Nanostructure Science and Technology

Series Editor

David J. Lockwood, FRSC
National Research Council of Canada
Ottawa, ON, Canada

Nanostructure science and technology now forms a common thread that runs through all physical and materials sciences and is emerging in industrial applications as nanotechnology. The breadth of the subject material is demonstrated by the fact that it covers and intertwines many of the traditional areas of physics, chemistry, biology, and medicine. Within each main topic in this field there can be many subfields. For example, the electrical properties of nanostructured materials is a topic that can cover electron transport in semiconductor quantum dots, self-assembled molecular nanostructures, carbon nanotubes, chemically tailored hybrid magnetic-semiconductor nanostructures, colloidal quantum dots, nanostructured superconductors, nanocrystalline electronic junctions, etc. Obviously, no one book can cope with such a diversity of subject matter. The nanostructured material system is, however, of increasing significance in our technology-dominated economy and this suggests the need for a series of books to cover recent developments.

The scope of the series is designed to cover as much of the subject matter as possible – from physics and chemistry to biology and medicine, and from basic science to applications. At present, the most significant subject areas are concentrated in basic science and mainly within physics and chemistry, but as time goes by more importance will inevitably be given to subjects in applied science and will also include biology and medicine. The series will naturally accommodate this flow of developments in the sciences and technology of nanostructures and maintain its topicality by virtue of its broad emphasis. It is important that emerging areas in the biological and medical sciences, for example, not be ignored as, despite their diversity, developments in this field are often interlinked. The series will maintain the required cohesiveness from a judicious mix of edited volumes and monographs that while covering subfields in depth will also contain more general and interdisciplinary texts.

Thus the series is planned to cover in a coherent fashion the developments in basic research from the distinct viewpoints of physics, chemistry, biology, and materials science and also the engineering technologies emerging from this research. Each volume will also reflect this flow from science to technology. As time goes by, the earlier series volumes will then serve as reference texts to subsequent volumes.

Srikanta Moharana · Duncan H. Gregory ·
Ram Naresh Mahaling
Editors

Emerging Nanodielectric Materials for Energy Storage

From Bench to Field

Springer

Editors
Srikanta Moharana
Department of Chemistry, School
of Applied Sciences
Centurion University of Technology
and Management
R. Sitapur, Paralakhemundi, Odisha, India

Duncan H. Gregory
School of Chemistry
University of Glasgow
Glasgow, UK

Ram Naresh Mahaling
School of Chemistry
Jyoti Vihar, Sambalpur University
Burla, Odisha, India

ISSN 1571-5744 ISSN 2197-7976 (electronic)
Nanostructure Science and Technology
ISBN 978-3-031-40937-0 ISBN 978-3-031-40938-7 (eBook)
https://doi.org/10.1007/978-3-031-40938-7

This Springer imprint is published by the registered company Springer Nature Switzerland AG
The registered company address is: Gewerbestrasse 11, 6330 Cham, Switzerland

Preface

Nanocomposites are typically made up of a host material and one or more guest materials. This book offers a comprehensive discussion of the synthetic methods that can be utilised to prepare them as well as the physicochemical and electrical characteristics of nanocomposites. The volume pays particular attention to the surface alternation of inorganic core-shell materials with nanometric dimensions, which assists in the development of an understanding of how nanocomposites function. More specifically, the book demonstrates in great detail how a broad interface between the polymer and the inorganic fillers governs the characteristics of nanocomposite dielectrics. Nanodielectrics are a new class of promising dielectric materials designed at the nanometric scale for potential applications in advanced electronic devices such as energy storage.

This book on *Emerging Nanodielectric Materials for Energy Storage: From Bench to Field* is intended to place an emphasis on the most advanced and fundamental phenomena pertinent to current global energy challenges. The scale of these challenges demands new technologies, and the design of new classes of nanodielectrics with applications in several key areas (which will be highlighted) will be at the centre of this new technological drive. Also highlighted will be the various facile processing techniques employed to fabricate nanodielectric materials and their composites with attractive properties like mechanical strength and flexibility that can complement electrical performance for end-use applications in the form of flexible thin-film capacitors, for example. The book thus endeavours to provide simultaneous fundamental insight into properties, performance and utility from bench to field. It places a tool for the birth and growth of new technological solutions in the hands of graduate students and both industrial and academic researchers including material scientists, engineers, physicists and chemists.

Since the discovery of nanocomposite dielectrics, scientists have learnt to produce several kinds of dielectric materials with enhanced qualities. As a result, the interest in nanocomposite dielectrics and their use has grown more widespread. For this interest in dielectrics to proliferate further, it is essential to have an understanding of the most recent developments in the field. It is hard to address all of the potential applications of nanocomposite dielectrics in a single book, owing to the tremendous progress that has been made in nanocomposite dielectric research. In this book, we have provided an overview of the current research being conducted in nanodielectrics of a specific field, and our subsequent books in this series will cover a great deal of additional material. This book comprises fifteen chapters based on recent nanodielectric materials for energy storage research. Chapter 1 "The Past, Present and Potential Future of Dielectric Nanomaterials" describes the energy, electronics and sensor sectors. This chapter mainly focuses on the properties and application of dielectric nanomaterials. Chapter 2 "Synthesis Approaches of Nanodielectric Materials" deals with various synthesis techniques towards nanodielectric materials. Chapter 3 "Synthesis, Microstructural and Dielectric Characterization of Nanodielectrics" describes the synthesis, microstructural and dielectric properties of many of the most important materials. Chapter 4 "Role and Prospects of Polymer Based Nanomaterials in the Dielectric World" provides a relatively brief overview and outlook of polymer-based nanomaterials with applications in various fields. Chapter 5 "Recent Progress in Nanodielectric Composites and Their Applications" presents a current and topical summary of different nanocomposite dielectrics along with their benefits. Chapter 6 "Effect of Nanofillers Reinforced Polymer Blends For Dielectric Applications" gives a concise introduction to the properties and dielectric applications of polymer blends and nanocomposites incorporating nanofillers. Chapter 7 "Metal and Metal Oxide-Based Polymeric Nanodielectrics For Energy Storage: Opportunities and Challenges" focuses on the properties of metal oxide-based polymeric nanocomposite dielectrics, whereas Chap. 8 "Fabrication and Properties of Dielectric Elastomer Based Nanocomposites" takes a broad view of the synthesis and properties of elastomeric nanocomposites. Chapter 9 "Physical and Chemical Properties of Inorganic-Polymer Nanodielectrics and Their Applications" explores the progress that has been made in identifying and interpreting the physical and chemical properties of nanocomposite dielectrics formed from inorganic fillers and polymeric hosts together with the applications that are most pertinent. Chapter 10 "Surface Engineering of Graphene Based Polymeric Composites for Energy Storage Devices" deals with the prospects for surface alteration of graphene-based polymer composites, their properties and how they find application in energy storage. In Chap. 11 "Core-Shell Structured Nanomaterials for High Performance Dielectric Applications", the authors describe how core-shell structures have become important for dielectric applications and how one might synthesise such nanomaterials. Chapter 12 "Transition Metal Oxide Based Nanomaterial for Advanced Energy Storage" offers a detailed discussion on transition metal oxide-based nanomaterials for practical applicability in the field of advanced energy storage, while Chap. 13 "Fluoropolymer Based Nanodielectrics for Energy Storage Application" takes a specific look at the growing number of fluoropolymer-based nanocomposite dielectrics that have been reported so far and their increasing influence in energy storage applications. Chapter 14 "Application of Organic–Inorganic Nanodielectrics for Energy Storage" introduces some of the most advanced energy storage applications offered by hybrid organic-inorganic nanocomposite dielectrics, and then lastly Chap. 15 "Metal Oxide Nanofiller Introduced Polymer Based Nanocomposite Dielectrics for

Advanced Energy Storage Devices" demonstrates the use of metal oxide ceramics in the doping of polymeric matrix dielectric nanocomposites. Overall, the book deals with many of the hosts of different aspects of nanodielectric composite research. Despite the vast breadth and depth of the area, we hope that you agree that the book provides a conveniently concise outline of the polymer (material) science and technology involved. We have high hopes that once the reader has had a chance to appreciate this volume, she or he will be equipped with sufficient information to proceed further into the otherwise daunting expanse of nanocomposite dielectric research. Each chapter of this book is comprehensively referenced, citing many examples of the most cutting-edge research from across the globe. In this respect, this volume acts as an atlas for further exploration, which should be of equal use to professionals in academia and industry, as to beginners in the field in navigating the far reaches of the nanodielectric world.

R. Sitapur, India — Srikanta Moharana, Ph.D.
Glasgow, Scotland — Duncan H. Gregory, Ph.D.
Burla, India — Ram Naresh Mahaling, Ph.D.

Contents

About the Editors

Editors and Contributors

Dr. Srikanta Moharana is presently working as Assistant Professor at the Department of Chemistry, School of Applied Sciences, Centurion University of Technology and Management, Odisha, India. He obtained his Ph.D. and M.Phil. degrees in Chemistry from the School of Chemistry, Sambalpur University, India. He has received his M.Sc. degree in Chemistry from National Institute of Technology, Rourkela. He has published more than 45 research papers in international peer-reviewed journals, 19 book chapters and many more counting. He has awarded **Prof. G. B. Behera Best Ph.D. thesis award** under the banner of Orissa Chemical Society, Odisha, India. He has delivered two invited talks and presented his work at various national and international conferences. Currently, two students are continuing Ph.D. under his guidance. He has more than 4 years of teaching and 4 years of research experience. His research experience as well as research interests lies in graphene, carbon nanotubes and ceramic-based polymer nanocomposites synthesis and characterisation for advanced energy storage applications. He is Life Member of Orissa Chemical Society.

Prof. Duncan H. Gregory is West CHEM chair of Inorganic Materials, University of Glasgow. He was previously EPSRC Advanced Fellow, Lecturer, then Reader in Materials Chemistry at the University of Nottingham. He is currently Visiting Professor at Kyushu University and was Vice President of the Royal Society of Chemistry (RSC) Materials Chemistry Division Council (2009–2014). His research interests focus on the discovery of new solids including sustainable energy materials (e.g. Li batteries, fuel storage, thermoelectrics), inorganic nanomaterials and the solid-state chemistry of non-oxides. His research also embraces the sustainable production of materials including the microwave synthesis and processing of solids. He has published more than 180 papers in peer-reviewed journals, is Editor-in-Chief of the journal *Inorganics* and Associate Editor of *Materials for Renewable and Sustainable Energy*. He was the winner of the RSC Sustainable Energy Award in

2009 and was awarded the Institute of Materials, Minerals and Mining (IOM3) Kroll Medal in 2019.

Prof. Ram Naresh Mahaling at present working in School of Chemistry, Jyoti vihar, Sambalpur University, Odisha, India (BHARAT). He has served Sambalpur University in the capacity of Assistant Professor and Associate Professor prior to join as full Professor. He has more than 13 years of teaching and 17 years of research experience and has published more than 60 research papers in peer-reviewed journals and authored seven book chapters. His current areas of research include the synthesis of perovskite-type metal oxide ($BaTiO_3$, $BiFeO_3$, $CaTiO_3$, etc.)-based polymer composite materials for electronic applications and energy storage, nanocomposites and their structure-properties relationship. He has visited the Institute of Polymer Research Dresden, Germany, and University of Orleans, France, for two-and-half years as Postdoctoral Fellow. He has received research grants from different funding agency, such as UGC, New Delhi; DST SERB, New Delhi, and Government of Odisha. So far he has guided fourteen (14) M.Phil. students, and two Ph.D. Currently 5 students are working under him for his Ph.D. He has delivered **seven invited talks** in different national and international conferences.

Contributors

Adeel Afzal School of Chemistry, University of the Punjab, Lahore, Pakistan

P. K. Akhilesh Institute of Condensed Matter and Nanosciences, Université-catholique de Louvain, Ottignies-Louvain-La-Neuve, Belgium

K. P. Andryushin Research Institute of Physics, Southern Federal University, Rostov-on-Don, Russia

Sarfaraz Ansari Nanostructured Composite Materials Laboratory, Indian Institute of Technology (Indian School of Mines), Dhanbad, Jharkhand, India

Kavya Pulagam Srinivasa Babu Department of Chemistry, CHRIST (Deemed to be University), Bengaluru, India

Minal Bafna Department of Physics, Agrawal PG College, Jaipur, India

Nipun Bafna Mechanical and Industrial Engineering Department, IIT Roorkee, Roorkee, India

Meera Balachandran Department of Chemical Engineering and Materials Science, Amrita School of Engineering, Amrita Vishwa Vidyapeetham, Coimbatore, India;
Centre of Excellence in Advanced Materials and Green Technologies (CoE-AMGT), Amrita School of Engineering, Amrita Vishwa Vidyapeetham, Coimbatore, India

V. Balasubramani Department of Analytical Chemistry, University of Madras, Chennai, India

Rabia Batool School of Chemistry, University of the Punjab, Lahore, Pakistan

Jayanta Bauri Nanostructured Composite Materials Laboratory, Indian Institute of Technology (Indian School of Mines), Dhanbad, Jharkhand, India

L. Biswal Department of Physics, School of Applied Sciences, KIIT Deemed to be University, Bhubaneswar, Odisha, India

K. P. Chaithra Department of Chemistry, CHRIST (Deemed to be University), Bengaluru, India

Subhendu Chakroborty Department of Basic Sciences, IITM, IES University, Bhopal, Madhya Pradesh, India

S. Chandraleka Department of Chemistry, R.V. Government Arts College, Chengalpattu, India

Arunkumar Chandrasekhar Nanosensors and Nanoenergy Lab, Department of Sensors and Biomedical Technology, School of Electronics Engineering, Vellore Institute of Technology, Tamil Nadu, Vellore, India

Ram Bilash Choudhary Nanostructured Composite Materials Laboratory, Indian Institute of Technology (Indian School of Mines), Dhanbad, Jharkhand, India

Barnali Dasgupta Ghosh Department of Chemistry, Birla Institute of Technology Mesra, Ranchi, India

Farah Deeba SS Jain Subodh P G College, Jaipur, India;
School of Applied Sciences, Suresh GyanVihar University, Jaipur, India

Sudhanshu Dwivedi S.S. Jain Subodh P.G. (Autonomous) College, Jaipur, India

Ankita Subhrasmita Gadtya Department of Chemistry, School of Applied Sciences, Centurion University of Technology and Management, R.Sitapur, Paralakhemundi, Odisha, India

Anupam Ghosh Department of Management, Birla Institute of Technology Mesra, Ranchi, India

Tajamal Hussain School of Chemistry, University of the Punjab, Lahore, Pakistan

Ankur Jain School of Applied Sciences, Suresh GyanVihar University, Jaipur, India;
Center for Renewable Energy and Storage, Suresh GyanVihar University, Jaipur, India

Vinod Kashyap Department of Chemistry, National Institute of Technology Tiruchirappalli, Tiruchirappalli, Tamil Nadu, India;
Department of Materials Engineering, Indian Institute of Science, Bangalore, India

Pragati Kumar Nano Materials and Device Lab, Department of Nanoscience and Materials, Central University of Jammu, Jammu, J & K, India

Sanjeev Kumar Nanostructured Composite Materials Laboratory, Indian Institute of Technology (Indian School of Mines), Dhanbad, Jharkhand, India

M. R. Kuppusamy Department of Chemistry, R.V. Government Arts College, Chengalpattu, India

Priyambada Mallick School of Applied Sciences, Centurion University of Technology and Management, Bhubaneswar, Odisha, India

Anchit Modi Department of Basic Sciences, IITM, IES University, Bhopal, Madhya Pradesh, India

Srikanta Moharana Department of Chemistry, School of Applied Sciences, Centurion University of Technology and Management, R.Sitapur, Paralakhemundi, Odisha, India;
School of Applied Sciences, Centurion University of Technology and Management, Bhubaneswar, Odisha, India

Adnan Mujahid School of Chemistry, University of the Punjab, Lahore, Pakistan

Anindita Mukherjee Department of Chemistry, Birla Institute of Technology Mesra, Ranchi, India

Debashish Nayak Nanostructured Composite Materials Laboratory, Indian Institute of Technology (Indian School of Mines), Dhanbad, Jharkhand, India

Joshi Harsh Nitinkumar Department of Chemical Engineering and Materials Science, Amrita School of Engineering, Amrita Vishwa Vidyapeetham, Coimbatore, India;
Centre of Excellence in Advanced Materials and Green Technologies (CoE-AMGT), Amrita School of Engineering, Amrita Vishwa Vidyapeetham, Coimbatore, India

Pravati Panda Department of Basic Sciences, RIE, Bhubaneswar, Bhubaneswar, India

Prasanth Raghavan Materials Science and NanoEngineering Lab (MSNE-Lab), Department of Polymer Science and Rubber Technology, Cochin University of Science and Technology (CUSAT), Cochin, India;
Department of Materials Engineering and Convergence Technology, Gyeongsang National University, Jinju, Republic of Korea

Navyasree Reghu Department of Chemical Engineering and Materials Science, Amrita School of Engineering, Amrita Vishwa Vidyapeetham, Coimbatore, India;
Centre of Excellence in Advanced Materials and Green Technologies (CoE-AMGT), Amrita School of Engineering, Amrita Vishwa Vidyapeetham, Coimbatore, India

L. A. Reznichenko Research Institute of Physics, Southern Federal University, Rostov-on-Don, Russia

Lipeeka Rout Department of Chemistry, Madanapalle Institute of Technology & Science, Madanapalle, Andhra Pradesh, India

Sushrisangita Sahoo Research Institute of Physics, Southern Federal University, 344090 Rostov-on-Don, Russia;
Department of Material Science and Engineering, Tuskegee University, Tuskegee, AL, 36088 USA

Anupam Sahoo Department of Chemistry, Sri Krushna Chandra Gajapati (Autonomous) College, Paralakhemundi, Odisha, India

Bibhuti B. Sahu Basic Science and Humanities Department, Nalanda Institute of Technology (NIT), Bhubaneswar, Odisha, India

P. Sakthivel Centre for Materials Science, Department of Physics, Science and Humanities, Faculty of Engineering, Karpagam Academy of Higher Education, Coimbatore, Tamil Nadu, India

R. Sasikumar Department of Physical Chemistry, University of Madras, Chennai, India

Santosh Ku Satpathy School of Applied Sciences, Centurion University of Technology and Management, Bhubaneswar, Odisha, India

Nupur Saxena Department of Physics, Indian Institute of Technology (IIT), Jammu, Jammu, India;
Organisation of Science Innovations and Research, Bah, India

Khurram Shehzad School of Micro-Nano Electronics, ZJU-Hangzhou Global Scientific and Technological Innovation Center, Zhejiang University, Zhejiang, China

Sukhvant Singh Department of Chemistry, National Institute of Technology Tiruchirappalli, Tiruchirappalli, Tamil Nadu, India;
Centre for Sustainable Technologies (CST), Indian Institute of Science, Bengaluru, India

T. M. Sridhar Department of Analytical Chemistry, University of Madras, Chennai, India

D. Sridharan Vioma Motors Pvt Ltd., Mumbai, Maharashtra, India

Sumit Department of Chemistry, Indian Institute of Technology, Delhi, India

Sangita Kumari Swain Centre of Excellence, Berhampur University, Berhampur, Odisha, India

Sukanta Kumar Swain Department of ECE, Indian Institute of Information Technology, Ranchi, Jharkhand, India

Debajani Tripathy Department of Chemistry, School of Applied Sciences, Centurion University of Technology and Management, R.Sitapur, Paralakhemundi, Odisha, India

T. P. Vinod Department of Chemistry, CHRIST (Deemed to be University), Bengaluru, India

Alexandru Vlad Institute of Condensed Matter and Nanosciences, Université-catholique de Louvain, Ottignies-Louvain-La-Neuve, Belgium

Abhinav Yadav Research Institute of Physics, Southern Federal University, Rostov-on-Don, Russia

Muhammad Zahid Department of Chemistry, University of Agriculture Faisalabad, Faisalabad, Pakistan

Chapter 1
The Past, Present, and Potential Future of Dielectric Nanomaterials

S. Chandraleka, V. Balasubramani, R. Sasikumar, M. R. Kuppusamy, T. M. Sridhar, Pragati Kumar, and Nupur Saxena

Abstract Dielectric nanomaterials with a relatively high dielectric loss tangent are used in various fields such are transformers, solar cells, transistors, capacitors, energy storage devices, microwave, and nanophotonic applications. An abundance of bonded surface charges is present on the surface of these nanomaterials that change their functional properties. Dielectric nanomaterials help to improve the dispersion of nanomaterials by applying them as thin film layers or as composites. These dielectric nanomaterials provide less space charge and a lower electric field by forming a multilayer composite and extending the life of electronic devices and electric motors. In this chapter, the past, present, and potential future of dielectric nanomaterials and their surface modification with polymers or several composites such are nanodielectric of metals and metal oxides including their characteristics are highlighted. The importance of dielectric nanomaterials and flexible nanocomposite materials with their wide range of applications for energy storage, electronics, sensors, and their potential deliverables as futuristic devices are presented along with their future scope.

Keywords Dielectric nanomaterials · Flexible dielectric nanocomposites · Energy · Electronics · Sensors

S. Chandraleka · M. R. Kuppusamy
Department of Chemistry, R.V. Government Arts College, Chengalpattu 603001, India

V. Balasubramani · T. M. Sridhar (✉)
Department of Analytical Chemistry, University of Madras, Chennai 600025, India
e-mail: tmsridhar23@gmail.com

R. Sasikumar
Department of Physical Chemistry, University of Madras, Chennai 600025, India

P. Kumar
Department of Nanoscience and Materials, Central University of Jammu, Samba, Jammu, J & K 181143, India

N. Saxena
Organisation of Science Innovations and Research, Bah 283104, India
e-mail: n1saxena@gmail.com

S. Moharana et al. (eds.), *Emerging Nanodielectric Materials for Energy Storage*, Nanostructure Science and Technology, https://doi.org/10.1007/978-3-031-40938-7_1

1.1 Introduction of Dielectric Nanomaterials and Their Origin

Interest in dielectric devices grew in the early 1990s with the discovery that certain thin elastomeric films, such as silicon films, can withstand large electrical stresses and generate large mechanical forces. It is interesting to take a look at the past, present, and potential current developments, current challenges, and future advances in dielectric nanomaterials related to the properties of dielectric nanomaterials [1].

A number of commercial products based on dielectric nanomaterials technology have already reached the market. This effect lags far behind other transducer technologies such as electromagnetism, but the history and potential of the technology are linked to antennas, loudspeakers, microphones (which convert waves into sound signals), and thermometers. It continues to expand from transponders, wireless communications, echo scanning, electromagnetic sensors, and energy storage. Polymer-based nanocomposites are usually chosen for energy storage devices because of their dielectric properties that enhance device performance [2]. Solid–liquid and gaseous dielectric nanomaterials are used along with metal dopants for various applications, whose properties vary with respect to their corresponding applications have made strides in the field of nanodielectrics with advances in the synthesis methodology to modify the structure of chemical nanomaterials applied in electronic applications. The combination of chemical constituents of nanocomposites brings out the dielectric characteristics.

The challenges exist with the use of nanoparticles as they are prone to agglomeration, and there are issues with regard to stability and viscosity in addition to polymer matrices. This results in the need to carry out surface modification of nanoparticles with organic dyes or other organic or electromagnetic nanomaterials and is presented in Fig. 1.1. Changes to the surface of the nanomaterials are a viable method to alter the dielectric properties of nanoparticles with polymers, improving their energy storage and capacity behavior over a long history [3]. This has led to the development of nanodielectrics properties with a wide range of materials and nanocomposites that have reactive flexibility of each dielectric nanocomposite. These nanocomposites also come with challenges, unique properties, advantages, and limitations.

The dielectric behavior of the nanocomposites selected in this chapter includes a wide variety of metals, widely used in energy storage, solar cells, batteries, and many other applications. Modification of dielectric nanoparticles results in changes to the interfacial structure that bring about the reduction in the interfacial force between particles, and the performance of dielectric nanocomposites is degraded. The interfacial charge density provides a stimulus effect on the dielectric properties of the modified nanocomposites [4]. The interfacial location of cationic groups and electron-withdrawing functional groups may reduce the overall dielectric loss and improve the dielectric breakdown strength. Ceramics, plastics, mica, glass, distilled water, dry air, vacuum, nitrogen, and helium are among the natural sources of dielectric nanomaterials. This would lead to a better evaluation of all types of dielectric

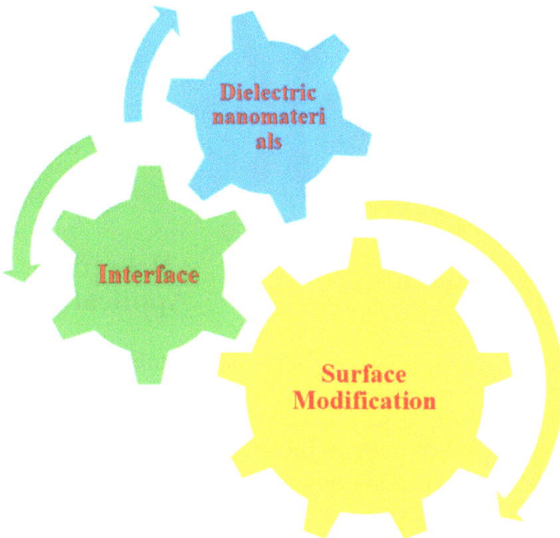

Fig. 1.1 Overview of nanodielectric materials

nanomaterials, and a better understanding of their long-term impact, stability, and potential for global energy storage and human health can be explored [5].

Normally, the dielectric constant of nanoparticles generally increases, and changes can be attributed to the interface relationship between the metal nanoparticles and polymer, which can improve the homogeneity of the composites. High dielectric parameters make it suitable for applications such as conductive paints and sensors. Polymer matrix dielectric non-materials were introduced in the twenty-first century. The rapid growth of life sciences and electrochemistry is impacted by the introduction of carbon-based materials, semiconducting metal oxides, hydroxyapatite (chemical constituent present in natural bone), and many dielectric nanomaterials [6]. Non-dielectric materials of the past have fueled innovation in portable electronics manufacturing and electrical insulation. With their high thermal conductivity, dielectric nanomaterials have found their way into almost all fields, resulting in higher energy storage compared to nanomaterials with polymer matrix dielectric nanomaterials [7]. Their potential applications of nanomaterials-based dielectrics are given in Fig. 1.2.

Their multiferroic properties at room temperature make them ideal candidates for future potential applications in memory devices. These devices are required from scientific calculators to supercomputers with varying sizes and speeds. Life science research is one of the future developments of dielectric nanomaterials involved in the molecules that will lead to upcoming areas of application of nanotechnology-based devices. Although the above dielectric parameters can be used for past and present nanomaterials applications and devices, and the future potential of nanodielectric nanomaterials has differences mainly in the parameters involved in interfacial effects of molecular ordering. The disadvantage of currently used insulating composites is the disintegration of electrical properties [8, 9].

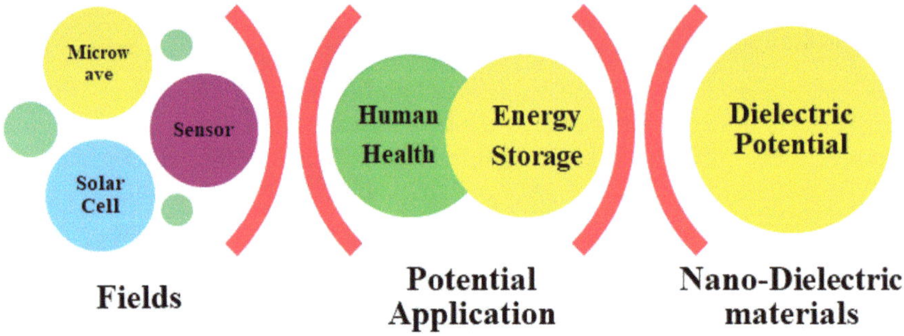

Fig. 1.2 Potential applications of nanodielectric materials

The shortcomings of the present applications offer us the scope for further modifications or the addition of metals to nanoparticles that are required to improve the strength of current dielectric nanodevices. Polymers act as fillers, and the interactions between fillers and nanomaterials impact the physico-chemical and mechanical properties of the polymers paving the way for future improvements in dielectric and electrical insulation. Polymer nanocomposites properties are further modified with inorganic fillers to bring out significant changes in dielectric properties widely used as energy storage materials [10].

1.2 Properties of Dielectric Nanomaterials and Their Applications in the Present Scenario

Dielectric nanomaterials exhibit properties that are required for the fabrication of highly flexible materials. Flexible polymer matrix nanodielectrics have been used in emerging or developing international research fields as it offers scientists the opportunity to present different perspectives, methods, and results [11]. It is presently applied to energy systems, life sciences, medical technology, robotics, machine learning, a host of biomedical devices like artificial skin and dresses with sensors, and electrical isolation among several others. Polydopamine (PDA) is used as a surface adhesive material in combination with polymers, semiconductor materials, inorganic ceramics, and rare metals [12]. Next, the PDA modifiers are commonly applied with nanomaterials, changing the dielectric nature with the use of small molecules like organosilanes, phosphonic acids, dopamine, and carboxylic acids. These characteristic primary organic molecules or nanomaterials prevent aggregation. The leakage current and dielectric damage of these nanomaterial-based devices can be brought down. Further, the dielectric breakdown property is enhanced by the modified surface of the metal oxide which presents an electron-withdrawing functional group with an electropositive phenyl configuration at the interface. After dopamine surface modification, the nanofiller dispersion and its dielectrics in the nanocomposites have enhanced their performance. Surface modification has led to the development of cost-effective methods to improve dielectric devices.

The properties of the nanoparticles present in nanomaterials are affected by the changes in interface charge density which in turn impacts the performance of dielectric composites. Dielectric nanomaterials and their composites are attractive for applications in dielectric capacitors as the carbon-based energy and fuel resources are being over-exploited for energy requirements, and its pollution levels have harmed the environment depriving us of clean air. Renewable energy technologies based on solar, wind, biofuel, geothermal, and tidal power are the leading alternatives to replace carbon-based energy resources [13]. A major challenge for renewable energy today is achieving efficient conversion and storage. The dielectric of capacitors with surface-modified nanomaterial brings about a change in the levels of dispersion due to flexible dielectric nanomaterials. Dielectric nanoparticles with relatively high dissipation factors are chosen due to the abundance of bound surface charges present on their surfaces. The dielectric properties of the fabricated nanocomposites are evaluated by recording the capacitance of the surface-modified nanocomposites as a function of frequency to determine the real (ε') and imaginary (ε") components of the complex permittivity [14].

Dielectrics are non-metallic substances with high resistivity, insulation resistance, and negative temperature coefficient of resistance. Other hands it is a non-conducting substance that retains an electric charge, and it has free electrons for current to flow. In the presence of an electric field, no charge flows through the material. The charge shifts slightly from its average equilibrium position due to dielectric polarization in the presence of an electric field. Further polarization results in the flow of positive charges toward the field, while negative charges move in the reverse way of the field. This property brings about changes in the internal electric field leading to the enhanced capacity to accumulate electrons inside materials, and this property result enhances the energy storage applications [15].

The dielectric properties of nanomaterials have changed in their fundamental molecular properties that are capable of imminent electron migration even though the energy gap is very large for dielectric nanomaterials. This in turn brings about a change in polarization when it is subjected to a peripheral electric field. Further, a negative temperature coefficient of resistance is obtained that could increase the conductivity, the resistivity of the semiconductor material decreases with increasing temperature, and the insulation resistance is high which allows objects of a certain size on the surface of metal plate. Being a poor insulator and an excellent conductor of electricity, it retains a large amount of charge for a long period. Dielectric nanomaterials have high resistivity, and the electrostatic force between electrons and nuclei is very strong. Electrical breakdown is observed with these materials as a disadvantage with these insulating materials. Finally, it is the mechanism of nanometric interfacial and molecular ordering processes that triggers the electrical breakdown in these insulating materials which need to be studied. To overcome the drawbacks, the present applications need some more changes or addition to enhance the current nanodielectric devices' strength [16].

Hydrogels synthesized by polymerization using dielectric nanomaterials as dopants are excellent candidates for 3D cell models due to their biocompatible structures. Further, it provides useful properties for biosensors and tunable drug delivery applications. Electronic device applications are based on the light intensity, chemical structure, and dielectric characterization of dyes—methylene-blue (MB), rhodamine-B (RB), and MB/RB dyes along with $NiFe_2O_4$ nanoparticle-based hydrogels as a function of frequency and bias, viscosity, and zigzag motion have approximately equal parts of the same absorbing medium absorbing equal parts of incident light in terms of suitability for biocompatibility for biosensor based applications. Hydrogels doped with organic dyes exhibit low adsorption coefficients such as MB, RB, and MB/RB, whereas dielectric nanoparticles exhibit high adsorption coefficients as a result of the dipole, Fermi direct distribution, which increases electron transfer [17]. Hydrogels doped with $NiFe_2O_4$ nanoparticles, the small size and high ratio of emitted photons to absorbed photons offer new scope for biomedical imaging of cancer and other target molecules in-vivo in the human body.

Organic dyes have been incorporated into hydrogels framed works to detect the membrane biology of the cells and image them with doped nanoparticles. These materials should have increased dielectric capacity on the application of electrical potential that results in partial ionization at elevated temperatures for applications in e-cars, e-bikes, space, and high-temperature furnaces industry. Polymers incorporated with dielectric nanomaterials show superior results of excellent power density and excellent rate capabilities. Power density indicates how quickly a device can deliver energy than batteries such as magnesium, lithium, and zinc ion batteries. One of the major problems of today's scenario in aircraft engines is that it requires some form of cooling to prevent engine damage. Normally, these cooling devices break down resulting in degradation of the critical electronic-based control systems due to disabling of cooling systems. Here, there is a requirement for materials that can withstand temperatures of up to 250 °C for power electronics and heat sources which causes damage to electronics systems [18]. Therefore, polymer matrix nanoparticles provide excellent cooling even under a high electric field which is a property to be incorporated in dielectric capacitors to withstand all weather conditions. The various types of dielectric nanomaterials currently being used along with their potential applications are summarized in Table 1.1.

1.3 Nanodielectric of Metals

The metal and polymer composites are solid-based dielectric nanomaterials that show extremely improved properties with respect to surface structure, the interaction between particles, and the size of nanomaterials. Nanodielectric metals are commonly expensive to manufacture as it involves the use of a large magnetic spin moment and strong ferromagnetic coupling due to high Curie temperature. Metals are solids, and they can have anisotropic behavior where nanodielectric metals have

Table 1.1 Types of dielectric nanomaterials currently being used along with their potential applications

Types of dielectric nanomaterials	Dielectric constant value at kHz	Application
Pure PVC	3.2	High thermal conductivity metal industry
Metal nanoparticles	1.2	Energy, sensors, and superconductors
Polydopamine	3.2	Energy, solar cell, and biomedical
Metals	1.3	Energy, sensors, biomedical
Hydrogels	1.5	Biomedical, drug delivery, and 3D cells models
Metal oxides	1.5	Energy, sensors, and biomedical
Polymer nanocomposites	3.2	High-voltage cables
Carbon-based nanocomposites	2.8	Solar, wind, biofuel, geothermal, and tidal power
Organic nanocomposites	2.5	Membrane biology and thermal barrier coating industry

large magnetocrystalline anisotropy which means it does not have the extrinsic property of metals which has ferromagnetic behavior and is not dependent on the particle size and structure, but the quantum of energy required for magnetization in each direction is high. The crystal lattice angles dictate these directions, as it is essential that the orbital motion of electrons coupled to the electric field of the crystal would provide the first-order contribution to the magnetocrystalline anisotropy. Studies have reported that gold, germanium, zinc, and some other metal nanoparticles have shown that the dielectric behavior changes contribute the device applications [5, 19]. These nanodielectric properties could be applied to interdisciplinary sciences involving chemistry, biology, physics, and materials science to fabricate these devices.

1.3.1 Gold

Gold has characteristics of high tensile strength which is responsible for developing uniform coatings with polymer matrix, and the obtained surface structures are ideal for the required applications. Trace levels of gold are used to cure the disease in medicinal applications so that it can kill or destroy the harmful cells present in the human body. The smaller levels of gold are well tolerated by the human body. Gold nanoparticles present in the composite with the polymer matrix are suitable materials for health care, cosmetics, tissue engineering, and most importantly medical diagnostics and sensors applications. Gold plays a vital application on both sides of electrochemical-based devices as well as in the process involving biomedical aspects of drug delivery systems [20]. The reason this gold is an ideal material for many applications is due to its enormous dielectric property of gold nanoparticles and its antibacterial activity that favors the development of smart drug delivery systems. Gold-containing nanoparticles have high dielectric constant and dielectric loss which enhances the dielectric properties when compared to the pure polymeric matrix [21].

The vibrant optical property of nanomaterials enhances the device application and also has strong catalytic properties for all types of industrial applications. The presence of gold in the composite with polymer matrix enhances the dielectric constant due to the composition of polymer materials as they act as conductive fillers. This type of high-energy storage material is more needful for the future to replace the need for deficiency of nonrenewable sources. Society needs storage devices to store more energy for the future with high dielectric permittivity (ε') at the low applied voltage, and this can be developed with gold nanoparticles [4].

The polymers used have different chemical compositions, and these changes affect the dielectric constant of nanomaterials; for example, the copolymer which is formed with two different monomers react together to form more flexible properties. Already gold has more tensile strength compared with the transition elements present in the periodic table and also as a composite with copolymers it enhances more flexible properties that interconnect with the conductive fillers to produce more efficient energy-stored capacitors.

1.3.2 Metal Oxides

Zinc oxide is a transition metal oxide with the formula of ZnO which is commonly used for several applications from plastics to rechargeable batteries as ZnO nanostructures can be used as anode materials in lithium-ion batteries. Materials with low frequency and temperature dependence are well suited for the fabrication of devices. Nano copper (II) oxide (CuO) has a superior dielectric constant and is independent of the temperature of the device. M-type hexagonal ferrite with the chemical formula $M\,Fe_{12}O_{19}$ (M = Ba, Sr, Pb) has excellent oxidation resistance, high magnetic material resistance behavior to variations in the magnetic field, and has the magnetic field strength capacity to demagnetize a completely magnetized material [22]. These nano oxides have the ability to provide information about the remanence and magnetic energy of the products. These characteristics of dielectric nanomaterials can be used as sensors, radar adsorptive materials, and several others. The rare earth metal doped in glass matrix results in enriching the optical properties. The silicon solar cell shows a high refractive index due to the plasmonic effect of gold and silver nanoparticles which are the reason for the rare earth metal growth on the glass matrix. This kind of material is widely used in ultrafast device response [22].

1.4 Applications of Dielectric Nanomaterials in Energy, Electronics, and Sensors

The wide range of applications of nanodielectrics is illustrated in Fig. 1.2, and this technology has been applied in many areas. The prominent areas are space charge suppression, charge storage, partial discharge (PD) resistance, high thermal conductivity, and the biomedical field which lead to the significant role of nanodielectrics in sectors like energy (supercapacitors and batteries), electronics, and sensors. The most prominent applications of nanodielectrics are as follows:

1.4.1 High Voltage Direct Current (HVDC) Cable

Nanodielectrics possess excellent dielectric properties and can moderate the internally developed electric field. In the case of conventional dielectrics, the space charge accumulation can destroy the distribution of the electric field and thereby degrading the polymer chain gradually. The polymer-nanoparticle composite (PNC) is a revolutionary material system that prohibits space charge accumulation and has been proven as a boon to researchers in this field. Many researchers have explored different PNCs. The incorporation of poly(stearyl methacrylate)-grafted SiO_2 into crosslinked polyethylene (XLPE) was studied by Zhang et al. [23] and showed the efficient suppression of space charge that led to limit the distortion in the internal field by 10.6% at room temperature over a broad externally applied DC field ranging from 30 to 100 kV/mm. Chu et al. [24] have used carbon-doped TiO_2 nanofibers in P(VDF-HFP) nanocomposites and studied their dielectric and energy storage properties. The incorporation of MgO nanoparticles in the MgO/low-density polyethylene (LDPE) nanodielectrics resulted in the enhancement of the average trap depth that led to space charge accumulation near the electrodes [25]. Ganea et al. [26] reported the effect of incorporating TiO_2 nanoparticles in bis-imidazolium salt with two cyanobiphenyl groups and dodecyl sulfate counterion (BIC). They have showed the conductivity enhancement with an increase in temperature and doping concentration at a constant frequency as shown in Fig. 1.3a–c. Besides, the characteristic relaxation time was found to decrease with an increase in the TiO_2 nanoparticles' concentration. The effect of the coupling agent was studied by Wang et al. [27] on the electrical properties of Al_2O_3/LDPE nanodielectrics and observed that the modified Al_2O_3 exhibits better space charge suppression than that of pristine one. Chen et al. [28] explained the charge dynamics in nanodielectric materials using deep trap formation by introducing nanoparticles in the material. In another work, space charge characteristics of polypropylene (PP)-based nanodielectrics were investigated by Zhou et al. [29] for its application as a recyclable insulating material for HVDC cable (Fig. 1.3).

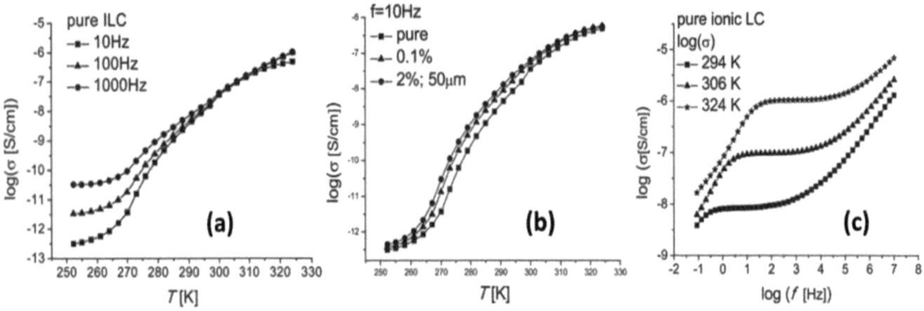

Fig. 1.3 **a** Temperature dependency of the electric conductivity σ' for BIC, **b** Electric conductivity versus temperature at a fixed frequency, $f = 10$ Hz, for the samples: BIC, BIC-01, and BIC-2, and **c** Conductivity spectra at $T = 296$ K, $T = 306$ K, and $T = 324$ K for BIC. Adapted with permission from ref. [26], Copyright: Elsevier, 2020

1.4.2 Nanodielectrics for Energy Storage Applications

Capacitors are the main energy storage components in electrical and electronic circuits as well as sensors [30]. These are of various types, viz. thin film, ceramic (film and laminated), electrolytic (aluminum and titanium), and the most sought supercapacitors type. Among these, the largest are supercapacitors and electrolytic which are manufactured by many manufacturers like Panasonic, Kemet, Murata, Vishay, etc. [31]. The energy storage capacitors seek huge applications in defense for transportation through vehicles, aircraft, and ships. For significant improvement of proficiency of energy storage in the capacitor, the dielectric material is required to possess high energy density [32]. Capacitors are highly required in power circuits for conditioning. It provides a constant current flow in the electrical circuit by regularizing the fluctuations in the current. The capacitors are more important for the circuits which are used for a wide range of temperature and frequency [33]. Energy-storing capacitors are highly important in power electronic circuitries, AC, and DC filters. Huai et al. [34] analyzed the importance of electrolytic and PNC film capacitors in DC-link filters for their high energy density, current, and voltage regulation efficiency, reliability, and stability provided to the circuit. Nanodielectric materials possess high dielectric constant and exhibit high breakdown field strength, hence are an important part of capacitors. The nanodielectrics are advantageous in enhancing the storage density of capacitors and supporting the miniaturization of the capacitor. The excellent dielectric breakdown properties, anti-aging properties, temperature, frequency independence, and reliability are some of the issues that need to be explored in the near future [35].

1.4.3 High Thermal Conductivity of Nanodielectrics

Nanodielectrics with high thermal conductivity are highly required for electrical as well as electronic circuits and systems. For applications employing high-voltage fields, the dissipation due to leakage current is enormous when the conducting voltage is enhanced. Particularly, for miniaturized devices, the compactness of the device will lead to more generation of heat in integrated circuitries.

Most of the literature on nanodielectrics focuses only on the thermal conductivity of the material. However, thermal expansion and breakdown strength studies are equally important and play a key role in the microelectronic packaging of integrated circuits. High thermal conductivity can be achieved by incorporating nano-sized fillers in polymer host matrices [36]. These fillers can create thermally conductive pathways like whiskers and filaments in the host matrix and modify the thermal resistance at the contacts [37].

There are various fillers that have been used to enhance the thermal conductivity of dielectric materials. As mentioned in an earlier section, these are classified as metal nanoparticles like Ag, Cu [38] inorganic nanoparticles like Si_3N_4, BN, AlN, SiC, MgO, SiO_2, and Al_2O_3) [39–41], and carbon-based materials [42, 43]. Sengwa et al. [39] reported a comparative study of different metal oxides fillers like Al_2O_3, SnO_2, TiO_2, and ZnO) in PVDF blended with poly (ethylene oxide) (PEO) for their dielectric properties and concluded that the incorporation of Al_2O_3 nanoparticles significantly improved the dielectric permittivity of the PNC as shown in Fig. 1.4. Boron nitride nanosheets (BNNSs) have also been proven to be a great influencer in improving the thermal conductivity of PNCs and explored by Seyhan et al. [44] and Sun et al. [45]. Additionally, the thermal conductivity of the polymer was reported to be highly dependent on the geometric distribution of the BNNSs in the host in the former case, whereas Ag nanoparticles were also used in the latter case to reduce the interfacial thermal resistance between two adjacent BNNSs. In another report, Hong et al. [46] suggested a different approach for enhancing the thermal conductivity and mechanical stretching of the PNC by incorporating a 3D hexagonal BN network in the polymer. Graphite was used as a filler for synthesizing nanodielectric PNC by Zhang et al. [47].

1.4.4 Nanodielectrics for Rotating Machines

The rotating machine systems include electric motors and generators. Nanodielectrics play a crucial role here too to restrain partial discharge (PD) to occur. Rotating machines are highly useful to run many appliances, smaller equipment, machines, etc. And PD may accelerate the degradation of polymer sequences and can cause major breakdowns of appliances or machines particularly for high-voltage devices. In high-voltage machines like pellet on and other electrostatic accelerators, homogeneous

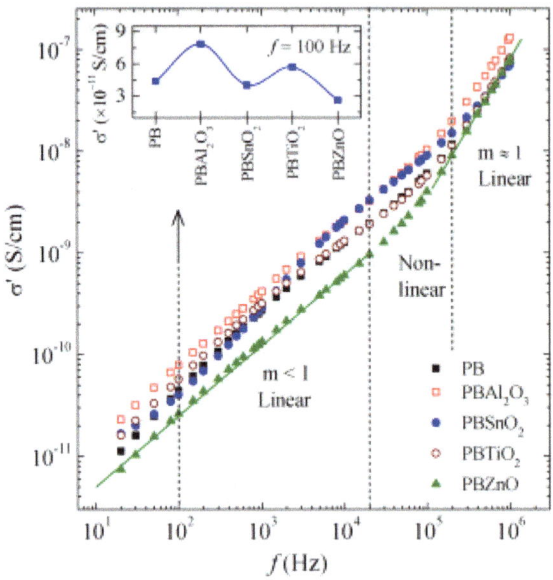

Fig. 1.4 Thermal conductivity spectra of the 75PVDF/25PEO blend (PB) film and the 75PVDF/25PEO–5 wt% nanofillers-based different PNC films ($PBAl_2O_3$, $PBSnO_2$, $PBTiO_2$, and PBZnO), at 27 °C. The solid lines drawn on the conductivity spectrum of PBZnO film represent linear fits of the low-frequency data as well as high-frequency data. Inset shows the comparative plot of thermal conductivity at 100 Hz for different PNC films. Adapted with permission from ref. [39], Copyright: Elsevier, 2020

distribution of electric field is hard to realize, and hence, there are high chances for surface corona discharge thereby breakdown down the machine.

An investigation on the property of PD resistance in nanodielectrics is highly desirable to understand the quality and its performance for prolonged high-voltage applications. There are three key parameters that decide the efficiency of a nanodielectrics material. These are PD current, erosion depth, and surface roughness. Here, again the incorporation of nanofillers like fumed silica (SiO_2) in XLPE [48] and silicate in polyamide [49] can be a boon to strengthen the bonding by resisting energetic charges moving in the polymer. This way, it is possible to shield the polymer and enhance the life of the equipment.

1.4.5 Nanodielectrics for Modulation and Sensing

In a recent breakthrough work, Tan et al. [50] reported the induction of dynamically controllable quasi-bound state in the continuum (QBIC) with ultrahigh quality factor in a symmetric metallic metal surface at terahertz (THz) frequencies using a nanodielectric or semiconductor layer, having thickness 1000 times less than the resonant wavelength (λ/1000), Additionally, the germanium nanostrips work as microchannels for their promising application as BIC-based refractive index sensor. The nanostrips provide 200% transmission intensity modulation when subjected to photoexcitation with very fast recovery (~ps). The sensitivity of the refractive index sensor was reported as −24.5 GHz/RIU with a 500-nm-thick superstrate as depicted in Fig. 1.5a–c.

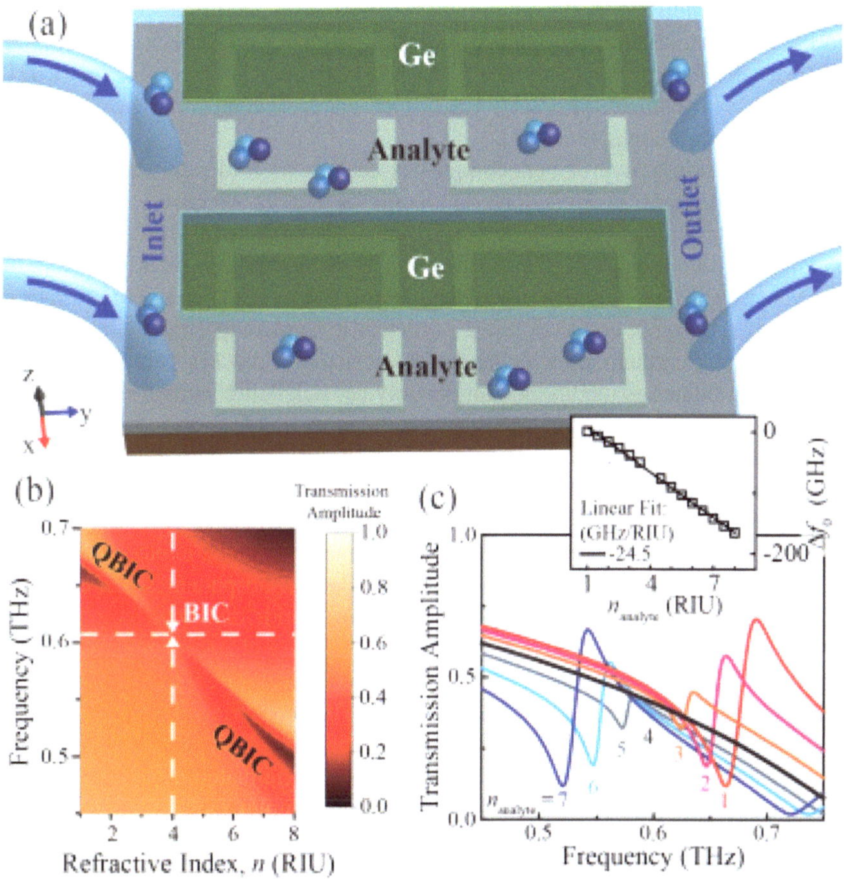

Fig. 1.5 Terahertz refractive index BIC sensor. **a** Illustration of the device with analyte flowing through the Ge strip microchannels. **b** Simulated transmission amplitude spectra showing the collapse and revival of the BIC resonance mode with changing the refractive index of an analyte placed in microchannels separated by the Ge strips of the device of Fig. 1.2a. The refractive index of Ge is 4 (vertical dashed line), and the calculated BIC resonance (from COMSOL) is at 0.61 THz (horizontal dashed line). **c** Transmission amplitude spectra of the QBIC resonance for a superstrate thickness (Ge strip and analyte) of $h = 500$ nm and different refractive indices of the analyte. The inset shows the frequency spectral shift as a function of the analyte refractive index. The device sensitivity in terms of frequency shift per refractive index unit is obtained from the slope through a linear fit. Adopted with permission from ref. [50], Copyright 2021

1.4.6 Advanced Applications of Dielectric Nanomaterials in Various Fields

The solid dielectric material such as ceramic substances has proven superior dielectric properties which is the major requirement for the development of supercapacitor applications. These solid materials should have characteristics of definite shape and structure. Solid dielectric nanomaterials depend on the distance between two grains and their size followed by the bonding between particles. Ferromagnetic substances doped into ceramic materials enhance the performance of supercapacitors. Chromium nanoparticles incorporated into the crystalline structure of superconducting (CuTl)-1223 matrix to obtain (Cr)x/(CuTl)-1223 superconductor composites are an example of supercapacitor application [51].

The dielectric nature of nanocomposites enables us to deploy them in energy storage applications as capacitors. The common factors which affect the energy storage levels and its discharge properties in nanocomposite materials are an accumulation of nanofillers and phase separation in the composites as this leads to the two constituent phases which are separated apart. This leads us to concentrate on improving the energy conversion rate of dielectric nanocomposites to influence the potential to enrich them to be incorporated as capacitive devices. Silanes have been the best candidates as modifiers with the only setback being its hydrolytic stability that restricts its development. The other candidate available is phosphonates which contain phosphorus and are also an insurmountable weakness in its detrimental properties that affect our environment. To overcome these challenges, there is a need for surface modifiers such are low-molecular-weight carboxylic acids and catechol which are present in dopamine and polydopamine and can be explored as surface modifiers. This surface modification focuses on silanes, phosphonates, and dopamine as chemical agents to modify the surfaces of metal oxide dielectric nanomaterials. Chemical modification of nanomaterials brings about a change in the functioning capacity of dielectric properties which include breakdown strength, energy density, and dielectric constant of the nanocomposites. The dielectric calculations of these nanomaterials depend on the frequency and temperature applied to the system as it fluctuates, and the parameters which affect the dielectric materials include conductance (G), capacitance (c), tangent loss (tanδ), and complex dielectric constant ($\varepsilon r'' = \varepsilon' r - i\varepsilon'' r$) varies. These parameters can be calculated by using an impedance analyzer. If temperature increases, the $\varepsilon r''$ are also increases as most energy is lost from the sample, but it is inversely proportional to the frequency given to the system.

Specific polymer matrices are used to potentially determine the final performance for optimizing nanocomposites and the properties of dielectric nanomaterials which are responsible for the various applications and the ways to overcome them have to be explored [51]. Dielectric-polymer nanocomposites are most suitable, but the choice of chemical reactions or chemical reagents used to functionalize these nanomaterials in the polymeric matrix determines the potential performance in various applications. Surface modification techniques have been widely used to modify nanomaterials for dialectics, but there is scope for further research in this field.

Several classifications have been used to classify dielectric nanomaterials as they differ in their dimensions, breakdown strength, surface area, composition, porosity, nature of ceramic inorganic materials, negative temperature coefficient of resistance, high insulating resistance, and polarization. MgxCa (0.90-x)Zn0.10Fe_2O_4 nanoparticle is an example of a surface-modified system for microwave applications [52]. Permittivity and dissipation factor, loss tangent, and dielectric constant are the properties that need to be evaluated to test the suitability of nanomaterials for microwave application. These properties indicate the role of metal nanoparticles on the dielectric properties of the developed nanocomposites. The presence of polymer matrix helps to clarify the applicability of metals/microelectronics industry such as transistors, capacitors and resistors, solid-state devices like solar cells, semiconductor laser, electrochemical sensing, and electricity resistance.

Parallel plate capacitors can only store a limited amount of energy before dielectric breakdown occurs. Because of the setting of the electric field in between two parallel plates are connected across the battery by external charge. This is used to measure the dielectric properties of materials; on the other hand, the impedance analyzer was used to measure the dielectric loss with respect to changes in frequency. It explained that dielectric loss is the energy loss that occurs when heating a dielectric material with a varying electric field or AC circuit alternately charges and discharges every half cycle. Dielectric nanomaterial especially gold has higher dielectric loss with polymer matrix concerning all frequency ranges from impedance analyzer. The dielectric loss of nanomaterials is directly proportional to the polarization resulting in a higher dielectric constant, and dielectric loss can be attributed to the maximum polarization of dipoles at low and high frequencies. The dipole polarization is minimal and the dielectric constant and Dielectric loss is reduced. The dielectric loss can be calculated by using the below formulae

$$\varepsilon * (\omega) = \varepsilon(\omega) - \mathrm{i}\varepsilon'(\omega)$$

where ε is a real number (permittivity), and ε′ is an imaginary number (dielectric loss) Part of the dielectric constant of a material.

1.5 Flexible Dielectric Nanocomposites

Flexible polymer-based nanocomposites (PNCs) have attracted researchers for frontier research on flexible electronics, sensors, and energy devices. Flexible matrix provides the folding and bending opportunity to device foldable electronics [40, 53–56], fabrication of sensors [50], energy devices [54–59], electromagnetic shielding [60], etc.

As far as flexible energy harvesters are concerned, Kar et al. [59] devised a highly efficient self-cleaning piezoelectric nanogenerator (PSNG) based on 2D SnO_2 nanosheet and PVDF polymer nanocomposite. The schematic representation of the fabrication process of PSNG is depicted in Fig. 1.6a–e. They adopted a simple

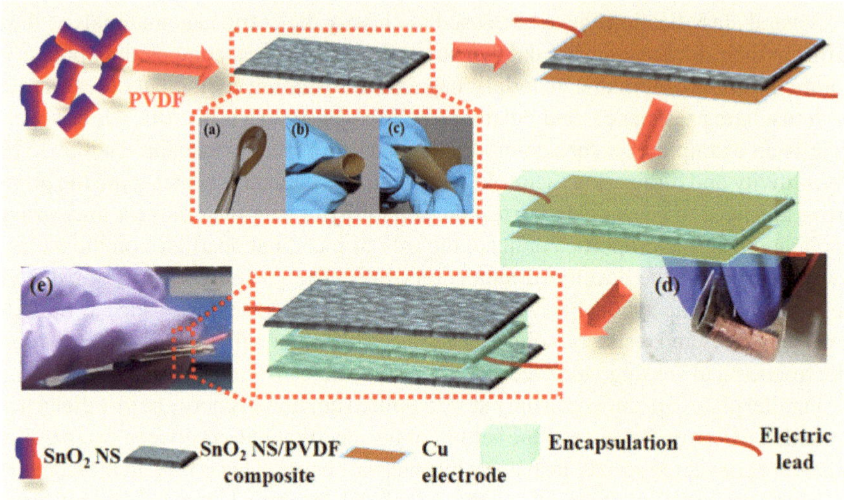

Fig. 1.6 Schematic representation of the process of fabrication for PSNG, Images show **a** bending, **b** rolling, and **c** twisting of the sample S-5.0. Digital images of PSNG **d** before and **e** after covering with SnO_2 NS/PVDF. Adopted with permission from ref. [59], Copyright: Elsevier, 2019

approach of hydrothermal to prepare 2D SnO_2 nanosheets and their composites with PVDF in different concentrations of SnO_2 were synthesized using the facile solution casting method in thickness $\sim 100 \pm 5$ μm. So formed self-standing PNC films are easy to bend, fold, twist, and roll as shown in Fig. 1.6a–c.

A summary of this entire work is schematically represented in Fig. 1.7a. The PSNG is capable of energy generation (illumination of LED panel) by body movement, sound energy of the string of a guitar, air blowing, and the self-cleaning property of 2D SnO_2 nanosheet-PVDF nanocomposites. Figure 1.7b shows the real-time applications of this PSNG under bio-mechanical pressures, particularly with human body movements like tapping by finger, twisting by the wrist, pressure by heel and toe, etc. The PSNG is capable and efficient to illuminate a panel of 85 blue and yellow LEDs directly and instantly without any capacitor or storage through a rectifier circuit connection using gentle imparting of human fingers. Other ways to generate power are depicted in Fig. 1.7c–f. The power generation using this PSNG was investigated under air blowing by an air-blower (Fig. 1.7g) and sound using a speaker of 3W (Fig. 1.7h).

Polymer-based flexible nanodielectrics are immensely beneficial for biomedical and futuristic robotics as well. The electroactive polymers (EAPs) exhibit a change in shape or size if subjected to an external bias. Advanced robotics find numerous applications of these EAPs as artificial skin to the human-like robots. EAPs are preferred owing to their fantastic dielectric properties, flexibility, robustness, lightweight, faster response as well as economic [61].

Numerous efforts have been carried out to realize this futuristic and revolutionary research. Particularly, capacitive type sensors are extremely useful for different

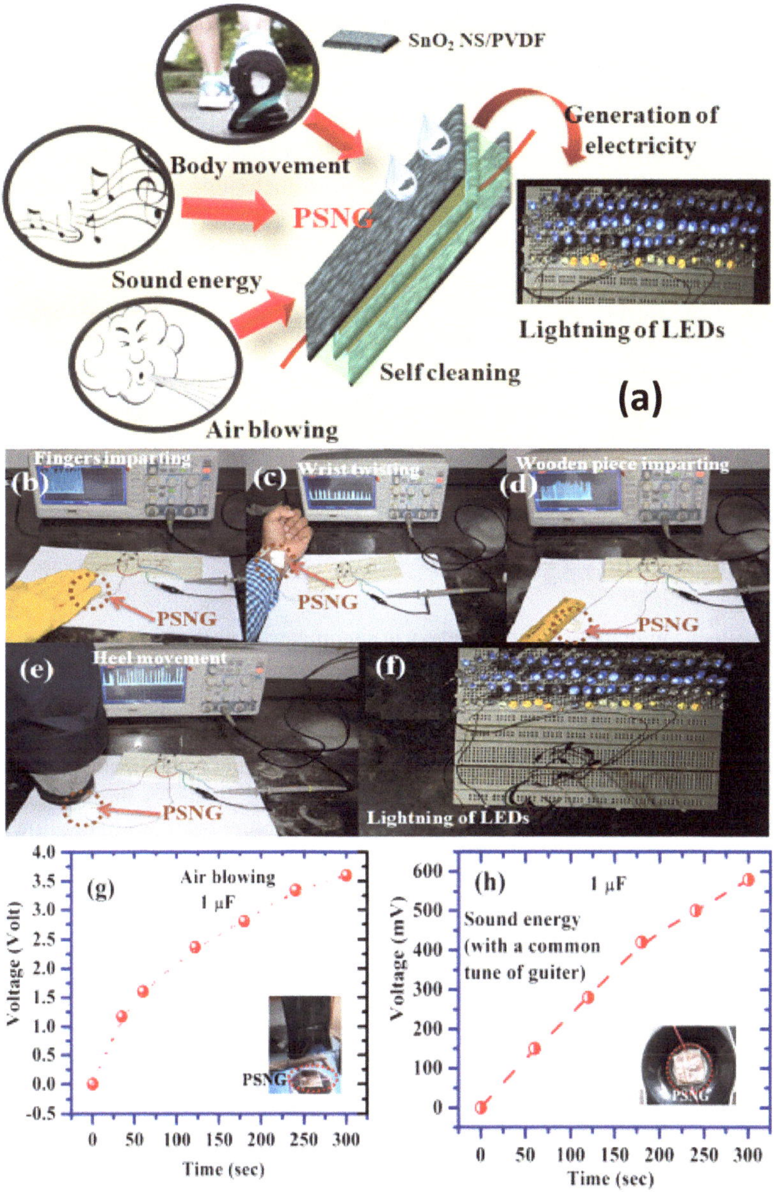

Fig. 1.7 **a** Schematic representation of the function of the flexible and self-cleaning energy harvester. Digital images of piezoelectric power generation (rectified output voltage) by PSNG due to **b** fingers imparting, **c** wrist twisting, **d** wooden piece imparting, **e** heel pressure. **f** Digital image of instant illumination of 85 numbers of commercial LEDs by simple finger imparting onto the upper surface of PSNG. Charging capacitor (1μF) by power generated from the PSNG via **g** air blowing, and **h** sound (a common tune of guitar) from a speaker using a full wave rectifier. Capacitor voltage reached around 3.5 V for air blowing of250s duration. Adopted with permission from ref. [59], Copyright: Elsevier, 2019

biomedical applications owing to their low power consumption, environment friendliness, and sustainability in adverse conditions. Their promising applications are health monitoring can be done by wearable sensors, electronic skin (e-skin) for artificial intelligent appliances, etc. [62]. Dielectric elastomers (DE) are a special class of EPAs, which are designed by sandwiching a thin and soft layer of polymer between two compliant electrodes and are extremely promising for making muscles of human-like robots. In this row, simulations of human facial expressions on the artificial skin were experimented with by Kwak et al. [63]. Additionally, the electromechanical properties of Des are extensively useful as touch sensors and functional sensors as well as the interconnects for the artificial outer skin of the robot to the circuits inside [64]. In other work, multi-walled CNTs were used as fillers in DEs to prepare nanocomposite. Further, their electrical and shape properties were examined, and it was observed that less power is required with better deformation properties.

1.5.1 Challenges with Present Dielectric Nanomaterials and Future Perspectives

The challenge with these types of dielectric nanomaterials is the separation of conductive particles of porous ferrite or dielectric nanomaterials samples, the value of the dielectric parameter shows a very small change, and it should obey Koop's phenomenology and Maxwell–Wagner polarization ease to show that the dielectric dispersion pattern was also elucidated by Maxwell–Wagner polarization, following Koop's phenomenology. It states that conductive particle shows the separation of the porous ferrite or dielectric nanomaterials. Nanocomposites sample allow very small changes in the numbers of the dielectric properties which are breakdown value, dielectric loss tangent and permittivity, resistance, and conductance which result from band gap variation due to grain boundary. Maximum electrical permittivity values arise due to the presence of thin grain boundaries to keep away Koop's hypothesis and the Maxwell–Wagner model of interfacial polarization. The connection between frequency and AC conductivity is explained with Koop's theory for different nanodielectric materials. Here, grain boundaries that possess increased resistance at low frequencies result in maintaining constant conductivity. The rise in AC conductivity values occurs in the high-frequency region where higher conductivity of the grains when compared to the grain boundaries of the corresponding dielectric spectrum is dominated by polarization processes that start from the grain boundaries of ferrite and dielectric nanoparticles. The thickness of the grain boundaries determines the performance of the dielectric constant as they are inversely proportional. The Maxwell–Wagner model at low frequencies is widely used to define grain boundary polarization standardized materials. Johnsher's power law is applied to heterogeneous disordered nanodielectric solid materials or multiphase materials built up with interfaces and defects which are responsible for the electrical properties of ferrites. Koop's theory has been also used to explain the correlation between AC conductivity and frequency for different ferrites.

Integrating the sensor scattering property of silicon nanoparticles and the electrons present inside the atom shows a magnetic dipole due to its spinning nature around the nuclei of silicon nanoparticles to escape thermal forfeiture. Anapole mode is an anon-emitting radio rays source that exists when electric and circulating dipole is present on the surface of an imaginary torus along the plane (toroidal dipole) suitable for refractometric sensors. The nanoparticles which have large dipole show a high dielectric constant. Refractometric sensors are currently used in the field of food and cell membrane application in the biomedical field. Organic dye-doped and nanoparticles including Fe and Ni combination particles doped in hydrogel exist with high dielectric constant values respective to their frequency range by impedance hence proved by using Koops theory and Maxwell–Wagner approach.

In general, Maxwell–Wagner polarization reduces the efficiency of nanomaterials. To avoid polarization, the cyclability of the connections of the lithium-ion batteries should be improved. Rare-earth metal doping of $LiR_xMn_{2-x}O_4$ nanoparticles improved their structural stability and increased their dielectric constant for memory devices. The dielectric barrier discharge method (DBD) is diverse from another synthesis of C_3N_4-Mn_3O_4 nanoparticles nonthermal plasma formed in between the two electrodes separated on the application of AC potential. DBDs include the ionization of the surrounding air followed by an H_2S gas atmosphere, producing a variety of reactive species for gas detection in sensors. Reducing the size of the nanoparticles increases the surface area resulting in increased sensitivity and faster responses in both response time and recovery time.

Dielectric nanomaterials have created an impact on the currently manufactured devices for dielectric applications from health care to communications engineering. This has paved the way for the future generation of novel smart technologies for harnessing nonrenewable energy resources from sensors to microelectronic devices as given below in Fig. 1.8.

1.6 Conclusions

Nanomaterials have significantly enhanced the properties of dielectric devices, and it is currently applied across all industries and in device manufacturing. The twenty-first century is the move toward protecting our natural fossil resources and living in a green environment free of pollution. The demonstration of e-vehicles as a challenge to the present carbon-based fuels had increased our dependence on nanodielectric materials for developing smart sensors and fast charging tools. Renewable energy resources have to be optimally used to build a sustainable future for our people on earth in the coming decades where the requirement of reusable materials for batteries, better storage capacity, and monitoring their performance remotely with sensor and artificial intelligence technologies would be the next challenge. Nanodielectrics are now in the forefront of frontier technologies which scientists are working up to technical

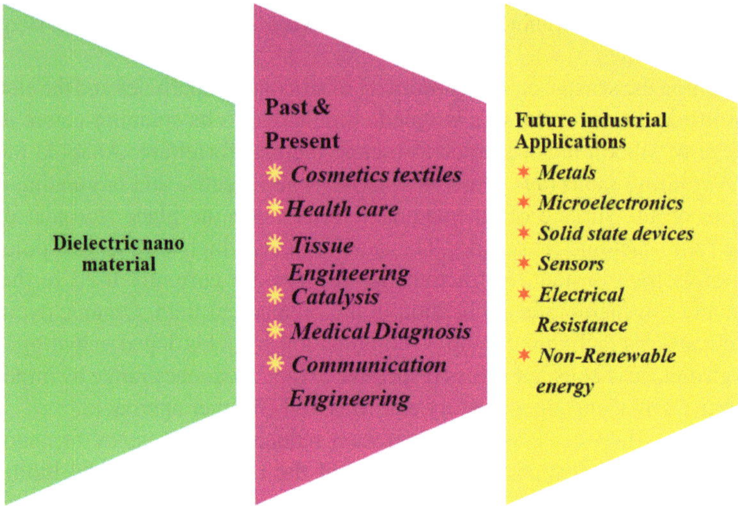

Fig. 1.8 **a** Schematic representation of past, present, and future applications of dielectric nanomaterial

produce products and devices using flexible polymer-matrixes, doped metal semiconducting oxides, and hybrid nanomaterials. The advancement in the development of nanomaterials for dielectric applications would further lead to the development of a sustainable economy by harnessing of nature through green energy routes and containing pollution.

Acknowledgements Ms. Chandralekha acknowledges the financial support of the Government of Tamil Nadu fellowship Under Department of Collegiate Education (Ref. No. 31350/1/2020). The authors acknowledge RUSA-MHRD (COMMUNICATION No: C3/RI & QI/PF4-Appoint/Theme-2/ Group-1/2021/150 for the financial support and facilities under UGC SAP DRS-I (Ref. No. F.540/16/DRS-I/2016 (SAP-I)), Department of Analytical Chemistry, University of Madras, Guindy Campus, Chennai-600025, Tamil Nadu, India.

References

1. Abadias G, Chason E, Keckes J, Sebastiani M, Thompson GB, Barthel E, Doll GL, Murray CE, Stoessel CH, Martinu L (2018) Stress in thin films and coatings: current status, challenges, and prospects. J Vac Sci Technol, A: Vac, Surf Films 36(2):020801
2. Blatsi C, Patsidis AC, Psarras GC (2022) Dielectric properties and energy storage of hybrid/boron nitride/titanium carbide/epoxy nanocomposites. J Compos Sci 6(9):259
3. Mahmood A, Naeem A, Mahmood T (2017) High-k polymer nanocomposites for energy storage applications. Polymer dielectrics: properties and applications of, p 23
4. Khan I, Saeed K, Khan I (2019) Nanoparticles: properties, applications and toxicities. Arab J Chem 12(7):908–931

5. Niu Y, Wang H (2019) Dielectric nanomaterials for power energy storage: surface modification and characterization. ACS Appl Nano Mater 2(2):627–642
6. Sundar U, Lao Z, Cook-Chennault K (2020) Enhanced dielectric permittivity of optimized surface modified of barium titanate nanocomposites. Polymers 12(4):827
7. Singh D, Singh N, Garg A, Gupta RK (2019) Engineered thiol anchored Au-$BaTiO_3$/PVDF polymer nanocomposite as efficient dielectric for electronic applications. Compos Sci Technol 174:158–168
8. Wahab YA, Fatmadiana S, Naseer MN, Johan MR, Hamizi NA, Sagadevan S, Akbarzadeh O, Chowdhury ZZ, Sabapathy T, Al Douri Y (2020) Metal oxides powder technology in dielectric materials. In: Metal oxide powder technologies. Elsevier, pp 385–399
9. Song JH, Min SH, Kim SG, Cho Y, Ahn SH (2022) Multi-functionalization strategies using nanomaterials: a review and case study in sensing applications. Int J Precis Eng Manuf-Green Technol 9(1):323–347
10. Lewis TJ (1994) Nanometric dielectrics. IEEE Trans Dielectr Electr Insul 1(5):812–825
11. Pandey JC, Singh M (2021) Dielectric polymer nanocomposites: past advances and future prospects in electrical insulation perspective. SPE Polym 2(4):236–256
12. Zhong SL, Dang ZM, Zhou WY, Cai HW (2018) Past and future on nanodielectrics. IET Nanodielectrics 1(1):41–47
13. Xu C, Song Y, Han M, Zhang H (2021) Portable and wearable self-powered systems based on emerging energy harvesting technology. Microsyst Nanoeng 7(1):1–14
14. Tafete GA, Abera MK, Thothadri G (2022) Review on nanocellulose-based materials for supercapacitors applications. J Energy Storage 48:103938
15. Panwar NL, Kaushik SC, Kothari S (2011) Role of renewable energy sources in environmental protection: a review. Renew Sustain Energy Rev 15(3):1513–1524
16. Jeevanandam J, Barhoum A, Chan YS, Dufresne A, Danquah MK (2018) Review on nanoparticles and nanostructured materials: history, sources, toxicity and regulations. Beilstein J Nanotechnol 9(1):1050–1074
17. Puiggalí J, del Valle LJ, Katsarava R (2019) Other miscellaneous materials and their nanocomposites. In: Nanomaterials and polymer nanocomposites, pp 353–398
18. Coşkun R, Okutan M, Öztürk M, Yalçın O (2019) Experimental model to describe the dielectric response of different dye and nanoparticles doped hydrogels for biological cell membranes and biological systems. J Mol Liq 296:112072
19. Hassan YA, Hu H (2020) Current status of polymer nanocomposite dielectrics for high-temperature applications. Compos Part A Appl Sci Manuf 138:106064
20. Amina SJ, Guo B (2020) A review on the synthesis and functionalization of gold nanoparticles as a drug delivery vehicle. Int J Nanomed 15:9823
21. Anik MI, Mahmud N, Al Masud A, Hasan M (2022) Gold nanoparticles (GNPs) in biomedical and clinical applications: a review. Nano Select 3(4):792–828
22. Mutlu S, Metin E, Yuksel SA, Bayrak U, Nuhoglu C, Arsu N (2021) In-situ photochemical synthesis and dielectric properties of nanocomposite thin films containing Au, Ag and MnO nanoparticles. Eur Polymer J 144:110238
23. Zhang L, Khani MM, Krentz TM (2017) Suppression of space charge in crosslinked polyethylene filled with poly (stearyl methacrylate)-grafted SiO_2 nanoparticles. Appl Phys Lett 110:130903
24. Chu H, Fu C, Xu J, Li W, Qian J, Nie W, Ran X (2020) Carbon-doped inorganic nanoassemblies as fillers to tailor the dielectric and energy storage properties in polymer-based nanocomposites. Mater Des 188:108486
25. Zhang L, Zhou YX, Tian JH (2014) Experiment and simulation of space charge suppression in LDPE/MgO nanocomposite under external DC electric field. J Electrostat 72:252–260
26. Ganea CP, Cîrcu V, Manaila-Maximean D (2020) Effect of titanium oxide nanoparticles on the dielectric properties and ionic conductivity of a new smectic bis-imidazolium salt with dodecyl sulfate anion and cyanobiphenyl mesogenic groups. J Mol Liq 317:113939
27. Wang SJ, Zha JW, Wu YH (2015) Preparation, microstructure and properties of polyethylene/alumina nanocomposites for HVDC insulation. IEEE Trans Dielectr Electr Insul 22(6):3350–3356

28. Chen G, Li ST, Zhong LS (2015) Space charge in nanodielectrics and its impact on electrical performance. In: IEEE 11th International conference on the properties and applications of dielectric materials (ICPADM), Sydney, NSW, Australia, July, pp 36–39
29. Zhou Y, Hu J, Dang B (2017) Effect of different nanoparticles on tuning electrical properties of polypropylene nanocomposites. IEEE Trans Dielectr Electr Insul 24(3):1380–1389
30. Cao Y, Irwin PC, Younsi K (2004) The future of nanodielectrics in the electrical power industry. IEEE Trans Dielectr Electr Insul 11(5):797–807
31. Angel C, Ramon R, Victoria J (2016) Entropy characterisation of overstressed capacitors for life time prediction. J Power Sources 336:272–278
32. Rafael VR, Sheng C, Vladimir P (2012) Electrical and electron paramagnetic resonance spectroscopy characterization of Mn-doped nanostructured TiO_2 for capacitor applications. J Power Sources 210:21–25
33. DeCerbo JN, Bray KR, Merett JN (2015) Analysis of multilayered, nitrogen-doped aluminium oxide and hafnium oxide dielectric films for wide-temperature capacitor applications. Thin Solid Films 590:71–75
34. Huai W, Dennis A, Frede B (2015) Degradation testing and failure analysis of DC film capacitors under high humidity conditions. Microelectron Reliab 55:2007–2011
35. Daniel T, Yang C, Enis T (2013) Nanofiller dispersion in polymer dielectrics. J Mater Sci 4:6–15
36. El-Hag AH, Simon LC, Jayaram SH (2006) Erosion resistance of nano-filled silicone rubber. IEEE Trans Dielectr Electr Insul 13:122–128
37. Huang X, Jiang P, Tanaka T (2011) A review of dielectric polymer composites with high thermal conductivity. IEEE Electr Insul Mag 27:8–16
38. Jiang PK, Chen J, Huang XY (2017) Research status of thermally conductive but electrically insulating polymer nanocomposites (in Chinese). High Voltage Eng 43(9):2791–2799
39. Sengwa RJ, Dhatarwal P, Choudhary S (2020) A comparative study of different metal oxide nanoparticles dispersed PVDF/PEO blend matrix-based advanced multifunctional nanodielectrics for flexible electronic devices. Mater Today Commun 25:101380
40. Wu K, Fang JC, Ma JR. Achieving a collapsible, strong, and highly thermally conductive film based onoriented functionalized boron nitride nanosheets and cellulose nanofiber. ACS Appl Mater Interfaces 9:20035–30045
41. Kozako M, Okazaki Y, Hikita M (2010) Preparation and evaluation of epoxy composite insulating materials toward high thermal conductivity. International conference on solid dielectrics, Potsdam, Germany, July, pp 1–4
42. Sun RH, Yao H, Zhang HB (2016) Decoration of defect-free graphene nanoplatelets with alumina for thermally conductive and electrically insulating epoxy composites. Compos Sci Technol 137:16–23
43. Wu ZJ, Gao S, Chen L (2017) Electrically insulated epoxy nanocomposites reinforced with synergistic core-shell SiO_2@MWCNTs and montmorillonite bifillers. Macromol Chem Phys 218:1700357
44. Seyhan AT, Goncu Y, Durukan O (2017) Silanization of boron nitride nanosheets (BNNSs) through microfluidization and their use for producing thermally conductive and electrically insulating polymer nanocomposites. J Solid State Chem 249:98–107
45. Sun JJ, Yao YM, Zeng XL (2017) Preparation of boron nitride nanosheet/nanofibrillated cellulose nanocomposites with ultrahigh thermal conductivity via engineering interfacial thermal resistance. Adv Mater Interfaces 4:1700563
46. Hong HJ, Kwan SM, Lee DS (2017) Highly flexible and stretchable thermally conductive composite film by polyurethane supported 3D networks for boron nitride. Compos Sci Technol 152:94–100
47. Zhang XM, Zhang JJ, Zhang XL (2017) Toward high efficiency thermally conductive and electrically insulating pathways through uniformly dispersed and highly oriented graphites close-packed with SiC. Compos Sci Technol 150:217–226
48. Tanaka T, Bulinski A, Castellon J et al (2011) Dielectric properties of XLPE/SiO_2 nanocomposites based on CIGRE WG D1.24 cooperative test result. IEEE Trans Dielectr Electr Insul 18(5):1484–1517

49. Kozako M, Fuse N, Ohki Y et al (2004) Surface degradation of polyamide nanocomposites caused by partial discharges using IEC(b) electrodes. IEEE Trans Dielectr Electr Insul 11(5):833–839
50. Tan TCW, Srivastava YK, Ako RT, Wang W, Bhaskaran M, Sriram S, Al-Naib I, Plum E, Singh R (2021) Active control of nanodielectric-induced THz Quasi-BIC in flexible metasurfaces: a platform for modulation and sensing. Adv Mater 33(27):2100836
51. Siddabattuni S, Schuman TP, Dogan F (2013) Dielectric properties of polymer–particle nanocomposites influenced by electronic nature of filler surfaces. ACS Appl Mater Interfaces 5(6):1917–1927
52. Rahman MA, Islam MT, Hossain A, Singh MJ, Albadran SM, Soliman MS, Samsuzzaman M (2022) Preparation of new flexible antenna based on sol–gel synthesized $Mg_xCa_{(0.9-x)}Zn_{0.10}Fe_2O_4$ nanoparticle for microwave imaging applications. J Market Res 20:3579–3591
53. Ambrosio R, Carrillo A, Mota ML, de la Torre K, Torrealba R, Moreno M, Vazquez H, Flores J, Vivaldo I (2018) Polymeric nanocomposites membranes with high permittivity based on PVA-ZnO nanoparticles for potential applications in flexible electronics. Polymers 10:1370
54. Kumar S, Supriya S, Kar M (2018) Enhancement of dielectric constant in polymer-ceramic nanocomposite for flexible electronics and energy storage applications. Compos Sci Technol 157:48–56
55. Morsi MA, Rajeh A, Al-Muntaser AA (2019) Reinforcement of the optical, thermal and electrical properties of PEO based on MWCNTs/Au hybrid fillers: nanodielectrics materials for organoelectronic devices. Compos Part B 173:106957
56. Choudhary S, Sengwa RJ (2018) ZnO nanoparticles dispersed PVA–PVP blend matrix based high performance flexible nanodielectrics for multifunctional microelectronic devices. Curr Appl Phys 18:1041–1058
57. Luo H, Zhou X, Ellingford C, Zhang Y, Chen S, Zhou K, Zhang D, Bowen CR, Wan C (2019) Interface design for high energy density polymer nanocomposites. Chem Soc Rev 48:4424–4465
58. Yan J, Liu M, Jeong YG, Kang W, Li L, Zhao Y, Deng N, Cheng B, Yang G (2019) Performance enhancements in poly(vinylidene fluoride)-based piezoelectric nano- generators for efficient energy harvesting. Nano Energy 56:662–692
59. Kar E, Bose N, Dutta B, Banerjee S, Mukherjee N, Mukherjee S (2019) 2D SnO_2 nanosheet/PVDF composite based flexible, self-cleaning piezoelectric energy harvester, energy convers. Manage 184:600–608
60. Tsonos C, Soin N, Tomara G, Yang B, Psarras GC, Kanapitsasa A, Siores E (2016) Electromagnetic wave absorption properties of ternary poly(vinylidene fluoride)/magnetite nanocomposites with carbon nanotubes and grapheme. RSC Adv 6:1919–1924
61. Liu L, Zhang C, Luo M (2017) A biologically inspired artificial muscle based on fiber-reinforced and electropneumatic dielectric elastomers. Smart Mater Struct 26:085018
62. Chhetry A, Yoon H, Park JY (2017) A flexible and highly sensitive capacitive pressure sensor based on conductive fibers with a microporous dielectric for wearable electronics. J. Mater. Chem. C 5:10068–10076
63. Kwak JW, Chi HJ, Jung KM (2005) A face robot actuated with artificial muscle based on dielectric elastomer. J Mech Sci Technol 19(2):578–588
64. Yoon SG, Chang ST (2017) Microfluidic capacitive sensors with ionic liquid electrodes and CNT/PDMS nanocomposites for simultaneous sensing of pressure and temperature. J Mater Chem C 5:1910–1919

Chapter 2
Synthesis Approaches for Nanodielectric Materials

Vinod Kashyap, Sukhvant Singh, Sumit, and Lipeeka Rout

Abstract It is highly critical to acquire desirable performance and durability of materials for various engineering applications. To overcome these issues, in recent years, studies on nanodielectric material are found to be one of the most dynamic fields of research among scientists. A multi-component nanostructure that can lead to the change of several dielectric properties is termed as nanodielectric. These materials provide extensively high-dielectric properties due to which they can act as a suitable material for energy storage device applications, in particular for capacitors. Nanodielectric materials were associated with high-dielectric and insulating properties. Such properties of nanodielectric material have commonly been observed in nanocomposites and ceramics materials. More recently, nanoparticles were being incorporated in various metal oxide and polymer support to achieve the enhanced dielectric property of the materials. Fundamentally, the basic principle of nanodielectric materials lies on a large internal area of the surface. Owing to their wide range of properties, it is highly necessary to make the nanodielectric material using the proper synthetic procedure. The most important factor to achieve the dielectric property of the material is to synthesize the nanocomposite or to incorporate the nanoparticle with uniform size distribution. A well-dispersed nanoparticle can enhance the dielectric property of the material. Since many variables can affect the dielectric property such as particle size, dispersion, and aspect ratio, synthesis of these materials is of

V. Kashyap · S. Singh
Department of Chemistry, National Institute of Technology Tiruchirappalli, Tiruchirappalli, Tamil Nadu, India

V. Kashyap
Department of Materials Engineering, Indian Institute of Science, Bangalore, India

S. Singh
Centre for Sustainable Technologies (CST), Indian Institute of Science, Bengaluru, India

Sumit
Department of Chemistry, Indian Institute of Technology, Delhi, India

L. Rout (✉)
Department of Chemistry, Madanapalle Institute of Technology & Science, Madanapalle, Andhra Pradesh, India
e-mail: lipika.pooja@gmail.com

S. Moharana et al. (eds.), *Emerging Nanodielectric Materials for Energy Storage*, Nanostructure Science and Technology, https://doi.org/10.1007/978-3-031-40938-7_2

high priority and highly important. Keeping the above in mind, the current chapter will give a thorough study of various synthetic methods which can be utilized for the synthesis of suitable nanodielectric material. In the current chapter, the authors have explained various available methods for the synthesis of nanodielectric material.

Keywords Nanodielectric materials · Sol–gel · Hydrothermal method · Electrospinning · Spray pyrolysis

2.1 Introduction

Presently, it is imperative to increase the use of energy and renewable sources due to the rising energy demand and the depletion of fossil resources [1]. The development of renewable energy generation is one of the major scientific interest in order to achieve the productivity of energy consumption [1]. The development of nanomaterials fills two primary needs: the establishment of their functioning and their applications in the practical field [2]. Since the introduction of DC electricity and the electric bulb by Thomas Edison in the late nineteenth century, electric power has become an effective source of energy for the human world [2].

For their prospective applications, dielectric materials with high dielectric constant and low dielectric loss have drawn a lot of attention [3, 4]. Nanodielectric materials were found to have superior dielectric properties and possess immense interest owing to their potential applications as capacitor materials [5, 6]. These capacitor materials can be further utilized for electronic devices, gate dielectrics, electroactive materials, management of grid energy by power electronics, and also in pulsed power systems [7–9]. High-performance dielectric materials also have been utilized in various industrial applications such as computers, telecommunications, defense, and aerospace [10–12]. In general, the development of pulse power is in much demand, because of the requirement of much energy over a longer period and which further can release it quickly, thus increasing the available instantaneous power [13, 14]. The capacity of electrostatic energy relies upon induced electric field polarization of dielectric materials, which further need high strength of breakdown and high dielectric permittivity.

Previously, few researchers' reported that compared to available conventional ceramic-based dielectric materials, composites based on polymers were found to have better efficiency, good flexibility, minimal expense, and lightweight to become a standard material in the microelectronics industries [15, 16]. They observed that flexible polymer composites with a high dielectric constant are profoundly attractive, and this high dielectric constant of polymer nanocomposites is of great interest for various practical applications [17]. Similarly, over the past few decades' successful commercialization was made for organic light-emitting diodes (OLEDs) using organic electronics. One of the emerging device building blocks is organic field-effect transistors (OFETs) which have shown higher efficiency compared to amorphous silicon thin-film transistors [18], essentially utilizing the properties of flexible and printable materials. In recent years, applications of OFET in backplane displays [19], biosensors [20], and electronic devices were broadly investigated [21, 22]. Moreover, it is believed that in near future these types of materials with high synthetic tenability, multifunctional properties, and high interface sensitivity are most likely to expand the scope of technological development [23].

Additionally, other forms of nanodielectric materials are available which have been used for numerous industrial applications [24]. In the current instance, biaxially oriented polypropylene (BOPP) is broadly applied as a dielectric material for the storage of electrical energy, because of its low dielectric loss, high breakdown strength, and easy way of manufacturing. Recently, poly (vinylidene fluoride) (PVDF) and its copolymers were found to be the most prevalent dielectric material for the application of capacitors with high energy density [25]. Also, a great deal of effort has gone into creating PVDF-based polymers by mixing, blending, coating, spinning, and nanocomposite production. Among these methods, nanocomposites have demonstrated advantages for enhancing the impact of the inorganic nanofillers on the dielectric permittivity [26].

To develop such materials that are very flexible, those having a high dielectric constant, and are less expensive, efforts have been made to synthesize these materials using various synthetic techniques. There are several techniques available for the preparation of numerous forms of nanodielectric materials. Every synthetic strategy has advantages and disadvantages depending on the reaction time, particle size, thickness of the film, etc.

Keeping all the above in mind, in the present chapter, our prime focus is to discuss the novel and cost-effective synthetic routes of nanodielectric materials. There are several available methods for the preparation of materials with excellent dielectric properties. A schematic of available synthesis methods for nanodielectric materials synthesis are shown in Fig. 2.1.

2.2 Introduction to Various Synthetic Routes for Nanodielectric Materials Synthesis

It is vitally important to create and design materials for use in electronic applications. While examining the dielectric characteristics of nanodielectric, the production of the materials was frequently overlooked. The physical characteristics of the raw materials employed in the preparation process are what give rise to the physical characteristics of the composite material, it has been noted [27]. To achieve various properties and various forms of nanodielectric materials for a diverse range of applications, it is very important and essential to synthesize those materials with suitable techniques and under optimized conditions.

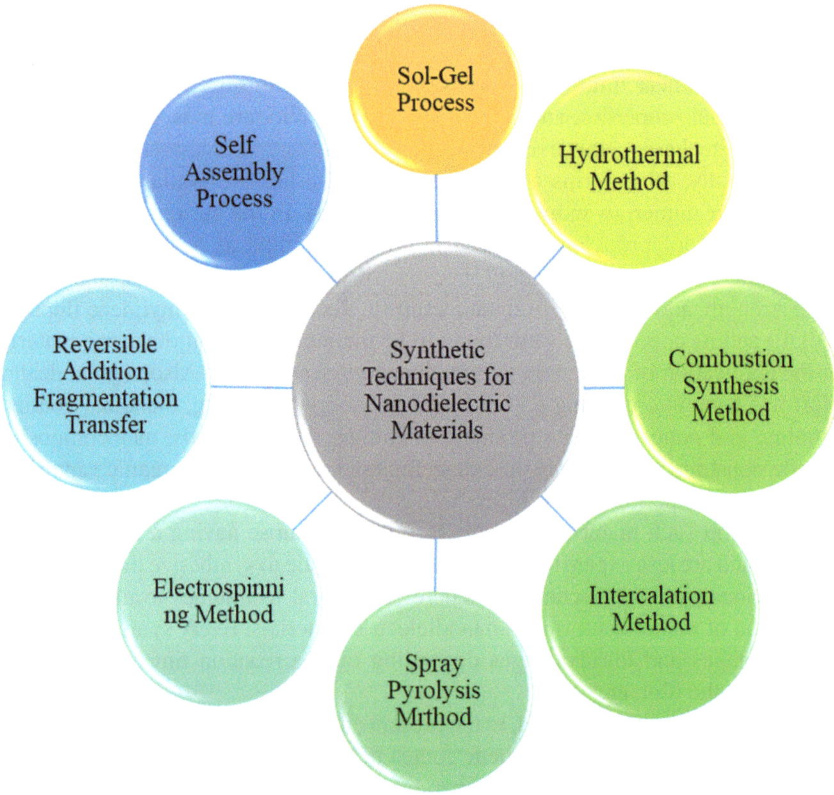

Fig. 2.1 Various synthetic techniques for nanodielectric materials synthesis

In recent years, highly cost-effective synthetic techniques have been adapted to create various forms of nanodielectric materials such as polymer-based, nanoparticle-based, and ceramic-based [28]. The enormous synthetic techniques for nanodielectric materials have attracted a lot of research interest, as they provide material with a high dielectric constant which is highly required for various applications. Recently, researchers have tried to overcome the difficulties associated with the development of nanodielectric with effective dielectric properties for devices and applications. Mostly, nanodielectric properties are observed in nanocomposites and ceramic materials [29]. In recent years, it has been found that in ceramics form nanostructures were obtained to large extent. Notably, it has been observed that nanocrystalline powder can be used as starting material for the synthesis of nanodielectric materials. The main focus was being made to synthesize nanostructure materials with dielectric property using novel concepts and approaches. There was various novel approached which has been utilized to obtain nanodielectric materials [30, 31]. The basic principle of materials nanodielectric mostly relies on its large internal surface area. In order to engineer these materials, one needs to select a specific aspect ratio, apply internal modifications, and change their morphology. Nanodielectric can also be fabricated using organic oxide materials. The traditional method for making dielectric materials with nanoparticles in them required combining polymer materials with nanoparticle precursors to create a composite dielectric material. Till date, various novel synthetic routes have been utilized to form highly effective nanodielectric materials for several industrial applications. The details of each synthetic techniques have been discussed in the later part of the chapter.

2.3 Sol–gel Technique for Nanodielectric Materials Synthesis

A sol–gel synthetic method is a bottom-up approach in which components of atomic dimensions are assembled to form nanomaterials [32]. The benefit of this technique is that particle size may be controlled. It is a method of wet chemical synthesis used to produce fibers, thin films, and powders of ceramic, dielectric, and glass materials [33–35]. A sol is a solid ion or particle suspension in a solvent, either colloidal or molecularly. When the solvent in the sol starts to evaporate and the particles left behind start to join up in a continuous network, a gel is created [36]. In this method, metal alkoxides and chlorides are typically utilized as precursors. During the synthesis technique, pH must be managed because this process is sensitive to it, and it will help to prevent precipitation and create a homogeneous gel.

The following steps occurred in the general sol–gel process (Fig. 2.2):

1. *Hydrolysis*—Metal precursors undergo hydrolysis by water in presence of catalysts such as HCl or ammonia.

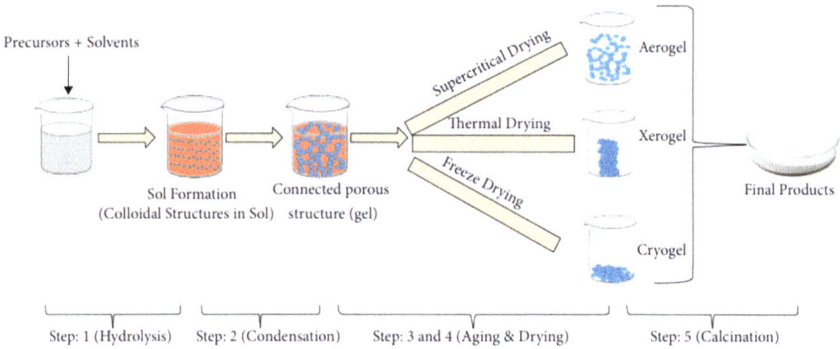

Fig. 2.2 Schematic diagram of different stages of the sol–gel process: from precursor to aerogel [1–5, 32]. Adapted with permission from ref. [32]. Open access under a CC BY 4.0 license. Copyright © 2021 Dmitry Bokov et al.

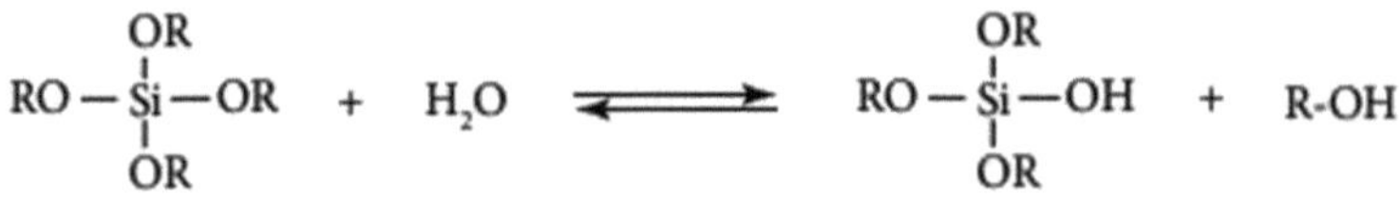

2. *Condensation*—Condensation reactions that either produce water or alcohol are used to polymerize siloxane to create bonds. Further, this results in the formation of monomers, dimers, cyclic trimers, and higher-order rings.

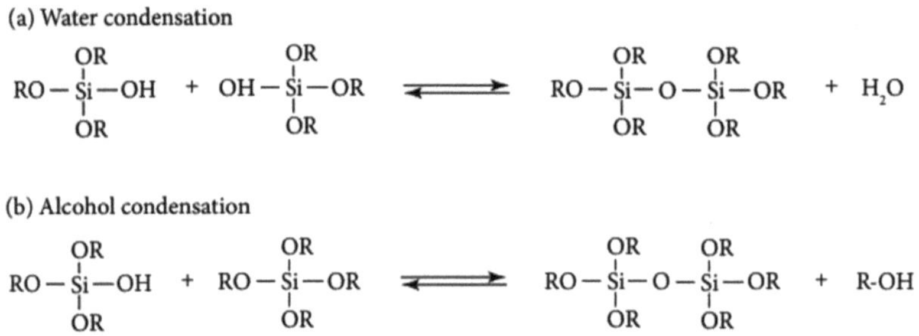

3. *Growth of material*—The molecules start to aggregate with an increase in the number of siloxane bonds where they create a network, and when they dry, gel forms. The network contracts as the water and alcohol are driven away. The formation of spherical nanoparticles occurs at pH values greater than thee $\frac{H_2O}{Si}$ the ratio which ranges from 7 to 5.

High-purity and highly efficient nanodielectric materials can be prepared at room temperature using the sol–gel method. Nanodielectric materials with high efficiency have a wide range of applications in optical fibers and high-performance electrical insulation [32]. Dielectric materials are also used in semiconductor materials and capacitors which can store charge. Dielectric materials have applications in instrumental analysis. The instruments use dielectric material in diodes in the detectors and radiant power monitor devices.

Recently, in 2020, N. Murali and co-workers synthesized cobalt ferrite nanodielectric materials by utilizing the sol–gel method [37]. They have investigated the cobalt ferrite's magnetic, dielectric, and structural characteristics. They discovered that the magnetic and dielectric properties of nanomaterials are greatly influenced by several important factors, including particle size, crystallinity, composition, and site occupancy, and the earlier parameters can be effectively managed by adjusting the calcination temperature and calcination time. They have noticed that as the calcination temperature rises, the materials' coercivity diminishes. Additionally, they have researched $CoFe_2O_4$'s dielectric characteristics in the 100 to 5 MHz frequency range.

Many functional groups have been grafted onto metal oxide surfaces using the sol–gel procedures, and their dielectric properties have been further examined by a small number of researchers [38]. Dierkes and co-workers investigated how charge transfer of (PP)/(ethylene-octene) copolymer (EOC)/silica nanodielectrics was affected by 3 aminopropyltriethoxysilane (APTES) functionalized nano-silica [39]. They have incorporated APTES over the silica surface to generate an amine functional group which is responsible to modify the electronic properties and interfacial interaction on the silica-polymer surface. According to a few reports, it has been observed that surface modification can enhance the nucleation rate and crystallinity of the nanodielectrics. Most recently, in 2022, sol–gel technology was used by Ghamarpoor et al. to create silica nanoparticles based on vinyl [40]. They used the sol–gel technique to graft vinyltriethoxysilane (VTES) over silica for this purpose. The grafting was required to create strong interaction between the surface of the silica and nitrile rubber (NBR) Matrix.

A lot of interest has been paid to thin-film nanodielectric in addition to powder-based and polymer-based nanodielectric. Aoki et al. have reported the synthesis of HfO_2 nano-films using the sol–gel method ad further studied their dielectric properties. They stated that the sol–gel method and post-annealing can be used to create a high-quality metal oxide nano-film. In this thin-film manufacturing procedure, metal oxide precursors were first made to chemisorb on the silicon wafer in an organic solvent to create a homogeneous layer. Next, the alkoxide group was hydrolyzed. Later, the deposited film was annealed for a few cycles to get uniform nanodielectric thin film [41]. Similar to this, Jian et al. published in 2005 on the sol–gel produced $ZnFe_2O_4$–SiO_2 composite thin film's dielectric behavior [42].

A straightforward method called sol–gel can be utilized to create high-purity products with a high level of production efficiency. Using this method, optical components with complex shapes can be synthesized. The process can design and control chemical homogeneous composition. This technique can be used to create porous and rich materials with organic and polymeric properties, as well as thin layers of amorphous materials [43]. Precursors are highly reactive due to the solution-phase process. The structure of a material can be altered by modifying factors like pH, temperature, etc. This method is capable of sintering at low temperatures, typically between 200 and 600 °C. The sol–gel method is versatile as it can produce aerogels, xerogels, ceramic materials, nanopowders, nanorods, nanostructures thin films, etc. Apart from the merits, the sol–gel process has some demerits too. Nanoparticles (NPs), sometimes, get contacted during agglomeration. Processing time is a long, and large production of NPs is difficult. The use of toxic chemical solvents is harmful to both humans and the environment [44]. One of the major problems is the purification of NPs after the synthesis. Separation of the NPs from the solvent consumes a lot of time and generates toxic residues.

2.4 Hydrothermal Synthesis of Nanodielectric Material

In a hydrothermal method, nanoparticles are created through hydrolysis processes that take place at high temperatures and high vapor pressure [45]. In this process, materials are obtained by the crystallization of phases under high temperatures and pressure. The method can be used at room temperature and below critical pressure [46]. This technique is dependent on the solubility of minerals in hot, high-pressure water. Factors affecting the hydrothermal synthesis process are the initial pH of the solution, duration, temperature, and pressure in the autoclave. Typically, an autoclave is used to perform the hydrothermal technique [47]. In this process, a metal precursor solution is mixed in a hydrolyzing solution and kept at a high temperature for 10–24 h in an autoclave, then cooling it down to ambient temperature for crystallization. The resulting crystals are collected using filtration, washing, and drying. Large particle-size nanocrystals can be synthesized using the hydrothermal method which is useful to increase the dielectric constant of the material and hence increases dielectric properties of the synthesized materials. Nanomaterials synthesized using this method tend to withstand high frequencies and high temperatures and therefore can be used in semiconductor devices [48].

In recent years, many researchers have utilized the hydrothermal method for the synthesis of various forms of nanodielectric materials and also studied their dielectric properties for electronic applications. For example, the hydrothermal process was used to create silver nanowires that were then coated with a carbonaceous shell as shown in Fig. 2.3. The synthesis was carried out by using silver nitrate as a metal precursor, Cetyl trimethyl ammonium bromide (CTAB), and glucose [49]. Silver nanowire/manganese dioxide nanowire/poly(methyl methacrylate) (AgNW/ MnO_2NW/PMMA) was synthesized, according to Zerrati et al. Here, they have utilized two types of synthetic techniques to get the final nanocomposite. For the synthesis of Ag, nanowires were synthesized via the hard-template technique, whereas the hydrothermal method was used for the synthesis of MnO_2 nanowires. They observed a high dielectric permittivity of 64 at 8.2 GHz was noted in the synthesized nanocomposite [50].

Similarly, hydrothermal synthesis in supercritical water has an advantage over the conventional method as the rate increases by 103 times under critical conditions. Supercritical water provides control of crystal phase, particle size, and morphology. Titanium dioxide NPs synthesized using this method have high crystallinity and high surface area due to which the dielectric properties of TiO_2 NPs are enhanced. For the synthesis, a slurry of titanium tetra propoxidein an aqueous medium was kept in a gold-tube reactor inside the autoclave. The synthesis is carried out at 400 °C at 25 MPa pressure for 24 h. This system is referred to as the batchwise system [51].

High-dielectric materials are used in diodes, solar cells, transistors, and gates. According to Ahmad et al., the hydrothermal process they used to create cubic ZrO_2 was far more stable at relatively high temperatures. Zirconia has potentially high dielectric properties due to which it has extensive use in gates. They discovered that solid zirconia NPs have a large surface area and function as p-type semiconductors. Additionally, they observed that the dielectric coefficient, which is influenced by the synthesis temperature, crystalline structure, and particle size, determines the band gap of these materials. For the synthesis, a solution of zirconium oxynitrate as a limiting reagent is prepared in NaOH and double distilled water. The solution was then put into a hydrothermal flask, and the precipitate was later cleaned and dried for 1.5 h at 110 °C before being ground [52].

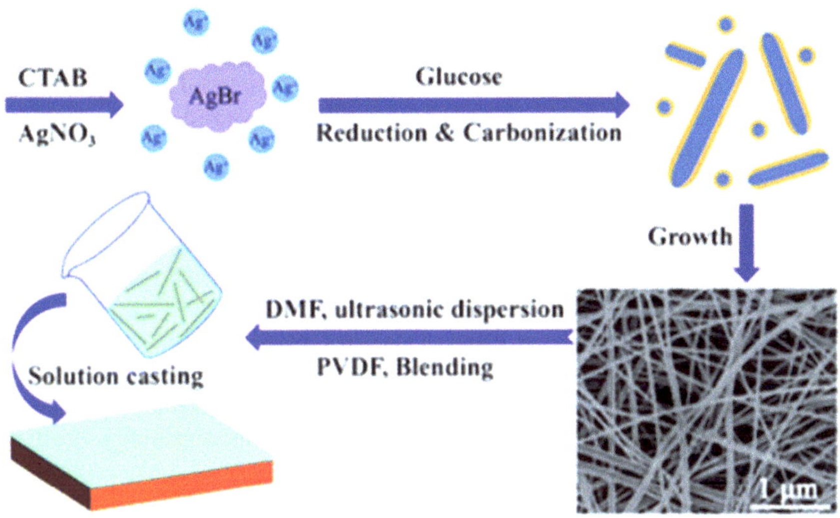

Fig. 2.3 Schematic diagram of the synthesis of Ag@Cnano cables [49]. Adapted with permission from ref [49] Copyright (2018) (Royal society of chemistry)

One of the advantages of the hydrothermal synthesis approach is its propensity to produce unstable or less stable materials at high temperatures. With this technique, one can produce large crystals with good purity. The method is easy and simple with a low cost of synthesis, but the initial setup cost is high [53]. The material's loss is minimal with a high product yield. Besides this, they have a few disadvantages which include longer reaction time, and it is difficulty to observe the crystals as they grow during the synthesis process inside the autoclave as the autoclave is fully equipped with Teflon.

2.5 Combustion Synthesis of Nanodielectric Materials

One of the most utilized processes for fabricating a variety of oxide materials is solution combustion synthesis (SCS). K.C. Patil and others invented and perfected this technique in India in the 1980s [54]. When urea and $Al(NO_3)_3$. $9H_2O$ were heated fast to a temperature of 500 °C in a muffle furnace (preheated), the first unintentional synthesis of α-alumina occurred. A large amount of white material that was recognized as α-Al_2O_3 was produced during the reaction, and the mixture foamed and burned with an incandescent flame. Concepts from propellant chemistry were applied to properly learn about the combustion reaction's high exothermic nature. The energy released during the reaction is at its highest, and there is no carbon residue left once the combustion is finished. To create pigments, phosphors, and catalysts, the host oxide matrix can be doped with the appropriate amount of metal ions. By using this technique, pink alumina was produced by adding an appropriate amount of chromium nitrate to the redox mixture of urea and aluminum nitrate, which uniformly doped the alumina matrix about 1% of Cr. Later, hydrazides were employed in place of urea to produce a wide range of oxides. Patil's group employed hydrazides such as malonic acid dihydrazide, oxalyldihydrazide, maleic hydrazide, and carbohydrazide [55]. Other researchers investigated various fuels, including glycine, alanine, hexamine, citric acid, etc.

Due to their high carrier mobilities, outstanding thermal stability, mechanical flexibility, and optical transparency, metal-oxide semiconductors and dielectrics, particularly in amorphous phases, constitute potential materials for next-generation electronics. The development of metal oxide films, particularly those for thin-film transistors, is widely performed via sol–gel processes. However, the necessary stages of sol–gel condensation, densification, and contamination removal are commonly needed for processing temperatures of > 500 °C, which are incompatible with affordable glasses and conventional flexible polymeric substrates. A unique variant of the SCS approach was recently developed by Marks and co-researchers at Northwestern University to fabricate thin-film transistors for next-generation electronics. To improve the preparation conditions and attributes of the synthesized films, many studies have already been published. We predict that, in addition to the production of electrical devices, combustion-based deposition techniques may have a significant impact on the creation of dielectric nanomaterials and electrochemical devices [56]. Controlling the type of combustion allowed for the modification of solid combustion product particle size, and by employing fuels that produce (i) the rate at which the fuel burns is slow, (ii) a decrease in exothermicity (making the reaction smolders instead of flam), and (iii) an increase in the number of gaseous products (H_2O, N_2, CO_2, etc., in which the heat is dissipated). Unlike volume combustion, linear combustion appears to produce oxide powders that are nanoscale in size. Additionally, it was found that metal acetates enhanced the synthesis of nanoscale oxides.

By using the solution combustion synthesis method, Kim et al. demonstrated a straightforward and relatively low-temperature (at 200 °C) synthesis technique for the construction of metal oxide thin-film transistors.In_2O_3, Zn-Sn–O, In-Zn–O, and In–Sn–O are materials used to create metal oxide thin-film transistors (ITO), and they incorporated NH_4NO_3 that had been dissolved in 2-methoxy ethanol with metal sources including $In(NO_3)_3$, $Zn(NO_3)_2$, and $SnCl_2$ to produce the precursor solution. To control the pH of the solution, they add ammonia solution, and precursor solutions were maintained for 72 h before film casting. Then, silicon wafers were spin-coated with solutions containing 0.05 M of total metal concentration, and they were heated for 30 min at 150–400 °C. Solution combustion synthesis-derived transistors with thin-film layers was found to be highly efficient with saturation mobility of 6.0 cm^2/V s. The authors claim that this property is far better than the mobility of amorphous silicon, which is now employed widely in large-area electronic applications (below 1 cm^2/Vs) [57]. The schematic of the device structure has been presented in Fig. 2.4.

Krengvirat and Sreekantan reported bismuth titanate ceramics which is considered as well knows nanodielectric material, using an intermediate fuel agent-assisted

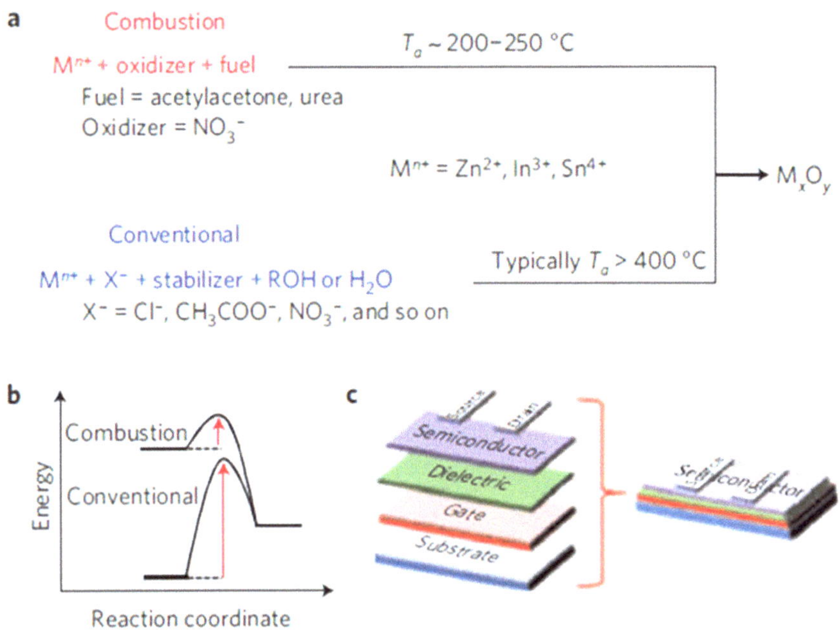

Fig. 2.4 Low-temperature solution-processing principles for metal oxides and fabricated device structures. **a** Depiction of the two different synthetic approaches. **b**, Energetics of combustion synthesis-based processes versus conventional processes. **c**, Top-contact bottom-gate TFT device structure used in this study [57]. Adapted with permission from ref. [57] Copyright (2018) (Springer Nature)

self-combustion synthesis (SCS) technique to replace praseodymium [58]. Similarly, Sekar et al. reported (Pb- BNb)O_3 (where B is Fe, Ni, Mg, or Zn), a lead-based niobate material based on two different combustion synthesis techniques. In the first combustion synthesis technique, they ignited the solutions of metal nitrates, NH_4NO_3, tetraformaltrisazine, and $Nb_2(C_2O_4)_5$ (Niobium oxalate) in a furnace (which is preheated) through a single-step fabrication approach. In the second combustion synthesis method, solution combustion was used in a two-step process to produce PbO (lead monoxide), $FeNbO_4$, $MgNb_2O_6$, $NiNb_2O_6$, and $ZnNb_2O_6$. They then combined PbO with metal niobates in the following stage, calcined the mixture, and the result was the desired lead-based ferroelectric ceramics. Microstructural investigations revealed that both synthesis methods create powders with good particle structures that can be sintered into compact samples with a relative density of around 97% [59]. They noticed that the ceramics created using the two-step synthesis process demonstrated higher dielectric properties in comparison with materials created using a single-stage synthesis strategy.

The solution combustion synthesis method's key benefits include: (i) the procedure is speedy and requires just basic equipment, (ii) since in this process redox reaction has exothermic nature, it can save energy by using the exothermicity; (iii) that method is allowing the modification in the composition of the product, structural modifications, also that method allows homogeneity as well as stoichiometry modifications of the product; (iv) that method allows to the formation of highly pure materials; (v) to create industrially usable materials, it enables incorporation of the necessary amount of impurity ions or dopants in the oxide hosts; (vi) that it allows metastable phases to be stabilized (γ-Fe_2O_3, anatase TiO_2, etc.); and apart from that (vii) it provides the fabrication of almost any size and shape of the obtained materials, for example, micron to nano size and spherical to hexagonal shape. Magnetic oxides, pigments, phosphors, high Tc cuprates, materials for solid oxide fuel cells, metal oxide catalysts, and durable materials are among the materials mentioned. (e.g., t-ZrO_2/Al_2O_3).

The solution combustion synthesis-derived powders are particularly efficient for removing pollutants from the environment. One-step combustion has been used to successfully deposit nano-sized crystalline ionic catalysts on ceramic cordierite honeycombs.

2.6 Electrospinning Method for Nanodielectric Material Synthesis

An electrostatically charged jet of molten polymer or a polymeric solution is used in the electrospinning (ES) technique to create continuous polymeric or inorganic fibers that can be anywhere between a few nanometers to a few micrometers in size. As depicted in Fig. 2.5, the procedure is carried out via a needle with a polymer solution streaming via it coupled to a grounded electrode for collecting and an HV DC generator [27].

The HV source is used to electrostatically charge a polymer solution, which then jets out of the needle tip as a hanging drop. The drop is distorted by the strong electric field between the grounded electrode and the needle until it becomes the shape of a Taylor cone. The thin fluid polymer jet is created and drawn in the direction of the metal collector when the charged drop experiences an electrostatic force that is greater than surface tension and exceeds a certain level of electric potential. The charged jet is then released. Objects are stretched and accelerated by the electric field and go through an instability process known as whipping instability; the spiral path taken by the fibers speeds up the stretching process, leading to fiber thinning as the solvent evaporates (or the melt polymer solidifies). During this instability process, fibers with diameters in the region of a few hundred nanometers are produced [60]. The jet's uneven motion results in the random deposition of fibers in the nonwoven nanoporous material type materials on the collector. Complex manufacturing procedures are required for the synthesis of high-quality nanocomposites. Electrospinning is now the most used technique in the field of electrical power engineering for creating nanodielectric materials based on polymers. The performance of composites depends on how evenly the fillers are distributed throughout the polymer. The two

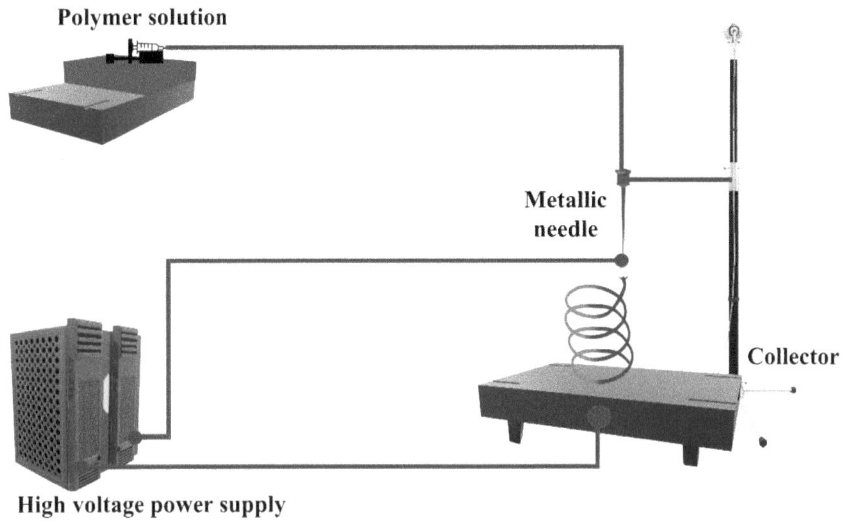

Fig. 2.5 Electrospinning apparatus

most common dispersion techniques utilized today are the solid-phase approach and the liquid-phase method. The solid-phase process, which is straightforward, simple to use, and convenient for large-scale production, under mechanical pressure, combined the matrix and fillers. However, the filler/matrix interface adhesion is poor, and the filler dispersion is not sufficiently uniform. The liquid-phase method, also known as the chemical method, has a higher interface adhesion and a far greater ability to spread fillers. For improving material characteristics, appropriate synthesis and preparation processes are also crucial in addition to material design and dispersion. In the electrospinning process, for instance, it is highly desirable for the development of composite qualities that the fibrous fillers can be layered sequentially along the direction of the fiber orientation to construct the network structure, retain the directional interconnection structure well, and produce no visible interface void or pore [61].

Li et al. reported a polymer-based nanodielectric material with high energy densities for electrical energy storage applications. They create uniform $BaTiO_3$ nanofibers (BT nfs) with a high aspect ratio, which are used as fillers in a poly(vinylidene fluoride-hexafluoropropylene) (PVDF-HFP) nanocomposite after being electrospun and surface-modified using poly(vinyl pyrrolidone). A straightforward solution-cast approach was used to create the nanocomposite films. They discovered that, compared to the polymer matrix and the nanocomposites with BT nfs, the dielectric constant and the energy density of the nanocomposites are significantly increased with a little loading of BT nfs [62]. Similarly, in 2017, Lin and co-workers prepared high-temperature resistance and high thermally stable core–shell barium titanate@silicon dioxide (BT@SiO_2) nanofibers with polyimide using the electrospinning technique as presented in Fig. 2.6. They have reported that the dielectric

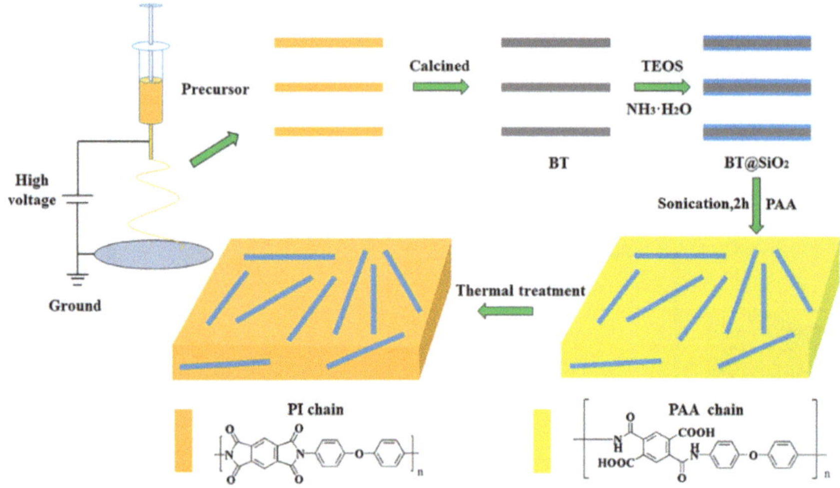

Fig. 2.6 Fabrication procedures of BT@SiO_2/PI nanocomposite films [63]. Adapted with permission from ref [63] Copyright (2017) (Elsevier)

permittivity of nanocomposite film increases significantly as the presence of thin SiO_2 layers was utilized to restrict the high dielectric loss from the nanofibers [63].

The ES method is thought to be straightforward, affordable, and useable in an industrial setting and creates extremely porous nanocomposite with excellent inorganic nanoparticle dispersion. For example, high-performance polymeric nanocomposite separators for Li-ion batteries could be made using ES. In another study, silicone rubber was mixed with nano-silica using the ES approach to lessen particle aggregation [64]. Nano-silica was electrospun into silicone fibers, which were then integrated into a silicone rubber matrix. Comparing ES to high-shear melt compounding, it has been demonstrated that a greater volume percentage of nano-silica may be disseminated more evenly into silicone rubber. In addition to this, a few limitations of the ES method include the use of toxic organic solvents, the difficulty to maintain uniform porosity of materials, and also, difficult to synthesize nanomaterials on large scale.

2.7 Intercalation Method for Nanodielectric Materials

Mostly, the intercalation method is used to synthesize polymer-based nanodielectric materials. PCNs, or polymer-clay nanocomposites, have received a lot of interest due to their exceptional properties, which include high dielectric properties, high thermal, and gas permeability, and pharmaceutical properties like optical, electrical, and mechanical at the nanoscaled level as compared to their pristine polymers. Dielectric properties of PCNs (acrylonitrile butadiene styrene and poly methyl methacrylate) with altering amounts (up to 10 wt%) of organically modified montmorillonite (OMMT) in the frequency range of 100 Hz^{-1} to MHz showed significant greater dielectric constant, and compared to pure polymers, exfoliated nanocomposites have greater dielectric loss values owing to the increased interfacial area between the clay layers and the polymer, which is driven by Maxwell–Wagner polarization. Pure polymers were shown to have higher dielectric constant and dielectric loss values in the same frequency range as intercalated nanocomposites, which is explained by the reduced mobility of polymer chains in the nanoscale pathways of OMMT [65].

These discoveries can be explained in terms of unique microstructures created in nanocomposite films after clay has completely laminated, exfoliated, and distributed in the polymer matrix with exfoliated polymer nanocomposites. The exfoliated OMMT and polymer matrix form a substantial interfacial area. In an applied electric field, different conductivities in the two phases lead to a significant interfacial or space charge polarization near the interface, due to trapped charges, and this phenomenon is known as Maxwell–Wagner polarization. As the OMMT content in exfoliated nanocomposites rises, the area at the interface between the polymers and OMMT expands, and as a result, the dielectric constant rises. Lower real and imaginary components of dielectric permittivity are displayed in intercalated nanocomposites than those displayed by the corresponding pristine polymers because the polymer chains are restricted between OMMT layers.

Polymer–clay nanocomposites (PCNs) are also known as organic–inorganic nanocomposites, and a suitable synthetic route can be utilized to control their useful properties [66]. The intercalation approach for the preparation of nanodielectric materials is most widely utilized for the preparation of PCNs; e.g., the clay, Nylon-6 hybrid was produced by the intercalation process and polymerization, and it is the prior nanocomposite developed by the Toyota group [67]. In general, the intercalation process is achieved by incorporating polymers within the layers of clay having dimensions in the range: from the thickness of a few nanometers to ~100 nm, length of 0.5 μm to as much as mm, and clay includes natural and synthetic clays, kaolinite, micas, and further layered inorganics [28] depending upon the exfoliated and intercalated hybrid techniques [28]. Moreover, when intercalate was requisite, organic materials were made to immerse within the layers of clay which further expands inside the component mixtures. Further, it is made to exfoliate the complete layers of clay materials that were distinct from one another but at the same time tied up inside the matrix of the organic component [66]. This strategy is a reversible technique that includes a collection of substances with layered structures. In PCN materials, three distinct phases occur from hydrogen bonding interactions between the clay intercalated/exfoliated at the nanoscale and the polymer (an organic substance) (an inorganic material). These steps are as follows:

(1) Microphase-separated composites, in which clay and polymer matrix are still incompatible,

(2) Intercalated clay assemblies, where polymer molecules infiltrate among clay galleries,
(3) Exfoliated structures, wherein the polymer matrix is filled with individual clay nanoplatelets [66].

Furthermore, intercalation or exfoliation is more feasible for a continuous in-line process than spherical inorganic nanodielectric, which can be produced in the lab more conveniently through a batch technique.

PEO is an etheric-linked simple chain polymer, contrary to PVA, which is a polymer with a backbone made of carbon chains and hydroxyl groups connected to carbons of methane. These polymer matrices composite (PMC) are frequently used in conjunction with montmorillonite clay for the development of organic–inorganic nanocomposite (NC) materials because they have a wide range of H-bond forming potential for the hydroxyl group (OH) of PVA monomer units and the ether oxygen of PEO monomer units. MMT clay is typically made up of linked silica and alumina sheets with an aspect ratio (length/width) of 100–1000 nm and a thickness of about 1 nm. To allow polar solvents and polymers to easily access the interlayer galleries, the cation quantity ratio can be changed depending on the source of the minerals, and its cation exchange mechanism can exhibit more hydrophilic organophilic MMT. This causes the exfoliation and intercalation of the polymer-layered silicate nanostructures.

In 2010, Sengwa et al. synthesized PEO-PVA blend-MMT clay nanocomposites using an intercalation process, and further, they presented the dielectric characteristics of the material in which their dielectric analysis showed that by incorporating 0.5–2 weight percent MMT clay into the PVA-PEO blend, the ϵ' value may be adjusted to be close to 1 [66]. As a result, the polymer-layered silicate nanostructures' end matrix is exfoliated and intercalated, enabling the use of these structures as low-dielectric constant nanodielectric with improved mechanical and thermal properties in low-frequency microelectronic technology. They created a PEO-PVA blend-MMT clay nanocomposite. In their research, they describe how they intercalated and exfoliated colloidally stable MMT clay in the suspension of the PVA-PEO blend matrix over a subsequent 24 h. Subsequently, adding clay hydrocolloids to 60 mm stainless steel rings, the PVA-PEO blend-MMT clay nanocomposite was cast and allowed to dry for a week at room temperature. Meanwhile, in the intercalation process, the miscibility of PVA and PEO is facilitated by the rise in polymer chain segmental motion relaxation time with increasing MMT clay concentration up to 2% by weight which is primarily due to intercalated and exfoliated nanostructures of MMT clay sheets allowing polymer chains to be connected by H-bonds.

Recently, exfoliated or Intercalated Montmorillonite (MMT)—PANI nanocomposites (PANC) using emulsion intercalation were reported by Hundekal et al. [68]. They have chosen to utilize a layered silicate known as natural montmorillonite (MMT) as a filler. With an increase in frequency, it is evident that the values of the dielectric constant and dielectric loss decrease. This drops significantly sharply after the onset frequency of the AC conductivity. All PANC nanocomposites' dielectric permittivity significantly increased at lower frequencies (less than 100 Hz), and this demonstrates that in this frequency range, the ionic conduction mechanism predominates.

For the PANI synthesized in this work, a high dielectric constant was observed, which is significantly greater than that stated in the literature [69, 70]. This might be a result of the PANI exhibiting nanocrystalline domains instead of microdomains simply because it was exposed to mechanical shear. They have observed that hyper-electronic polarization and a significant polar on delocalization may be contributing to the high dielectric response for the nanocrystalline PANC at a lower frequency (100 Hz–1 kHz). They have demonstrated that the use of swelling agents in combination with significant shear forces is the most efficient technique of exfoliation or intercalation, and the versatility of clay molecule orientation in the solid state offers a wider route for conduction in the composite.

In 2008, Akat and his co-workers reported a novel synthetic route for the preparation of PNCs which is considered efficient nanodielectric materials by employing a chain transfer agent that is intercalated in the process of traditional free radical polymerization [71]. This method of synthesis is extremely simple, and a facile quaternization procedure yields the interactive chain transfer agent from commercially accessible substances. Irrespective of the structure of the monomer, the above-mentioned technique can be used to polymerize free radicals from a variety of monomers since chain transfer events involving thiol groups are ubiquitous.

A few advantages associated with this technique are that it eliminates the need for various organic during the synthesis of nanocomposite materials. If the filler surface is transformed in such a way that the forces anchoring the particles or platelets together are very weak, the polymer can intercalate between the interlayers. Heat, radiation, diffusion of a suitable initiator, a catalyst, or organic initiator attached by cation exchange inside the interlayer can be used to start polymerization before the swelling phase. Apart from this, the synthesis process has some demerits which include the need for high temperatures but that is undesirable for polymer materials [72]. The desired monomer, when polymerized, was incapable of producing adequate shear stress to effectively exfoliate and achieve adequate dispersion of the nanoclay particles in the polymer matrix [73]. Also, it will work only for low viscous polymers [74].

2.8 Spray Pyrolysis of Nanodielectric Materials

Spray pyrolysis is a technique that produces thin films by simply misting a solution onto a heated surface, where a chemical product is produced when two or more components react. At the deposition temperature, the chemical reagents are chosen so that: other than the desired molecule, the byproducts are typically volatile. The technique has been used for a long time to create a transparent SnOx electrical conductor to glass and is advantageous for the deposition of oxides. Since Chamberlin & Skarman's pioneering work on CdS films for application in solar cells in 1966, there have been numerous investigations reported in this area [75].

A lot of effort has been made in the last several decades to synthesize $BaTiO_3$ powders that are nano-sized and have great phase purity and narrow particle size distribution. Sol–gel processing and co-precipitation of mixed metal oxalates are two examples of solution-based approaches that have been used to design various kinds of nanoparticles [76], including the hydrothermal approach [77]. Spray pyrolysis is being utilized reliably to develop ceramic particles that are between microns and submicron in size. Since it is an effective, continuous method that generates spherical particles with good phase purity and regulated size distributions [78]. Spray pyrolysis yields particles through the evaporation, pyrolysis, and crystallization of atomized liquid droplets [79]. Consequently, on a nanometer scale, all the components of a droplet are mixed homogeneously, and thus, the resulting particles exhibit remarkable phase purity and an outstanding stoichiometric composition for multicomponent oxides. The procedure has a limitation because there was not enough time for solute diffusion or particle densification, which results in a high rate of evaporation and the bulk of the synthesized particles having hollow and porous structures. To control particle morphology, extensive research has been done [80, 81]. A schematic representation of the spray pyrolysis method is presented in Fig. 2.7

Chemical spray deposition processes are categorized based on the type of reaction. As the solvent initially evaporates in process A, a droplet adheres to the surface, providing a solid residue that may still react after it dries [82]. In process B, the dry solid changes the surface because the solvent evaporates before the droplet reaches the surface, where the breakdown occurs. In process C, as the droplet gets closer

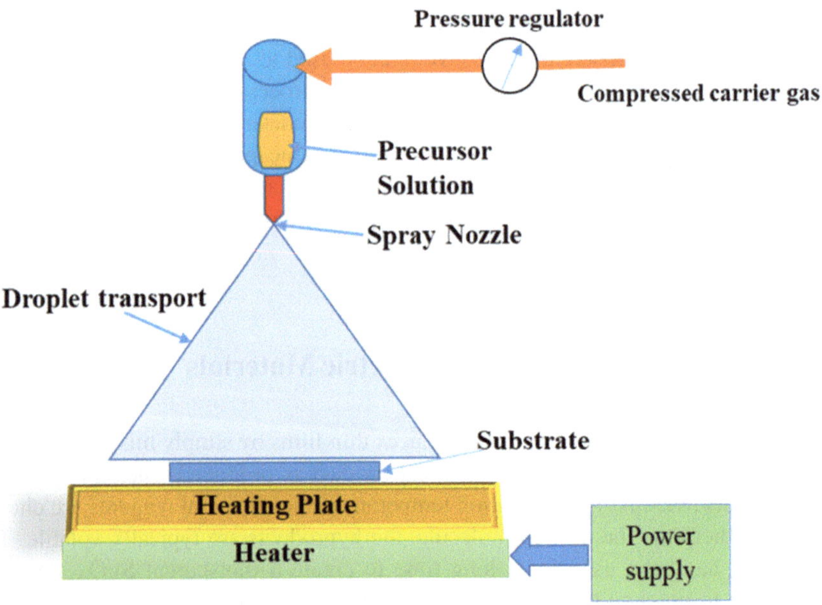

Fig. 2.7 Schematic diagram for spray pyrolysis system

to the substrate, the solvent evaporates; the solid melts and transforms into vapor, which then diffuses to the substrate, where it undergoes a heterogeneous process. (They refer to it as a genuine CVD technique.). In process D, a chemical reaction happens in the vapor phase at the maximum temperatures whenever the metallic compound evaporates before reaching the substrate [82]. The ideal temperature, solution concentration, carrier gas flow rate, droplet radius, nozzle-to-substrate distance, solution flow rate, and, for continuous processes, substrate motion are the key factors in all operations. The chemical makeup of the carrier gas and/or environment should be included in this list, as well as, most notably, the substrate temperature. Type A or Type B studies on spray pyrolysis depositions are the most frequent [82].

2.8.1 *Types of Spray Pyrolysis Methods*

Several attempts have been made to synthesize nanoparticles utilizing the spray pyrolysis technique, such as low-pressure spray pyrolysis [83], electrospray pyrolysis [84], and salt-assisted aerosol decomposition are reported [85]. Kang and Park introduced a novel aerosol method called the filter expansion aerosol generator (FEAG), and they reported that it can effectively produce nano-sized ZnO particles [78]. Another author, Xia et al. have also reported a novel spray pyrolysis synthesis method to efficiently fabricate nanoparticles, known as "salt-assisted spray pyrolysis" [85]. In addition to this, many researchers have utilized the spray pyrolysis technique for the preparation of highly efficient nanodielectric material and have presented their dielectric properties for various applications.

In 2012, Choi et al. prepared nanoscaled $Ba_{0.7}Sr_{0.3}TiO_3$ using spray pyrolysis, to create hollow, thin-walled particles for dielectric properties. In synthesis, strontium nitrate [$Sr(NO_3)_2$], titanium tetra-isopropoxide, and barium nitrate [$Ba(NO_3)_2$] are used as the precursors for the synthesis (TTIP). The hydrolyzed TTIP was peptized with a small amount of nitric acid [HNO_3] to create a clear solution, and ethylenediaminetetraacetic acid (EDTA) and citric acid were utilized as chelating agents as well as gas evolution additives to increase the hollowness of the particles of precursor. The concentrations of citric acid [$(CH_2CO_2H)_2$], EDTA, and metal components were set at 0.1 M. Precursor powders were processed via spray pyrolysis at 900 °C, and the air carrier gas flow rate was set at 40 L min^{-1}. Post-treatments after spray pyrolysis were carried out for 2 h in an environment of air at temperatures of 900, 1000, and 1100 °C. After being pelletized at 250 kg f cm^{-2} to a diameter of 10 mm, the powders were sintered for two hours at 1300 °C. Using the surfaces of the pellets as an electrode, the silver paste was applied to evaluate the dielectric characteristics. Figure 2.8 illustrates the equipment layout and the process of the nano-sized $Ba_{0.7}Sr_{0.3}TiO_3$ powders developed through spray pyrolysis [86].

Ko and co-workers have also reported on spray pyrolysis by applying the same procedure as described in Fig. 2.8 to create precursors of $BaTiO_3$ nanopowders with varying quantities of Ag. The pellet surfaces were coated with Ag paste to create

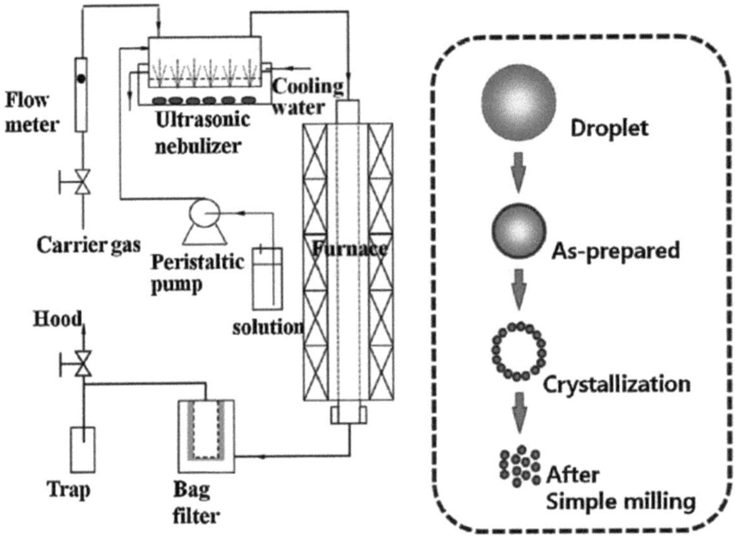

Fig. 2.8 Diagram illustrating the spray pyrolysis process and the method by which ceramic particles of nanoscale are formed [86]. Adapted with permission from ref. [86] Copyright (2012) (Elsevier)

an electrode, which was then used to create electrical contact. An LCR meter was utilized to measure the dielectric characteristics [87].

In another article, authors have reported spray pyrolysis fabrication of Al_2O_3, TiO_2, $Al_{2x-1}{\cdot}Ti_xO_y$, and ZnO nanodielectric materials. Aerosols comprising 50 mg/mL aluminum chloride ($AlCl_3$) and titanium dioxide solutions spray-coated at 420 °C onto commercially available indium tin oxide (ITO) coated glass (sheet resistivity Rs. 15 Ohms/) were applied using a pneumatic airbrush maintained at a vertical distance of around 30 cm (1:5). The spray coating of the precursors and blends' aerosols took place for 30 s, with a 20-s interruption to enable the vapors to settle. They have shown that after successful synthesis of the nanodielectric materials, they were spray-coated on ITO substrate to obtain an optimized thickness. Bottom-gate top-contact (BG-TC) transistors were developed using aluminum (Al) as source material for the manufacturing of thin-film transistors (TFTs) and the spray-coated glass/ITO/composite/ZnO stacks were used to thermally evaporate drain (S/D) electrodes with a shadow mask under a high vacuum. Device characterization was carried out at room temperature and under a high vacuum [88].

Similarly, Adamopoulos et al. reported the fabrication of a dielectric material zinc oxide (ZnO), ZrO_2, and Li-ZnO deposition through the spray pyrolysis process. They have presented transistor fabrication; ITO glass substrates were used to construct bottom-gate top-contact transistors. At a substrate temperature of around 400 °C, high-k ZrO_2 dielectric coatings were sprayed over ITO electrodes. Then, a layer of undoped/Li-doped ZnO with a thickness approximately of 35 nm was sprayed. Through a shadow mask and a high vacuum, aluminum source/drain electrodes successfully thermally evaporated (10^{-6} mbar) after the semiconducting films were deposited by using a spray pyrolysis process [89]. Jung et al. demonstrate the preparation of barium titanate nanodielectric material with tetragonal crystal structures by flame spray pyrolysis method. $BaCO_3$ and titanium tetra-isopropoxide (TTIP) were the substrate materials utilized in the fabrication of the barium titanate ($BaTiO_3$) powder. In the report, they observed that barium titanate powder synthesis via the spray pyrolysis method can function as an efficient dielectric material for various practical applications. Figure 2.9 depicts the flame spray pyrolysis apparatus schematic representation, and Fig. 2.10 represents the synthesis mechanism of the nanoscaled barium titanate ($BaTiO_3$) powders by flame spray pyrolysis process [90].

Spray pyrolysis thin-film coating of khaya gum biopolymer for electrical and dielectric properties was also reported. The primary source for the production of khaya gum was the exudates of the Senegalese khaya trees. Using an agate mortar and pestle, the collected gum was crushed into a powder and then sieved. The solution was then prepared by first refining the KG powder and then dissolving it in 20 ml of distilled water. After that, using a hot plate and spray pyrolysis, it was homogenized and sprayed onto the glass for four minutes at 200 °C and glass slides that had been coated with indium tin oxide (ITO). Aluminum (Al) having a thickness of 100 nm

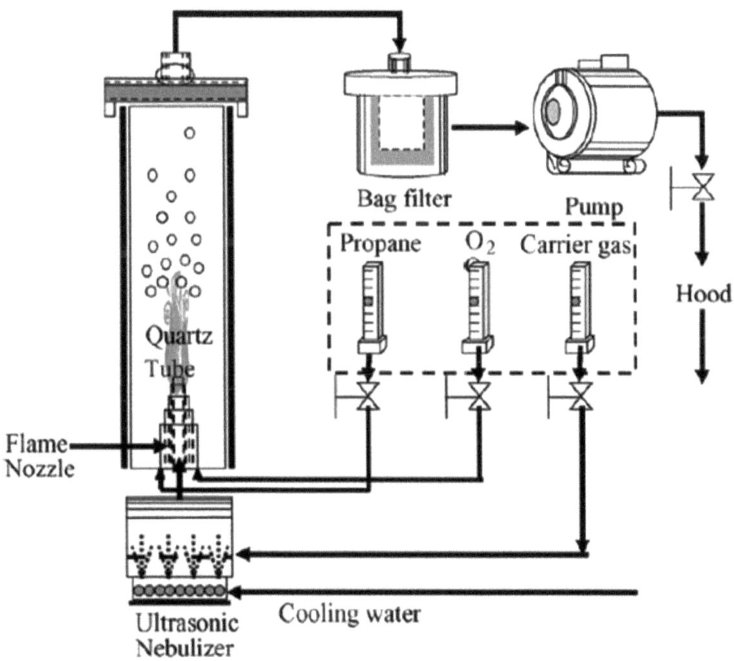

Fig. 2.9 Schematic diagram of flame spray pyrolysis [90]. Adapted with permission from ref. [90] Copyright (2008) (Elsevier)

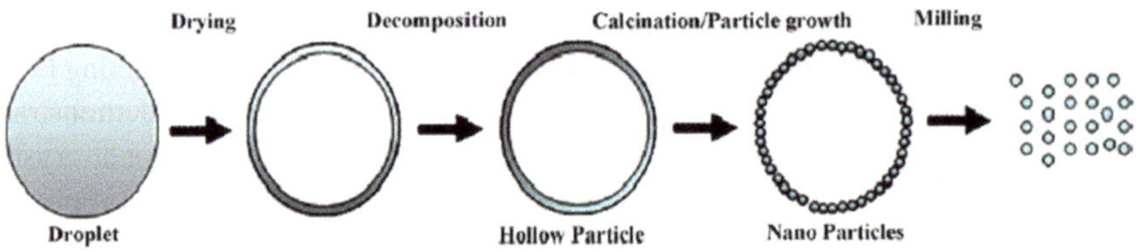

Fig. 2.10 Synthesis mechanism of the nano-sized $BaTiO_3$ powder by flame spray pyrolysis [90]. Adapted with permission from ref. [90] Copyright (2008) (Elsevier)

was thermally evaporated by a shadow mask with an area of 0.01 cm^2 to perform capacitance measurements. They calculated the capacitance values using several frequencies [91].

The following are some advantages of spray pyrolysis: (i) Open-air procedure; (ii) The opportunity to monitor the deposition technique (iii) Economical viable and ongoing operation; (iv) There is no requirement for high-quality regents as precursors; (v) Have higher production rate; (vi) High surface area allows crystal size to be controlled, and (vii) Compositional homogeneity present in fabricated materials. Spray pyrolysis has certain drawbacks also, including (i) a very poor yield produced from the powder materials; (ii) It is necessary to convert the sulfides into an oxide; and (iii) There are several difficulties with temperature optimization [92].

2.9 Reversible Addition-Fragmentation Transfer (RAFT) Process for Nanodielectric Materials

Due to their potential applications, such as in small electronic devices, capacitor materials, pulsed power systems, and power electronics for grid energy management, novel polymeric nanocomposite materials with improved dielectric characteristics are of great interest [93, 94]. Polymer nanodielectric materials synthesized by adding inorganic nanoparticles into polymer matrices have obtained significant attention because they promise the best characteristics of polymer and ceramic dielectrics and are among the most promising techniques for producing high-performance dielectrics [94]. The approach includes the incorporation of conductive polymer along with semiconductive fillers (such as carbon nanotubes and conductive polymers) and high-permittivity inorganic fillers (e.g., Barium titanate, i.e., $BaTiO_3$) or polymers with robust breakdown strength [95]. Ferroelectric polymers including PVDF-based homo- and copolymers and polyimides have been included in polymer nanocomposite matrices. The ceramic filler-based nanocomposites of conductive polymer (PNCs) can incorporate the features of polymers and ceramics and demonstrates a novel category of material that shows flexibility, easy processability, and of low cost [96]. Recently, significantly great approaches have been made toward the fabrication of core–shell nanoparticles, which involve the modification of nanoparticles incorporating polymer shells utilizing several synthetic methods (e.g., grafting from, grafting to, etc.). To develop single or multiple-component nanocomposite systems, most of the modified nanoparticles can be used.

To date, many polymer nanocomposites (PNCs) have been developed, However, certain challenging issues need to be addressed, e.g., homogenous nanoparticle dispersion, strengthening the adhesion of the polymeric particle interface, and optimizing dielectric loss. Consequently, several approaches have been used to deal with these challenges by altering the interface of PNCs through the addition of organic molecules to nanoparticles, along with coupling agents, phosphoric acids, surfactants, and others. However, these techniques still have constraints since surface modifications do not make a direct contribution to accomplishing the objective of enhanced dielectric properties. It has been established that the synthesis of core–shell structured dielectric of polymer nanocomposites (PNCs) via the "grafting from" technique is a promising strategy for this class of materials [97].

Tang and co-workers in 2013, developed a novel polymer nanocomposite using nanodielectric materials containing conjugated oligothiophene side chains. They depicted that the polymers synthesized by the RAFT method possess significant dielectric permittivity, which is comparable to poly (vinylidene fluoride) and significantly greater than most of the other polymers (such as 2–5 for polystyrene and polypropylene, etc.). Polyvinylidene fluoride (PVDF) in particular, polymers exhibit exceptional consistency over a wide frequency range, from 100 Hz to several MHz, with negligible dielectric loss (0.02). This functionality is due to fast-relaxing nano dipoles and significantly higher polarizable related to nanoscaled crystalline domains (<2 nm) incorporated into oligothiophene side groups. Using barium titanate ($BaTiO_3$, BT) nanoparticles and polymers containing oligothiophene, they soon created core–shell nanoparticles with a monolayered structure for use in dielectric polymer nanocomposites. Under controlled/living in situ surface-initiated RAFT polymerization, the oligothiophene-incorporating polymer was grafted from BT nanoparticles [98].

RAFT polymerization has become apparent as one of the best optimistic approaches due to its facile chemical experimental conditions, functional group tolerability, and versatility to a variety of vinyl monomers. In contrast to atom transfer radical polymerization (ATRP) and nitroxide-mediated polymerization, reversible addition-fragmentation transfer polymerization (NMP) relies on the degenerative chain transfer technique rather than the reversible termination of propagating radicals. A chain transfer agent (CTA) that contains thiocarbonylthio is often used by reversible addition-fragmentation transfer (RAFT) to accomplish its deactivation–activation equilibrium. The radical concentration is not significantly affected by the activation or deactivation process, an external initiator is required, and initiation proceeds as in a standard radical polymerization process. A typical RAFT polymerization involves the following four steps: initiation, addition-fragmentation, reinitiation, and equilibration. In the subsequent initiation step, the thiocarbonylthio molecule receives the propagating radical (Pn•).

(1) to generate intermediate radical
(2) followed by gradual fragmentation to generate another thiocarbonylthio group

(3) with a newly produced radical (R•)
(4) Thus, the radical (R•) may react with the newly formed thiocarbonyl group or with a monomer to produce a new propagating radical (Pm•), which can then start the polymerization process. After the subsequent consumption of the initially injected CTA, the degenerative chain transfer process between propagating (Pn or Pm) and dormant chains establish the "primary equilibrium" of activation–deactivation (i.e., the end of the "initialization period").

2.9.1 Merits and Demerits of the RAFT Method

It is believed that reversible addition-fragmentation chain transfer (RAFT) polymerization is a reliable method. Polymerization can be achieved in a wide diverse range of solvents, including water, at various temperatures, with a high functional group tolerance and without the need for metals. Majority of monomers susceptible to radical polymerization can be utilized to prepare confined copolymers or polydispersity polymers using this technique. It is feasible to enhance the conversion of RAFT polymerizations and attain economically viable polymerization rates. In heterogeneous media (emulsion, mini emulsion, and suspension), polymerizations can be carried out robustly. Monomers, solvents, and initiators are compatible with a diverse range of functionalities. Accessible and capable of exhibiting high purity are stars, microgel, blocks, supramolecular assemblies, hyperbranched structures, and other complex architectures. As of 2014, almost all monomer classes susceptible to radical polymerization are incorporated by the diverse range of RAFT agents that are commercially available. Besides this, the RAFT method also has some demerits, such as a specific RAFT agent is only appropriate for a specific set of monomers, and its synthesis often necessitates a multistep synthetic procedure followed by purification. Because the thioester moiety gradually breaks down to produce tiny sulfur compounds, RAFT agents can be unstable over extended periods, be highly colored, and have a pungent scent. For some applications, the resultant polymers presence of sulfur and color may also be undesirable; however, this can, to a certain extent, be eliminated using further chemical and physical purification techniques [99]. This remarkable versatility, however, cannot be attained without carefully considering the RAFT agent and reaction parameters. Inappropriate RAFT agent selection for the monomer(s) and/or reaction conditions is commonly responsible for reported RAFT polymerization challenges (retardation, ordinary than expected control). RAFT agents that perform effectively in one set of circumstances may not always be the best approach.

2.10 Self-assembly Process for Nanodielectric Materials

Organic, inorganic, and polymeric semiconductors that can be processed using a solution have been produced during the last 20 years as a result of factors like printability, the potential for large-area depositions, the ability to manufacture devices at a low cost, and compatibility with substrates that are versatile mechanically [100]. In addition to their mechanical adaptability and flexibility, hybrid materials offer the environmental, optical, and electrical resilience of inorganic materials. Several groups have combined hybrid materials made of self-assembled monolayers (SAMs) and ultrathin MO layers, such as those made of HfOx, AlOx, and ZrOx, to obtain extraordinarily low operating voltage OTFTs. Multi-layered organic–inorganic hybrid materials are necessary for potential and high-quality multilayers adopting the simplest and most reliable way in conventional thin-film transistors and other applications. High-quality hybrid films could be made using vapor-phase fabrication methods for organic/inorganic hybrid materials. However, they usually need high- or medium-vacuum deposition machinery (such as chemical/physical vapor deposition or atomic layer deposition), and it is not immediately clear how to incorporate them into large-volume coating techniques. By layer-by-layer solution-based deposition of precisely defined precursor molecules, a wide variety of functional materials can be produced with a high level of precision and structural control at the molecular level. A variety of such assembly's chemistry has been observed, frequently based on metal–ligand coordination or siloxane. One of these material systems is the family of silane-based nanodielectrics known as self-assembled nanodielectrics (SANDs), which has promising features for a variety of optoelectronic applications, including TFTs. Strong siloxane connections that alternate between organic and inorganic multilayers make up SANDs. Various self-assembly methods can be used to create high-performance multi-layered gate dielectrics for TFT applications [101]. Moreover, the functions of these dielectrics efficiently represent both organic and inorganic semiconductor devices and can be optimized by numerous self-assembled monolayers (SAMs) to boost the TFT efficiency even more.

In general, the molecular self-assembly method describes the spontaneous assembly of precursor molecules to produce nanostructured materials, which can be accompanied depending on their environments and the inherent characteristics of the molecules [102]. To modify the structures which emerge after the self-assembly method, we have to depend on the following deceptive approach: Firstly, the precursor molecule's framework can be utilized to modify the direction of intermolecular interactions, and secondly, carefully choosing the ambient temperature. The spontaneous assembling of molecules into well-defined, stable aggregates with non-covalent bonds at equilibrium conditions is referred to as molecular self-assembly. This method emerges as a novel approach in synthetic chemistry, having the potential of producing nonbiological assemblies having dimensions of 1 to 102 nm (with a molecular mass of 104 to 1010 Dalton's). The above process helps in synthesizing structures with comparable size to those which can be synthesized by microlithography and other methodology of microfabrication [103].

Thin-film transistors (TFTs) made from unusual materials have attracted interest recently because they can be used to make inexpensive electronic devices like printed RF-ID cards, sensors, and flexible electronic displays [104]. Strong capacitances and low processing temperatures (such as 200 °C) are provided by solution-processable self-assembled nanodielectrics (SANDs), which are composed of structurally distinct and substantial layers of alternately alternating inorganic oxide (such as ZrOx and HfOx) and organic (such as stilbazolium) layers. SANDs have produced outstanding device metrics when incorporated with a variety of semiconductors; they have never been incorporated into the most technologically sophisticated top-gate thin-film transistor architecture. Recently, solution-phase methods for the low-temperature development of highly efficient semiconductor films and dielectrics were synthesized by utilizing the self-assembly process of organic–inorganic hybrid nanodielectric (SANDs) films. The potential benefits are also limited by the diversity of the SAND structure, which can be modified to take into account both the organic and inorganic components in addition to the vast variety of suitable unconventional semiconductors.

Tobin J. Marks et al. demonstrated the development pathways of solution-processed zirconia self-assembled nanodielectric material (Zr-SAND) on template-stripped Al substrates. Based on the ultrasmooth (r.m.s. roughness 0.4 nm) nature of the resulting Zr-SAND on aluminum structures, template-stripped metal substrates show the same remarkable electronic uniformity (capacitance 700 nF cm^2 and leakage current < 1 Ma cm^{-2} at -2 MV cm^{-1}) and growth of multi-layered Zr-SAND on Si while demonstrating superior temperature and response toward the voltage capacitance. They reported that the synthesized nanodielectric material is essential to carry out comprehensive transport measurements in developing transistor technologies with SAND including for developing future technologies in flexible electronics or integrated circuits (ICs) [105].

The growth of Zr-SAND is shown schematically in which (a–d) layer-by-layer assembly of the zirconia self-assembled nanodielectric (Zr-SAND) multilayer structure is shown in the following Fig. 2.11. In a separate article, researchers utilized the solution-processed a-IGZO with the four-layer Hf-SAND to create top-gate thin-film transistors (TFTs), which exhibit exceptional bias stress stability and electron mobility (SAT = 19.4 and Vth = (10–100.83 V), subthreshold slopes (SS = 293 mV/dec), and gate leakage currents A. Structures for self-assembled nanodielectric (SAND) gate dielectrics [106].

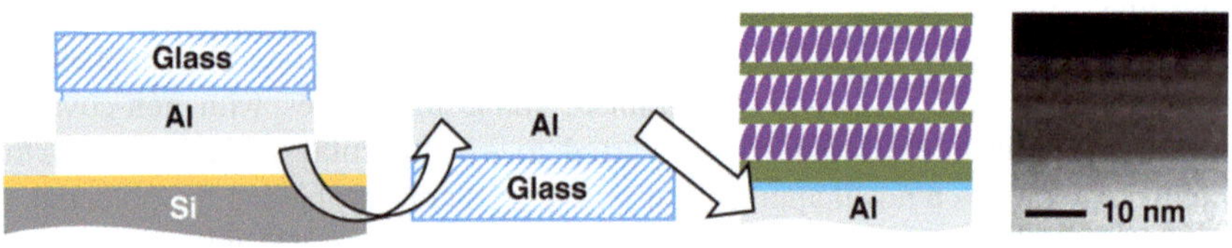

Fig. 2.11 Zirconia self-assembled nanodielectrics (Zr-SAND) solution-processed on aluminum substrates using template-stripped [106]. Adapted with permission from ref. [106] Copyright (2021) (American chemical society)

For the synthesis of nanostructures, solution-based self-assembly techniques have several advantages, such as the ability to control the characteristics of the nanostructures by altering the concentrations of constituent molecules, accessibility of this approach for effectively regulating the size and shape of nanostructures, and simple and economic viable experimental processes. The self-assembly technique yields silicate materials that are prone to aggregation. There are numerous significant challenges with the self-assembly technique because it is relatively a novel technique in nanotechnology. Unfortunately, it is difficult to handle self-assembly on a large scale, and to be extensively applied, we will need to ensure significantly high degrees of repeatability at these scales. The basic principles of atomistic and macroscale processes might differ dramatically from those of nanostructures, and the thermodynamic and kinetic mechanisms of self-assembly are poorly understood. In self-assembling systems, equilibrium time scale and kinetic rates are affected by concepts of thermal motion and capillary action that are less evident.

2.11 Conclusion and Perspectives of Every Synthesis Method for Nanodielectric Materials

Due to their widespread use in industry, particularly in telecommunication and electronic applications, nanodielectric materials. Moreover, different types of nanodielectric materials expand the possibility of realizing numerous significant applications. The nanoparticles have been produced using a variety of unique synthetic methods in order to achieve greater dielectric characteristics. In this chapter, we address alternative synthetic techniques for the creation of nanomaterials with greater dielectric characteristics while keeping the aforementioned in mind. Various synthetic routes have their own merits and demerits which have been well explained. Depending on the type of application of the nanomaterials, researchers have chosen the synthetic procedure, such as thin film, nanocomposite, polymer-based nanocomposite, and metal nanoparticle-based nanodielectric materials. In the sol–gel synthesis approach, nanosized particles can be prepared with good stoichiometric control and ultrafine particles with narrow size distribution. Also, it takes short time for processing at a lower temperature. The hydrothermal method takes place in a closed system with high pressure and temperature. This process provides large-sized particles for various applications. To synthesize these film-based dielectric materials, spray pyrolysis method can be used, which provides limited size distribution of particles with a uniform coating. In the spray pyrolysis technique, a solution is made to spray on the heated substrate to create a thin film, which further can be used for various industrial applications. Polymer-based nanodielectric can be synthesized using the RAFT method. RAFT method includes the addition of inorganic fillers with high permittivities such as barium titanate, conductive or semiconductive fillers like carbon nanotubes, and other conductive polymers, to those polymers with high breakdown strength. The combustion method provides fast synthesis with the consumption of low energy and gives products with high purity. Control over particle size using the combustion method by choosing appropriate fuel that produces a slow burning rate and less exothermicity. Polymer-clay nanocomposite can be synthesized using the intercalation method. In the intercalation technique, the polymer was incorporated within the layers of clay having few nanometer dimensions. Polymer incorporation largely affected by the intercalated clay structure which is further responsible for the high electrical and dielectric properties of the nanocomposite. Silane-based nanodielectrics are referred to as self-assembled nanodielectrics (SANDs). In the SAMs technique, nanostructured materials were formed by spontaneous assembly of precursor molecules. This method provides a stable and structurally well-defined aggregate associated with non-covalent bonds. Nanofibers with higher dielectric properties can be synthesized by using electrospinning techniques. The electrospinning method is a comprehensive and simple method to get ultrafine fibers from various materials. The applicability of various synthetic routes has provided various challenges for the synthesis of nanomaterials with electronic properties. In the future, these synthetic routes can be further modified to produce other forms of nanodielectric for various industrial applications.

References

1. Dang ZM, Yuan JK, Yao SH, Liao RJ (2013) Flexible nanodielectric materials with high permittivity for power energy storage. Adv Mater 25(44):6334–6365
2. Tan DQ (2020) Review of polymer-based nanodielectric exploration and film scale up for advanced capacitors. Adv Funct Mater 30(18):1808567
3. Tian F, Lei Q, Wang X, Wang Y (2012) Investigation of electrical properties of LDPE/ZnO nanocomposite dielectrics. IEEE Trans Dielectr Electr Insul 9:763–769
4. Yang K, Huang X, Zhu M, Xie L, Tanaka T, Jiang P (2014) Combining RAFT polymerization and thiol–ene click reaction for core–shell structured polymer@ $BaTiO_3$ nanodielectrics with high dielectric constant, low dielectric loss, and high energy storage capability. ACS Appl Mater Interfaces 6(3)
5. Simon P, Gogotsi Y (2008) Materials for electrochemical capacitors. Nat Mater 7(11):845–854
6. Thongbai P, Yamwong T, Maensiri S (2009) Electrical responses in high permittivity dielectric (Li, Fe)-doped NiO ceramics. Appl Phys Lett 94(15):152905
7. Brochu P, Pei Q (2012) Dielectric elastomers for actuators and artificial muscles. In: Electroactivity in polymeric materials. Springer, Boston, MA
8. Wang Y, Zhou X, Chen Q, Chu B, Zhang Q (2010) Recent development of high energy density polymers for dielectric capacitors. IEEE Trans Dielectr Electr Insul 17(4):1036–1042
9. Ortiz RP, Facchetti A, Marks TJ (2010) High-k organic, inorganic, and hybrid dielectrics for low-voltage organic field-effect transistors. Chem Rev 110(1):205–239
10. Wang Q, Zhu L (2011) Polymer nanocomposites for electrical energy storage. J Polym Sci B: Polym Phys 49(20):1421–1429
11. Cao Y, Irwin PC, Younsi K (2004) The future of nanodielectrics in the electrical power industry. IEEE Trans Dielectr Electr Insul 11(5):797–807
12. Sanida A, Stavropoulos SG, Speliotis T, Psarras GC (2021) Evaluating the multifunctional performance of polymer matrix nanodielectrics incorporating magnetic nanoparticles: a comparative study. Polymer 236:124311
13. Slenes KM, Winsor P, Scholz T, Hudis M (2001) Pulse power capability of high energy density capacitors based on a new dielectric material. IEEE Trans Magn 37(1):324–327

14. Hardy CG, Islam MS, Gonzalez-Delozier D, Ploehn HJ, Tang C (2012) Oligoaniline-containing supramolecular block copolymer nanodielectric materials. Macromol Rapid Commun 33(9):791–797
15. Sikalidis CE (2011) Advances in ceramics: electric and magnetic ceramics, bioceramics, ceramics and environment. BoD–Books on Demand
16. Pan Z, Zhai J, Shen B (2017) Multilayer hierarchical interfaces with high energy density in polymer nanocomposites composed of $BaTiO_3$@ TiO_2@ Al_2O_3 nanofibers. J Mater Chem 5(29):15217–15226
17. Li J, Claude J, Norena-Franco LE, Seok SI, Wang Q (2008) Electrical energy storage in ferroelectric polymer nanocomposites containing surface-functionalized $BaTiO_3$ nanoparticles. Chem Mater 20(20):6304–6306
18. Horowitz G, Hajlaoui R, Bouchriha H, Bourguiga R, Hajlaoui M (1998) The concept of threshold voltage in organic field-effect transistors. Adv Mater 10(12):923–927
19. Kleemann H, Schwartz G, Zott S, Baumann M, Furno M (2020) Megahertz operation of vertical organic transistors for ultra-high resolution active-matrix display. Flex Print Electron 5(1):014009
20. Ta J, Sun W, Lu L (2022) Organic small molecule semiconductor materials for OFET-based biosensors. Biosens Bioelectron 216:114667
21. Ji D, Li T, Hu W, Fuchs H (2019) Recent progress in aromatic polyimide dielectrics for organic electronic devices and circuits. Adv Mater 31(15):1806070
22. Singh TB, Sariciftci NS (2006) Progress in plastic electronics devices. Annu Rev Mater Res 36:199–230
23. Kim CH, Kymissis I (2017) Graphene–organic hybrid electronics. J Mater Chem C 5(19):4598–4613
24. Chen Q, Shen Y, Zhang S, Zhang QM (2015) Polymer-based dielectrics with high energy storage density. Annu Rev Mater Res 45(1):433–458
25. Feng M, Zhang C, Zhou G, Zhang T, Feng Y, Chi Q, Lei Q (2020) Enhanced energy storage characteristics in PVDF-based nanodielectrics with core-shell structured and optimized shape fillers. IEEE Access 8:81542–81550
26. Liu Y, Gao J, Wang Y, Zhou J, Cao L, He Z, Zhang Y, Tang C, Zhong L (2019) Enhanced temperature stability of high energy density ferroelectric polymer blends: the spatial confinement effect. Macromol Rapid Commun 40(21):1900406
27. Andritsch T, Fabiani D, Vazquez IR (2013) Nanodielectrics-examples of preparation and microstructure. IEEE Electr Insul Mag 29(6):21–28
28. Tanaka T, Montanari G, Mulhaupt R (2004) Polymer nanocomposites as dielectrics and electrical insulation-perspectives for processing technologies, material characterization and future applications. IEEE Trans Dielectr Electr Insul 11(5):763–784
29. Anandraj J, Joshi GM (2018) Fabrication, performance and applications of integrated nanodielectric properties of materials—a review. Compos Interfaces 25(5–7):455–489
30. Saha S, Chakravorty D (2006) One-dimensional organic giant dielectrics. Appl Phys Lett 89(4):043117
31. Chen X, Saito T, Yamada H, Matsushige K (2001) Aligning single-wall carbon nanotubes with an alternating-current electric field. Appl Phys Lett 78(23):3714–3716
32. Bokov D, Turki Jalil A, Chupradit S, Suksatan W, Javed Ansari M, Shewael IH, Valiev GH, Kianfar E (2021) Nanomaterial by sol-gel method: synthesis and application. Adv Mater Sci Eng 2021
33. Znaidi L, Illia GS, Benyahia S, Sanchez C, Kanaev A (2003) Oriented ZnO thin films synthesis by sol–gel process for laser application. Thin Solid Films 428(1–2):257–262
34. Wang N, Magdassi S, Mandler D, Long Y (2013) Simple sol–gel process and one-step annealing of vanadium dioxide thin films: synthesis and thermochromic properties. Thin Solid Films 534:594–598
35. Li L, Yalcin B, Nguyen BN, Meador MAB, Cakmak M (2009) Flexible nanofiber-reinforced aerogel (xerogel) synthesis, manufacture, and characterization. ACS Appl Mater Interfaces 1(11):2491–2501

36. Abid N, Khan AM, Shujait S, Chaudhary K, Ikram M, Imran M, Haider J, Khan M, Khan Q, Maqbool M (2021) Synthesis of nanomaterials using various top-down and bottom-up approaches, influencing factors, advantages, and disadvantages: a review. Adv Colloid Interface Sci 102597
37. Mammo TW, Murali N, Kumari CV, Margarette S, Ramakrishna A, Vemuri R, Rao YS, Prasad KV, Ramakrishna Y, Samatha K (2020) Synthesis, structural, dielectric and magnetic properties of cobalt ferrite nanomaterial prepared by sol-gel autocombustion technique. Phys B: Condens Matter 581:411769
38. Gunji N, Komori Y, Yoshitake H (2016) Double functionalization with mercaptopropyl and vinyl groups of the surface of silica nanoparticles and its application to tire rubber. Colloids Surf A Physicochem Eng 511:351–356
39. Mahtabani A, Rytöluoto I, Anyszka R, He X, Saarimäki E, Lahti K, Paajanen M, Dierkes W, Blume A (2020) On the silica surface modification and its effect on charge trapping and transport in pp-based dielectric nanocomposites. ACS Appl Polym Mater 2(8):3148–3160
40. Ghamarpoor R, Jamshidi M (2022) Synthesis of vinyl-based silica nanoparticles by sol–gel method and their influences on network microstructure and dynamic mechanical properties of nitrile rubber nanocomposites. Sci Rep 12(1):1–15
41. Aoki Y, Kunitake T, Nakao A (2005) Sol–gel fabrication of dielectric HfO_2 nano-films; formation of uniform, void-free layers and their superior electrical properties. Chem Mater 17(2):450–458
42. Jiang H, Liu H, Yu H, Gao F, Liu J-M, Nan C (2005) Dielectric behaviors of $ZnFe_2O_4$–SiO_2 composite thin films prepared by sol-gel method. Int J Mod Phys 19(15–17):2682–2686
43. Srivastava M, Mishra P, Gupta VK, Srivastava N (2020) Green synthesis of nanomaterials for bioenergy applications. Wiley
44. Parashar M, Shukla VK, Singh R (2020) Metal oxides nanoparticles via sol–gel method: a review on synthesis, characterization and applications. J Mater Sci: Mater Electron 31(5):3729–3749
45. Adschiri T, Kanazawa K, Arai K (1992) Rapid and continuous hydrothermal crystallization of metal oxide particles in supercritical water. J Am Ceram Soc 75(4):1019–1022
46. Raghavendra M, Jagadish K, Srikantaswamy S, Pradeep T, Gnana Prakash A, Ravikumar H (2022) Effect of CeO_2 nanoparticles on dielectric properties of PVB/CeO_2 polymer nanodielectrics: a positron lifetime study. J Mater Sci: Mater Electron 33(2):1063–1077
47. Dewangan R, Asthana A, Singh AK, Carabineiro SA (2021) Control of surface functionalization of graphene-metal oxide polymer nanocomposites prepared by a hydrothermal method. Pol Bull 78(8):4665–4683
48. Das S, Das S, Sutradhar S (2017) Effect of Gd^{3+} and Al^{3+} on optical and dielectric properties of ZnO nanoparticle prepared by two-step hydrothermal method. Ceram Int 43(9):6932–6941
49. Chen Z, Li H, Xie G, Yang K (2018) Core–shell structured Ag@ C nanocables for flexible ferroelectric polymer nanodielectric materials with low percolation threshold and excellent dielectric properties. RSC Adv 8(1):1–9
50. Zeraati AS, Arjmand M, Sundararaj U (2017) Silver nanowire/MnO_2 nanowire hybrid polymer nanocomposites: materials with high dielectric permittivity and low dielectric loss. ACS Appl Mater Interfaces 9(16):14328–14336
51. Hayashi H, Hakuta Y (2010) Hydrothermal synthesis of metal oxide nanoparticles in supercritical water. Materials 3(7):3794–3817
52. Ahmad T, Shahazad M, Phul R (2017) Hydrothermal synthesis, characterization and dielectric properties of zirconia nanoparticles. Mater Sci Eng Int J 1:100–104
53. Gan YX, Jayatissa AH, Yu Z, Chen X, Li M (2020) Hydrothermal synthesis of nanomaterials. vol 2020. Hindawi
54. Patil K (2008) Chemistry of nanocrystalline oxide materials: combustion synthesis, properties and applications. World Scientific
55. Patil KC, Aruna S, Mimani T (2002) Combustion synthesis: an update. Curr Opin Solid State Mater Sci 6(6):507–512

56. Varma A, Mukasyan AS, Rogachev AS, Manukyan KV (2016) Solution combustion synthesis of nanoscale materials. Chem Rev 116(23):14493–14586
57. Kim M-G, Kanatzidis MG, Facchetti A, Marks TJ (2011) Low-temperature fabrication of high-performance metal oxide thin-film electronics via combustion processing. Nat Mater 10(5):382–388
58. Krengvirat W, Sreekantan S, Ahmad-Fauzi M, Chinwanitcharoen C, Kawamura G, Matsuda A (2012) Control of the structure, morphology and dielectric properties of bismuth titanate ceramics by praseodymium substitution using an intermediate fuel agent-assisted self-combustion synthesis. J Mater Sci 47(9):4019–4027
59. Sekar MM, Halliyal A (1998) Low-temperature synthesis, characterization, and properties of lead-based ferroelectric niobates. J Am Ceram Soc 81(2):380–388
60. Ramakrishna S (2005) An introduction to electrospinning and nanofibers. World scientific
61. Jiang J, Shen Z, Cai X, Qian J, Dan Z, Lin Y, Liu B, Nan CW, Chen L, Shen Y (2019) Polymer nanocomposites with interpenetrating gradient structure exhibiting ultrahigh discharge efficiency and energy density. Adv Energy Mater 9(15):1803411
62. Li Z, Liu F, Yang G, Li H, Dong L, Xiong C, Wang Q (2018) Enhanced energy storage performance of ferroelectric polymer nanocomposites at relatively low electric fields induced by surface modified $BaTiO_3$ nanofibers. Compos Sci Technol 164:214–221
63. Wang J, Long Y, Sun Y, Zhang X, Yang H, Lin B (2017) Enhanced energy density and thermostability in polyimide nanocomposites containing core-shell structured $BaTiO_3$ @ SiO_2 nanofibers. Appl Surf Sci 426:437–445
64. Bian S, Jayaram SH, Cherney EA (2010) Use of electrospinning to disperse nanosilica into silicone rubber. In: 2010 annual report conference on electrical insulation and dielectic phenomena. IEEE, pp 1–4
65. Anwar N, Ishtiaq M, Shakoor A, Niaz NA, Rizvi TZ, Qasim M, Irfan M, Mahmood A (2021) Dielectric properties of polymer/clay nanocomposites. Polym Polym Compos 29(6):807–813
66. Sengwa R, Choudhary S, Sankhla S (2010) Dielectric properties of montmorillonite clay filled poly (vinyl alcohol)/poly (ethylene oxide) blend nanocomposites. Compos Sci Technol 70(11):1621–1627
67. Kojima Y, Usuki A, Kawasumi M, Okada A, Kurauchi T, Kamigaito O (1993) Synthesis of nylon 6–clay hybrid by montmorillonite intercalated with ϵ-caprolactam. J Polym Sci, Part A: Polym Chem 31(4):983–986
68. Vellakkat M, Kamath A, Raghu S, Chapi S, Hundekal D (2014) Dielectric constant and transport mechanism of percolated polyaniline nanoclay composites. Ind Eng Chem Res 53(43):16873–16882
69. Yan X, Goodson T (2006) High dielectric hyperbranched polyaniline materials. J Phys Chem B 110(30):14667–14672
70. Joo J, Long S, Pouget J, Oh E, MacDiarmid A, Epstein A (1998) Charge transport of the mesoscopic metallic state in partially crystalline polyanilines. Phys Rev B 57(16):9567
71. Akat H, Tasdelen MA, Du Prez F, Yagci Y (2008) Synthesis and characterization of polymer/clay nanocomposites by intercalated chain transfer agent. Eur Polymer J 44(7):1949–1954
72. Pavlidou S, Papaspyrides C (2008) A review on polymer–layered silicate nanocomposites. Prog Polym Sci 33(12):1119–1198
73. Pizzatto L, Lizot A, Fiorio R, Amorim CL, Machado G, Giovanela M, Zattera AJ, Crespo JS (2009) Synthesis and characterization of thermoplastic polyurethane/nanoclay composites. Mater Sci Engg C 29(2):474–478
74. Thomas S, Maria HJ (2016) Progress in rubber nanocomposites. Woodhead Publishing
75. Chamberlin R, Skarman J (1966) Chemical spray deposition process for inorganic films. J Electrochem Soc 113(1):86
76. Kim S, Lee M, Noh T, Lee C (1996) Preparation of barium titanate by homogeneous precipitation. J Mater Sci 31(14):3643–3645
77. Hayashi K, Yamamoto T, Sakuma T (1996) Grain orientation dependence of the PTCR effect in niobium-doped barium titanate. J Am Ceram Soc 79(6):1669–1672

78. Kang YC, Park SB (1997) Effect of preparation conditions on the formation of primary ZnO particles in filter expansion aerosol generator. J Mater Sci Lett 16(2):131–133
79. Chiu H-H (2000) Advances and challenges in droplet and spray combustion. I. Toward a unified theory of droplet aerothermochemistry. Prog Energy Combust Sci 26(4–6):381–416
80. Roh HS, Kang YC, Park SB (2000) Morphology and luminescence of $(GdY)_2O_3$: Eu particles prepared by colloidal seed-assisted spray pyrolysis. J Colloid Interface Sci 228(2):195–199
81. Lee D, Kang Y, Park H, Ryu S (2003) VUV characteristics of BaAl {sub 12} O {sub 19}: Mn {sup $^{2+}$} phosphor particles prepared from aluminum polycation solutions by spray pyrolysis. J Alloys Compd 353
82. Mooney JB, Radding SB (1982) Spray pyrolysis processing. Annu Rev Mater Sci 12(1):81–101
83. Lenggoro IW, Okuyama K, de la Mora JF, Tohge N (2000) Preparation of ZnS nanoparticles by electrospray pyrolysis. J Aerosol Sci 31(1):121–136
84. Kang YC, Park SB (1995) A high-volume spray aerosol generator producing small droplets for low pressure applications. J Aerosol Sci 26(7):1131–1138
85. Xia B, Lenggoro IW, Okuyama K (2002) Nanoparticle separation in salted droplet microreactors. Chem Mater 14(6):2623–2627
86. Choi SH, Ko YN, Lee J-K, Kang YC (2012) Dielectric properties of nano-sized $Ba_{0.7}Sr_{0.3}TiO_3$ powders prepared by spray pyrolysis. Ceram Int 38 (5):4029–4033
87. Ko YN, Choi SH, Kang YC (2013) Nano-sized Ag–$BaTiO_3$ composite powders with various amount of Ag prepared by spray pyrolysis. J Eur Ceram Soc 33(7):1335–1341
88. Afouxenidis D, Mazzocco R, Vourlias G, Livesley PJ, Krier A, Milne WI, Kolosov O, Adamopoulos G (2015) ZnO-based thin film transistors employing aluminum titanate gate dielectrics deposited by spray pyrolysis at ambient air. ACS Appl Mater Interfaces 7(13):7334–7341
89. Adamopoulos G, Thomas S, Wöbkenberg PH, Bradley DD, McLachlan MA, Anthopoulos TD (2011) High-mobility low-voltage ZnO and Li-doped ZnO transistors based on ZrO_2 high-k dielectric grown by spray pyrolysis in ambient air. Adv Mater 23(16):1894–1898
90. Jung D, Hong S, Cho J, Kang Y (2008) Nano-sized barium titanate powders with tetragonal crystal structure prepared by flame spray pyrolysis. J Eur Ceram Soc 28(1):109–115
91. Seck M, Mohammadian N, Diallo AK, Faraji S, Erouel M, Bouguila N, Ndiaye D, Khirouni K, Majewski LA (2020) Organic FETs using biodegradable almond gum as gate dielectric: a promising way towards green electronics. Org Electron 83:105735
92. Majumder S (2022) Synthesis methods of nanomaterials for visible light photocatalysis. In: Nanostructured materials for visible light photocatalysis. Elsevier, pp 47–113
93. Bhattacharya SK, Tummala RR (2000) Next generation integral passives: materials, processes, and integration of resistors and capacitors on PWB substrates. J Mater Sci Mater Electro 11(3):253–268
94. Barber P, Balasubramanian S, Anguchamy Y, Gong S, Wibowo A, Gao H, Ploehn HJ, Zur Loye H-C (2009) Polymer composite and nanocomposite dielectric materials for pulse power energy storage. Materials 2(4):1697–1733
95. Wang S, Huang X, Wang G, Wang Y, He J, Jiang P (2015) Increasing the energy efficiency and breakdown strength of high-energy-density polymer nanocomposites by engineering the $Ba_{0.7}Sr_{0.3}TiO_3$ nanowire surface via reversible addition–fragmentation chain transfer polymerization. J Phys Chem C 119(45):25307–25318
96. Yang K, Huang X, Xie L, Wu C, Jiang P, Tanaka T (2012) Core–shell structured polystyrene/$BaTiO_3$ hybrid nanodielectrics prepared by in situ RAFT polymerization: a route to high dielectric constant and low loss materials with weak frequency dependence. Macromol Rapid Commun 33(22):1921–1926
97. Qiao Y, Yin X, Wang L, Islam MS, Benicewicz BC, Ploehn HJ, Tang C (2015) Bimodal polymer brush core–shell barium titanate nanoparticles: a strategy for high-permittivity polymer nanocomposites. Macromolecules 48(24):8998–9006
98. Qiao Y, Islam MS, Han K, Leonhardt E, Zhang J, Wang Q, Ploehn HJ, Tang C (2013) Polymers containing highly polarizable conjugated side chains as high-performance all-organic nanodielectric materials. Adv Funct Mater 23(45):5638–5646

99. Moad G, Chong Y, Postma A, Rizzardo E, Thang SH (2005) Advances in RAFT polymerization: the synthesis of polymers with defined end-groups. Polymer 46(19):8458–8468
100. Arias AC, MacKenzie JD, McCulloch I, Rivnay J, Salleo A (2010) Materials and applications for large area electronics: solution-based approaches. Chem Rev 110(1):3–24
101. Ashurbekova K, Ashurbekova K, Botta G, Yurkevich O, Knez M (2020) Vapor phase processing: a novel approach for fabricating functional hybrid materials. Nanotechnology 31(34):342001
102. Whitesides GM, Grzybowski B (2002) Self-assembly at all scales. Science 295(5564):2418–2421
103. Whitesides GM, Mathias JP, Seto CT (1991) Molecular self-assembly and nanochemistry: a chemical strategy for the synthesis of nanostructures. Science 254(5036):1312–1319
104. Ha Y-G, Emery JD, Bedzyk MJ, Usta H, Facchetti A, Marks TJ (2011) Solution-deposited organic–inorganic hybrid multilayer gate dielectrics. Design, synthesis, microstructures, and electrical properties with thin-film transistors. J Am Chem Soc 133(26):10239–10250
105. Everaerts K, Zeng L, Hennek JW, Camacho DI, Jariwala D, Bedzyk MJ, Hersam MC, Marks TJ (2013) Printed indium gallium zinc oxide transistors. Self-assembled nanodielectric effects on low-temperature combustion growth and carrier mobility. ACS Appl Mater Interfaces 5(22):11884–11893
106. Stallings K, Smith J, Chen Y, Zeng L, Wang B, Di Carlo G, Bedzyk MJ, Facchetti A, Marks TJ (2021) Self-assembled nanodielectrics for solution-processed top-gate amorphous IGZO thin-film transistors. ACS Appl Mater Interfaces 13(13):15399–15408

Chapter 3
Synthesis, Microstructural and Dielectric Characterization of Nanodielectrics

Sudhanshu Dwivedi

Abstract Nanostructured dielectric materials are a special class of high-quality electronic-grade materials having confinement of at least one of its dimensions at the nanoscale for advanced nanoelectronics, nanodevice, optical, and many other potential applications. Nanodielectric materials became significantly important with the progression of the miniaturization of device dimensions. Currently, transistor fabrication technology has seen many advancements reaching almost atomic scales that inculcate very high quality and ultrathin gate dielectric layers. High capacitance nanodielectric gate materials are extremely important in high-throughput thin-film transistors (TFTs) for potentially hysteresis-free and low-voltage action. Moreover, organic–inorganic hybrid nanodielectric materials are the most-sought-after for flexible electronics on account of mechanical bendability. Organic nanodielectric materials are highly desired for advanced low-voltage operating carbon nanotube (CNT)-based TFTs and for complementary logic gates applications. Another fact is the ease of deposition of dielectric solution over plastic substrates for flexible nanoelectronic applications. Other categories of nanodielectric materials include carbon black composited with polymers to form nanocomposites, silica aerogel, epoxy microcomposites enhanced with nanosized clays, crystalline dielectric materials, and nanometric dendrimers. Next-generation applications can be classified into increased stress, higher temperature, stress-grading, and surface modification submissions. Engineered nanodielectric layers are of paramount importance for self-cleaning or easy-to-clean surfaces for outdoor applications in cars and windows, for example. The main feature of a high static contact angle provides for a nonwetting surface. Next-generation highly efficient insulation materials are the need of the hour that is free from unambiguous dielectric breakdown character including surface flashover between the dielectric surface and vacuum or surrounding environment. Nanoscale fillers, nanodielectric materials, and nanoparticles have been synthesized and worked on for state-of-the-art applications. This chapter is a concise effort to present information about nanosized dielectric materials, their synthesis methods, and their state-of-the-art applications.

S. Dwivedi (✉)
S.S. Jain Subodh P.G. (Autonomous) College, Jaipur, India
e-mail: sudhanshu.dwivedi@gmail.com

S. Moharana et al. (eds.), *Emerging Nanodielectric Materials for Energy Storage*, Nanostructure Science and Technology, https://doi.org/10.1007/978-3-031-40938-7_3

Keywords Nanotechnology · Dielectric materials · Theory of dielectrics · Nanodielectrics · Nanofillers · Polymer matrix

3.1 Introduction

The term "nanodielectrics" entails a multicomponent nanostructure possessing dielectric properties that are engineered in a way such that it is at least one of the dimensions is confined in the nanoregime typically less than 100 nm [1–4]. Fréchette [1] points out two different definitions of nanodielectrics as follows, "Nanodielectrics are those materials composed of multicomponent dielectrics confined in nanodimensions, so that its dielectric properties are altered as compared to the bulk phase". "Nanodielectric can be defined as materials that exhibit altered properties due to nanoconfinement within 100 nm of at least one of its dimensions". This definition is closer to a specific feature of nanostructured materials showing modified novel properties at the nanoscale.

Michel F. Fréchette states in the IEEE Electrical Insulation Magazine that he coined the word nanodielectrics entailing a combination of nanotechnology and dielectrics [1]. This concept of nanostructured dielectric materials came to the forefront in form of published literature after 2001. In the first experimental report, current–voltage characteristics were markedly modified by the use of a nano-additives-based mixed compound in the case of a ZnO varistor.

Nanodielectric materials exhibit completely distinguished performance as exemplified by the above-stated definitions and should not be confused with micron-sized agglomerated structures possessing distinct properties due to the inclusion of defects or even mild enhancement in dielectric properties. In addition to the chief feature of nanoconfinement in form of a nanoparticle, nanocrystal, nanocomposites, nanopowders, or mixed nanomaterials, the fundamental process should allow for the dielectric behavior to prevail. Michel F. Fréchette recognizes four different types of nanodielectrics that are nanoclusters of varying aspect-ratios, multilayered nanostructured dielectric films, nanocoating of only a few nanometers thickness, and nanophases. In a sense, self-assembly lies at the core of the development of a nanodielectric material along with the potential of erecting a structure based on building blocks and interfaces.

Nanodielectric materials can be utilized in a range of applications that includes space charge suppression, high-density energy storage, nonlinear field-grading applications, high thermal conductivity applications, and even in advanced applications in biomedical systems [1, 4–7]. Nanodielectric materials with nonlinear field-grading properties mean that the dielectric properties can be varied as a function of the electric field-induced stress [5]. Information acquired as a result of nonlinear field grading can be used in the distribution of electric fields over the dielectric surfaces homogeneously. This is particularly useful in the prevention of the formation of electric stress zones in high-voltage applications. Sometimes, nonlinear field grading can be performed intentionally to induce nonlinear electrical characteristics by incorporation of nanofillers into the polymer matrix. Nonlinear field-grading characteristics are also useful for outdoor high-voltage applications relying on high-voltages. Another state-of-the-art application of nanodielectrics is in integrated circuitry (IC) which has scaled down to as low as 3 nm node in the fabrication of next-generation devices. As a result, highly advanced nanodielectric materials are the need of the hour for effective electrical and thermal insulation as more heat per volume will be generated. Additionally, super-advanced nanodielectrics as gate materials are required to avoid electron tunneling phenomenon due to confinement of thickness in the 1–3 nm range [8–10]. This is highly desirable to control the increment of open-circuit current or leakage current because of the operation of nanodevices in the high-voltage regime. The development of thinner insulation material with much enhanced dielectric strength will ensure significantly reduced dielectric layer material for the same voltage level in a much more compact integrated circuitry nearing approximately atomic dimensions for a single node. Nanofillers-polymer composite materials are significantly useful because of low costs, flexible features, high breakdown strength, and high permittivity. Further, the incorporation of inorganic nanomaterials in a polymer base avoids not only the incorporation of space charges but accumulation as well. Space charges are detrimental to the local electric field distribution that degrades the efficiency of the involved polymer as a function of time under high voltages of the direct current operation. Enis Tuncer and Isidor Sauers have reviewed the nanoparticle as fillers in a polymer matrix along with their applications in a detailed manner [11]. Dielectric materials are of prime commercial value in the sense that lesser insulation material for the same voltage level with a positively modified breakdown can result in the development of much more efficient, cost-effective, and weight-reduced applications including power capacitors, transformers, and even power cables. Voltage endurance is an important parameter in the case of solid insulation materials that depict the time taken in a breakdown on the application of an electric field that is lower in magnitude in comparison with the stress generated due to a field leading to instant failure. Nanodielectrics can also improve direct current (DC)-based transmission cable systems. Nanodielectric material systems can improve mechanical features in form of tensile strength, elongation at the point of fracture, hardness, resistance to wettability and temperature variations, and electrical properties in form of enhancement of tolerance level against electrical arcing and reduced dielectric permittivities. Nanocomposite-enameled wires were subjected to repetitive surge voltage in a study forwarded by Okuba et al. and Hayakawa and Okubo which estimated a breakdown lifetime longer by a factor of 1000 as compared to bulk-phased enameled wires under similar surge voltage application [12]. Nanodielectrics can also be promising for high-temperature dielectric applications that include capacitors, compact transformers, and generators. Nanodielectrics are also under consideration for use in cryogenic applications with a focus on thermoplastics, such as polyvinyl alcohol (PVA), *poly*methyl methacrylate (PMMA), or even a thermoset. A remarkable feature is that high permittivity particles on compositing with low permittivity surfactant produce a material that possesses low dielectric permittivity.

For insulation applications, nanodielectrics must possess sound features of high electric strength, effective relative permittivity along with low dielectric loss ($\boldsymbol{\delta}$) of the tailored material medium as per the application, and high tolerance level against material erosion, high stability of the optimized nanostructure design for prolonged periods in a reduced nano-entropy configuration, stable behavior as a function of temperature, or even on prolonged hours of operation including effective mechanical properties. Effective relative permittivity along with low dielectric loss ($\boldsymbol{\delta}$) need to be engineered in a nanodielectric material according to the requirements. In terms of commercial applications, the switchover from the current module of bulk insulation materials to nanodimensional materials with a brand-new set of properties is a giant step. All original equipment manufacturers (OEM) are fearful of making a changeover because of dreadful commercial implications. The important features of nanodielectric materials are inclusive of highly efficient voltage endurance, high fidelity against discharge-driven deformations, superior dielectric behavior with distention for intrinsic space charge collection, and increment in alternating current (AC) breakdown strength with positive consequences of enhancing the operational electric load that become significant at high temperatures under application of divergent fields along with direct voltage. This can ensure a highly concise mechanical design of electric machines along with thermal features.

3.1.1 Theory of Dielectric Materials

Dielectrics belong to the broader class of materials that possess the property of polarizability under the application of an electric field. The formation of dipolar structures allows the containment of an electrostatic field within them [13]. Insulators are also dielectric materials that prevent the flow of leakage currents or open-circuit currents in devices. Dielectrics have become extremely important in integrated circuitry (IC) since shrinking of device architectures to almost atomic scales (1–3 nm node) has compelled the development of ultrathin high-$\boldsymbol{k}$ dielectric materials that can not only provide effective gating but prevent leakage currents as well. Under the effect of an electric field, positive charges are along the direction of the electric field ($\boldsymbol{E}$) while negative charges are aligned in a direction opposite to the direction of the electric field ($\boldsymbol{E}$). In this process, dielectric material remains neutral as a whole, although, electrons get displaced from their mean vibrating positions in length scales smaller than atomic sizes. No significant charge displacement is observed at the macro-scale similar to the conductors to cause higher conductivities. G. C. Psarras has provided details of the theory of dielectrics in his book chapter on "Fundamentals of Dielectric Theories" in a very nice manner [13].

Polarization happens due to the distribution of charged species in a dielectric with a vector pointing from the midpoint of negatively charged species toward the midpoint of positively charged species. The developed electric dipole moment (μ) due to the absolute value of positive and negative charges with distance r between the two centers is given as follows,

$$\mu = \mathbf{q} * \mathbf{r} \tag{3.1}$$

The total dipole moment can be calculated by summing the above product over a range of different dipoles. Polarization (***P***) is defined as the total electric moment per unit volume (**V**) of the dielectric material, which is given as,

$$\mathbf{P} = \frac{\mathbf{M}}{\mathbf{V}} \tag{3.2}$$

Polarization is an important parameter that practically resembles charge surface density with units of $\mathbf{C\,m^{-2}}$ and dimensions as $[\mathbf{L^{-2}TI}]$. Polarization can be defined as the average sum of ***n*** electric dipole moments contained in a unit volume which can be expressed as,

$$\boldsymbol{P} = \left\langle \sum_{j=1}^{n} \mu_i \right\rangle \tag{3.3}$$

In an isotropic material, under the effect of zero electric fields ($\mathbf{E} = 0$), permanent electric dipole moments are arranged randomly. On application of an **E**, permanent electric moments are along the direction of ***E***, while induced dipole moments ($\boldsymbol{\mu}_{\mathbf{ind}}$) are also developed as outlined above. Hence, dipole moment as a whole is expressed as,

$$\boldsymbol{\mu}_{\mathbf{Total}} = \boldsymbol{\mu}_{\mathbf{i}} + \boldsymbol{\mu}_{\mathbf{ind}} \tag{3.4}$$

And, in that case, total polarization is,

$$\boldsymbol{P} = \left\langle \sum_{j=1}^{n} \boldsymbol{\mu}_{\mathbf{Total}} \right\rangle \tag{3.5}$$

The coefficient of polarization or polarizability ($\boldsymbol{\alpha}$) is an important dielectric parameter for dielectrics that give the ability of atoms or molecules to get polarized.

In the case of linear dielectrics, polarizability can be defined as the total dipole moment developed in a dielectric under the application of an external electric field (**E**) or better total dipole moment produced per unit of electric field (**E**).

Hence,

$$\boldsymbol{\alpha} = \frac{\boldsymbol{\mu}_{\mathbf{Total}}}{\boldsymbol{E}} \tag{3.6}$$

In the above equation, ***E*** is the acting electric field and not the field applied externally. Polarizability possesses dimensions of volume and can be estimated for dielectric materials with different types of atoms.

Electric displacement vector ($\vec{D}$) with $\varepsilon_0 = 8.854 * 10^{-12} F/m$ is expressed as follows,

$$D = \varepsilon_0 E + P \tag{3.7}$$

Noticeable point is that polarization relates to induced charges over edge surfaces of the sample in a parallel plate capacitor filled with a dielectric material, electric displacement pertains to independent charges at the electrode surfaces, while electric field vector (**D**) depends on both independent and induced charges.

The relative dielectric constant (ε) of a material is defined as follows,

$$\text{Relative dielectric constant}(\varepsilon) = \frac{Electric\ Permittivity\ of\ Material(\varepsilon_m)}{Electric\ Permittivity\ of\ Vacuum(\varepsilon_0)}$$

Relative dielectric constant (ε) in terms of electric displacement vector (D) is expressed as,

$$\varepsilon = \frac{D}{\varepsilon_0 E} = 1 + \frac{1}{\varepsilon_0}\frac{P}{E} \tag{3.8}$$

Here, $\frac{P}{\varepsilon_0 E}$ is electric susceptibility that provides the degree of induced electric polarization due to the application of an electric field.

Dielectrics can be classified into polar and nonpolar dielectrics. Polar dielectrics possess permanent electric molecular dipoles since the line joining the midpoints of positive and negative charge carriers is not along a single direction. Permanent molecular dipole moments persist even in the absence of the application of an electric field. The essence lies in the asymmetrical nature of the molecules constituting molecular dielectrics with the value of dipole moment increasing on increasing the difference in electronegativity of the constituent atoms. On the other hand, symmetrical molecules in nonpolar molecular dielectrics are aligned, so that the midpoints of positive and negative charge carriers coincide to cause a loss of nonsymmetrical nature. Molecular dipole moments tend to be oriented in an externally applied electric field. Perfect molecular orientation happens when the imposed force due to an externally applied field overcomes the thermal energy of molecules. Dipoles tend toward incremental alignment with an increase in the strength of the external electric field while the ordering becomes randomized on increasing the temperature based on additionally imparted thermal energy. Atoms do not possess permanent electric dipole moments because of being spherically symmetric. However, under the effect of an externally applied field, their electron clouds acquire certain orientations leading to the development of induced dipole moments. There may be interactions happening between permanent electric moments and induced dipoles that may onset a distortion in the perfect alignment of dipoles in the direction of an electric field.

Technically, polarization can be classified into distortional and orientational polarizability.

(i) Distortional polarizability is also known as deformational polarizability. Distortion polarizability can be classified commonly into electronic and atomic or ionic polarizability based on relative deformations or distortions of electronic charges or between ions generated due to an externally applied electric field. It is to be noted that electronic polarizability keeps on increasing as a function of the increment in the number of electrons. The genesis lies in the fact that electrons in outer shells of atoms with many electrons experience a lesser degree of binding forces with nuclei imparting significant contribution to α-values. Atomic or ionic polarizability is a consequence of displacements of atoms or ions from their equilibrium positions under the effect of an externally applied electric field. Atomic or ionic polarizability progresses much slower in comparison with electronic polarization due to the higher mass of ions or atoms than electrons. However, both electronic and atomic polarizability has low relaxation times making them "fast" processes. Another important parameter is that deformational polarization has no dependence on temperature, and hence, no energy loss is reflected as a function of temperature.

(ii) Orientational polarizability is also called as dipolar polarizability. Permanent electric dipole moment is exhibited due to structural molecules or groups in polar dielectric materials. Randomly oriented dipoles get perfectly aligned in the direction of an externally applied electric field. In the structural configuration of solids, the relative displacement of charges with localized attachment to atoms or molecules leads to distortional and orientational polarizability. Space charge is an important element in dielectrics that portray additional charge carriers moving short distances across the dielectric and imparting overall polarization and leading to an enhancement in the capacitance of dielectric material. An important characteristic is a constriction in movement or traveling paths due to trapping at the grain boundaries or interfaces in the dielectric material. Space charge contribution can be significant that becomes indistinct or approximately impossible to filter out from other polarizability contributions. Hence, space charge or interfacial polarizability (α) appears as a fourth term in the total polarizability of a material in addition to electronic, atomic, or ionic and orientational polarizabilities. As pointed out previously, polarization attempts to overcome thermal energy-based contribution to kinetic energy (K.E.) of molecules and hence is a function of the applied temperature.

Determination of the static value of dielectric permittivity or dielectric constant as well as polarization is a direct function of dipole moments of a polar dielectric. Peter J. W. Debye adopted Langevin's theory of statistic orientation to derive an analytical explanation of permanent dipole moments of a polar dielectric material. As pointed out in this theory, the cumulative effect of two opposing actions generated due to an externally applied electric field that aligns dipoles in preferred orientations and thermal agitation-driven randomization of dipoles results in a statistical equilibrium. Here, the definite rotation of molecules as well as their corresponding relative interactions are neglected in the derivation of statistical equilibrium. The assumption is based on factors that include very slowly varying externally applied electric field to achieve polarizations of all different types in an equilibrium state, permanent dipole moment ($\boldsymbol{\mu}$) is free from effects of temperature as well as applied fields, dipole axes can adopt any alignment for the direction of externally applied field in an initial phase, Boltzmann statistical distribution of permanent electric dipole moments in a direction to that of the applied field, and isotropic nature of the dielectric material. Based on above-mentioned assumptions, Debye formulated the theory of dielectrics by doing away with complexities resulting in dielectric materials due to close interactions of molecules attributed to dense packing fractions.

3.1.2 Nanodielectrics

The development of nanodielectric technology relies heavily on morphological features of nanomaterials that can acquire granular microstructure, one-dimensional (1D) nanofiber, and two-dimensional (2D) nanoplatelets. Such morphologically varying nanostructures can be used as fillers for spatial distribution in a polymer matrix creating an interface of two distinguished materials. This interfacial zone has a strong bearing on the overall properties reflected by these typified polymer-nanoparticles mixed nanocomposites. Advanced tailoring of nanodielectric properties is possible by suitable modifications in the microstructure of the nanoparticle along with tuning of filler percentage and spatial distribution in the polymer matrix. Zhong et al. have reviewed polymer-nanoparticle composite materials as nanodielectrics in detail [6].

Nanodielectric materials are mainly of two types nanostructured ceramics and nanocomposites. The use of nanocrystalline materials as starting precursor material leads to the development of a highly efficient final product. A simplistic approach is to mingle nanostructured particles in a polymer matrix for the preparation of polymer nanocomposites. However, highly efficient approaches need to be followed for the development of nanometric dielectric materials. Nanodielectric particles of very small sizes show an agglomeration tendency. The great interest in nanodielectric materials lies in the large surface area available in nanosized surfaces, better functionalization opportunities offering high loading sites, and highly modified properties in contrast to their bulk counterparts. Spontaneous dielectric polarization as a function of the externally applied electric field happens in nanodielectric materials [14]. Nanosized dipoles are formed in the nanostructured dielectric material at an instant under an external electric force. Generally, the electric polarization of perfect insulator materials is termed dielectric polarization. Dielectric polarization can be explained based on the following three mechanisms of electronic, atomic or ionic, and orientational polarization,

(i) Spatial shift of electron cloud about the nuclei that induce the formation of atomic dipoles developing electronic polarization ($\boldsymbol{P}_e$),

(ii) Dislocation of various atomic nuclei from their initial relative positions in an inverse space to induce molecular dipoles developing atomic or ionic polarization ($\boldsymbol{P}_{\mathbf{a}}$),
(iii) Alignment of permanent molecular dipoles under the influence of an electric field force to develop orientational polarization ($\boldsymbol{P}_{\mu}$).

The resultant dielectric polarization vector ($\vec{\boldsymbol{P}}$) is the actual sum of all the above-mentioned three contributing polarization vectors.

$$\left(\vec{P}\right) = \overrightarrow{P_e} + \overrightarrow{P_a} + \overrightarrow{P_\mu} \tag{3.9}$$

Electronic polarization and atomic polarization as well as depolarization are very fast processes in dielectric materials in comparison with orientational polarization as well as disorientation polarization.

Engineering of nanodielectric materials requires suitable tailoring of nanostructure shape and sizes, morphological features, design of structures with appropriate aspect ratios, and relevant interfacial modifications. Nanofillers can be suitably divided into classes of **0D**, **1D**, **2D,** and **3D** types with each specific type of nanofiller improving on the particular properties of mechanical, electrical, thermal, and optical properties. Quantum dots belong to the category of **0D** quantum confined nanofillers. Similarly, nanoshells, nanolaminas, and nanoplates belong to **1D** confined nanofillers. Nanotubes, nanowires, and nanofibers are **2D** confined nanostructures. Nanosized $\mathbf{SiO_2}$ forms a good example of **3D** confined quantum structures.

Polymer nanocomposites (PNCs) can be synthesized by the application of three major PNC components of the polymer matrix, nanomodified particles along with transitional constituents, such as surfactants and intercalants [15–20]. A range of polymers is available in the commercial market that can be exploited as a matrix for the synthesis of PNCs. Different polymers including polyvinyl alcohol (PVA) [21, 22], polyethylene glycol (PEG) [23], and even *di*-block copolymers (poly-styrene-poly-ethylene oxide) [24, 25] can be used for the preparation of PNCs. Polymers need to be processed with nanomodified particles and transitional constituents carefully in a thermal environment below their glass transition temperature ($\boldsymbol{T}_{\mathbf{g}}$). Earlier thermosets were used frequently for PNCs which were subsequently replaced by thermoplastics [5, 26–28]. Polyolefins (**POs**) belong to the category of thermoplastic polymers possessing intricate organic molecular structures with relatively low melting points (M.P), semi-crystalline, cost-effective, less resistant to degradation against UV-radiation, mechanical breakage, and oxidation chemical means along with the temperature of processing not to exceed 250 °C [28]. Clay-containing PNCs can be synthesized by reactive, solution, and melt-compounding methods. The reactive method involves in situ polymerization technology, solution method is focused on creating a dispersion of organoclay in a solution dissolved with polymer, while melt-compounding consists of formation by compounders [28]. Clay platelets as well as polymers are of the separate chemical constitution with polymers being hydrophobic and largely immiscible with hydrophilic clays, mostly silicates. Compatibilization is a two-step process forming a diffused ionic layer in a region surrounding the clay platelet along with the dispersion of extra ionic bunches in the material medium [28]. The interface between clay and matrix offers a range of complex reactions. Organic molecules are physisorbed followed by solidification resulting in enhancement of overall net solid content. Significant enhancement of electric modulus along with barrier characteristics due to totaling of a small quantity of effectively distributed clay particles is caused because of increment in solid content at the interface. This may also halt the UV-driven degradation process. The inclusion of clay particles enhances the degree of crystallinity along with providing efficient nucleation sites. An important parameter is the optimization of type and quantity of intercalant as well as compatibilizer that may cover the clay particles effectively and halt effective nucleation. The inclusion of an optimized content of clay in PNC is an important parameter that shows the most effective performance in the 2–4 wt% range of organoclay. The degradation of clays may be driven catalytically by transition metal ions as contaminants. Organoclay does not have a sound effect on dielectric properties when added in small quantities but has a pronounced effect on mechanical as well as barrier characteristics.

3.1.3 *Interface Chemistry*

Interface chemistry plays a major role in each of the steps involved in the synthesis methodology of nanocomposites. Glass transition temperature ($\boldsymbol{T}_g$) in polymers is a classical illustration of the involved interfacial chemistry in nanocomposites [29]. Simulation or modeling or even experimental study of interfacial chemistry-related characteristics is an important factor that can be tailored effectively by varying simulation or experimental conditions. A plethora of literature has been presented on interface chemistry in nanocomposites. Alcoutlabi and McKenna concluded in their report that the present theoretical accounts to explain $\boldsymbol{T}_g$ are not sufficient to describe the distinct features at the nanoscale [28, 30]. The defect diffusion model inculcated factors of concentration of defects, percolation fraction, defect lattice geometry correlation length, and defect–defect interaction enthalpy [28]. This model on nanoconfined $\boldsymbol{T}_g$ was presented by Bendler et al. and focused on a calculable association between the percolation ratio obtained by fixed to transportable areas and $\boldsymbol{T}_g$ [31]. Lewis et al. have pointed out that nanocomposite presents an interface between nanofiller and polymer matrix and it has nothing to do with a simplistic two-component phasic system [32, 33]. As a result, the incorporation of a minimum of a single interphase region becomes of prime importance to get a nanoproduct in which structure–property relationships and involved chemistries become distinct in contrast to the two initial bulk phases that react primarily to produce the nanocomposite. A significant distinction of this system is that nanoparticles possess the enormously exposed surface area to contribute to the interphase region in a nanocomposite.

Complex permittivity ($\boldsymbol{\varepsilon}$) is an important parameter and dielectric response to externally applied AC electric fields studied in terms of complex permittivity.

Complex permittivity consists of a real part (ε') and an imaginary part (ε''). The real part of complex permittivity (ε') entails energy stored in a dielectric material, while the imaginary part of complex permittivity (ε'') talks about different mechanisms that correlate to the dissipation of energy. Tanaka gave an initial modeling theory that considered a nanocomposite formed by the dispersion of nanoparticles in a polymer matrix. In a model based on this approach as shown in Fig. 3.1, nanoparticles are encircled by three different layers of separate structural geometry covered by a diffuse Gouy–Chapman charge layer [34]. In the core lies the nanoparticle, while other layers lie one above the other as shown in the model given below. Adjacent two layers close to the core nanoparticle in the model, consisting of immobilized species that bear a strong interacting relationship with their inner neighbor, are covalently attached to it. An enhanced free volume including chain mobility lies at the core of this system confined in an area between the immobilized shells and undisturbed matrix summing up with other factors to generate a distinct effect on the dielectric nature of the formed three-phase system.

The rule of mixing the composite material consisting of two systems A and B portrays that the properties of the finally derived composite material will lie in between the values shown by precursor materials A and B.

Raetzke and Kindersberger have discussed the impact of interphase in terms of fraction based on the diameter of the particle and thickness of interphase signifying the role of physical or even chemical bonding at the interphase, acting radius, and correlated material properties [29]. As pointed out above, nanoparticles as fillers possess a surrounding area covered with polymer forming the interphase. This surrounding layer around the nanoparticles or interphase has a thickness typically in the range of 0.5–10 nm. Polymeric chains contained in the surrounding matrix are bound to the particle surface by physical or chemical bonds that may possess orientations as shown in Fig. 3.2a and b. The interphase structure depends on a range of parameters that include filler or type of nanoparticles, surface treatment meted-out, type of matrix material, and even the crystalline or semi-crystalline nature of interphase that plays a deciding role. Properties of interphase are completely different from surrounding matrix material because of distinct interphase structure and typical bond formation with the particle surface.

Attachment of polymeric chains over nanofiller surfaces happens by chemical bonds that depend on polar or nonpolar characteristics of the two bonding species. An interesting factor is the conversion of surface groups from nonpolar to polar characteristics due to surface treatment. Ehrenstein has pointed out that chemical bonds of polymeric chains possess bond energies between 40 and 800 kJ mol^{-1} with bond lengths typically lying in the range of 0.075 to 3 nm [35]. On the other hand, bond energies in physical bonds are typically between 0.2 and 25 kJ mol^{-1} for bond lengths between 0.5 and 0.8 nm.

An explanation of interphase phenomena in terms of a multicore model entails involving all three layers, so that the contribution from interphase becomes significant along with nanofiller surfaces impacting the polymeric chains. Different nanofiller concentrations will produce interphase made up of different fractions. For the establishment of a simplistic explanation, it is pertinent to consider the three separate

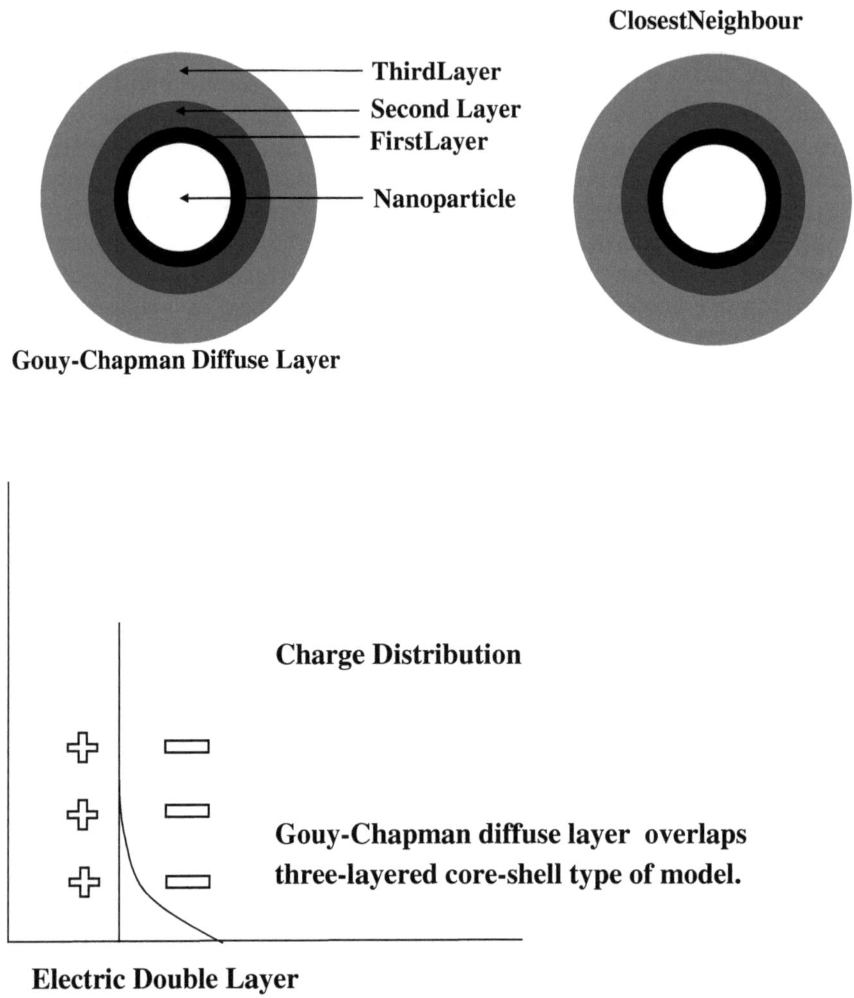

Fig. 3.1 Schematic representation of the interfacial model for core–shell structured nanoparticles for advanced nanodielectric materials [34]

layers as a single interphase layer possessing thickness, *j*. Bond lengths and atomic diameters are the parameters important for the determination of thickness *j* of interphase in the multicore model. Typically, the first layer possesses a thickness in the range of 1–1.5 nm in which polymeric functional groups are chemically bonded to the nanofiller surfaces, and consists of contribution from nanofiller surfaces as well, with a typical value of 0.4 nm. The thickness of the initially attached organic layer in a tight-binding approach lies in the range of ~0.3 nm–1.0 nm, which typically is dependent upon the type of polymer. The second layer has polymeric chains attached to nanofiller surfaces or the first layer as per the multicore model. In the second layer,

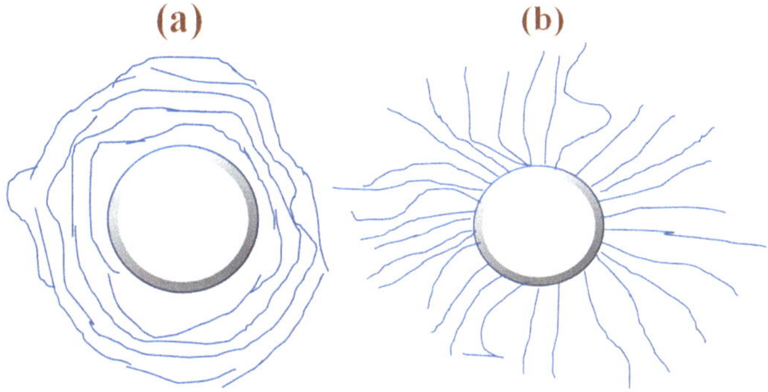

Fig. 3.2 Schematic representation of **a** nanofiller surrounded by polymer matrix in such that **a** polymer chains are arranged parallel to the nanofiller surface, and **b** polymeric chains oriented radially about the nanofiller surface [29]

the sterical arrangement of polymeric functional groups is affected mainly and facilitates an ordered arrangement. In the parallel arrangement of polymeric chains about the nanofiller surface, as shown in Fig. 3.2a, the little number of functional groups constitute mere thickness of the second layer that is dependent on smaller bond lengths of $\mathbf{H_2}$-bonds as well as dipole bonds. It signifies that the thickness of the second layer depends on the degree of strength of the bonds. The thickness of the interphase layer can be estimated by the diameter of polymeric chains typically in the range of ~0.3 nm to 1.0 nm along with bond lengths between 0.5 and 0.8 nm. A typical thickness between 1.5 and 8 nm prevails for about 2–4 entangled polymeric chains in the second layer. In the case of radial attachment of polymeric groups to the nanofiller surface, as shown in Fig. 3.2b, the second layer can adopt higher thicknesses based on polymeric chain lengths.

In contrast to the first and second polymer layers, the third layer is affected by the orientation of polymeric chains in the second layer, and usually, the thickness is between 1 and 3 nm. With all the above three-layered structure considerations, the thickness ($\boldsymbol{j}$) of interphase lies in the range of 3–12 nm.

Interphase structure greatly impacts the mechanical, dielectric, electrical and thermal conductivity, and prevention of material degradation. The mobility of polymeric chains affects dielectric properties, such as dielectric loss (**tan δ**). On increasing interphase fractions, the permittivity of dielectric material can show drastic changes.

Raetzke and Kindersberger have given a simplified model for the estimation of interphase faction at different fractions of nanofillers [29]. Assumptions for this model are outlined below,

(i) All particles are assumed to be spherical and uncapped in interphase having nonvariable thickness $\boldsymbol{j}$.
(ii) All particles are assumed to be equal in size with a constant diameter "$\boldsymbol{d}$".
(iii) Homogeneous dispersion of particles happens in dispersing medium or in the matrix.

In a cubic configuration with homogeneous dispersion, as shown in Fig. 3.3a, eight particles share each corner in the cubic geometry along with one particle at the body center of the cubic structure. Let the center-to-center distance of the two closest lying particle centers be $\boldsymbol{r_0}$. This type of particle arrangement in a cubic configuration can accommodate a maximum nanofiller concentration of 68%.

Estimation of interphase fraction or $\boldsymbol{f_j}$ can be performed by knowing about nanofiller concentration $\boldsymbol{V_f}$, while interphase thickness or $\boldsymbol{j}$ is another important parameter. Four different conditions can be considered in this regard,

$$\text{(i)}\quad \frac{2}{\sqrt{3}}(\boldsymbol{d}+2\boldsymbol{j}) \le \boldsymbol{r}_0 \tag{3.10}$$

At low concentrations of nanofillers, interparticle distance is significantly high, and hence, no overlapping of neighboring interphases happens. Considering spherical volumes to be $\boldsymbol{V_d}$ and $\boldsymbol{V_{d+2j}}$ for diameters $\boldsymbol{d}$ and $\boldsymbol{d+2j}$, respectively, let the interphase volume be represented as $\boldsymbol{V_j}$, then,

$$V_j = V_{d+2j} - V_d = \frac{1}{6}\pi\left[(d+2j)^3 - d^3\right] \tag{3.11}$$

Now, the estimation of interphase fraction $\boldsymbol{f_i}$ to nanofiller concentration $\boldsymbol{f}$ can be given as,

$$\frac{\boldsymbol{f}_{j,1}}{\boldsymbol{f}} = \frac{2\boldsymbol{V}_j}{\boldsymbol{V}_d} = \frac{\frac{1}{6}\pi\left[(\boldsymbol{d}+2\boldsymbol{j})^3 - \boldsymbol{d}^3\right]}{\frac{1}{6}\pi\boldsymbol{d}^3} \tag{3.12}$$

$$\boldsymbol{f}_{j,1} = \boldsymbol{f}\left[\left(1+\frac{2\boldsymbol{j}}{\boldsymbol{d}}\right)^3 - 1\right] \tag{3.13}$$

$$\text{(ii)}\quad (\boldsymbol{d}+2\boldsymbol{j}) \le \boldsymbol{r}_0 < \frac{2}{\sqrt{3}}(\boldsymbol{d}+2\boldsymbol{j}) \tag{3.14}$$

In this range, interphase fraction enhances with nanofiller concentration, so that closely lying interphases overlap.

$$\boldsymbol{f}_{j,2} = \boldsymbol{f}\left[\left(1+\frac{2\boldsymbol{j}}{\boldsymbol{d}}\right)^3 - 1 - 8\left(\frac{1}{2}+\frac{\boldsymbol{j}}{\boldsymbol{d}}-\frac{\sqrt{3}}{4}\frac{\boldsymbol{r}_0}{\boldsymbol{d}}\right)^2\left(2+\frac{4\boldsymbol{j}}{\boldsymbol{d}}+\frac{\sqrt{3}}{2}\frac{\boldsymbol{r}_0}{\boldsymbol{d}}\right)\right] \tag{3.15}$$

$$\text{(iii)}\quad \frac{2\sqrt{2}}{3}(\boldsymbol{d}+2\boldsymbol{j}) < \boldsymbol{r}_0 < (\boldsymbol{d}+2\boldsymbol{j}) \tag{3.16}$$

High overlapping of closely lying interphases happens in the above-specified range. Under this condition, interphase concentration reduces because nanofillers

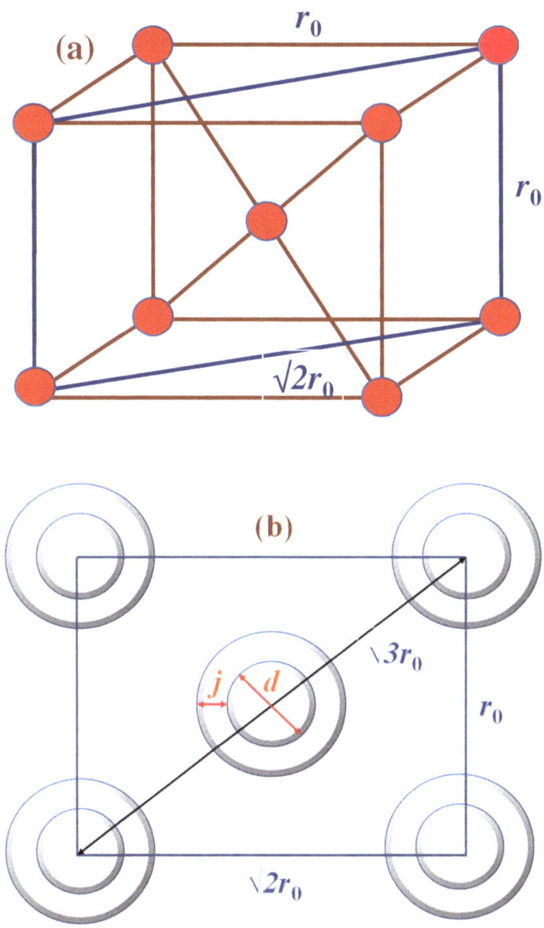

Fig. 3.3 **a** For homogeneous dispersions of particles, a cubic geometry can be assumed for a perfect arrangement of particles such that eight particles share each corner of the cube similar to the cubic crystal geometry in face-centered cubic (*fcc*) configuration. This can be described as the fundamental geometry of the arrangement of particles that are dispersed homogeneously in a polymer matrix. In addition, a body-centered cubic (bcc)-type configurational structure is also visible such that a particle is placed at the center of the body of the cube, so that a maximum of 68% of nanofiller concentration by volume can be accommodated [29]. **b** Nonoverlapping particles are represented schematically here since nanofiller particles are placed at considerable distances attributed to low nanofiller concentrations used in the formation of nanocomposites. As a result, interphases of closely-lying particles do not partake in the formation of common regions. Interphase fraction is supposed to increase linearly with an increment in nanofiller concentration [29]

begin to shift toward closely lying interphases with an increment in nanofiller concentration (Fig. 3.4).

$$f_{j,3} = f\left[\left(1+\frac{2j}{d}\right)^3 - 1 - 8\left(\frac{1}{2}+\frac{j}{d}-\frac{\sqrt{3}}{4}\frac{r_0}{d}\right)^2\left(2+\frac{4j}{d}+\frac{\sqrt{3}}{2}\frac{r_0}{d}\right)\right.$$
$$\left.-6\left(\frac{1}{2}+\frac{j}{d}-\frac{r_0}{2d}\right)^2\left(2+\frac{4j}{d}+\frac{r_0}{d}\right)\right] \tag{3.17}$$

(iv) $r_0 < \frac{2\sqrt{2}}{3}(d+2j)$ (3.18)

If interphase fraction dominates in the overall structural configuration, the effect of incremental nanofiller concentration leads to a decremental interphase fraction (Fig. 3.5).

$$f_{j,3} = 1 - f \tag{3.19}$$

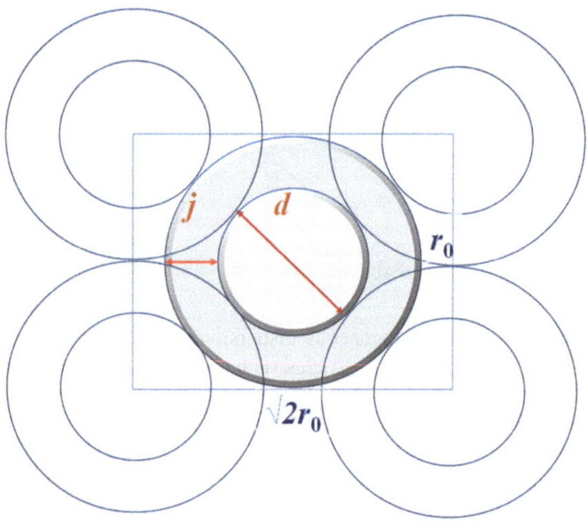

Fig. 3.4 Schematic representation of the diagonal element of fundamental geometry of particles reflecting overlapping of interphases under the third condition [29]

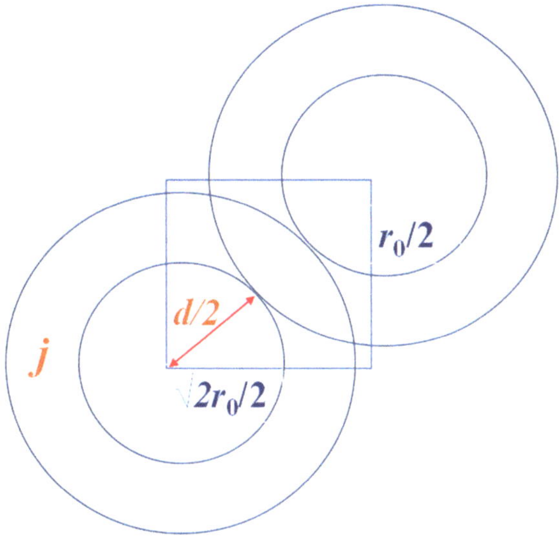

Fig. 3.5 Schematic representation of the interphase-based formation of the complete polymeric layer [29]

3.2 Synthesis, Microstructural, and Dielectric Characterization of Nanodielectrics

Different techniques employed for nanoparticle synthesis are also used for the preparation of dielectric nanoscale materials. This is inclusive of simple solution techniques, co-precipitation methods, simple mixing of nanoparticles in a polymer environment for the formation of polymer nanocomposites preparation of nanoparticles under vacuum, hydrothermal synthesis, sol–gel method, sonochemical synthesis of nanoparticles, microwave-assisted synthesis, template-based synthesis approach, and application of biological templates for synthesis of nanoparticles and physical methods that include pulsed laser ablation.

3.2.1 Simple Solution Techniques for Nanodielectric Particles

Solution-based mixing of nanofillers with polymers in a solution along with the melt-compounding method in the context of broader mechanical mixing of two components is a well-established process. Mechanical mixing is a common process in industrial production that entails the management of a diverse physical system to produce a resulting homogeneous system in the configuration of unit operation. Mass or heat transfer happens due to the mechanical agitation of multiple phases. In the simple solution chemistry or mechanical mixing method, polymers are usually dissolved in a solution followed by the dispersion of nanofillers in this polymer solution to synthesize nanocomposites in the solution. This nanocomposite solution is subjected to evaporation to get solidified nanocomposites in a mixed configuration.

Shear forces in the simple solution chemistry method are of a much lower degree due to the imposition of mechanical agitation of a lesser degree. This is usually performed by placing the mixed components solution on a mechanical stirrer under constant magnetic stirring under the effect of temperatures if so required. However, the applied temperature must be kept below its $\boldsymbol{T}_g$ to avoid the transformation of the polymeric phase. Due to this fact, external parameters play a vital role in the dispersion of nanofillers in the polymer solution, such as surface loading of nanoparticles, surface modification by organic molecules, and presence of binders. The advantage of this method involves high control over the produced nanocomposites since constituents can be mixed in desired percentages and mechanical mixing. Another benefit is to have better control over the shape, size, and morphological features of nanoparticles that are produced separately as one of the constituents. Ajay Vasudeo Rane and others have given details of clay-polymer nanocomposites prepared by mechanical mixing of polymers with both pristine and modified clays [36]. Solid–solid phase mixing and solid–liquid phase mixings are the stereotyped methods in the formation of clay-polymer nanocomposites. Solid–solid mixing in this case is further classified into convective and intensive mixing. Convective mixing generates a random state eventually with the potential risk of demixing constituents due to segregation caused by variations in shape, size, and density. Convective mixing is not suitable for the mechanical mixing of materials possessing cohesive nature mostly exercised in the case of very finely divided particles or even wet compounds. Convective or transportation forces are of a very mild nature that is not sufficient to overcome relatively strong cohesive forces acting among particles. As a result, energy-based mixing or impact employment or application of shear stress may be followed to negate the effect of cohesive interparticle forces. Four different synthesis methods of clay-polymer nanocomposites include in situ template synthesis, solution intercalation, in situ intercalative polymerization, and melt intercalation [36]. Melt intercalation in conjunction with a shear mixer is the preferred method for the synthesis of clay-polymer nanocomposites based on thermoplastics and elastomers. Solution intercalation is applied for the synthesis of clay-polymer nanocomposites in which different thermosets, thermoplastics, and elastomers based on solubility parameters are dissolved in the specified solvents followed by intercalation with solvent-dissolved nanoclay particles. At the industrial scale, mechanical mixing is performed in batches inline or dynamic mixing with mixers possessing the capability to operate at 1500 or 1800 rpm.

Ren et al. have compared the dielectric properties of pristine polyetherimide (**PEI**) polymeric materials with those of polyetherimide polymer composited with hafnium oxide (**HfO_2**) nanoparticles synthesized mainly by the solution casting method [37]. Solutions were cast over glass slides followed by drying at a suitable temperature for a specific period to evaporate the solvent. Dried solution casted films were peeled-off and dried again to obtain 9 to 12 μm thick films.

Nanocomposites enhanced with **HfO_2** nanoparticles show an improved dielectric constant along with a reduction in high-field current density. A difference in morphological features of the pristine **PEI** and **PEI-HfO_2** nanocomposites was observed from scanning electron micrographs (SEM) as shown in Fig. 3.6.

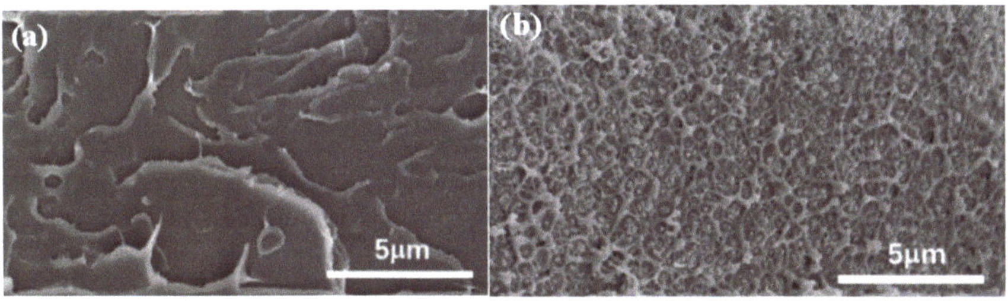

Fig. 3.6 SEM images of pristine **PEI** and **HfO_2**-enhanced nanocomposites in **PEI** matrix [37]. Morphological changes are visible at the microscale of 5 μm with a network formation in the **HfO_2**-enhanced matrix in comparison with a rather larger layered-type formation in pristine PEI composites [37]. Reprinted with permission from ref. [37]. Copyright (2021) (Elsevier)

For pristine **PEI** and **PEI/HfO_2** nanocomposites with various nanofiller contents, a comparison of ***D-E*** loops was performed at $\boldsymbol{E} = 300$ MV.m^{-1} as shown in Fig. 3.3a and b. It was inferred that electric displacement (***D***) is steadily improved as a function of the induction of **HfO_2** nanofillers in the **PEI** matrix along with more than 95% efficiency of the charge–discharge cycle. At nanofiller contents of 1 vol%, 3 vol%, 5 vol%, 7 vol%, and 9 vol%, ***D***-value increased to 1.03×10^{-2}, 1.07×10^{-2}, 1.11×10^{-2}, 1.14×10^{-2}, and 1.22×10^{-2} C m^{-2} for nanocomposites from 9.77×10^{-3} C m^{-2} of pristine **PEI** at 300 MV m^{-1}.Thus, the ***D***-value reflects a trend similar to that of dielectric constant since ***D*** changes in a linear manner as a function of $\boldsymbol{\varepsilon}_r(\boldsymbol{D} = \boldsymbol{\varepsilon}_0\boldsymbol{\varepsilon}_r\boldsymbol{E})$ in the case of linear dielectrics Fig. 3.7.

Figure 3.8 shows a schematic of the **PEI/HfO_2** nanocomposites with nanoparticles at the core and first, second, and third layers as the bonding, bound, and loose layers typically in a Guoy–Chapman electrical double-layer configuration [37]. Band

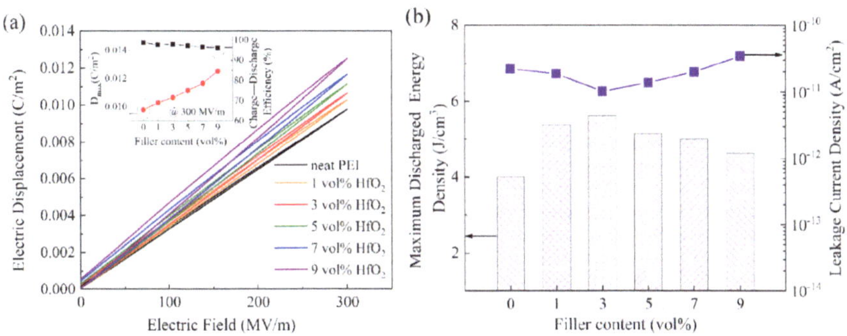

Fig. 3.7 **a** *D-E* loops of pristine **PEI** and **PEI/HfO_2** nanocomposites derived at $E = 300$ MV m^{-1}. Inset shows the D_{max} and charge–discharge efficiency in % of pristine **PEI** and **PEI/HfO_2** nanocomposites as a function of nanofiller content at $E = 300$ MV m^{-1}. **b** Maximum discharged energy density (J cm^{-3}) and leakage current density (A cm^{-2}) were derived as a function of nanofiller content in vol% for pristine **PEI** and **PEI/HfO_2** nanocomposites [37]. Reprinted with permission from ref. [37]. Copyright (2021) (Elsevier)

diagram with HOMO and LUMO bands can also be seen along with this core–shell nanoparticle interfacial model. Charge incorporation from electrodes and charge transport including charge trapping and de-trapping in pristine **PEI** and **PEI/HfO_2** nanocomposites can be seen in Fig. 3.10. High degree of thermal and electrical stimulation prevails only at high temperatures and electric fields, respectively, and hence, conduction mechanisms based on charge density in pristine **PEI** are significantly boosted to overcome the potential barrier conveniently. However, an opposite mechanism happens in **PEI/HfO_2** nanocomposites, and charge carriers are trapped in holes or positive vacancies formed at the interface of **HfO_2** nanofillers and polymer matrix. Insertion of wide band gap **HfO_2** dielectric nanofillers impedes the electrical conduction mechanism as documented by significantly reduced leakage current density.

3.2.2 Co-precipitation Method

Co-precipitation is a synthesis method in which compounds soluble in a solution are abolished during the precipitation process [38]. More than one compounds become soluble instantly to make happen precipitation when dissolved in a single solution phase. However, the degree of contamination becomes of a very high degree. Generally, co-precipitation is the process of bringing down involved materials that are dissolved in normal conditions. Precipitation can be understood as the process of formation of insoluble solid-phase particles from a super-saturated solution on the dissolution of a substance. Precipitant is the chemical reagent that causes the formation of the insoluble solid phase.

Precipitation happens in a much faster manner in a strongly super-saturated solution. The super-saturated solution has its concentration exceeding its solubility owed to the formation of a mixture of solvents, evaporation of solvents, and associated temperature changes. Precipitation involves the development of a solid phase that forms an interface with the constituent solution. Interface chemistry gets involved here inculcating free energy connected with the dissolution reaction including relative surface energy formation between the two phases. Free energy of dissolution reaction enhances the degree of randomness and hence entropy of the system in an exothermic or endothermic reaction mechanism. The nicety of the co-precipitation method is that no precipitate product will be obtained in case no appropriate nucleation sites are found associated with a lesser favorable energy.

On the other hand, post-precipitation is a method in which precipitation of the unwanted phase happens initially after the precipitation of the required compound. The degree of contamination is low. On this formed precipitate, a precipitation layer is formed over this initially formed layer.

Co-precipitation usually involves the instant happening of nucleation or beginning of precipitate formation, growth of solid phase, and coarsening along with agglomeration in the solid content. Regarding the formation of nanocomposites, Ostwald ripening, and other secondary processes tailor the shape, size, morphological features,

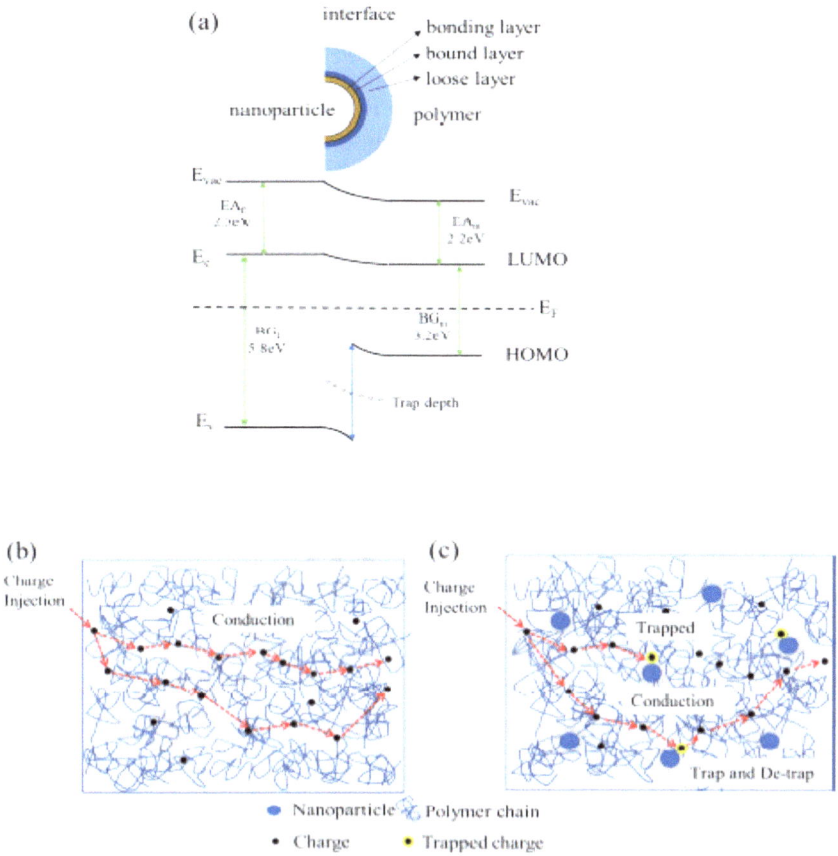

Fig. 3.8 Band diagrams for the core–shell nanoparticle interfacial model along with HOMO and LUMO bands. E_{vac} is the vacuum energy level, E_F is the Fermi energy level, E_C is the conduction band, E_V is the valence band, E_{Af} is electron affinity, and BG_F is the band gap of nanofiller, while E_{Am} and B_{Gm} are electron affinity and bandgap of the polymer matrix, respectively [37]. Reprinted with permission from ref. [37]. Copyright (2021) (Elsevier)

and associated properties of the resulting products. Co-precipitation of nanocomposites is a widely used process for the formation of metals from aqueous solutions or obtained from nonaqueous solutions by reduction, electrochemical reduction, formation of oxides from aqueous and nonaqueous solutions, or even microwave-sonication-assisted co-precipitation methods. Some of the benefits of co-precipitation in the formation of nanocomposites are fast synthesis procedures along with control of shape, size and composition, modification of particle surfaces along with homogeneous character, offering synthesis facility at low temperatures, energy efficiency, and no use of organic solvents as a medium. Disadvantages are inclusive of nonapplicability to noncharged species, taking a bit of time to synthesize, associated irregularity in batch-to-batch production, the inclusion of impurities, and nonconformity with reactants that possess highly varying precipitation rates.

Vadivel et al. formed magnetic nanosized particles of chromium (**Cr**)-displaced cobalt ferrite ($\mathbf{CoFe_{2\text{-}x}Cr_xO_4}$) at different **Cr** concentrations of 0.0, 0.1, 0.2, and 0.3 by co-precipitation followed by annealing for 3 h at T = 600 °C [39]. Multi-component particles were single phasic in a cubic spinel geometry with an average crystallite size of 15–23 nm. Dielectric studies revealed that both dielectric constant (ε) and dielectric loss (ε) were analyzed to be higher in comparison with the pristine $\mathbf{CoFe_2O_4}$ nanoparticles. Fluorescence spectral analysis displayed UV emissions in the strong red zone as well as in the weak blue zone. Figure 3.9a and b show the variation of dielectric constant (ε') and dielectric loss (ε'') as a function of frequency in various **Cr**-substituted $\mathbf{CoFe_2O_4}$ compounds. It can be seen that the dielectric constant (ε') increases at reduced f-values.

Variation of dielectric constant (ε') as a function of logarithmic (f) at different temperatures of 40, 60, 80, 100, 120, 140, 160, 180, and 200 °C is shown for (a) $\mathbf{CoFe_2O_4}$, (b) $\mathbf{CoFe_{1.9}Cr_{0.1}O_4}$, (c) $\mathbf{CoFe_{1.8}Cr_{0.2}O_4}$, and (d) $\mathbf{CoFe_{1.7}Cr_{0.3}O_4}$ nanoparticles in Fig. 3.10a–d. This is observable that the dielectric constant (ε') increases from 8.0×10^3 to ~ 1×10^4 at room temperatures on increasing **Cr** from 0.0 to 0.2 and 0.3. The effect of temperature is that the value of the dielectric constant (ε') increases with increasing temperature. Dielectric loss (ε'') for the four samples under the same operating temperatures is shown in Fig. 3.11a–d. Dielectric loss (ε'') at room temperature for the four samples shows erratic behavior with value decreasing from ~ 1.4×10^4 to 8.0×10^3 in the case of $\mathbf{CoFe_{1.9}Cr_{0.1}O_4}$, again rising to 1.0×10^4 for $\mathbf{CoFe_{1.8}Cr_{0.2}O_4}$, and finally rising to 2.0×10^4 for $\mathbf{CoFe_{1.7}Cr_{0.3}O_4}$. However, all the samples show comparatively higher values of dielectric loss for the operational temperature of 200 °C.

Krishna et al. used chemical co-precipitation for the synthesis of lanthanum oxide nanoparticles by employing organic and biological capping agents [40]. The organic capping agent was ethylene-diamine-tetra-acetic acid (EDTA) and the biological capping agent was starch and deoxy-ribo-nucleic-acid (DNA). Lanthanum oxide nanoparticles are obtained by annealing the carbonates at temperatures of 450 to 700 °C. Dielectric permittivity was estimated from drawing plots of capacitance as a function of applied frequencies for the two operational temperatures. Dielectric permittivity was found to possess enhanced values at low operational frequencies and reduced values at high operational frequencies. Under AC conduction, maximum AC conductivity is achieved at high operational frequencies.

Joshi et al. co-precipitated nickel ferrite nanoparticles of single-phase cubic spinel crystal geometry in a space group of Fd3m possessing an average crystallite size of 8 to 20 nm based on annealing temperature [41]. Strong temperature dependence was observed at all the operational frequencies for dielectric permittivity, dielectric loss, and AC conductivity measurements. Frail polaron hopping was discovered between $\mathbf{Fe^{3+}/Fe^{2+}}$ ions for dominant ac-conduction. Optical bandgap was estimated to lie in the range of 1.27 eV to 1.47 eV for $\mathbf{NiFe_2O_4}$ by UV–visible diffuse spectra and established as an indirect bandgap material. Dielectric constant (ε') and dielectric loss (**tan**$\boldsymbol{\delta}$) were studied at room temperature as shown in Fig. 3.12. Dielectric constant

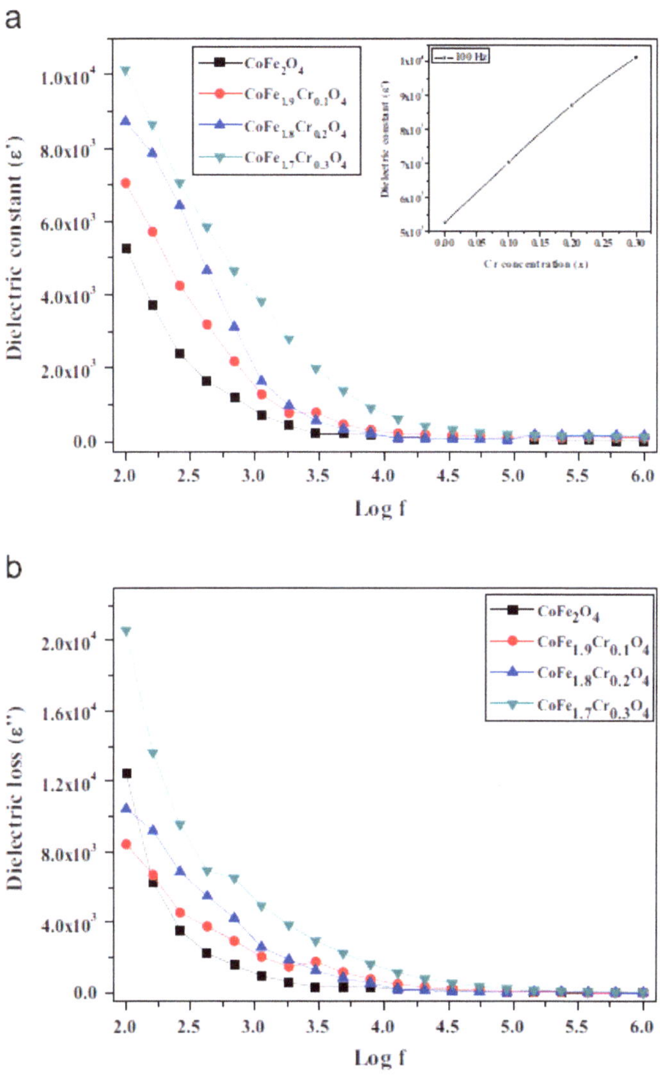

Fig. 3.9 **a** Dielectric constant (ε') and **b** dielectric loss (ε'') as a function of logarithmic (f) in $CoFe_2O_4$, $CoFe_{1.9}Cr_{0.1}O_4$, $CoFe_{1.8}Cr_{0.2}O_4$, and $CoFe_{1.7}Cr_{0.3}O_4$ [39]. Reprinted with permission from ref. [39] Copyright (2014) (Elsevier)

showed frequency dispersion in the lower frequency range that became constant nature for frequencies beyond 10 kHz. The dielectric constant value fell sharply in the low-frequency region and was found to increase gradually at higher operational frequencies. Electronic, ionic, and orientational polarizations as well as space charge were the reasons for high dielectric constant values in the region of low frequencies.

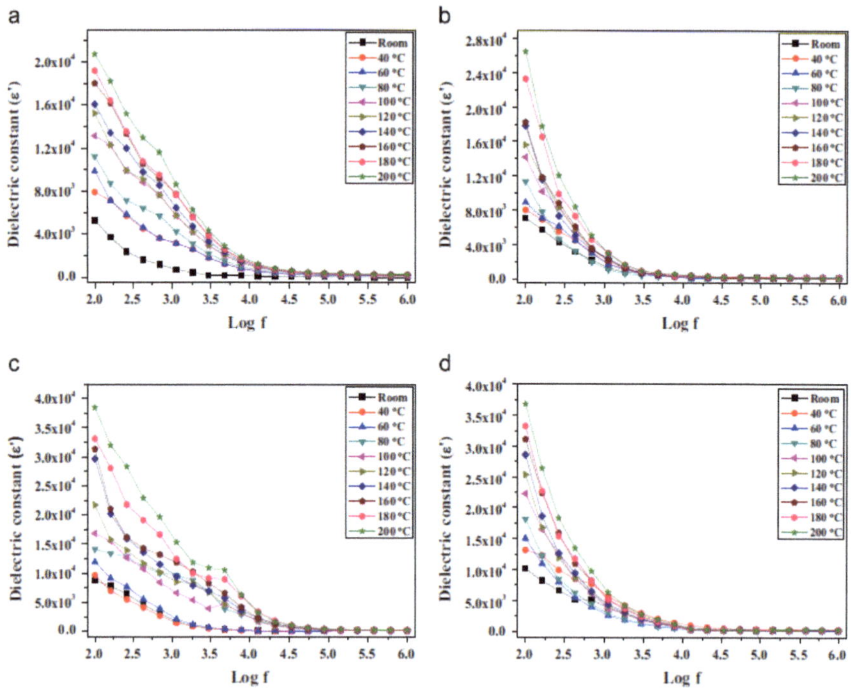

Fig. 3.10 Dielectric constant (ε') as a function of logarithmic (f) in **a** $CoFe_2O_4$, **b** $CoFe_{1.9}Cr_{0.1}O_4$, **c** $CoFe_{1.8}Cr_{0.2}O_4$, and **d** $CoFe_{1.7}Cr_{0.3}O_4$ at different temperatures of 40, 60, 80, 100, 120, 140, 160, 180, and 200 °C [39]. Reprinted with permission from ref. [39], Copyright (2014) (Elsevier)

Space charge polarization was the dominating factor at high temperatures and low frequencies.

Dielectric constant ($\boldsymbol{\varepsilon}'$) and dielectric loss ($\boldsymbol{\delta}$) were also drawn as a function of temperature for the sample calcined at 550 °C for 2 h as shown in Fig. 3.13. Curves were acquired at different operational frequencies of 10 kHz, 25 kHz, 50 kHz, 75 kHz, 100 kHz, and 1 MHz. The normal behavior of magnetic semiconductor ferrite shows an initial increment in dielectric constant as a function of temperature. The formation of free localized dipoles increases the dielectric constant ($\boldsymbol{\varepsilon}'$) due to supplemental thermal energy. These free localized dipoles tend to line up along a direction in which the field is applied. Enhancement in temperature causes the creation of lattice phonons that undergo interaction with electrons for electron–phonon scattering.

Figure 3.14 shows the variation of AC conductivity $\boldsymbol{\sigma}_{\mathbf{ac}}$ with angular frequency at different temperatures of 40, 60, 80, 100, 120, 140, 160, 180, and 200 °C [41]. It can be observed that $\boldsymbol{\sigma}_{\mathbf{ac}}$ shows dispersion characteristics at higher operating frequencies with a variable gradient. Eventually, all $\boldsymbol{\sigma}_{\mathbf{ac}}$ curves meet each other on higher operating frequencies. This shows the presence of several temperature induced as well as multiple relaxation processes. This electrical conduction can be expressed by a phenomenal expression in which the hopping of charge carriers is the major

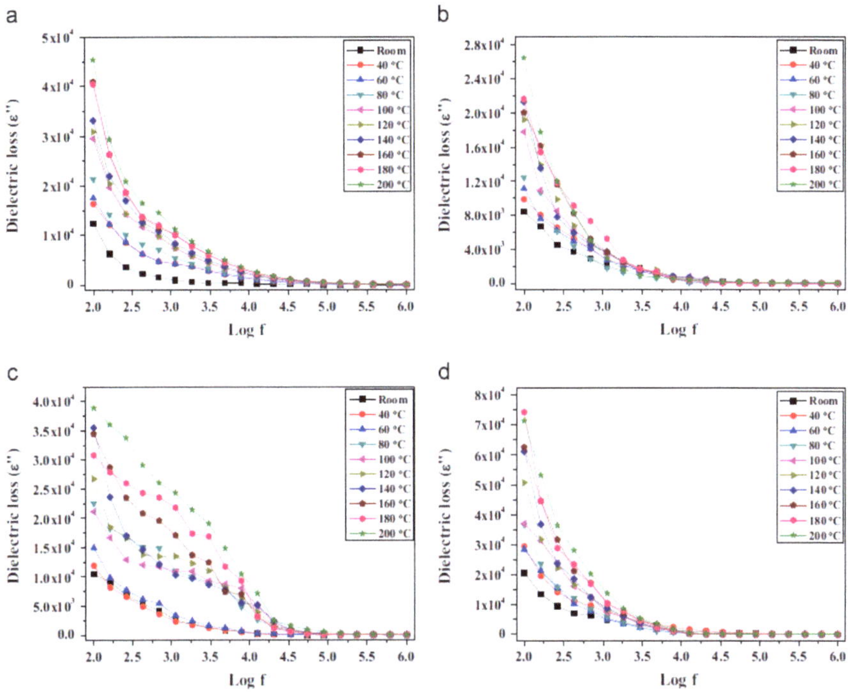

Fig. 3.11 Dielectric loss (ε'') as a function of logarithmic (f) in **a** $CoFe_2O_4$, **b** $CoFe_{1.9}Cr_{0.1}O_4$, **c** $CoFe_{1.8}Cr_{0.2}O_4$, and **d** $CoFe_{1.7}Cr_{0.3}O_4$ at different temperatures of 40, 60, 80, 100, 120, 140, 160, 180, and 200 °C [39]. Reprinted with permission from ref. [39] Copyright (2014) (Elsevier)

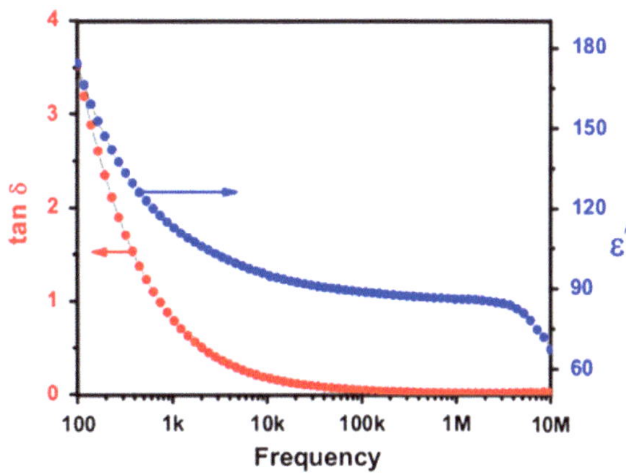

Fig. 3.12 Dielectric constant and loss as a function of frequency for the grounded sample and calcined for 2 h at 550 °C [41]. Reprinted with permission from ref. [41], Copyright (2014) (Elsevier)

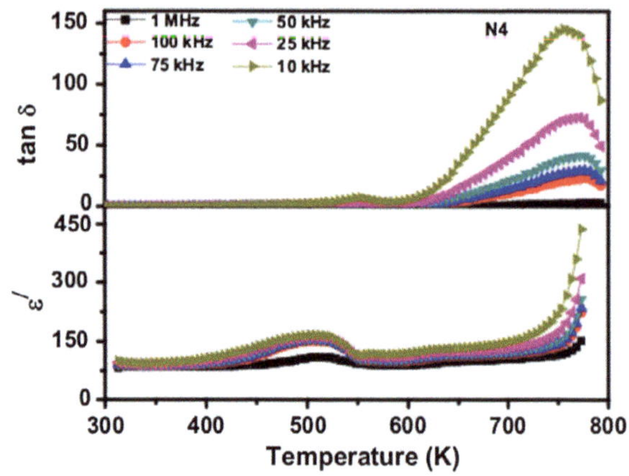

Fig. 3.13 Dielectric constant and dielectric loss as a function of temperature under different operational frequencies for the sample calcined at 500 °C for 2 h [41]. Reprinted with permission from ref. [41], Copyright (2014) (Elsevier)

mechanism. This relationship is called Jonscher's universal power law,

$$\sigma_{\mathbf{ac}} = \sigma_{\mathbf{dc}} + \mathbf{K}\omega^{\mathrm{s}} \tag{3.20}$$

Here, $\boldsymbol{K}$ is a constant that depends on temperature, $\boldsymbol{s}$ occupies a value between 0 and 1, and $\boldsymbol{\omega}$ is the angular frequency of the AC field. The value of $\boldsymbol{s}$ was found to increase as a function of temperature which can be explained based on small polaron hopping.

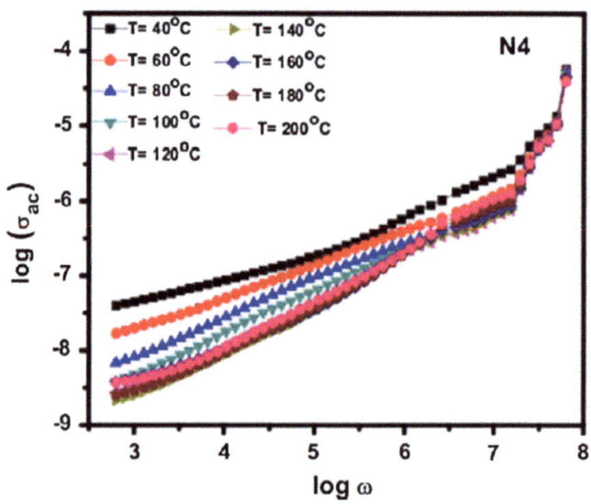

Fig. 3.14 AC-conductivity plotted as a function of angular frequency of the AC applied field in logarithmic scales at different temperatures of 40, 60, 80, 100, 120, 140, 160, 180, and 200 °C [41]. Reprinted with permission from ref. [41], Copyright (2014) (Elsevier)

3.2.3 Hydrothermal Process

Hydrothermal synthesis involves chemical reactions of precursor substances kept in an air-tight sealed reactor heated to create temperatures and pressures above the normal atmospheric ranges [42]. When a reaction takes place at elevated temperatures and pressures in an organic solvent, it is more commonly referred to as solvothermal synthesis.

The advantage of the hydrothermal synthesis method is that it can be used to dissolve most of the materials in an appropriate solvent by thermal treatment and creating pressures in the vicinity of the critical point. This synthesis method is an effective method for the production of metastable or intermediate-phase compounds. A range of flexibility is offered in this synthesis method by tailoring parameters of reaction time, temperature, and change of surfactant, solvent, and precursors. Another benefit is to obtain products with reduced M.P., a tendency to undergo pyrolysis, and possessing high vapor pressures. A high disadvantage of the hydrothermal process is that it does not offer any flexibility to observe and study reaction mechanisms, rates, and other parameters in situ because of the happening of reactions under high pressures in the sealed reactor.

Köseoğlu et al. hydrothermally synthesized $\mathbf{Mn_{0.2}Ni_{0.8}Fe_2O_4}$ nanoparticles assisted with polyethylene glycol (PEG) [43]. Temperature-dependent properties as a function of frequency (1 Hz to 3 MHz) showed dielectric dispersion that is based on Koop's theory. Koop's theory relies on the Maxwell–Wagner model focused on homogeneous double structure with reference to interfacial polarization [44].

Jiang et al. have shed light on nanoscale $\mathbf{BaTiO_3}$ as well as **polymer-**$\mathbf{BaTiO_3}$ nanocomposites with tailored properties in form of a review [45]. Polymer-based nanocomposites of $\mathbf{BaTiO_3}$ have several advantages over pristine $\mathbf{BaTiO_3}$ nanocrystals. Hydrothermal as well as solvothermal processes of $\mathbf{BaTiO_3}$ form the core of this detailed review article. In a work performed by Phromviyo et al. **Ag** nanoparticles were hydrothermally synthesized by applying Aloe vera-plant extracted solution in form of a surface stabilizer and reducing agent to produce **Ag@Ale-NPs** [46]. This was followed by the preparation of **Ag@Ale-NPs/PVDF** polymer nanocomposites by liquid-phase backed dispersion as well as hot-pressing techniques. Modified surfaces of **Ag@Ale-NPs** were confirmed by X-ray photoelectron spectroscopy (XPS). Dielectric properties were studied at two different filler compositions of 0.18 and 0.22 in volume fractions. At the volume fraction of 0.18, high dielectric permittivity ($\boldsymbol{\varepsilon}$) of $\approx$92.5 along with highly reduced tangential loss of 0.049 was discovered at 1 kHz and room temperature. On increasing the nanoparticle volume fraction to 0.22, a substantially higher value of $\boldsymbol{\varepsilon}$ of $\approx$257.2 became evident along with a low loss of $\mathbf{tan\delta}$ $\approx$0.26. Interfacial polarization along with the micro capacitance effect in the **PVDF** matrix was the factors responsible for outstanding dielectric properties of the synthesized **Ag@Ale-NPs/PVDF** polymer nanocomposites.

SEM images of surface-modified nanoparticles of **Ag@Ale-NPs** by reacting Aloe vera extract in a reactor for 3 h, 6 h, and 12 h in hydrothermal synthesis is shown in Fig. 3.15 [46].

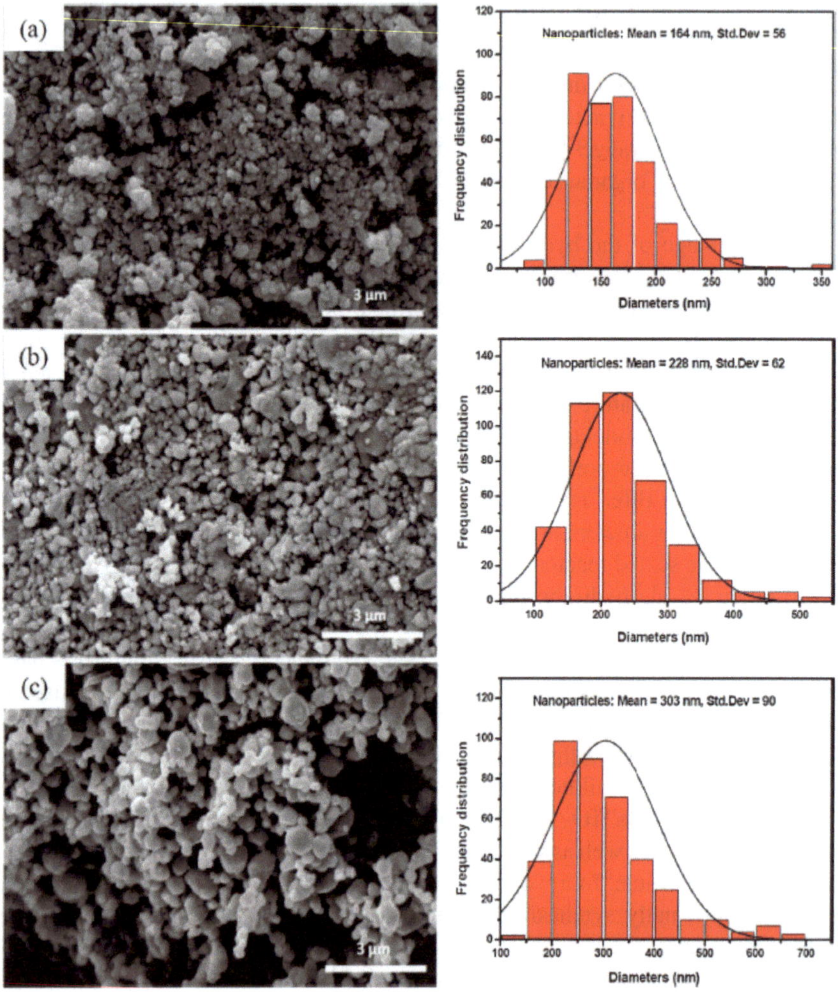

Fig. 3.15 Morphological features along with particle size distributions of **a** 164 nm with S.D. = 56, **b** 228 nm with S.D. = 62, and **c** 303 nm with S.D. = 90 as visible in SEM images of surface-modified Ag@Ale-NPs prepared with the help of Aloe vera plant extracts reacted for three different times of 3 h, 6 h, and 12 h [46]. Reprinted with permission from ref. [46], Copyright (2018) (Elsevier)

Microstructure of nanocomposites was also observed by SEM imaging in a largely fractured cross-sectional area for filler volume fractions of $f_{\mathbf{Ag@Ale}} = 0.08$ and $f_{\mathbf{Ag@Ale}} = 0.22$ [46]. Microstructure consisted of random distribution of **Ag@Ale-NPs**in a surrounding PVDF polymer matrix in a cluster configuration more specifically on increasing filler volume fraction. Cluster configuration indicated the high tendency of **Ag@Ale-NPs** to agglomerate due to interparticle cohesive forces. Information on agglomeration is essential for establishing a correlation between microstructure and derived dielectric propertiesFig. 3.16.

Fig. 3.16 Morphology of **Ag@Ale-NPs** in PVDF polymer in a cross-section under different filler volume fractions of $f_{\mathbf{Ag@Ale}} = 0.08$ & $f_{\mathbf{Ag@Ale}} = 0.14$ [46]. Reprinted with permission from ref. [46], Copyright (2018) (Elsevier)

The tangential dielectric loss was estimated to change from 0.039 for a filler volume fraction of 0.22 to 0.049 with a filler volume fraction of 0.18 under an operational frequency of 1 kHz. Cluster formation was detected for the filler volume fraction of 0.18 with each of the clusters separated by a polymer layer of **PVDF** resulting in mild enhancement of the percolation threshold. However, the presence of a barrier polymer layer prevented the formation of weak conductive links between different conductive clusters. Hence, dielectric tangential loss (**tanδ**) at lower frequencies was maintained at lower values. Lower operating frequencies are an approximation to direct current (**DC**) conductivities. Free electrons generated from conductive clusters tend to assemble at the interface of the PVDF barrier layer between different loose nanoparticle agglomerates. This gives rise to the effect of interfacial polarization. Low tangential loss (**tan δ**) of polymeric nanocomposites was ascribed to micro capacitor formation because of two close-lying nanoparticles clusters acting as microelectrodes along the direction of the applied electric field separated by a dielectric polymer barrier layer (Fig. 3.17).

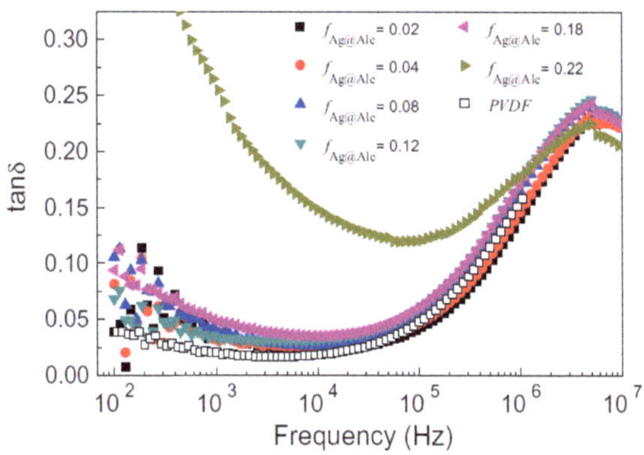

Fig. 3.17 Tangential loss (tanδ) plotted as a function of frequency in Hz for different filler volume fractions of $f_{\mathbf{Ag@Ale}} = 0.02$, 0.04, 0.08, 0.12, 0.18, and 0.22 at room temperature [46]. Reprinted with permission from ref. [46], Copyright (2018) (Elsevier)

3.2.4 Sol–gel

Sol–gel method involves the formation of a sol or solution that slowly but eventually progresses in form of a diphasic system reflecting a behavior similar to gels [47]. Gel-resembling diphasic system consists of a solid as well as a liquid phase. Sol–gel-derived particles can possess surface morphological features in the geometry of discrete or distinct particles and extend to more regular networks of polymeric chains. Sol–gel method possesses the capability to produce ultrafine divided distinct particles in ceramic powders of single as well as multicomponent constitution at the nanometer scale. Metal nanoparticles are usually derived by the sol–gel process by mixing the precursor substances in the liquid phase, formation of stable sol that is normally transparent by hydrolysis and polycondensation reactions, and further aging of sols leads to the development of a 3D structural network followed by drying, calcination, and sintering.

Advantages of sol–gel are inclusive of the (i) formation of a homogeneous product at relatively low temperatures, (ii) better control over shape, size, porosity, and (iii) metal-oxide or ceramic particles in a polymer matrix form an α-β matrix for enhanced applications, and (iv) suitable methodology for the formation of multicomponent particles with the possibility to include dopant atoms.

Yadav et al. used a honey-mediated sol–gel combustion method for the synthesis of *fcc*-spinel ferrite $\mathbf{CoFe_2O_4}$ nanoparticles subsequently calcined at 500, 700, 900, and 1100 °C [48]. The dielectric constant ($\boldsymbol{\varepsilon}$) of $\mathbf{CoFe_2O_4}$ nanoparticles was plotted as a function of frequency, and it decreased with increment in frequency. At lower operating frequencies that are nearly identical to DC conductivity, $\boldsymbol{\varepsilon}$-values lie on the higher side depicting effects associated with space charges, ionic charges, and interfacial polarization. Contrarily, $\boldsymbol{\varepsilon}$-values became practically independent of applied frequencies because formed electric dipoles were not capable of catching up with the fast-varying electric field lines as a function of time. Again, Koop's theory is of paramount importance here as outlined above relying on dielectric formation in which two heterogeneous medium layers of the Maxwell–Wagner type are considered. The dielectric module consists of insulating grain boundaries separating the conductive grains in a granular or polycrystalline or multicrystalline configuration. At low operational frequencies, grain boundaries play a major role as compared to the grain itself, so that $\boldsymbol{\varepsilon}$-values become effectively high. Moreover, the hopping mechanism of charge carriers takes place, so that charge carriers pile up at the insulating boundary because of high resistances and thus producing polarization. This polarization or accumulation of charge carriers is not observed at higher operating frequencies since the charge carriers can not pursue the AC electric field at higher frequencies. Dielectric constant varied as a function of cationic distribution depending on granular size and type of microstructure. Dielectric loss suffers from a lag or phase difference in polarization about the AC electric field. This may be caused due to microstructural attributes, such as grain boundaries, imperfections, defects or dislocations, or distortions in unit cell lattice volume. The hopping frequency whereby becomes equal to the frequency of the applied AC field, and oscillating ions begin receiving maximized electrical energy. This is reflected in form of a power loss peak in the plot of **tan δ** versus frequency (**f**). The mathematical expression for maximum tangential loss in dielectric materials is given as,

$$\omega\tau = \mathbf{1} \tag{3.21}$$

$$\omega = \mathbf{2}\pi f \tag{3.22}$$

The mathematical relationship between relaxation time ($\boldsymbol{\tau}$) and jumping probability per unit time ($\boldsymbol{p}$) is given as,

$$\tau = \mathbf{1/2}p \tag{3.23}$$

$$\omega_{\mathbf{max}} = \mathbf{2}p \tag{3.24}$$

Debye relaxation is observed when the rate of hopping of jumping probability per unit time becomes equal to ac-frequency.

Vasudevan et al. applied the sol–gel method for the synthesis of poly(vinyl pyrrolidone) (PVP) doped with varying concentrations of titanium dioxide ($\mathbf{TiO_2}$) to obtain PVP-TiO_2 nanocomposites [49]. Dielectric properties were estimated for the nanocomposites in the range of operational frequencies of 1 kHz to 2 MHz in the vicinity of room temperature. The dielectric constant and tangential loss were estimated to be reduced at high operational frequencies of AC conduction. Conductivity values were 10^{-7} to 10^{-8} S cm^{-1}, which decreased steeply on increasing wt% of TiO_2 due to enhancement in randomness or reduction of homogeneity in microstructure pattern in the whole volume of the nanocomposite.

Singh et al. used the sol–gel method for the preparation of polysulfone-ZnO nanocomposites and studied frequency-dependent and temperature-dependent dielectric properties [50]. Dipolar polarization was solely deciphered to be responsible for the observed dielectric behavior, while interfacial polarization was partly. Dielectric relaxation was studied in terms of electric modulus to observe the effects of electrode polarization and resolving low-frequency relaxation mechanisms. Figures 3.9 and 3.10 show the real (M′) and imaginary (M″) parts of complex electric modulus of pure polysulfone and ZnO-loaded polysulfone nanocomposites at different operating temperatures of 30, 60, 90, and 120 °C (Fig. 3.18).

Mathematical equations for the calculation of real and imaginary parts of electric modulus are given as,

$$\boldsymbol{M'} = \frac{\boldsymbol{\varepsilon}'}{\boldsymbol{\varepsilon}'^2 + \boldsymbol{\varepsilon}''^2} \tag{3.25}$$

$$\boldsymbol{M''} = \frac{\boldsymbol{\varepsilon}''}{\boldsymbol{\varepsilon}'^2 + \boldsymbol{\varepsilon}''^2} \tag{3.26}$$

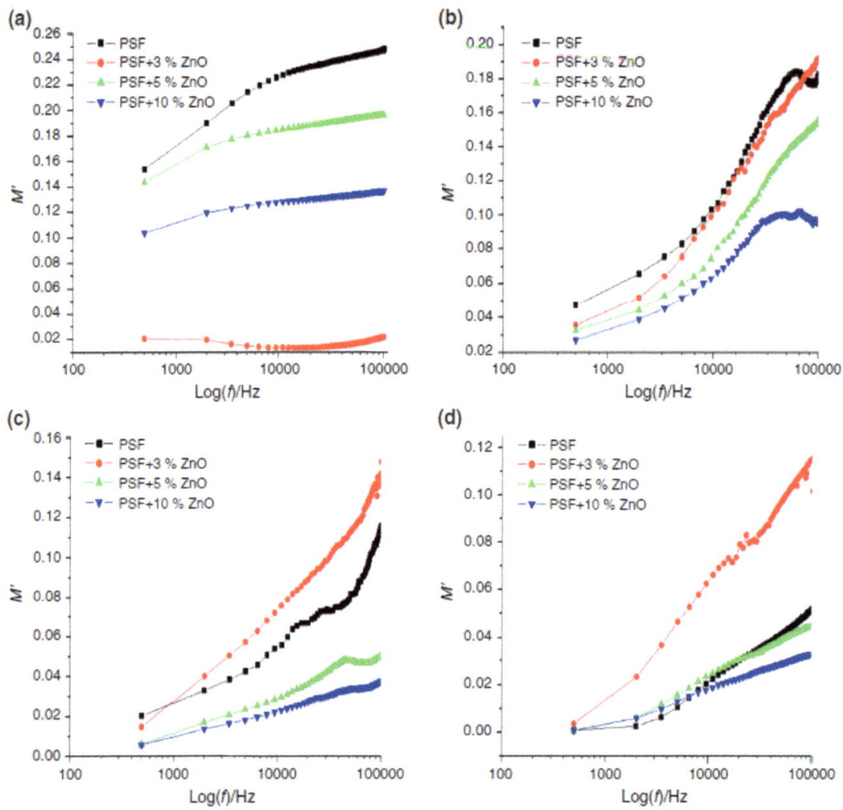

Fig. 3.18 Real part of electric modulus (M') plotted as a function of frequency (f) for pure and ZnO-loaded polymer nanocomposites at **a** $T = 30$ °C, **b** $T = 60$ °C, **c** $T = 90$ °C, and **d** $T = 120$ °C [50]. Reprinted with permission from ref. [50], Copyright (2015) (Springer)

It can be observed from Fig. 3.19 that the electric modulus of pristine and ZnO-loaded nanocomposites shows an increasing trend as a function of increasing frequencies at different operational temperatures. However, the imaginary part (***M"***) of pristine and ZnO-loaded nanocomposites do not show any peak in the curve at an operating temperature of 30 °C. Observance of sharp peaks at operational temperatures of 90 °C and 120 °C is representative of dielectric relaxation at higher operational frequencies reflecting very quick crystalline relaxation propagating with retentive symmetries. At higher temperatures of 120 °C, peaks of imaginary parts (***M"***) of dielectric relaxation shift toward higher frequencies. Good broadening of peaks is observed for ***M"*** at 90 °C for different loading **wt%** of nanofillers. However, the peak became sharp and peak-broadening decreased at 120 °C of operational temperature reflecting Maxwell–Wagner–Sillers (MWS) type crystalline relaxation in any asymmetric mode. Further, it was inferred that the peak reduced in a vertical orientation on increasing the wt% of nanofillers.

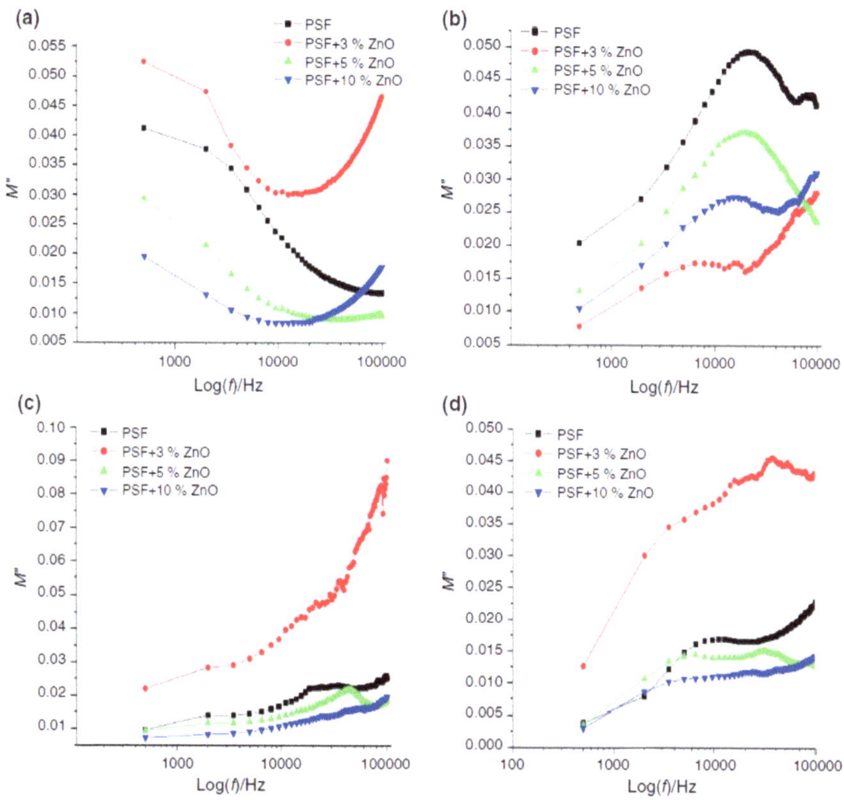

Fig. 3.19 Imaginary part of electric modulus (M″) plotted as a function of frequency (f) for pure and ZnO-loaded polymer nanocomposites at **a** $T = 30$ °C, **b** $T = 90$ °C, **c** $T = 120$ °C, and **d** $T = 180$ °C [50]. Reprinted with permission from ref. [50], Copyright (2015) (Springer)

Figure 3.20 shows the variation of AC conductivity (σ_{AC}) in S m^{-1} of the pristine PSF and ZnO-loaded nanocomposites at mass% of (i) 3%, (ii) 5%, and (iii) 10% at the specific frequencies of 20 kHz, 50 kHz, and 95 kHz derived at two different temperatures of 30 °C and 180 °C.

It can be observed from Fig. 3.20 that the σ_{AC} value is 3.61 × 10^{-7} S m^{-1} for pristine PSF, while it becomes 5.55 × 10^{-7} S m^{-1} for 10 mass% ZnO-loaded PSF at $f = 20$ kHz and $T = 30$ °C showing an increment of 53.74%. At $T = 180$ °C, σ_{AC} value was 1.56 × 10^{-5} S m^{-1} for pristine PSF which became 2.03 × 10^{-5} S m^{-1} for 10 mass % ZnO-loaded PSF at the same $f = 20$ kHz showing an increment of ~30%. It can also be observed easily that σ_{AC} value increased by an order of 100 as reflected by determined σ_{AC} values on switching $T = 30$ °C to $T = 180$ °C. This is close to that of a semiconducting type of behavior with ZnO itself being a wide-bandgap material. An increment in σ_{AC} values at $T = 180$ °C indicates that more free charge carriers become available at higher temperatures due to enhancement in kinetic energy (K.E.) because of an increased supply of thermal

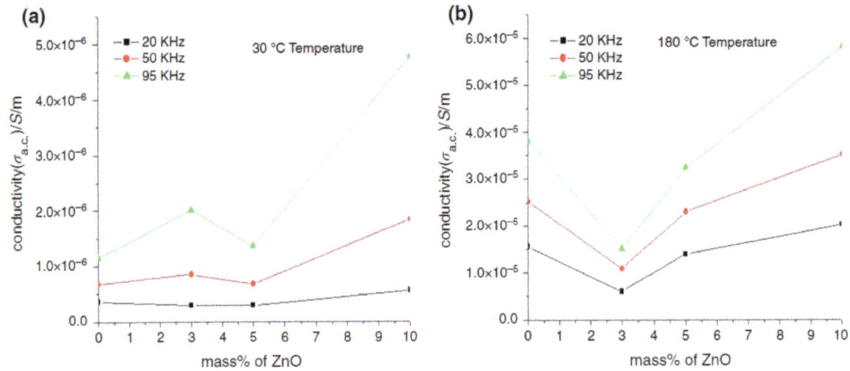

Fig. 3.20 AC conductivity (σ_{AC}) derived as a function of mass% of ZnO-loaded nanofillers at the operational temperatures of 30 °C and 180 °C [50]. Reprinted with permission from ref. [50], Copyright (2015) (Springer)

energy. Another reason is the reduced interparticle distance and increased probability of much-enhanced contacts developing higher probabilities of charge carrier transfer. This mechanism happens at different nanofiller concentrations to develop overlapped interfacial regions inside the nanocomposites. Dispersion of nanofillers in the polymer matrix is an important factor in the crossover of interfacial regions in the nanocomposites that leads to the formation of percolation of charge carriers in weakly conducting chains.

3.3 Conclusion

Nanodielectric materials offer countless opportunities for high-end applications based on nanoelectronics, nanosensors, and high-voltage-based electrical transmission devices as well as improvement in the performance of ceramic-based applications. Dispersion of nanofillers (spherically shaped nanoparticles, nanowires, nanotubes, etc.) in a matrix, usually a polymer, produces a nanocomposite material with much-enhanced properties. An interfacial region is formed at the boundary of the matrix and particle surface that possesses modified properties in comparison with both matrix and nanofiller. The design of nanodielectrics gives high consideration to properties of interphase in addition to that of nanofillers and matrix since interphase fraction is the substantial fraction in composite material. A major problem is that the majority of the peer-reviewed research reports do not include quantification of nanofiller dispersions. Further, concerning industrial applications, advanced research on optimized concentration of nanofillers in a matrix is required along with prolonged usage of derived nanocomposite materials in several cycles. In addition, the huge volume of consistent and reliable datasets is needed that can be applied to prepare models in a rather empirical approach along with providing due consideration to breakdown as well as recovered state after prolonged use in several cycles, a multiscale model to include parameters related to nanofiller type and functionalization, matrix, along with conditions of processing, and, last but not the least, involved costing as it has straight industrial consequences.

References

1. Frèchette MF (2013) What are nanodielectrics ? IEEE ElectrInsul Mag 29:8–11
2. IEEE (2010) Nanodielectrics: a panacea solving all electrical insulation problems? 10th IEEE Intl Conf Solid Dielectr 1–29
3. FrèchetteM F, Reed CW, Sedding H (2006) Progress, understanding and challenges in the field of nanodielectrics. IEEJ Trans Fund Mater 126:1031–1043
4. Stevens GC, Vaughan AS (2013) Nanodielectrics and their role in power transmission applications. In: Melhem Z (ed) Electricity transmission, distribution and storage systems, pp 206–241
5. Cao Y, Irwin PC, Younsi K (2004) The future of nanodielectrics in the electrical power industry. IEEE Trans Dielectr Electr Insul 11:797–807
6. ZhongS-L D-M, ZhouW-Y C-W (2018) Past and future on nanodielectrics. IET Nanodielectr 1:41–47
7. Bhugra VS (2020) Advanced magnetic and dielectric nanomaterials, Ph.D. Thesis, Victoria University of Wellington, pp 1–214
8. Marks T, Ye P, Facchetti A, Lu G, Lin H (2007) High performance field effect transistors with self-assembled nanodielectrics. US Patent (US2007/0284629 A1), pp 1–10
9. Stallings K, Smith J, Chen Y, Zeng L, Wang B, di Carlo G, Bedzyk MJ, Facchetti A, Marks TJ (2021) Self-assembled nanodielectrics for solution-processed top-gated amorphous IGZO thin film transistors. ACS Appl Mater Interf 13:15399–15408
10. Lee S, Han H, Kim C-H (2020) Nanodielectrics approaches to low-voltage organic transistors and circuits. Eur Phys J Appl Phys 91:20201(1–8)
11. Tuncer E, Sauers I (2010) Industrial applications perspective of nanodielectrics. In: Nelson J (ed) Dielectric polymer nanocomposites. Springer, Boston, MA
12. Okubo H, Nakamura Y, Inano H, Hayakawa N, Hiroshima S, Hirose T, Hamaguchi M (2007) Lifetime characteristics of nanocomposite enameled wire under surge voltage application. Ann Rep Conf Electr Insul Dielect Phenom 1–4
13. Psarras GC (2018) Fundamentals of dielectric theories. In: Dang Z-M (ed) Dielectric polymer materials for high-density energy storage. Elsevier
14. Jiang S, Jin L, Hou H, Zhang L (2019) Polymer-based nanocomposites with high dielectric permittivity. In: Polymer-based multifunctional nanocomposites and their applications, pp 201–243
15. Singh M, Apata IE, Samant S, Wu W, Tawade BV, Pradhan N, Raghavan D, Karim A (2021) Nanoscale strategies to enhance the energy storage capacity of polymer dielectric capacitors: review of recent advances. Poly Rev 62:211–260
16. Luo H, Wang F, Guo R, Zhang D, He G, Chen S, Wang Q (2022) Progress on polymer dielectrics for electrostatic capacitors application. Adv Sci 9:2202438(1–25)
17. Balaraman AA, Dutta S (2022) Inorganic dielectric materials for energy storage applications: a review. J Phys D: Appl Phys 55:183002(1–38)
18. Hassan YA, Hu H (2020) Current status of polymer nanocomposite dielectrics for high-temperature applications. Compos Part A: Appl Sci Manufact 138:106064(1–24)
19. Pandey JC, Singh M (2021) Dielectric polymer nanocomposites: past advances and future prospects in electrical insulation perspective. SPE Poly 2:236–256

20. Yuan J, Yao S, Poulin P (2016) Dielectric constant of polymer composites and the routes to high-k or low-k nanocomposite materials. In: Huang X, Zhi C (eds) Polymer nanocomposites. Springer, Cham
21. Reddy PL, Deshmukh K, Chidambaram K, Nazeer Ali MM, Sadasivuni KK, Kumar YR, Lakshmipathy R, Khadheer Pasha SK (2019) Dielectric properties of polyvinyl alcohol (PVA) nanocomposites filled with green synthesized zinc sulphide (ZnS) nanoparticles. J Mater Sci: Mater Elect 30:4676–4687
22. Roy AS, Gupta S, Sindhu S, ParveenA RPC (2013) Dielectric properties of novel PVA/ZnO hybrid nanocomposite films. Comp Part B: Engg 47:314–319
23. Li Y, Bi X, Wang S, Zhan Y, Liu H-Y, Mai Y-W, Liao C, Lu Z, Liao Y (2020) Core-shell structured polyethylene glycol functionalized graphene for energy-storage polymer dielectrics: combined mechanical and dielectric performances. Comp Sci Technol 199:108341(1–9)
24. Mao H, You Y, Tong L (2018) Dielectric properties of *di*-block copolymers containing a *poly*-arylene ether nitrile block and a *poly*-arylene ether ketone block. J Mater Sci: Mater Electr 29:3127–3134
25. Samant SP, Grabowski CA, Kisslinger K, Yager KG, Yuan G, Satija SK, Durstock MF, Raghavan D, Karim A (2016) Directed self-assembly of block copolymers for high breakdown strength polymer film capacitors. ACS Appl Mater Inter 8:7966–7976
26. Tanaka T, Imai T (2013) Advances in nanodielectric materials over the past 50 years. IEEE ElectrInsul Mag 29:10–23
27. Andritsch TM (2010) Epoxy based nanodielectrics for high voltage DC applications—synthesis, dielectric properties and space charge dynamics, Ph.D. Thesis, Delft University of Technology
28. International Science Nanodielectrics (2010) International conference on solid dielectrics, Postdam Germany, pp 1–29
29. Raetzke S, Kindersberger J (2006) The effect of interphase structures in nanodielectrics. IEEJ Trans Fund Mater 126:1044–1049
30. Alcoutlabi M, McKenna GB (2005) Effects of confinement on material behaviour at the nanometre size scale. J Phys: Condens Matt 17:R461–R524
31. Bendler JT, Fontanella JJ, Shlesinger MF, Wintersgill MC (2009) The defect diffusion model and the glass transition in nanoscale and bulk films. J Comput Theor Nanosci 6:1–5
32. Lewis TJ (2004) Interfaces are the dominant feature of dielectrics at the nanometric level. IEEE Trans Dielect Electr Insul 11:739–753
33. Lewis TJ (2005) Interfaces: nanometric dielectrics. J Phys D Appl Phys 38:202–212
34. Tanaka T (2005) Dielectric nanocomposites with insulating properties. IEEE Trans Dielect Electr Insul 12:914–928
35. Ehrenstein GW (1999) Polymer-Werkstoffe, Carl Hanser Verlag 2 Auflage
36. Rane AV, Kanny K, Abitha VK, Patil SS, Thomas S (2017) Clay-polymer composites: design of clay polymer nanocomposite by mixing. In: Jlassi K, Chehimi MM, Thomas S (eds) Clay polymer nanocomposites, pp 113–144
37. Ren L, Yang L, Zhang S, Li H, Zhou Y, Ai D, Xie Z, Zhao X, Peng Z, Liao R, Wang Q (2021) Largely enhanced dielectric properties of polymer composites with HfO_2 nanoparticles for high-temperature film capacitors. Comp Sci Technol 201:108528(1–32)
38. Kolthoff IM (1932) Theory of coprecipitation: the formation and properties of crystalline precipitates. J Phys Chem 36:860–881
39. Vadivel M, Ramesh Babu R, Sethuraman K, Ramamurthi K, Arivanandhan M (2014) Synthesis, structural, dielectric, magnetic and optical properties of Cr substituted $CoFe_2O_4$ nanoparticles by coprecipitation method. J MagnMagn Mater 362:122–129
40. Krishna PGA, Tharayil NJ (2019) Dielectric properties of lanthanum oxide nanoparticle synthesized using chemical coprecipitation method. AIP Conf Proc 2162:020079(1–9)
41. Joshi S, Kumar M, Chhoker S, Srivastava G, Jewariya M, Singh VN (2014) Structural, magnetic, dielectric and optical properties of nickel ferrite nanoparticles synthesized by coprecipitation method. J Mol Struct 1076:55–62

42. Yang G, Park S-J (2019) Conventional and microwave hydrothermal synthesis and application of functional materials: a review. Mater 12:1177(1–18)
43. Köseoğlu Y, Bay M, Tan M, Baykal A, Sözeri H, Topkaya R, Akdoğan N (2011) Magnetic and dielectric properties of $Mn_{0.2}Ni_{0.8}Fe_2O_4$ nanoparticles synthesized by PEG-assisted hydrothermal method. J Nanopart Res 13:2235–2244
44. Farea AMM, Kumar S, Batoo KM, Yousef A, Lee CG, Alimuddin (2008) Structure and electrical properties of $Co_{0.5}Cd_xFe_{2.5-x}O_4$ ferrites 464:361–369
45. Jiang B, Iocozzia J, Zhao L, Zhang H, Harn Y-W, Chen Y, Lin Z (2019) Barium titanate at the nanoscale: controlled synthesis and dielectric and ferroelectric properties. Chem Soc Revs 48:1194–1228
46. Phromviyo N, Chanlek N, Thongbai P, Maensiri S (2018) Enhanced dielectric permittivity with retaining low loss in poly(vinylidene fluoride) by incorporating with Ag nanoparticles synthesized via hydrothermal method. Appl Surf Sci 446:59–65
47. Kumar A, Yadav N, Bhatt M, Mishra NK, Chaudhary P, Singh R (2015) Sol–gel derived nanomaterials and it's applications: a review. Res J Chem Sci 5:1–8
48. Yadav RS, Kuřitka I, Vilcakova J, Havlica J, Masilko J, Kalina L, Tkacz J, Švec J, Enev V, Hajdùchovà M (2017) Impact of grain size and structural changes on magnetic, dielectric, electrical, impedance and modulus spectroscopic characteristics of $CoFe_2O_4$ nanoparticles synthesized by honey mediated sol-gel combustion method. Adv Nat Sci: Nanosci. Nanotechnol 8:045002(1–14)
49. Vasudevan P, Thomas S, Arunkumar KV, Karthika S, Unnikrishnan NV (2015) Synthesis and dielectric studies of poly(vinyl pyrrolidone)/titanium dioxide nanocomposites. IOP Conf Ser: Mater Sci Engg 73:012015
50. Singh PK, Gaur MS, Chauhan RS (2013) Dielectric properties of sol–gel synthesized polysulfone-ZnO nanocomposites. J Therm Analy Calor 111:743–751

Chapter 4
Role and Prospects of Polymer-Based Nanomaterials in the Dielectric World

Sushrisangita Sahoo, Abhinav Yadav, K. P. Andryushin, and L. A. Reznichenko

Abstract The development of reliable energy storage systems is the best strategy to resolve the global crisis of increased energy consumption in this modern high-tech world and exhaustion of fossil fuels for energy production. Electrostatic capacitors are one of the extensively used energy storage systems by the engineers and researchers among other such devices like supercapacitors and batteries due to its unique characteristics of superior charge density, ultrafast charge and discharge, high stability, and long-life time. These unique features made them suitable for distributed power systems, microelectronic circuits, electric vehicles, etc. But these electrostatic capacitors have the demerits of low energy density. Therefore, the main objective of researchers is now to achieve high energy density and energy storage efficiency along with the high-power density. The only possible way to attain the high energy density of the electrostatic capacitor is to tailor the features of dielectric materials. Both the ceramics and polymers were used individually in the dielectric layer of the electrostatic capacitors, while the both the individuals have some pros and cons. So, materials scientist combines both the ceramics and polymer to obtain the pro qualities of polymer like high breakdown strength and flexibility, and high dielectric constant of ceramics. Recently, ceramics in nanoform are chosen as the filler materials to the polymer matrix in the trend of device miniaturization. So, in this chapter, we will discuss about the various attempts made to design the polymer-ceramic nanocomposites to obtain the high dielectric constant and hence the energy density of the material.

Keywords Nanocomposites · Polymer · Filler · Dielectric permittivity · Breakdown strength

S. Sahoo (✉) · A. Yadav · K. P. Andryushin · L. A. Reznichenko
Research Institute of Physics, Southern Federal University, 344090 Rostov-on-Don, Russia
e-mail: sushri1990@gmail.com; ssahoo@tuskegee.edu

S. Sahoo
Department of Material Science and Engineering, Tuskegee University, Tuskegee, AL 36088, USA

S. Moharana et al. (eds.), *Emerging Nanodielectric Materials for Energy Storage*, Nanostructure Science and Technology, https://doi.org/10.1007/978-3-031-40938-7_4

4.1 Introduction

The current craze in portable and wearable electronics in the modern high-tech world triggered a growing interest of scientific communities toward the design and development of flexible piezo/triboelectric nanogenerators for energy harvesting and storage devices [1–4]. Polymer-based materials are the best option to choose as the base material for flexibility purpose as it can easily rollable, foldable, twistable, and bendable. Flexibility is not only the feature required for energy harvesting and storage devices, so we need to optimize other parameters for the device performance. On the other hand, we are all aware of the fact that the exhausted usage of fossil fuels and high consumption of energy in the modern high-tech world became a serious threat to our society. To avoid the consequences of this serious threat, we need to resolve these problems by (1) developing some cost-effective, ecofriendly, and efficient energy storage devices and (2) diverting our path from exhausting usage of nonrenewable energy resources to renewable energy sources (like wind, sun, sound thermal and mechanical energy, etc.). So, some of the researchers now put their efforts on energy conversion from environmental phenomena. Though the idea of energy conversion seems very simple, it is really challenging due to the intermittent nature of this energy sources [5–9]. Some other material researcher trying to resolve the storage of the conversion energy in a stable and controlled manner [10]. Among all the storage systems (such as flywheel energy storage, compressed air storage, and hydro storage), the electrochemical energy storage systems, fuel cells, and electrostatic capacitors are widely used by the researchers [11–13]. The two main subsystems of electrochemical storage systems are Li-ion batteries and supercapacitors, both depends on electrochemical processes to store energy, but both of them have different working mechanism yielding different charge storage properties [14]. However, the electrostatic capacitors usually store the energy through electric field induced polarization and depolarization process [15].

Li-ion batteries follow the diffusion controlled faradic reactions in the bulk electrode to store energy, i.e., the Li-ion travels from anode to cathode or cathode to anode during charging and discharging process, and hence, the process is slow [14]. The bulk storage mechanism of batteries causes its high energy density, whereas the lengthy process yields low power density and short life time. In contrast to batteries, supercapacitors stores energy either through the accumulation of electrostatic charge on the electrode–electrolyte surface (EDLC) or through fast and reversible redox reactions (pseudocapacitors) [16]. Supercapacitors exhibit high-power density and long-life span, but low energy density. In contrast to the above two electrochemical storage devices, the electrostatic (or dielectric) capacitors store energy through electric-field-induced polarization and depolarization process and have unique pro qualities of ultrahigh-power density, greater cycling stability, and long lifespan [17–19], but limited energy densities [20]. The energy density of these dielectric capacitors strongly depend on their dielectric permittivity and breakdown strength. Conventional dielectrics are either polymer or ceramics. Both of them have some merits and demerits for application purpose. The ceramics exhibit high dielectric permittivity and thermal stability, but have demerits of poor flexibility and low breakdown strength. However, the polymers have good flexibility and breakdown strength, but have low dielectric permittivity. Therefore, polymer nanocomposites were proposed, which incorporate the individual's merits by dispersing nanofillers in the polymer matrix [21–24]. Various types of polymer matrix used for energy storage devices including both ferroelectric and conducting polymer such as poly(vinylideneflouride) PVDF, poly(methylmethacrylate) (PMMA), polyetherimide (PEI), PVDF-based copolymers, polyimide (PI), polypyrrole (PPy), polyaniline (PANI), polyvinyl alcohol (PVA), and polytetrafluoroethylene (PTFE) [25–35]. The ultimate properties of the polymer nanocomposites are strongly affected or controlled by the features of nanofiller, nanofiller/polymer interfaces, interaction between the nanofillers, and spatial composite structure. Hence, it has been a great challenge for researchers to achieve simultaneous improvement in dielectric permittivity (dielectric and energy storage properties) as well as breakdown strength (mechanical properties). A continuous effort has been made by the material researchers to solve the technical challenges associated with the polymer nanocomposites and improve their performance by choosing the suitable types of fillers, surface modification, microstructure and alignment of filler, complex structures, etc., for the application perspective. Various parameters are designed such as the (1) selection of the polymer matrix, e.g., linear dielectric or ferroelectric polymers, in some cases even conductive polymer matrix was also chosen; (2) nanofillers with different electrical properties (conductor, semiconductor, and dielectric), shapes, size, and compositions; (3) surface engineering; (4) hierarchical structures with different orientation and morphology of fillers; (5) multilayered nanocomposites [36–41].

Though excellent reviews on polymer nanocomposites are summarized earlier, but the collective information regarding the fundamental mechanisms and the significance of dielectric properties of polymer nanocomposites for energy devices as well as the theoretical and experimental approach to explain the crucial parameters are discussed in this chapter. Section 4.2 covers the basic concepts of dielectric, i.e., dielectric polarization to understand the working mechanism of capacitive storage, the dielectric breakdown, and energy storage density. Section 4.3 covers the models for dielectric permittivity and breakdown strength, which are used to explain the experimental trend of the polymer nanocomposites. Section 4.4 describes the crucial parameters which affect the overall properties of the polymer nanocomposites. The last section describes the different attempts made by the researcher to enhance and optimize these crucial parameters for energy storage devices.

4.2 Some Basic Concepts of Dielectric Capacitors and Energy Storage

Dielectric capacitors stores energy through electric-field-induced polarization and depolarization process. Since polarization, breakdown, conduction, and dissipation strongly affect the different properties of a dielectric material, we have discussed these mechanisms for better understanding of the basic concepts.

4.2.1 Dielectric Polarization

Polarization mechanism can be explained based on the response of individual atoms, ions, and dipoles of a dielectric material under the action of external applied field. The dielectric material polarized through displacement of atoms, ions, or dipoles from their equilibrium position by the action of an external electric field causing electric dipole moment. Electronic, ionic, dipolar or orientational, and space charge polarizations are the four different types of polarization mechanisms. Electronic polarization is the response of individual atoms to the external applied electric field. Each material constitutes atoms, and there is a symmetric distribution of electronic cloud and nucleus resulting zero dipole moment in the absence of applied electric field. But these atoms can be polarized due to the displacement of the centers of electron cloud and nucleus under the action of electric field causing some nonzero dipole moment. Electronic polarization is observed in all dielectrics and exists up to optical frequency range (ultraviolet/visible), i.e., a very high frequency range due to involvement of electron. Similarly, the displacement of positive and negative ions from their equilibrium position results a net induced dipole moment under the influence of electric field, and the corresponding polarization is called as the ionic polarization. Ionic polarization is observed in those dielectric materials which have ionic bonding and exists up to infrared frequency. But some dielectrics possess some permanent dipole moments, and the application of electric field causes these dipole moments to align in one direction. Such kind of polarization is called orientational polarization. Orientational or dipolar polarization is observed in polar dielectrics and exists only up to microwave frequency due to slow response of dipoles to the applied field. Due to electrical inhomogeneity, defects, and imperfections in some inhomogeneous dielectric material, charges are accumulated on the surfaces of grain and grain boundary or near the electrode interfaces resulting space charge or interfacial polarization. The total polarization of the dielectric material is the contribution of all these four types of polarization [20, 42].

4.2.2 Dielectric Breakdown

Breakdown suggests about the failure of a dielectric material against a certain applied electric field which causes the loss of its insulating properties and the corresponding electric field called as the breakdown strength or dielectric strength of the material. The breakdown process of the dielectric material involves three mechanisms such as electromechanical breakdown, intrinsic breakdown, and thermal breakdown. The dominance of electrostatic compressive force over the mechanical compressive strength is the main reason for electromechanical breakdown of the dielectric material. The breakdown strength of electromechanical breakdown is generally expressed in terms of Young's modulus (Y) as [43, 44]:

$$E_b = 0.606\sqrt{\frac{Y}{\varepsilon_0 \varepsilon_r}} \tag{4.1}$$

The crossover of charge carriers (mainly electrons) from valence band to conduction band under the influence of very high electric field results large conduction currents and hence breakdown of dielectric material. This breakdown is known as intrinsic breakdown. The collision of highly accelerated electrons with atoms or molecules broke their covalent bonds under the application of high field and released more electrons and holes. This leads to a dielectric breakdown in a material, and it is known as avalanche breakdown. The thermal breakdown generally occurs in the dielectric material, when the heat produced through the conduction and leakage loss of the dielectric exceeds the dissipation of those energy through radiation and conduction.

4.2.3 Energy Storage

The energy density of polar or nonlinear dielectric materials can be defined as the integral of electric field (E) with respect to the electrical displacement (D) vector between the limit of remnant polarization (P_r) and maximum electrical displacement ($D_{\max}$) i.e. [19, 45],

$$U_e = \int_{P_r}^{D_{\max}} E dD \tag{4.2}$$

and the energy storage efficiency (η) can be defined in terms energy density (U_e) and energy loss (U_l) as the following equation:

$$\eta = \frac{U_e}{U_e + U_l} \tag{4.3}$$

The energy loss (U_l) represents the energy dissipated during the charge and discharge process of dielectric material. U_l also denotes the (1) energy consumption of all the dielectric materials during polarization and depolarization process and (2) unavoidable leakage loss.

The electrical displacement vector for these polar dielectrics can be expressed as:

$$D = P + \varepsilon_0 E \tag{4.4}$$

However, the energy density (U_e) for nonpolar or linear dielectric materials can be defined as the product of electrical displacement vector and electric field with some constant (i.e., $\frac{1}{2}$).

$$U_e = \frac{1}{2} DE \tag{4.5}$$

The electrical displacement vector for these nonpolar dielectric materials directly proportional to electric field and can be written as:

$$D = \varepsilon E = \varepsilon_0 \varepsilon_r E \tag{4.6}$$

ε_r, ε, and ε_0 are the relative dielectric constant and dielectric constants in medium and free space, respectively.

So, the above equation of energy density is deduced to [46, 47]:

$$\therefore U_e = \frac{1}{2} \varepsilon_0 \varepsilon_r E^2 \tag{4.7}$$

All these mathematical expressions clearly indicate the significant contribution of these critical parameters such as polarization, dielectric constant, breakdown strength, and energy loss to the energy density of the dielectric material. Any one of them can affect the energy storage performance of the material. High dielectric constant and high breakdown strength are required to obtain high energy density of nonpolar (linear) dielectric material. Similarly, high polarization (or high dielectric displacement) and high breakdown strength are required for high energy density of polar (or nonlinear) dielectric material. High dielectric constant is needed for high polarization in polar dielectrics. Therefore, dielectric constant and breakdown strength are most important parameter, and their value should be high to obtain enhanced energy density for both polar and nonpolar dielectrics. On the other hand, it is not only to achieve the high energy density, but we have to optimize the energy storage efficiency. So, the dielectric material should have low energy loss, leakage loss, and conductivity. Polymer nanocomposites incorporate the individual's pro factor such as high breakdown strength of polymer and filler's high dielectric constant (or polarization) to optimize the most essential factor energy density and consequently the dielectric energy storage performance. The high dielectric constant of these polymer nanocomposites can be obtained either from percolation polymer nanocomposites by addition of conductive fillers or from dielectric-reinforced polymer nanocomposites having intrinsic dielectric properties. Similarly, the breakdown strength has equal significance in energy storage, so the breakdown processes are also discussed in our following sections.

4.3 Models for Dielectric Permittivity and Breakdown Strength

As we have mentioned earlier, the polymer nanocomposites combined the pro quality of individual components such as the fillers which significantly contributes to the dielectric properties and the polymer matrix which provides the mechanical properties. The electrical properties of polymer nanocomposites exhibit completely different trends than the pure polymer and filler constituents. Therefore, several models are proposed to explain the dielectric properties of these polymer nanocomposites.

4.3.1 Models for Dielectric Permittivity

Initially, simple models such as series and parallel models are proposed by assuming continuous medium of polymer matrix with dielectric permittivity 'ε_p' filled with spherical fillers of dielectric permittivity 'ε_f', and both the filler and the polymer have no dielectric loss. The series and parallel model can be mathematically expressed as:

$$\text{Series Model}: \ \varepsilon_c = \frac{\varepsilon_p \varepsilon_f}{\varepsilon_p f_f + \varepsilon_f f_p} \tag{4.8}$$

$$\text{Parallel Model}: \ \varepsilon_c = \varepsilon_p f_p + \varepsilon_f f_f \tag{4.9}$$

where ε_c is the permittivity of the polymer composites, f_p and f_f represent the volume fraction of the polymer matrix and the filler, respectively.

Then, these simple models are modified by Sillars and Landau–Lifshitz which holds good for lower volume fraction of the filler, but limited to the conductivity of both filler and polymer matrix. The mathematical expression can be written as:

$$\varepsilon_c = \varepsilon_p \left[1 + \frac{3 f_f (\varepsilon_f - \varepsilon_p)}{2\varepsilon_p + \varepsilon_f} \right] \tag{4.10}$$

Maxwell–Garnet equation which is more accurate than the previous ones but have the limitation in resistivity of polymer matrix and filler can be expressed as [48, 49]:

$$\varepsilon_c = \varepsilon_p\left[1 + \frac{3f_f(\varepsilon_f - \varepsilon_p)}{(1 - f_f)(\varepsilon_f - \varepsilon_p) + 3\varepsilon_p}\right] \text{(For spherical fillers)} \tag{4.11}$$

and

$$\varepsilon_c = \varepsilon_p\left[1 + \frac{f_f(\varepsilon_f - \varepsilon_p)}{A(1 - f_f)(\varepsilon_f - \varepsilon_p) + \varepsilon_p}\right] \text{(for non spheroid fillers)} \tag{4.12}$$

$A = \frac{1}{3}$ is the depolarization factor, which indicates the deviation from spherical shape of the dispersed filler.

Then, Bruggeman proposed an improved model (expressed in the following equation) which is valid for relatively large volume fractions of spherical filler than the previous models and for agglomerated or closely packed filler particles.

$$\varepsilon_c = \varepsilon_f \frac{3\varepsilon_p + 2f_f(\varepsilon_f - \varepsilon_p)}{3\varepsilon_f - f_f(\varepsilon_f - \varepsilon_p)} \tag{4.13}$$

And the final Bruggeman equation can be written as [48, 50]:

$$\frac{(\varepsilon_f - \varepsilon_c)}{\varepsilon_c^{1/3}} = \frac{(1 - f_f)(\varepsilon_f - \varepsilon_p)}{\varepsilon_p^{1/3}} \tag{4.14}$$

Further, Jaysundere and Smith proposed a model by considering the interaction between the filler particles which cannot be neglected for nanosized fillers and also in large volume fractions of filler, hence, calculated the overall electric field by including the polarization of adjacent particles. This factor was neglected in previous models as all of them are valid for lower filler loadings and assumed less interaction of filler particles due to large distance between them. The Jaysundere–Smith equation can be written as [48, 51, 52]:

$$\varepsilon_c = \frac{\varepsilon_p f_p + \varepsilon_f f_f \frac{3\varepsilon_p}{(2\varepsilon_p + \varepsilon_f)}\left[1 + 3f_f\frac{(\varepsilon_f - \varepsilon_p)}{(2\varepsilon_p + \varepsilon_f)}\right]}{f_p + f_f\frac{3\varepsilon_p}{(2\varepsilon_p + \varepsilon_f)}\left[1 + 3f_f\frac{(\varepsilon_f - \varepsilon_p)}{(2\varepsilon_p + \varepsilon_f)}\right]} \tag{4.15}$$

Lichtenker proposed a model by modifying the initial idea of series and parallel model and calculated the formula for effective dielectric permittivity and conductivity of the polymer-based composites. The series and parallel models explained the lower and upper bound of the effective dielectric permittivity based on arrangement of the parallel layers of the composite perpendicular or along the field, respectively. The Lichtenker formula widen the lower and upper limit of the effective dielectric permittivity and can be written as [52, 53]:

$$\varepsilon_{\text{eff}}^{\alpha} = \varepsilon_p f_p^{\alpha} + \varepsilon_f f_f^{\alpha} \tag{4.16}$$

α is the parameter whose value explains the transition from anisotropy ($= -1$) to isotropy ($= 1$).

Another model was proposed to explain the dielectric properties of the polymer composite having conducting fillers and called as percolation model. Though insulating (inorganic) fillers mainly dispersed in polymer matrix to enhance the dielectric permittivity of the composite, the dielectric constant is not still high enough in application point of view even at high filler concentration and the high filler inclusion affects the mechanical properties of the host polymer. So, some researcher dispersed conducting fillers to achieve high dielectric constant of the composites. The high dielectric constant in these composites is the result of Maxwell–Wagner polarization. These composites also exhibit high loss and conductivity. Both the dielectric constant and conductivity of these composites follow a nonlinear trend above percolation threshold (f_c) which cannot be explained by the above-mentioned models. The equations for percolation theories can be written as [48, 49, 54]:

$$\varepsilon_c \propto \left(f_c - f_f\right)^{-q} \text{ for } f_f < f \tag{4.17}$$

$q = 0.8 - 1$ is the dielectric critical exponent.

Yamada model was proposed for polymer composites with higher loading of dielectric filler and the equation can be written as [55, 56]:

$$\varepsilon_c = 1 + \frac{\eta f_f\left(\varepsilon_f - \varepsilon_p\right)}{\eta\varepsilon_p + \left(\varepsilon_f - \varepsilon_p\right)\left(1 - f_f\right)} \tag{4.18}$$

η is the shape factor.

Yu et al. [56] fitted the experimental data of surface-modified $BaTiO_3$ nanofillers dispersed PVDF matrix composites with the Yamada model and find that the model simulation fitted well with the trend of experimental data of permittivity except the filler concentration at 60%.

Modified Kernel model was another model purposed which considers the interaction between the adjacent fillers in the polymer composites and the equation can be written as [53, 57]:

$$\varepsilon_c = \frac{\varepsilon_p f_p + \varepsilon_f f_f\left[\frac{3\varepsilon_p}{\varepsilon_f + 2\varepsilon_p}\right]\left[1 + \frac{3f_f\left(\varepsilon_f - \varepsilon_p\right)}{\varepsilon_f + 2\varepsilon_p}\right]}{f_p + f_f\left[\frac{3\varepsilon_p}{\varepsilon_f + 2\varepsilon_p}\right]\left[1 + \frac{3f_f\left(\varepsilon_f - \varepsilon_p\right)}{\varepsilon_f + 2\varepsilon_p}\right]} \tag{4.19}$$

4.3.2 Models for Breakdown Strength

Phase-field model, finite element simulation, stochastic simulation, and machine learning are the different type of breakdown simulations used to understand the dielectric breakdown mechanism of the polymer nanocomposites.

Dielectric breakdown model:

The dielectric breakdown model was proposed to understand the underlying mechanism of breakdown patterns in a polymer composite by assuming the homogeneous distribution of the dielectric fillers in the polymer matrix. The formed electrical trees in the polymer composite propagate through the dielectric medium during the dielectric breakdown, and the electrical field affects this tree channel growth. So, the mathematical relation between the probability (p) of tree channel growth and the electric field (E) can be expressed as [58]:

$$p(i, k \to k', k') = \frac{\left(E_{i,k'}\right)^{\eta}}{\sum\left(E_{i,k'}\right)^{\eta}} \tag{4.20}$$

where η is the exponent and $i,\ k'$ represents the tree channel growth sites.

Failure probability and fractal dimension are the two factors which can determine the electrical tree behaviors. The number of channels connected to the electrical tree are controlled by the factor called propagation time, which affect the failure probability. Similarly scaling behavior (C) is another function used to calculate the fractal dimension (D) using the following mathematical relation:

$$C_r = C_0 r^{D-2} \tag{4.21}$$

where r represents the radius of the circle which covers all the electrical sites.

The failure probability was also explained on the basis of Weibull distribution model. The Weibull statistics model can be mathematically expressed as [45, 59, 60]:

$$P = 1 - e^{-\left(\frac{E}{E_0}\right)^{\beta}} \tag{4.22}$$

where $P,\ E,\ E_0$, and β represent the cumulative probability of electric failure, experimental breakdown data, scale parameter, and shape parameter, respectively. The scale parameter suggests the field strength where there is a 63% probability for the sample to breakdown (characteristic E_b), and the shape parameter called as the Weibull modulus displays the dispersion of E.

Modified dielectric breakdown model:

The dielectric breakdown model was again modified for randomly distributed filler in the polymer matrix by introducing a new term "L_i" in the existing equation of probability as [58]:

$$p(i, k \to k', k') = \frac{\left(\frac{E_{i,k'}}{L_i}\right)^{\eta}}{\sum\left(\frac{E_{i,k'}}{L_i}\right)^{\eta}} \tag{4.23}$$

where L_i was represented as the pertaining breakdown channel length which can be expressed as:

$$L_i = L_0 - \frac{L_0}{a_i} \tag{4.24}$$

a_i is the distance between the conducting particle.

Wang et al. used the modified dielectric breakdown model to explain the improvement of breakdown strength of a sandwich polymer nanocomposites which constitute an array of ultrasmall metal particles (Ag, Au, or Pt) in between the layers of PVDF HFP.

Phase-field model:

Ceramics or dielectric fillers with high dielectric constant were dispersed in the polymer matrix to enhance the effective dielectric permittivity of the polymer nanocomposites. The accumulation of charges in the interfaces of ceramic fillers and the polymer matrix due to the huge difference in their dielectric constant also affects the local electric field at the interface and consequently deteriorates the breakdown strength of the polymer nanocomposites. Further, application of electric field above a critical value may initiate cracks and weak paths to propagate, resulting in a complete breakdown. Phase-field model was developed to describe the damage initiation and the progress in the polymer composites [61]. The dielectric breakdown of the polymer composites can be explored by obtaining the local field distributions and crack propagation details continuum phase-field model [62]. The influence of microstructure on the breakdown strength, energy density, and dielectric permittivity of the polymer composites can also be investigated from the simulation of this model. This model can also design the novel microstructures of desired energy density and breakdown strength. The influence of various stimuli (e.g., mechanical, electrical, and thermal stimulus) on breakdown behavior of polymer composites can be predicted by the continuous phase-field model. By considering the contribution of electric energy, phase separation energy, and gradient energy, one can develop the phase-field model to simulate the breakdown strength in polymer nanocomposites. The contribution of thermal energy which was not included in the phase-field model was incorporated to develop the electrothermal breakdown model, which predicts well about the breakdown strength of the 3D filler-based polymer nanocomposites.

Stochastic model:

Stochastic model was proposed to understand the breakdown phenomena by studying one of the most important features, i.e., the fractal properties of the branched discharges. According to this model, the growth probability depends on the local field (potential), and the mathematical expression can be written as [63]:

$$p(i, k \to i', k') = \frac{(\varphi_{i',k'})^\eta}{\sum (\varphi_{i',k'})^\eta} \tag{4.25}$$

The denominator represents all the possible growth processes.

Yue et al. [64] reported a modified stochastic model to study the growth of the breakdown path and the internal electric field distribution of the polymer composites under the applied electric field. The break down probability $P(r)$ was reported as:

$$P(r) = \frac{E(r)^2}{E_b(r)^2} / \sum \frac{E(r)^2}{E_b(r)^2} \tag{4.26}$$

where $E(r)$ and $E_b(r)$ represent the electric field of a local point and intrinsic breakdown strength, respectively. The denominator suggests the sum of all points where local electric field exceeds the breakdown strength. In addition to modified stochastic model, they also used the machine learning to study the breakdown strength of the polymer composites. They used three parameters, i.e., the dielectric permittivity, size, and content of the filler, to build the model for the breakdown process of the polymer composites.

Finite element simulations:

Wang et al. [65] investigated the local electric field distribution and its effect on breakdown strength of the polymer composites by finite element simulations, when high permittivity fillers like barium titanate were added on the polymer matrix. Similarly, the local field distributions and leakage current density of the double-layered composite of boron nitride surface-modified BCZT ferroelectric nanofillers and PVDF polymer were analyzed by the finite element simulations [66]. The electric field distribution in the layered structure of PVDF/PMMA polymer composites was analyzed from the finite element analysis [67]. The effect of surface-modified barium titanate nanoparticle by TiO_2 and polydopamine filler on breakdown strength of epoxy composite was studied by finite element simulations [68].

4.4 Crucial Parameters (Dielectric Permittivity, Dielectric Loss, Breakdown Strength, and Energy Density) of Polymer-Based Nanocomposites

The first crucial factor in polymer composite dielectric is the type of dielectric material which means the variation of electric displacement is different in different dielectric materials with application of electric field as the polarization and energy loss is different in polar and nonpolar (linear) dielectrics. There are mainly two types of dielectrics: (1) polar or nonlinear dielectrics and (2) nonpolar or linear dielectrics. Again, the polar or nonlinear dielectric subcategorized to ferroelectric (both normal and relaxor) and antiferroelectric. However, the nonpolar dielectrics include the linear dielectrics and paraelectric material. The variation of electric displacement for both polar and nonpolar dielectrics with application of electric field is illustrated in Fig. 4.1. It is clear from the figure that the nonlinear dielectrics have more pronounced energy loss, but they also exhibit high polarization and dielectric constant. Relaxor ferroelectric material possess high saturation polarization, low remnant polarization, and moderate breakdown strength, which made them potential material among all the dielectric. The efficiency of these relaxor materials can reach up to 90%, and the narrow hysteresis loop suggests the low energy dissipation in the form of heat during discharging. As we have mentioned in our previous Sect. 4.2.3, the energy storage density and efficiency of these polar dielectrics depend upon the remnant polarization and saturation polarization and low energy loss. However, the linear dielectrics (e.g., polymers or some composite) exhibits almost no energy loss, but their energy density was limited due to their low polarization and dielectric permittivity.

From the above discussion on energy density of capacitive storage, we found that the other important parameters are the dielectric permittivity and breakdown strength. Various types of nanofillers based on their electrical properties like dielectric/ ceramics, conducting, and semiconducting are incorporated in the polymer matrix to improve the dielectric permittivity of the polymer nanocomposites. More details regarding the inclusion of ceramics, conductive and semiconductive nanofillers for the enhancement of dielectric permittivity, and the energy storage properties are discussed in or following sections. However, in search for high permittivity of the polymer nanocomposites, researchers never tried to sacrifice the flexibility of the

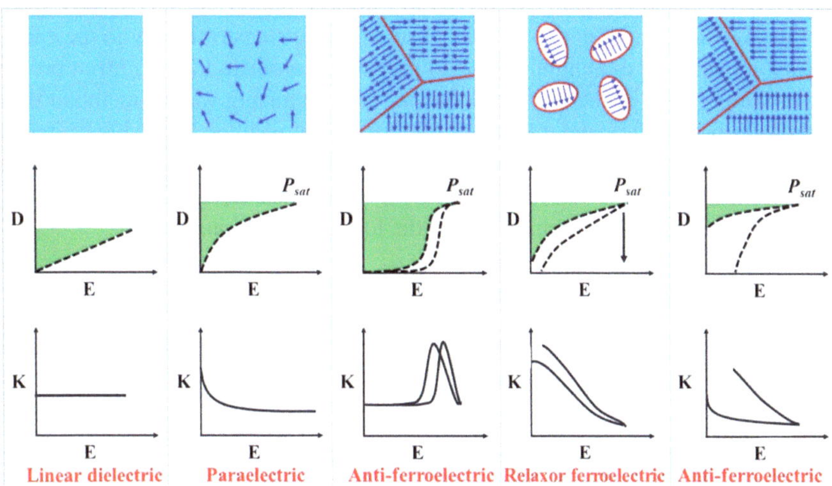

Fig. 4.1 Variation of electric displacement with applied electric field in linear and nonlinear dielectric. Reprinted with permission from Ref. [69]. Copyright 2019 Elsevier

polymer matrix. So, a small amount of filler incorporated in the polymer matrix to maintain the flexibility. In case of linear dielectrics, there is only induced polarization; whereas for polymer nanocomposites, the total polarization is the contribution of polarization of the nanofiller, the interaction between the fillers, and interface between the polymer matrix and nanofillers. The inclusion of nanofiller (whether it is conductive or insulating) in polymer matrix has a significant effect on overall properties of the polymer nanocomposites because the large surface area of nanomaterials create a large interaction zone among the nanofiller and the polymer matrix. In addition, the inclusion of nanofiller also changes the surface morphology of the polymer matrix. The small size of the nanofiller also modifies the internal field and space charge distribution of the polymer nanocomposites. The space charge distribution and the scattering effect due to the incorporation of nanofillers significantly affect the breakdown strength and dielectric permittivity of the polymer nanocomposites. Maxwell–Wager–Sillars theory was used to explain the interfacial space charge behavior of polymer nanocomposites [70]. Two different models, i.e., Lewis and Tanaka's model [71–73], are purposed to understand the effect of interface due to inclusion of nanosized filler in the polymer matrix, and the details of models, the influence of interface in different parameters, and overall properties of the composites are explained in Prateek et al. They also reviewed in detail the effect of (i) shape of conducting and nonconducting filler with different dimensions (0D, 1D, 2D), (ii) the 1D and 2D fillers with different aspect ratios on overall properties of the polymer nanocomposites. Another important parameter in the energy density equation is the breakdown strength, but it was studied that the enhancement of breakdown strength always accompanied by the decrease in dielectric permittivity. The inverse relationship between the breakdown strength and dielectric permittivity mainly attributed to the local electric field distribution in the polymer nanocomposites due to huge difference of individual dielectric permittivity of filler and matrix. We can neither ignore the breakdown strength nor the contribution of dielectric permittivity to the energy density, so we have to maintain these to parameter to obtain the optimized properties of the polymer nanocomposites. So, the one promising way to maintain these parameters is to reduce the difference of permittivity of filler and polymer matrix, and this can be obtained by adopting surface engineering, interfacial, and topological design. The examples of surface modification to maintain the breakdown strength of the polymer composites are discussed in the following section.

4.5 Different Approaches for Performance Improvement of Polymer Nanocomposites

(a) **By adding different types of fillers in polymer dielectric**

Different types of nanofillers such as conductors, semiconductors, and ceramics with different electrical properties are dispersed with the polymer matrix to tailor the dielectric and energy storage performance of the polymer nanocomposites.

Ceramics Fillers:

The most traditional way to achieve the high dielectric constant in the polymer nanocomposites is to incorporate ceramics filler such as $Pb(Zr_{0.5}Ti_{0.5})O_3$ (PZT) [74], $CaCu_3Ti_4O_{12}$ (CCTO) [75], $BaTiO_3$ (BT) [76], $Na_{1/2}Y_{1/2}Cu_3Ti_4O_{12}$ (NYCTO) [77], Ba_xSr_{1-x} TiO_3, $x$$NaNbO_3$- (1-$x$)$BaTiO_3$ (NN-BT) [78], $(Bi_{0.5}Ba_{0.25}Sr_{0.25})(Fe_{0.5}Ti_{0.5})O_3$ (BFO–BST) [79], and $0.78Bi_{0.5}Na_{0.5}TiO_3$–$0.22SrTiO_3$ (BNT-ST) [80] in the polymer matrix. Some of other recent examples are also discussed in the following paragraph.

Considering ultrahigh dielectric and polarization response of the ferroelectric ceramics near the morphotropic phase boundary (MPB), Xie et al. [81] chosen the MPB composition of the $Pb(Zr_{0.52}Ti_{0.48})O_3$ (PZT) nanocubes (NCs) as ceramic fillers for the P(VDF-CTFE) polymer matrix. But, as we know, the huge difference between the dielectric constant of the PZT ceramic and the PVDF-CTFE polymer would cause accumulation of charges in the interfaces between the ceramic and polymer, which also affect the local electric field at the interface and consequently the reduction of overall breakdown strength of the polymer nanocomposite. Again, the high conductivity of the ferroelectric ceramics filler will induce the high leakage current and loss of the polymer nanocomposite and hence affect the energy density. So, to avoid such problems, they tried to introduce a buffer coating layer of SiO_2 over the PZT ceramics. The dielectric constant of the polymer nanocomposites increased from 11 of pristine polymer to 15.8 at 1 kHz in the addition of 5 vol% of the filler. This core–shell structure PZT@SiO_2 embedded P(VDF-CTFE) nanocomposites possess a high energy density of 16.8 Jcm^{-3} at breakdown strength of 491 MVm^{-1}. Xie et al. [82] studied the effect of three different types of filler, i.e., semiconductive (SiO_2), conductive (Al_2O_3), and dielectric nanofiller ($BaTiO_3$) on dielectric and energy storage properties of polypropylene polymer matrix experimentally explained based on theoretical models and visualized simulation. Jaysundere–Smith, Maxwell–Wagner, and parallel models are used to explain the experimental data of the dielectric constant for all the composites. They found that the $BaTiO_3$/PP composites are the best option to obtain high dielectric constant with low loss among the others by keeping the same breakdown strength for all the composites.

Conductive or Metallic Fillers:

Conductive fillers include the pure metallic nanoparticles, conducting polymers, and carbon-based nanoparticles like Fe_2O_3 [83], NiO [84], Ni, carbon black, W and Ag [85, 86] carbon nanotube [87, 88], graphene, etc., were used for the polymer nanocomposites [89–91].

The dramatic increase of dielectric constant near the percolation threshold leads to the development of conductive filler-based polymer composites. Qi et al. [92] explained clearly that the size of the metal filler has a significant effect on the properties of polymer composites. They also mentioned the relation between the percolation probability to the size ratio of the filler to composite thickness and proposed that the metal fillers in the nanometer range will be the suitable option for polymer composites for practical applications. They added Ag nanoparticles on the epoxy polymer to form the metal-polymer nanocomposite which has the dielectric constant of 300 with a relatively low loss of 0.05. A highly aligned reduced graphene oxide (rGO)/ epoxy polymer nanocomposites exhibit an exceptionally high dielectric constant of 14,000 with 3 wt% of the filler concentration at1 kHz and explained on the basis of Maxwell–Wagner–Sillars polarization [93].

Semiconductor Fillers:

Apart from conducting and dielectric ceramics fillers, semiconducting nanofillers are also used to improve the dielectric constant of the polymer matrix and the enhanced dielectric constant in these semiconductor polymer nanocomposites is mainly observed near the percolation threshold. The pro qualities of semiconducting fillers include the appreciable band gap, low AC conductivity, relatively low dielectric constant compatible to the polymer matrix, and the nonlinear electrical characteristics.

The enhanced dielectric, energy storage performance of the semiconductor filler-based polymer nanocomposites, i.e., fullerene-type tungsten disulfide nanoparticle-PVDF, was recently reported by Guo et al. [94]. They observed the enhancement of polar electroactive phases with increase in filler content leading to enhanced dielectric, mechanical, and energy storage properties. The dielectric constant increases 1.5 to 4 times than the pristine polymer PVDF even in the moderate addition of filler, whereas the higher filler concentration exhibits dielectric constant more than 40 times with an increase in tan δ. The dielectric behavior with the filler content was explained with the percolation model and fitted with various model for the comparison as shown in Fig. 4.2.

Similarly, Jia et al. enhanced electrical polarization and improved dielectric constant of a semiconductor nanostructured filler (both flower and nanosheet) in the PVDF matrix [95]. A novel three-dimensional zinc oxide superstructures (flower and walnut like) filler was used by Wu et al. [96] to tailor the dielectric constant and breakdown strength of the PVDF polymer matrix. The polymer nanocomposites significantly enhance the dielectric constant to 19.4, 104.9, and 221.1 for commercial ZnO, walnut-like, and flower-like ZnO superstructures at 100 Hz, whereas their breakdown strengths are found to be 45, 40, and 42 kV/mm, respectively. Ji et al. [97] used nickel hydroxide $Ni(OH)_2$ nanoparticles in three different dimensions, i.e., nanosphere (0D), nanoplates (1D), and nanoflowers (3D) as fillers in the PVDF matrix, and the existence of -OH group will enhance the polar/electroactive β phase of PVDF. The -OH group will act as a surface layer above the Ni and do not need additional surface modification of the conductive filler. The hydroxyl group will provide a good interaction between the polymer matrix and filler. The maximum energy density of 17.3 J/cm3 at 421 kV/mm and dielectric constant of 16.3 at 100 Hz were observed for 5 wt% nanoplates-$Ni(OH)_2$ based composites in comparison with other nanofillers having different dimensions. It also suggests the filler dimension which also affects dielectric and energy storage properties. A semiconductor-based hybrid filler, i.e., $ZnO@MoS_2$ was used to modify the different properties of poly(vinylidene fluoride-chlorotrifluoroethylene-double bonds) [98]. The 2 wt% of this hybrid nanofiller (4 mol% of MoS_2 in ZnO) in the polymer matrix yields the

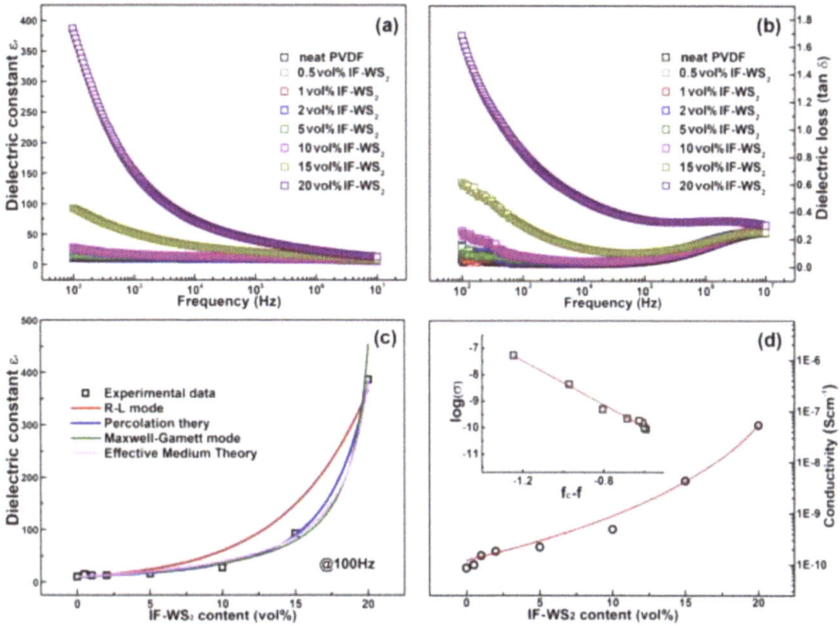

Fig. 4.2 Variation of dielectric constant with respect to frequency and filler content of fullerene-type tungsten disulfide nanoparticle-PVDF nanocomposites. Fitting of experimental data of dielectric constant with filler content by various model. Reprinted with permission from Ref. [94]. Copyright 2020 Elsevier

energy density of 7.2 Jcm^{-3} at 300 MVm^{-1}which is 70% higher compared to that of pristine polymer and possess high energy efficiency of 83%.

Jing et al. [99] prepared a polymer nanocomposite by incorporating a completely new type of filler, i.e., a high entropy oxide nanofiber $(Eu_{0.2}Bi_{0.2}Y_{0.2}La_{0.2}Cr_{0.2})_2O_3$ in the P(VDF-HFP)/PMMA polymer matrix. The 0.6 wt% of the filler concentration leads to highest breakdown strength of 509.4 kV mm^{-1} and energy density of 15.13 Jcm^{-3}. The enhanced breakdown strength in the nanocomposites is explained by the several factors such as polymorphic distortion caused by the high entropy in the lattice, and the movement of the long polymer was restricted by the pinning effect associated with the polymorphic distortion and reduces the loss in strong electric field and promotes the breakdown strength. The transmission of space charges is also blocked by the charge shielding layer between the nanofiber and the polymer, which again contributes to the enhanced breakdown strength.

Heterogenous fillers are also used in some literatures where both conductive and insulating fillers are used for the polymer nanocomposites. Ji et al. [100] used multi-walled carbon nanotube (MWCNTs) and barium titanate ($BaTiO_3$) fillers and PVDF polymer to prepare multilayered films with alternate distribution of layers. They also proposed a theoretical calculations/model for dielectric properties of heterogenous filler-based polymer nanocomposites by modifying the classical series model by considering the contribution of two additional factors, i.e., filler distribution and interfacial polarization.

4.6 By Structural Design and Surface Engineering

Recently, Zeng et al. [101] developed a novel polymer nanocomposite by incorporating a surface-modified $BaTiO_3$ nanoparticle with polyimide (PI) polymer shells in the polyetherimide (PEI) polymer matrix. The dispersion of core–shell structured PI@BTO nanofillers yields an improved the energy density to 15.55 Jcm^{-3} with charger–discharge efficiency (η) of 80% and enhanced breakdown strength of 600.7 MVm^{-1} in the PEI-PI@BTO nanocomposites. Similarly, Hu et al. [102] first designed a polymer nanocomposite by employing surface-modified core–shell nanofibers and then fabricated sandwiched structures by sandwiching these polymer nanocomposites layer in between the outer layer of host polymer matrix. They have chosen $Ba(Ti_{0.85}Zr_{0.15})O_3$ as core and $Ba(Ti_{0.65}Zr_{0.35})O_3$ as shell nanofibers for the core–shell structure, which was further modified by polydopamine (PDA). Then, after modification, they have added the surface-modified core–shell nanofiber to the PVDF polymer matrix in different filler concentration. Then, they fabricated three-layer sandwich structure by considering the polymer nanocomposite as the central layer in between the host PVDF outer layer. They observed the enhanced dielectric constant, breakdown strength, and energy density in lower filler content (2wt% of surface-modified core–shell structure) of both the single layer and trilayer nanocomposites. The energy density of the single and trilayer composites are found to be 8.2 and 9.9 Jcm^{-3}, respectively. The energy density and breakdown strength of the polymer nanocomposites are tailored by adding different polymer@$BaTiO_3$ nanoparticle core–shell structure on PVDF HFP matrix in Li et al. [103]. The $BaTiO_3$ is surface modified by the polymethyl methacrylate (PMMA), polymethylsulfonyl ethyl methacrylate (PMSEMA), and polyglycidyl methacrylate (PGMA) polymers to prepare the core–shell structure. They studied the polarity of different polymer shell on the energy storage performance of the polymer nanocomposite and found the enhanced energy density and breakdown strength for the PMMA polymer shell with lowest polarity. Pan et al. [104] reported the enhanced dielectric constant of 149.5 at 1 kHz, high breakdown strength of 250.6 MV m^{-1} with suppressed dielectric loss 0.19 of PVDF/MoS_2-PPy composites. The hybrid MoS_2-PPy nanofiller was fabricated by first exfoliating the molybdenum disulfide (MoS_2) with the cetyltrimethyl ammonium bromide (CTAB) solution and the exfoliated MoS_2 again followed by surface modification with conducting polymer polypyrrole (PPy). Xu et al. [68] recently reported a novel design of polymer nanocomposites by using a gradient dielectric filler means a double core–shell structure which not only improves the compatibility and dispersibility between the filler and polymer matrix but also improves the breakdown strength, nonlinear conductive, and dielectric properties. The double core–shell structure filler material constitutes $BaTiO_3$ nanoparticle as the core covered by TiO_2 as the first shell layer and then encapsulated by a second layer of polydopamine. Then, this double core–shell structure filler was dispersed in the epoxy polymer matrix to prepare the polymer nanocomposites. Then, again these polymer nanocomposites were exfoliated boron nitride nanosheets with parallel and random alignments. The dispersion of this double core–shell structure filler with 10 wt% and the exfoliation of the polymer nanocomposite significantly improves the breakdown strength of the pure polymer matrix epoxy by 55%, i.e., from 29.4 to 45.7 $kVmm^{-1}$. The nonlinear trend of dielectric and conductivity properties of these nanocomposites were explained on the basis of nonlinear effective medium theory (EMT) and the hopping transport model, respectively.

Li et al. [105] developed sandwich-structured ceramic-polymer nanocomposites, which constitute $BaTiO_3$ nanofiber-(PVDF-HFP) composites as the outer layer and MgO nanowires-(PVDF-HFP) composites in the central layer in an objective to improve the breakdown strength and high dielectric constant by the outer layer and central layer, respectively. They found best energy density of 15.55 Jcm^{-3} and charge–discharge efficiency of 68% at Weibull breakdown strength of 416 MVm^{-1} for the sandwich structure with 2 wt% MgO in the central layer and 20 wt% $BaTiO_3$ in the outer layers. Bai et al. [66] prepared a two-layer polymer composite, where the bottom layer constitutes the polymer composites of PVDF filled with boron nitride surface insulated $0.5(Ba_{0.7}Ca_{0.3})TiO_3$-$0.5Ba(Zr_{0.2}Ti_{0.8})O_3$ (BCZT) nanofibers and the top layer consists of the boron-nitride-filled PVDF composite. They obtained a high breakdown strength and energy density by tailoring the ratios of boron nitrides in the top layer and also the amounts of surface-insulated BCZT nanofibers in the bottom layer. The PVDF-boron nitrides nanocomposite in the top layer exhibits an ultrahigh breakdown strength of 730 MVm^{-1} for 0.8 wt% of fillers. The double-layer nanocomposites with the above composite in the top layer and the bottom layer with 1.6 wt% of BCZT@boron nitride nanofiber was found to be with energy density of 24.3 Jcm^{-3} at 718 MVm^{-1}. Luo et al. [106] designed a unique sandwich film in which the $Ba_{0.6}Sr_{0.4}TiO_3$ nanofibers network are completely interconnected and covered by the PVDF-HFP top and bottom layers. The sandwich film with 4 vol% of the nanofibers possesses an enhanced energy storage density of 9.46 J/cm^{-3}, improved dielectric constant of 24.8 at 1 kHz and higher breakdown strength of 2954 $kVcm^{-1}$. Here, instead of polymer-based nanocomposites in the middle layer, a direct nanofiber network was used for the sandwich structure to improve the dielectric constant, energy density, and breakdown strength. Zhang et al. [67] reported high breakdown strength of 486.05 MVm^{-1}, energy density of 8.65 Jcm^{-3}, and a high polarization of 4.48 μCcm^{-2} for a sandwich structure of PMMA/PVDF blended layer (for 30 mass% of PMMA) covered by the outer layer of polyetherimide. The designing of different layered structures by ordering the polymer nanocomposites layer (called hard layer) and pure polymer layer (soft layer) in five different arrangements and the influence of interfaces between the adjacent layers on different parameters like breakdown strength and energy density were studied by Wang et al. [107]. The $BaTiO_3$/poly(vinylidene fluoride) nanocomposites represent the hard layer and denoted by B, and the pure PVDF layer represents the soft layer denoted by P. They observed the layer structure with polymer nanocomposites as the outer layer, and pure polymer in the middle layer (i.e., B-P-B film) exhibits

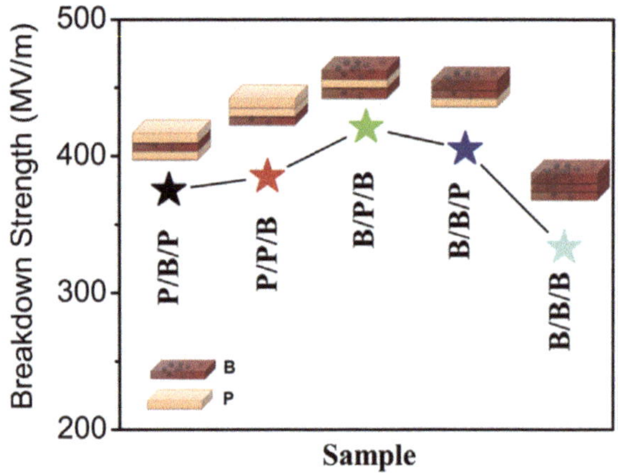

Fig. 4.3 Breakdown strengths of sandwich-structured films with different geometries. Reprinted with permission from Ref. [106]. Copyright 2021 Elsevier

the highest breakdown strength in comparison with others as shown in Fig. 4.3. They also prepared the single-layer polymer nanocomposites of PVDF and $BaTiO_3$ nanoparticles and compared the energy density and breakdown strength with the polymer nanocomposites. They found that both the energy density (7.02 Jcm^{-3}) and breakdown strength (390 $kVmm^{-1}$) of layered structure are higher than those of single-layer nanocomposites.

4.7 Conclusion and Perspectives

In comparison with the recent energy storage systems such as supercapacitors, batteries, and fuel cells, dielectric capacitors attracted much attention of contemporary materials researcher due to its unique qualities such as high-power density, fast cycling rate, and long lifespan. But the application potentiality of conventional dielectric capacitors (either ceramics or polymer) is restricted by alternate pro qualities of pure polymer and ceramics, i.e., pure polymer's application potentiality is limited due to its low permittivity and energy density, whereas the ceramics lack in flexibility and breakdown strength. These shortcomings of conventional dielectric capacitors are resolved by fabrication of polymer-ceramics composites through inclusion of ceramics filler in the polymer matrix. Again, for further improvement in the energy storage properties with high energy density, polymer nanocomposites were developed which incorporates the fillers in nanoform (0D, 1D, 2D nanofillers). The main focus of the researchers in recent trend to design and develop efficient energy storage system is to optimize all the parameters of dielectric to attain high energy density and efficiency in polymer nanocomposites. The dielectric properties of the polymer nanocomposites strongly depend on the physical, chemical, and mechanical properties of both the fillers and polymer matrix. So, selection of both the filler and polymer matrix is the most significant step to expect the desired parameters in the polymer nanocomposites. To obtain the high dielectric permittivity of the polymer nanocomposites, two different approaches have been made, i.e., addition of either (1) ceramics filler with high dielectric permittivity or (2) conductive filler associated with the percolation thresholds. But again, the addition of ceramics filler in the polymer matrix affects/deteriorates the breakdown strength of the polymer nanocomposites due to huge difference in their permittivity. So, to maintain both the breakdown strength and dielectric permittivity in polymer nanocomposites, surface modification of ceramics filler was carried out. Again, to reduce the difference of dielectric permittivity of polymer matrix and filler, conductive or semiconducting filler having comparatively lower permittivity was added to resolve the above issues. Both nanoparticles and nanofibers are added as fillers in the polymer matrix to prepare ternary composites with improved properties in comparison with binary composites. Novel structural design such as multilayered and core–shell structure has been developed to tailor the properties by interface minimization. Nanofillers (specifically 1D or 2D) with specific orientation or alignment found to be the best core filler in core–shell structured polymer nanocomposites to obtain enhanced dielectric constant. The difference in permittivity of polymer matrix and filler can be balanced by core–shell configuration and results a gradient dielectric permittivity. The development of different layered and laminated architecture also enables another way to improve in breakdown strength and energy density of the polymer nanocomposites.

Though there are many major breakthroughs in the design and development of polymer nanocomposites and tailoring their properties by novel approaches and architecture, still there are some unresolved quests to be explored in the future generation such as (1) the unsolved puzzle for simultaneous improvement breakdown strength and dielectric permittivity, (2) more attention needed to reduce the dielectric loss to improve the efficiency of energy storage, (3) new attempts and extensive analysis needed to alleviate the interfacial polarization, and (4) significant effort to design novel engineered architecture for effective energy storage systems.

Acknowledgements This research was funded by Ministry of Science and Higher Education of the Russian Federation: State task in the field of scientific activity, scientific project No. 0852-2020-0032 (BAZ0110/20-3-07IF).

References

1. Ghosh SK, Mandal D (2017) Bio-assembled, piezoelectric prawn shell made self-powered wearable sensor for noninvasive physiological signal monitoring. Appl Phys Lett 110:123701
2. Rahman W, Ghosh SK, Middya TR, Mandal D (2017) Highly durable piezo-electric energy harvester by a super toughened and flexible nanocomposite: effect of laponite nano-clay in poly(vinylidene fluoride). Mater. Res. Express 4:095305

3. Karan SK, Maiti S, Agrawal AK, Das AK, Maitra A, Paria S, Bera A, Bera R, Halder L, Mishra AK, Kim JK, Khatua BB (2019) Designing high energy conversion efficient bio-inspired vitamin assisted single-structured based self-powered piezoelectric/wind/acoustic multienergy harvester with remarkable power density. Nano Energy 59:169–183
4. Ding R, Zhange X, Chen G, Wang H, Kishor R, Xiao J, Gao F, Zeng K, Chen X, Sun XW, Zheng Y (2017) High-performance piezoelectric nanogenerators composed of formamidinium lead halide perovskite nanoparticles and poly(vinylidene fluoride). Nano Energy 37:126–135
5. Ryu H, Yoon HJ, Kim SW (2019) Hybrid energy harvesters: toward sustainable energy harvesting. Adv Mater 1:1802898
6. Vatansever D, Hadimani RL, Shah T, Siores E (2011) An investigation of energy harvesting from renewable sources with PVDF and PZT. Smart Mater Struct 20:055019
7. Karan SK, Maiti S, Agrawal DAK, Maitra A, Paria S, Bera A, Bera R, Halder L, Mishra AK, Kim JK, Khatua BB (2019) Designing high energy conversion efficient bio-inspired vitamin assisted single-structured based self-powered piezoelectric/wind/acoustic multienergy harvester with remarkable power density. Nano Energy 59:169–183
8. Liang Q, Yan X, Gu Y, Zhang K, Liang M, Lu S, Zheng X, Zhang Y (2015) Highly transparent triboelectric nanogenerator for harvesting water-related energy reinforced by antireflection coating. Sci Rep 5:1–7
9. Alam MM, Sultana A, Mandal D (2018) Biomechanical and acoustic energy harvesting from TiO_2 nanoparticle modulated PVDF nanofiber made high performance nanogenerator. ACS Appl Energy Mater 1:3103–3112
10. Yang Z, Zhang J, Kintner-Meyer MCW, Lu X, Choi D, Lemmon JP, Liu J (2011) Electrochemical energy storage for green grid. Chem Rev 111:3577–3613
11. Poullikkas A (2013) A comparative overview of large-scale battery systems for electricity storage. Renew Sustain Energy Rev 27:778–788
12. Luo X, Wang J, Dooner M, Clarke J (2015) Overview of current development in electrical energy storage technologies and the application potential in power system operation. Appl Energy 137:511–536
13. Ibrahim H, Ilinca A, Perron J (2008) Energy storage systems—characteristics and comparisons. Renew Sustain Energy Rev 12:1221–1250
14. Simon P, Gogotsi Y, Dunn B (2014) Where do batteries end and supercapacitors begin? Science 343:1210–1211
15. Yang M, Li Q, Zhang X, Bilotti E, Zhang C, Xu C, Gan S, Dang ZM (2022) Surface engineering of 2D dielectric polymer films for scalable production of high-energy-density films. Prog Mater Sci 128:100968
16. Simon P, Gogotsi Y (2008) Materials for electrochemical capacitors. Nat Mater 7:845–854
17. Sun Z, Ma C, Wang X, Liu M, Lu L, Wu M, Lou X, Wang H, Jia C (2017) Large energy density, excellent thermal stability, and high cycling endurance of lead-free $BaZr_{0.2}Ti_{0.8}O_3$ film capacitors. ACS Appl Mater Interfaces 9:17096–17101
18. Zhou M, Liang R, Zhou Z, Dong X (2018) Novel $BaTiO_3$-based lead-free ceramic capacitors featuring high energy storage density, high power density, and excellent stability. J Mater Chem C6:8528–8537
19. Hu J, Zhang S, Tang B (2021) Rational design of nanomaterials for high energy density dielectric capacitors via electrospinning. Energy Storage Mater 37:530–555
20. Cheng R, Wang Y, Men R, Lei Z, Song J, Li Y, Guo M (2022) High-energy-density polymer dielectrics via compositional and structural tailoring for electrical energy storage. iScience25, 104837
21. Gao F, Zhang K, Guo Y, Xu J, Szafran M (2021) (Ba, Sr)TiO_3/polymer dielectric composites–progress and perspective. Prog Mater Sci 121:100813
22. Huang X, Sun B, Zhu Y, Li S, Jiang P (2019) High-k polymer nanocomposites with 1D filler for dielectric and energy storage applications. Prog Mater Sci 100:187–225
23. Singh M, Apatal E, Samant S, Wu W, Tawade BV, Pradhan N, Raghavan D, Karima A (2022) Nanoscale strategies to enhance the energy storage capacity of polymeric dielectric capacitors: review of recent advances. Poly Rev 62:211–260

24. Zhang L, Liu Z, Lu X, Yang G, Zhang X, Cheng ZY (2016) Nano-clip based composites with a low percolation threshold and high dielectric constant. Nano Energy 26:550–557
25. Wanga H, Xie H, Wang S, Gao Z, Li C, Hu G, Xiong C (2018) Enhanced dielectric property and energy storage density of PVDF-HFP based dielectric composites by incorporation of silver nanoparticles-decorated exfoliated montmorillonite nanoplatelets. Compos A 108:62–68
26. Zh J-W, Zheng M-S, Fan B-H, Dang Z-M (2021) Polymer-based dielectrics with high permittivity for electric energy storage: a review. Nano Energy 89:106438
27. Sun L, Shi Z, Wang H, Zhang K, Dastan D, Sun K, Fan R (2020) Ultrahigh discharge efficiency and improved energy density in rationally designed bilayer polyetherimide-$BaTiO_3$ /P(VDF-HFP) composites. J Mater Chem A 8:5750–5757
28. Wang Y, Yao MG, Ma R, Yuan QB, Yang DS, Cui B, Ma CR, Liu M, Hu DW (2020) Design strategy of barium titanate/polyvinylidene fluoride-based nanocomposite films for high energy storage. J Mater Chem A 8:884–917
29. Bouharras FE, Raihane M, Ameduri B (2020) Recent progress on core-shell structured $BaTiO_3$@polymer/fluorinated polymers nanocomposites for high energy storage: synthesis, dielectric properties and applications. Prog Mater Sci 113:100670
30. Feng Y, Zhou YH, Zhang TD, Zhang CH, Zhang YQ, Zhang Y, Chen QG, Chi QG (2020) Ultrahigh discharge efficiency and excellent energy density in oriented core-shell nanofiber-polyetherimide composites. Energy Storage Mater 25:180–192
31. Guo MF, Jiang JY, Shen ZH, Lin YH, Nan CW, Shen Y (2019) High-energy-density ferroelectric polymer nanocomposites for capacitive energy storage: enhanced breakdown strength and improved discharge efficiency. Mater Today 29:49–67
32. Marwat MA, Xie B, Zhu YW, Fan PY, Ma WG, Liu HM, Ashtar M, Xiao JZ, Salamon D, Samart C, Zhang HB (2019) Largely enhanced discharge energy density in linear polymer nanocomposites by designing a sandwich structure. Compos Part A: Appl S 121:115–122
33. Lu X, Zou XW, Shen JL, Zhang L, Jin L, Cheng ZY (2020) High energy density with ultrahigh discharging efficiency obtained in ceramic-polymer nanocomposites using a non-ferroelectric polar polymer as matrix. Nano Energy 70:104551
34. Zhou Y, Wang Q (2020) Advanced polymer dielectrics for high temperature capacitive energy storage. J Appl Phys 127:240902
35. Cheng S, Zhou Y, Hu J, He J, Li Q (2020) Polyimide films coated by magnetron sputtered boron nitride for high-temperature capacitor dielectrics. IEEE T Dielect El In 27:498–503
36. Li Q, Han K, Gadinski MR, Zhang G, Wang Q (2014) High energy and power density capacitors from solution-processed ternary ferroelectric polymer nanocomposites. Adv Mater 26:6244–6249
37. Tang H, Sodano HA (2013) Ultra high energy density nanocomposite capacitors with fast discharge using $Ba_{0.2}Sr_{0.8}TiO_3$ nanowires. Nano Lett 13:1373–1379
38. Huang X, Jiang P (2015) Core-shell structured high-k polymer nanocomposites for energy storage and dielectric applications. Adv Mater 27:546–554
39. Hu P, Shen Y, Guan YH, Zhang XH, Lin YH, Zhang QM, Nan CW (2014) Topological-structure modulated polymer nanocomposites exhibiting highly enhanced dielectric strength energy density. Adv Funct Mater 24:3172–3178
40. Zhang G, Li Q, Allahyarov E, Li Y, Zhu L (2021) Challenges and opportunities of polymer nanodielectrics for capacitive energy storage. ACS Appl Mater Interfaces 13:37939–37960
41. JiangY ZhouM, ShenZ ZhangX, PanH LinYH (2021) Ferroelectric polymers and their nanocomposites for dielectric energy storage applications. APL Mater 9:020905
42. Shanmugasundram HPPV, Jayamani E, Soon KH (2022) A comprehensive review on dielectric composites: classification of dielectric composites. Renew Sustain Energy Rev 157:112075
43. StarkKH GartonCG (1955) Electric strength of irradiated polythene. Nature 176:1225–2126
44. Hu J, Zhang S, Tang B (2021) 2D filler-reinforced polymer nanocomposite dielectrics for high-k dielectric and energy storage applications. Energy Storage Mater 34:260–281
45. Xiong X, Zhang Q, Zhang Z, Yang H, Tong J, Wen J (2021) Superior energy storage performance of PVDF-based composites induced by a novel nanotube structural BST@SiO_2 filler. Composites: Part A 145:106375

46. Jian G, Jiao Y, Meng Q, Wei Z, Zhang J, Yan C, Moon K, Wong C (2020) Enhanced dielectric constant and energy density in a $BaTiO_3$/polymer-matrix composite sponge. Commun Mater 1:1–12
47. Ren L, Yang L, Zhang S, Li H, Zhou Y, Ai D, Xie Z, Zhao X, Peng Z, Liao R, Wang Q (2021) Largely enhanced dielectric properties of polymer composites with HfO_2 nanoparticles for high-temperature film capacitors. Compos Sci Technol 201:108528
48. Dang Z, Yuan J, Zha J, Zhou T, Li S, Hu G (2012) Fundamentals, processes and applications of high-permittivity polymer–matrix composites. Prog Mater Sci 57:660–723
49. Yang W, Yu S, Sun R, Du R (2011) Nano- and microsize effect of CCTO fillers on the dielectric behavior of CCTO/PVDF composites. Acta Mat 59:5593–5602
50. Nelsont SO, You T-S (1990) Relationships between microwave permittivities of solid and pulverized Plastics. J Phys D: Appl Phys 23:346
51. Thomas S, Deepu VN, Mohanan P, Sebastian MT (2008) Effect of filler content on the dielectric properties of PTFE/$ZnAl_2O_4$–TiO_2 composites. J Am Ceram Soc 91:1971–1975
52. Dash S, Choudhary RNP, Kumar A, Goswami MN (2019) Enhanced dielectric properties and theoretical modeling of PVDF–ceramic composites. J Mater Sci Mater Electron 30:19309–19318
53. Hossain ME, Liu SY, O'Brien S, Li J (2014) Modeling of high-k dielectric nanocomposites. Acta Mech 225:1197–1209
54. Zhang C, Chi Q, Dong J, Cui Y, Wang X, Liu L, Lei Q (2016) Enhanced dielectric properties of poly(vinylidene fluoride) composites filled with nano iron oxide-deposited barium titanate hybrid particles. Sci Rep 6:1–9
55. Yamada T, Ueda T, Kitayama T (1982) Piezoelectricity of a high-content lead zirconate titanate polymer composite. J Appl Phys 53:4328–4332
56. Yu K, Niu Y, Zhou Y, Bai Y, Wang H (2013) Nanocomposites of surface-modified $BaTiO_3$ nanoparticles filled ferroelectric polymer with enhanced energy density. J Am Ceram Soc 96:2519–2524
57. JayasundereN SV (1993) Dielectric constant for binary piezoelectric 0–3 composites. J Appl Phys 73:2462–2466
58. Hu H, Zhang F, Luo S, Chang W, Yue J, Wang CH (2020) Recent advances in rational design of polymer nanocomposite dielectrics for energy storage. Nano Energy 74:104844
59. Chen C, Xie Y, Liu J, Li J, Wei X, Zhang Z (2020) Enhanced energy storage capability of P(VDF-HFP) nanodielectrics by HfO_2 passivation layer: preparation, performance and simulation. Compos Sci Technol 188:107968
60. Tabhane GH, Giripunje SM, Kondawar SB (2021) Fabrication and dielectric performance of RGO-PANI reinforced PVDF/$BaTiO_3$ composite for energy harvesting. Synth Met 279:116845
61. Honga W, Pitike KC (2015) Modeling breakdown-resistant composite dielectrics. Procedia IUTAM 12:73–82
62. Sen ZH, Wang JJ, Lin Y, Nan CW, Chen LQ, Shen Y (2017) High-throughput phase-field design of high-energy-density polymer nanocomposites. Adv Mater 30:1704380
63. Niemeyer LP, Wiesmann HJ (1984) Fractal dimension of dielectric breakdown. Phys Rev Lett 52:1033
64. Yue D, Feng Y, Liu XX, Yin J-H, Zhang W-C, Guo H, Su B, Lei Q-Q (2022) Prediction of energy storage performance in polymer composites using high-throughput stochastic breakdown simulation and machine learning. Adv Sci 9:2105773
65. Wanga Z, Nelson JK, Hillborg H, Zhao S, Schadler LS (2013) Dielectric constant and breakdown strength of polymer composites with high aspect ratio fillers studied by finite element models. Compos Sci Technol 76:29–36
66. BaiH GeG, HeX ShenB, ZhaiJ PanH (2020) Ultrahigh breakdown strength and energy density of polymer nanocomposite containing surface insulated BCZT@BN nanofibers. Compos Sci Technol 195:108209
67. Zhang T, Sun Q, Kang F, Wang Z, Xue R, Wang J, Zhang L (2022) Sandwich-structured polymer dielectric composite films for improving breakdown strength and energy density at high temperature. Compos Sci Technol 227:109596

68. Xu H, Xie C, Gou B, Wang R, Zhou J, Li L (2022) Core-double-shell structured BT@TiO_2@PDA and oriented BNNSs doped epoxy nanocomposites with field-dependent nonlinear electrical properties and enhancing breakdown strength. Compos Sci Technol 230:109777
69. Huang X, Sun B, Zhua Y, Lib S, Jiang P (2019) High-k polymer nanocomposites with 1D filler for dielectric and energy storage applications. Prog Mate Sci 100:187–225
70. Rogti F, Ferhat M (2014) Maxwell Wagner polarization and interfacial charge at the multilayers of thermoplastic polymers. J Electrostat 72:91–97
71. Danikas MG (2010) On two nanocomposite models: differences, similarities and interpretational possibilities regarding Tsagaropoulos' model and Tanaka's model. J Electr Eng 61:241
72. Lewis TJ (2005) Interfaces: nanometric dielectrics. J Phys D: Appl Phys 38:202
73. Prateek TVK, Gupta RK (2016) Recent progress on ferroelectric polymer-based nanocomposites for high energy density capacitors: synthesis, dielectric properties, and future aspects. Chem Rev 116:4260–4317
74. Tiwari V, Srivastava G (2015) Structural, dielectric and piezoelectric properties of 0–3 PZT/PVDF composites. Ceram Int 41:8008–8013
75. Thomas P, Varughese KT, Dwarakanath K, Varma KBR (2010) Dielectric properties of Poly(vinylidene fluoride)/$CaCu_3Ti_4O_{12}$ composites. Compos Sci Technol 70:539–545
76. Fu J, Hou Y, Zheng M, Wei Q, Zhu M, Yan H (2015) Improving dielectric properties of PVDF composites by employing surface modified strong polarized $BaTiO_3$ particles derived by molten salt method. Appl Mater Interfaces 44:24480–24491
77. Kum-onsa P, Thongbai P (2020) Improved dielectric properties of poly(vinylidene fluoride) composites incorporating $Na_{1/2}Y_{1/2}Cu_3Ti_4O_{12}$ particles. Mater Today Commun 25:101654
78. Santos IA, Rosso JM, Cotica LF, Bonadio TGM, Freitas VF, Guo R, Bhalla AS (2016) Dielectric and structural features of the environmentally friendly leadfree PVDF/$Ba_{0.3}Na_{0.7}Ti_{0.3}Nb_{0.7}O_3$ 0–3 composite. Curr Appl Phys 16:1468–1472
79. Behera C, Choudhary RNP, Das PR (2017) Development of multiferroic polymer nanocomposite from PVDF and $(Bi_{0.5}Ba_{0.25}Sr_{0.25})(Fe_{0.5}Ti_{0.5})O_3$. J Mater Sci: Mater Electron 28:2586–2597
80. Ji SH, Cho JH, Jeong YH, Paik JH, Yun JD, Yun JS (2016) Flexible lead-free piezoelectric nanofiber composites based on BNT-ST and PVDF for frequency sensor applications. Sensors and Actuators A 247:316–322
81. Xie B, Wang T, Cai J, Zheng Q, Liu Z, Guo K, Mao P, Zhang H, Jiang S (2022) High energy density of ferroelectric polymer nanocomposites utilizing PZT@SiO_2nanocubes with morphotropic phase boundary. J Chem Eng 434:134659
82. Xie Z, Liu D, Xiao Y, Wang K, Zhang Q, Wu K, Fu Q (2022) The effect of filler permittivity on the dielectric properties of polymer-based composites. Compos Sci Technol 222:109342
83. Gonçalves R, Martins PM, Caparrós C, Martins P, Benelmekki M, Botelho G, Lanceros-Méndez S, Lasheras A, Gutierrez J, Barandiarán JM (2013) Nucleation of the electroactive β-phase, dielectric and magnetic response of poly(vinylidene fluoride) composites with Fe_2O_3 nanoparticles. J Non-Cryst Solids 361:93–99
84. Amoresi RAC, Felix AA, Botero ER, Domingues NLC, Falcão EA, Zaghet MA, Rinaldi AW (2015) Crystallinity, morphology and high dielectric permittivity of NiOnanosheets filling Poly(vinylidene fluoride). Ceram Int 41:14733–14739
85. XuHP DangZM (2007) Electrical property and microstructure analysis of poly(vinylidene fluoride)-based composites with different conducting fillers. Chem Phy Lett 438:196–202
86. Xu HP, XieHQ, Yang DD, Wu YH, Wang JR (2011) Novel dielectric behaviors in PVDF-based semiconductor composites. J Appl Polym Sci 122:3466–3473
87. Dang ZM, Wang L, Yin Y, Zhang Q, Lei QQ (2007) Giant dielectric permittivities in functionalized carbon-nanotube/electroactive-polymer nanocomposites. Adv Mater 19:852–857
88. Wang L, Dang ZM (2005) Carbon nanotube composites with high dielectric constant at low percolation threshold. Appl Phys Lett 87:042903

89. Yousefi LX, Zheng Q, Shen X, Pothnis JR, Jia J, Zussman E, Kim JK (2013) Simultaneous in situ reduction, self-alignment and covalent bonding in graphene oxide/epoxy composites Nariman. Carbon 59:406–417
90. Stankovich S, Dikin DA, Dommett GHB, Kohlhaas KM, Zimney EJ, Stach EA, Piner RD, Nguyen ST, Ruoff RS (2006) Graphene-based composite materials. Nature 442:282–286
91. Yousefi N, Gudarzi MM, Zheng Q, Aboutalebi SH, Sharif F, Kim JK (2012) Self-alignment and high electrical conductivity of ultralarge graphene oxide–polyurethane nanocomposites. J Mater Chem 22:12709–12717
92. QiL LI, ChenS SD, ExarhosG J (2005) High-dielectric-constant silver-epoxy composites as embedded dielectrics. Adv Mat 17:1777–1781
93. Yousefi N, Sun X, Lin X, Shen X, Jia J, Zhang B, Tang B, Chan M, Kim JK (2014) Highly aligned graphene/polymer nanocomposites with excellent dielectric properties for high-performance electromagnetic interference shielding. Adv Mater 26:5480–5487
94. Guo D, Cai K, Deng P, Si G, Sun L, Chen F, Ning H, Jin L, Ma J (2020) Structure tailorable triple-phase and pure double-polar-phase flexible IF-WS_2@poly(vinylidene fluoride) nanocomposites with enhanced electrical and mechanical properties. J Mater 6:563–572
95. Jia Q, Huang X, Wang G, Diao J, Jiang P (2016) MoS_2 nanosheet superstructures based polymer composites for high-dielectric and electrical energy storage applications. J Phys Chem C 120:10206–10214
96. Wu W, Huang X, Li S, Jiang P, Toshikatsu T (2012) Novel three-dimensional zinc oxide superstructures for high dielectric constant polymer composites capable of withstanding high electric field. J Phys Chem C 116:24887–24895
97. Ji W, Deng H, Sun C, Fu Q (2019) Nickel hydroxide as novel filler for high energy density dielectric polymer composites. Compos Sci Technol 172:117–124
98. Wen F, Zhu C, Li L, Zhou B, Zhang L, Han C, Li W, Yue Z, Wu W, Wang G, Zhang S (2022) Enhanced energy storage performance of polymer nanocomposites using hybrid 2D ZnO@MoS_2 semiconductive nano-fillers. J Chem Eng 430:132676
99. Jing L, Li W, Gao C, Li M, Fei W (2022) Enabling high energy storage performance in PVDF-based nanocomposites filled with high-entropy oxide nanofiber. Compos Sci Technol 230:109783
100. Ji W, Deng H, Fu Q (2017) Heterogeneous filler distribution in polymeric capacitor films: an efficient route to improve their dielectric properties. Compos Sci Technol 151:131–138
101. Zeng J, Yan J, Li BW, Zhang X (2022) Improved breakdown strength and energy storage performances of PEI-based nanocomposite with core-shell structured PI@$BaTiO_3$ nanofillers. Ceram Int 48:20526
102. Hu J, Liu Y, Zhang S, Tang B (2022) Novel designed core–shell nanofibers constituted by single element-doped $BaTiO_3$ for high-energy–density polymer nanocomposites. J Chem Eng 428:131046
103. Li H, Wang L, Zhu Y, Jiang P, Huang X (2021) Tailoring the polarity of polymer shell on $BaTiO_3$ nanoparticle surface for improved energy storage performance of dielectric polymer nanocomposites. Chin Chem Lett 32:2229–2232
104. Pan X, Wang M, Qi X, Zhang N, Huang T, Yang J, Wang Y (2020) Fabrication of sandwich-structured PPy/MoS_2/PPynanosheets for polymer composites with high dielectric constant, low loss and high breakdown strength. Compos A 137:106032
105. Li Z, Liu F, Li H, Ren L, Dong L, Xiong C, Wang Q (2019) Largely enhanced energy storage performance of sandwich-structured polymer nanocomposites with synergistic inorganic nanowires. Ceram Int 45:8216–8221
106. Luo W, Xu L, Zhang G, Zhou L, Li H (2021) Sandwich-structured polymer nanocomposites with $Ba_{0.6}Sr_{0.4}TiO_3$ nanofibers networks as mediate layer inducing enhanced energy storage density. Compos Sci Technol 204:108628
107. Wang Y, Hou Y, Deng Y (2017) Effects of interfaces between adjacent layers on breakdown strength and energy density in sandwich-structured polymer composites. Compos Sci Technol 145:71–77

Chapter 5
Recent Progress in Nanodielectric Composites and Their Applications

Joshi Harsh Nitinkumar, Navyasree Reghu, P. K. Akhilesh, Alexandru Vlad, Meera Balachandran, and Prasanth Raghavan

Abstract With the tremendous growth in electrical gadgets and electrification in recent years, there is an urgent need to utilise green/renewable energy and improve energy efficiency, energy conversion systems, and energy storage density. In the future, development in electrical power technologies will be aligned in the direction of higher power and frequency, wider bandwidth, faster speed, smaller form factor, higher temperature rating, lower cost, greater flexibility, and scalability. In this context, polymer nanocomposite dielectrics, also known as nanodielectrics, have garnered considerable scientific interest. Polymer nanodielectric combines the advantages of high dielectric constant nanofillers with high dielectric breakdown strength polymer matrix and offers the flexibility of engineering materials and structures to obtain desired and unique properties. This chapter discusses current developments in nanodielectric composites for energy storage applications with an emphasis on fundamental material concerns for two different types of polymer nanodielectrics: composites of polymer and ceramic and conductive nanoparticles. The role of tuneable nonlinear nanodielectrics in energy storage and harvesting is also discussed. The chapter highlights the applications, opportunities, and challenges associated with future developments in nanodielectrics.

J. H. Nitinkumar · N. Reghu · M. Balachandran (✉)
Department of Chemical Engineering and Materials Science, Amrita School of Engineering, Amrita Vishwa Vidyapeetham, Coimbatore 641112, India
e-mail: b_meera@cb.amrita.edu

Centre of Excellence in Advanced Materials and Green Technologies (CoE-AMGT), Amrita School of Engineering, Amrita Vishwa Vidyapeetham, Coimbatore 641112, India

P. K. Akhilesh · A. Vlad
Institute of Condensed Matter and Nanosciences, Universitécatholique de Louvain, 1348 Ottignies-Louvain-La-Neuve, Belgium

P. Raghavan
Materials Science and NanoEngineering Lab (MSNE-Lab), Department of Polymer Science and Rubber Technology, Cochin University of Science and Technology (CUSAT), Cochin 682022, India

Department of Materials Engineering and Convergence Technology, Gyeongsang National University, 501 Jinju-Daero, Jinju 52828, Republic of Korea

S. Moharana et al. (eds.), *Emerging Nanodielectric Materials for Energy Storage*, Nanostructure Science and Technology, https://doi.org/10.1007/978-3-031-40938-7_5

Keywords Polymer nanodielectrics · Dielectric properties · Ceramic nanoparticles · Conducting nanoparticles · Applications

5.1 Polymer Nanodielectrics: An Introduction

Polymer nanodielectric composites comprise a polymer matrix and a nanodimensional filler that has at least one dimension in the range of less than 100 nm. Polymer nanodielectrics combine the high breakdown strength and low loss of the polymer matrix with favourable dielectric properties like high dielectric constant of the nanofillers [1]. Due to their nanometric size and high specific surface area of nanoparticles that create large interfacial interactions, they exhibit exceptional properties compared to bulk materials [2]. A schematic representation of the polymer nanodielectric composite and microstructural appearance—correlation between a number of particles and size of particles and the nanoparticle–polymer interface is provided in Fig. 5.1. There are opportunities to tailor nanodielectrics to have a unique combination of properties by careful selection of materials along with engineering the dispersion and interface between the constituent materials. The interfacial region becomes significant as particle size reduces and constitutes a considerable fraction of the polymer dielectric, even at low concentration of the nanoparticle. Hence, it is important to design and optimise the dispersion of nanoparticle in the polymer matrix and the interfacial region to achieve desirable properties. There are various methods reported to prepare nanodielectrics. The popular preparation methods of nanodielectrics composites are depicted below in Fig. 5.2a.

The intrinsic interfacial region can be engineered by changes in matrix structure, mobility, and crosslinking of polymer chains while the extrinsic interfacial regions can be modified by ligands on nanoparticle surfaces resulting from functionalization [3]. Due to these features, nanodielectrics have immense potential in various fields ranging from high-voltage insulation to energy storage devices and sensors.

There are several properties of interest for polymer dielectrics viz. dielectric permittivity, dielectric breakdown strength, dielectric constant, energy density, partial discharge resistance, treeing lifetime, space charge density, tracking resistance, and conductivity [4, 5]. The major factors affecting the performance of nanodielectrics are presented in Fig. 5.2b. Over the past three decades, studies have been conducted on nanodielectrics with different polymers as matrices. They include epoxy resins, polyolefins, polyesters, polyimides, polyimide-amides, polyester-imide, and elastomers. Several nanoparticles are used in polymer nanodielectrics like layered silicates, silicon dioxide, titanium dioxide, aluminium oxide, silicon carbide, barium titanate, magnesium dioxide, zinc oxide, zirconium dioxide, manganese sulphoselenide, etc. [4–10]. The changes in the properties of nanodielectrics upon incorporation of nanofillers vary depending on the nanofiller properties, dispersion, and polymer–filler interaction. In general, it has been observed that the breakdown strength, tracking resistance, partial discharge resistance, tree inception, and growth time are enhanced while space charge accumulation is decreased. However, the effect

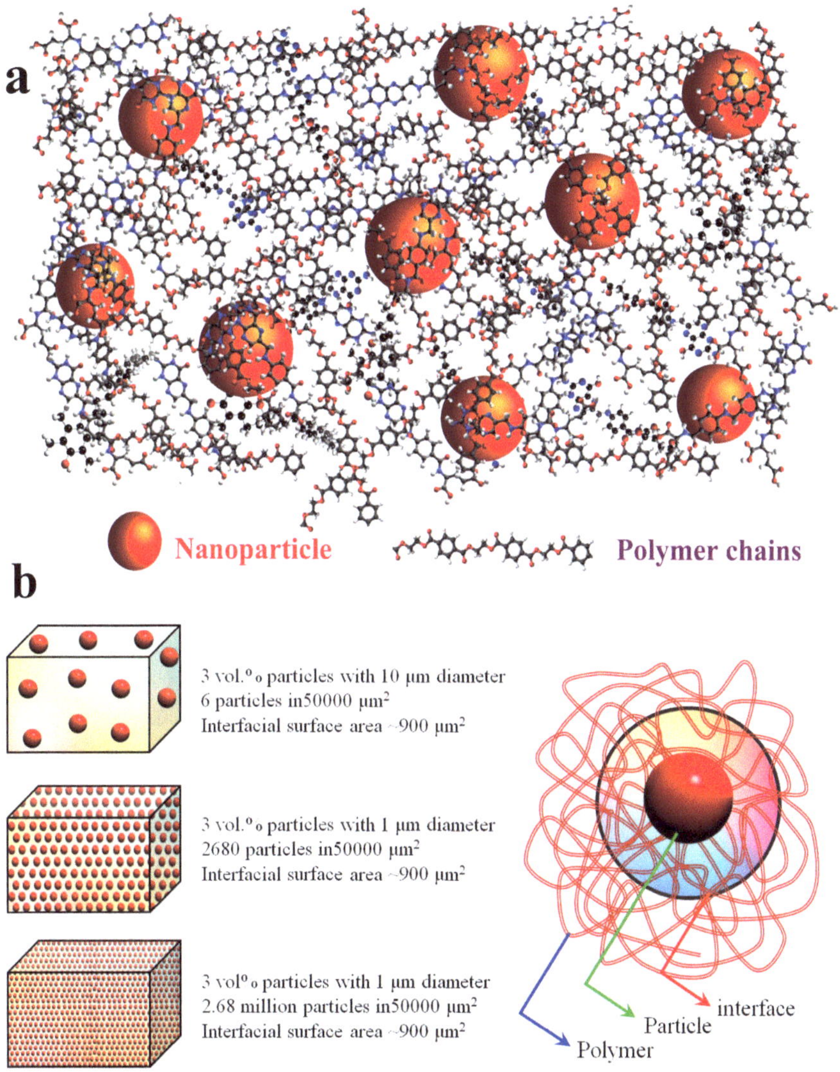

Fig. 5.1 **a** Schematic representation of polymer nanodielectric composite, **b** microstructural appearance—correlation between number of particles and size of particles and the nanoparticle—polymer interface

on properties like permittivity, dielectric loss, and DC conductivity are complicated and may increase or decrease depending on several factors [5]. This chapter examines the current developments in nanodielectric composites for energy storage applications with an emphasis on fundamental material concerns for two different types of polymer nanodielectrics: composites of polymer and ceramic and conductive

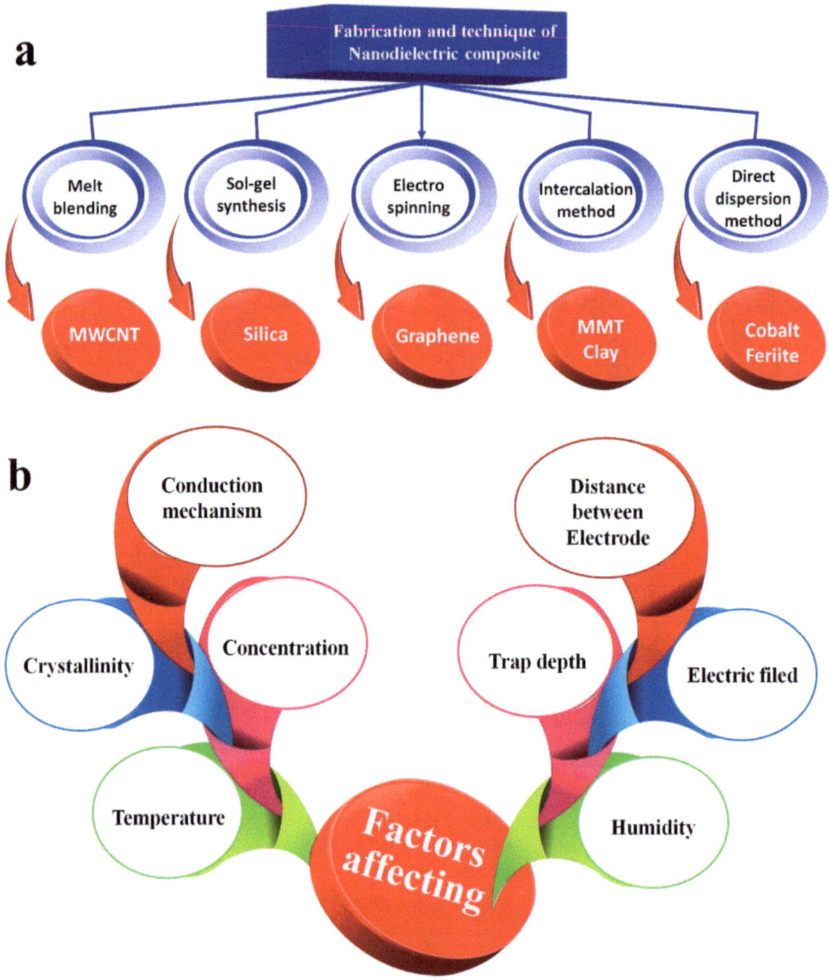

Fig. 5.2 **a** Various fabrication methods for synthesising nanodielectrics and **b** the major factors affecting the performance of nanodielectrics

nanoparticles. The applications, opportunities, and challenges associated with the future developments in nanodielectrics are also discussed.

5.2 Polymer Nanodielectrics Based on Ceramic Nanoparticles

A synergy between high breakdown strength of polymers and high permittivity of ceramic materials can be achieved by choosing specific ceramic materials having architecture surface morphology and particle size to be incorporated into the polymer matrix. Sintered bulk ceramics like ferroelectric $BaTiO_3$ possess relative permittivity (ε_r) higher than 1000 in value. By reducing the internal strain energy and electrostatic energy of jammed crystalline grains, bulk $BaTiO_3$ with polygrain has been shown to have ferroelectric domains of 180° orientation and domains of 90° for tetragonal phase [11].

In a study by Paniagua et al., phosphonic acid was adopted as a surface initiator for the formation of polystyrene and polymethylmethacrylate (PMMA) over the barium titanate (BTO) nanoparticles via atom transfer radical polymerization. This method of preparation resulted in $BaTiO_3$ as a core embedded in the grafted polymer as shown in Fig. 5.3a. The surface morphology of the resulting polymer nanodielectric composites is displayed in Fig. 5.3b–e. The performance evaluation at a field strength of ~220 V/μm showed that the one-component system, (PMMA-grafted-BTO) exhibited a maximum extractable energy density (2 J/cm^3), which was two times higher than that of the composite without covalent attachment or the pristine polymer. Due to the high permittivity, high breakdown strength, mechanical flexibility of the nanoparticles, and ease of processability due to the organic polymer, such materials have potential applications in hybrid capacitors [12].

The ability of a substance to store electrical energy in an electrical field is measured by a number called the dielectric constant, which is the relative permittivity of a dielectric material. Due to the absence of grain boundaries in ceramics, it is possible to enhance the dielectric characteristics of polymers by using carefully chosen nanoceramic fillers. The relationship between breakdown strength and dielectric permittivity of various dielectric materials is displayed in Fig. 5.4.

Higher ε_r values are caused by ionic polarisation between ferroelectric domains and movable domain walls in ferroelectric ceramics[2]. Because of the large domains, which lessen the number of domain walls, ε_r is not particularly high for bulk $BaTiO_3$ when grain size is above 1 μm. The greatest value of ε_r was reported to be ~5000 with a large number of relatively tiny domains, and when ε_r value continued to decrease, such as by 30 nm, it was reported to be ~1000 [11, 13, 14].

For $BaTiO_3$ particles that are nanosized, the ε_r value depends on both the synthesis methods used to create the nano-$BaTiO_3$ particles as well as the ferroelectric and paraelectric phases. Nanoparticles typically do not undergo the effects of sintering at high temperature, but they do exhibit a variety of crystallographic flaws, e.g. $BaTiO_3$ nanoparticles in the ferroelectric-tetragonal phase have a greater ε_r value compared to those in the paraelectric-cubic phase. Paraelectric $BaTiO_3$ nanoparticles with a size of less than 200 nm are typically produced using low temperature hydrothermal and sol–gel methods, whereas ferroelectric $BaTiO_3$ nanoparticles larger than 200 nm are typically produced using high-temperature pyrolysis and combustion methods. Since

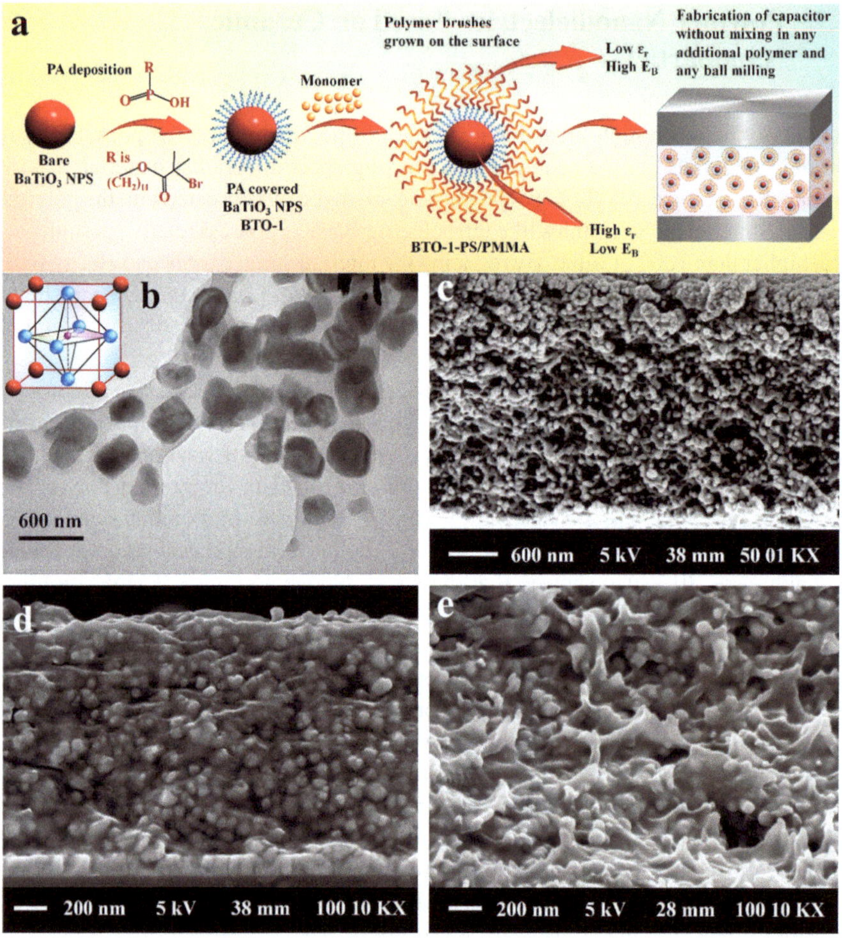

Fig. 5.3 **a** Scheme for the creation of a one-component, three-constituent hybrid dielectric that can be used to build capacitors. **b** TEM micrograph of BTO-1-PS composite [29% (v/v) BTO] (inset: crystal structure of $BaTiO_3$) and **c** cross section SEM image for composite 5 [19% (v/v) BTO], **d** cross-sectional SEM for composites BTO-1-PMMA and **e** BTO-1/cPMMA22 (both 22% (v/v) in BTO) [12]. Adapted with permission from ref. [12] Copyright (2014) (American Chemical Society)

it is practically impossible to quantify the ε_r value for such tiny particles directly, various mixing procedures have been used to separate nanoparticles from polymer composites. It is usually claimed that the Bruggeman mixing model, which takes into consideration particle–particle dipole interactions, provides a more accurate estimation of ε_r for $BaTiO_3$ nanoparticles than any other model [15]. The ε_r value was discovered to drop from 800 to 200 for particle sizes greater than 10 μm, and on further reduction in size, the ε_r value was reported to be ~50 for 5–20 nm $BaTiO_3$ nanoparticles [16–18].

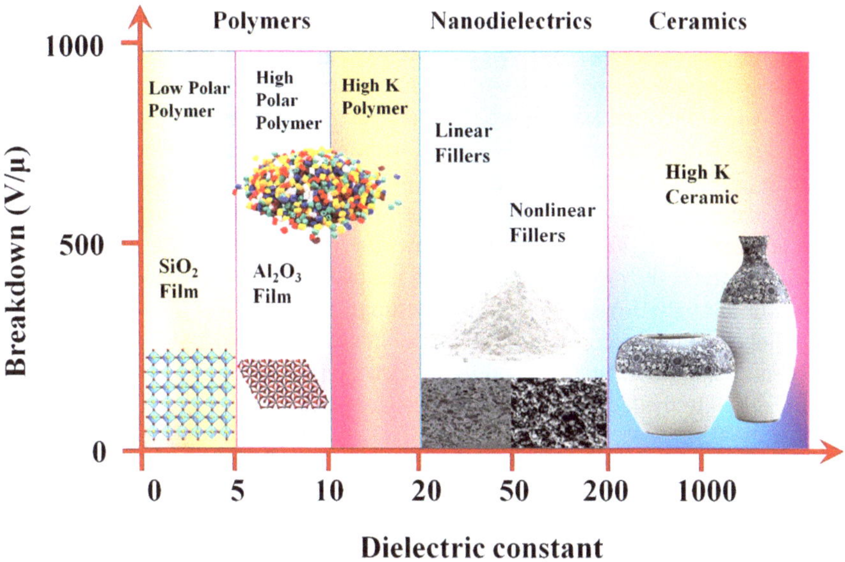

Fig. 5.4 Relationship between breakdown strength and dielectric permittivity (dielectric constant) of various dielectric materials

The crystalline grain and domain structure has a significant impact on the ε_r of perovskite nanoparticles. Multiple crystal grains are typically found in one phase of $BaTiO_3$ nanoparticles larger than 1–10 μm in size, and they typically exhibit the ferroelectric/tetragonal phase, which has 180° and 90° domains at room temperature. They have an extremely high ε_r value because 90° domain walls are present. Only single crystal grains were seen as particle size decreased from 0.2 to 1 μm because there was no particle jamming, and therefore, no need for them to minimise internal strain energy. The only 180° domain walls that were present in this kind of single crystal grain structure nullify the impact of the dipoles and cause the ε_r to decrease further [19–21].

To achieve uniform dispersion of nanoparticles without blending any homopolymers, polymethyl methacrylate (PMMA) was grafted with 70 nm $BaTiO_3$ nanoparticles in one experiment. In this experiment, it was found that ε_r increased with the $BaTiO_3$ volume fraction [22]. The maximal ε_r in the PVDF/BTO nanocomposite was discovered to be up to 60 vol%, which is three times greater than PVDF in value. ε_r began to decline above 60 vol% because of the spin-coated composite film's porosity. The results of the research conclude that compared to polymer matrices, addition of ceramic nanofillers into it enhances the ε_r by 3–5 times before percolation threshold is reached [23].

The 1D nanofillers, like nanowires, nanofibers, and nanotubes with higher aspect ratio (AR) were also tested for their effects when added to spherical ceramic nanofillers. It was found that these fillers boost at lower filler contents inside the ceramic nanofillers [24, 25]. It was discovered that filling the $BaTiO_3$ with larger aspect ratio nanowires increased the ε_r for PVDF at higher filler loadings [26].

The polar coupling contact is stronger in the longitudinal direction compared to the transverse direction (i.e., along the diameter) in high aspect ratio (AR) fillers, and as a result, ε_r is higher in the longitudinal direction compared to the transverse direction [26]. Because movable domain walls are present in ferroelectric/tetragonal $BaTiO_3$ nanotubes and nanowires, it has been found that their ε_r is higher than those of paraelectric/cubic nanotubes and nanowires. This suggests that the crystalline phase also affects the ε_rs of nanotubes and nanowires [27]. High aspect ratio fillers have a substantially lower percolation threshold than spherical nanoparticles [28]. A parallel capacitor circuit is used to model the overall capacitance when the high aspect ratio nanofillers are oriented to the electric field parallelly, whereas a series circuit is used to model when the nanofillers are perpendicular to electric field.

$$C = C_f + C_m$$

$$\frac{1}{C} = \frac{1}{C_f} + \frac{1}{C_m}$$

where C_f is capacitance for filler and C_m is capacitance for matrix

The parallel capacitor model gives significantly greater ε_r, i.e., when the fillers in the composites are oriented parallelly, it gives a greater ε_r compare to fillers oriented perpendicular to the applied electric field [29]. For ceramic filler, the ε_r's value is only slightly higher than base polymer when the percolation threshold value is not yet reached. However, above this percolation threshold value, the polymer/ceramic interface acts as a conductive pathway, significantly reducing E_b and η, making it only fit for low-voltage applications. Additionally, polymer nanocomposite materials with ceramic filler loadings greater than 10 vol% are frequently brittle and susceptible to mechanical failure, making them unsuitable for use in film capacitors.

5.2.1 Dielectric Loss

Low dielectric loss is crucial for any practical thin film capacitor application, and the equivalent series resistance provides the total of these various dielectric losses (ESR). As the ripple grows and the ESR rises, wound capacitors heat up significantly, shortening their lifetime [30]. For the deployment of novel polymer dielectrics for thin-film capacitance applications, a low dielectric loss achievement is more crucial than a high U_e since at higher temperatures ripple current heating is unfavourable.

Dielectric losses in polymer nanodielectrics are not only caused by the polymer matrix and ceramic fillers but also due to their interactions. Sources of dielectric losses depend on several polarisation mechanisms [31], including:

a. Dielectric loss owing to resonant (due to electronic and atomic polarization) and relaxation (due to orientational polarization);
b. Space charge conduction (due to ions and electrons);
c. Extrinsic and intrinsic ferroelectric switching of aligned electric domains, the former caused due to impurity ion losses and thermally activated electrons, the latter being a feature of dielectric itself.

Extrinsic dielectric losses in ceramic fillers are caused by lattice vibration in domains, which also causes ionic polarization, and intrinsic dielectric losses in domain walls caused by dipole vibration. Because of the high dipole mobility in the walls, relaxation loss for perovskite crystals occurs around about 1 GHz[13]. Lattice ionic polarisation occurs in ferroelectric domains at 20 GHz frequency [32] that is higher than the power frequency (1 Hz to 1 MHz). Ceramic fillers, therefore, exhibit relatively low intrinsic loss at the power frequency. When it comes to nanoceramic fillers, the leakage current is not significant until the percolation threshold is reached, but above this value, the significance of the leakage current because of the percolated polymer–ceramic particles interfaces under high electric field is enhanced. On the other hand, in case of bulk ceramics, the leakage current due to conduction of electrons originates from grain boundaries causing an extrinsic dielectric loss. The intrinsic and resonant losses occur in the optical (>10^{14} Hz) and infrared (10^{12}–10^{14} Hz) ranges, respectively. Since the power frequency range is within 10^2–10^6 Hz, little resonant loss is always seen. Since polar polymers inherently have larger relaxation dielectric loss than nonpolar polymers, the relaxation loss is placed right inside the power frequency range. When used for high-field poling, ferroelectric polymers like PVDF and its copolymers cause severe hysteresis or ferroelectric loss, which is why they are never used for electric energy storage [33, 34]. Both ionic and electronic currents are large in polymers above T_g/T_m temperature. The extrinsic dielectric loss due to impunity ion contamination and thermally activated electrons make them unfavourable candidate for applications of high-voltage capacitors. Compared to a liquid amorphous matrix, polymer crystals are better at preventing the conduction of electrons and ions [35]. Polar polymers typically have impurity ion concentration below trace levels but can be still contaminated at higher temperatures [36, 37].

The E_b for polymer/ceramic particle nanodielectrics is greatly reduced because of the strong interaction between high-κ (dielectric constant) ceramic fillers and polymer matrix. At lower frequencies, an impurity ion conduction loss in the polypropylene matrix of the PP/BTO@POSS nanocomposite was detected, which increased with temperature. The polarisation of thermally activated electrons in BTO nanoparticles is responsible for the rather broad and feeble peak at 1 kHz frequency that was seen. Remarkable hysteresis loss was noted for the PP/BTO@POSS at 20% nanocomposites with input electric fields greater than 50 MV/m. At the same electric field, the above-mentioned nanocomposite displays a higher U_e value than the pristine polypropylene, but the η considerably dropped as a result of the widened hysteresis loop. The conductivities and ε_rs of the PP matrix and the BTO fillers differ significantly. BTO has a high thermal conductivity because of its comparatively high ε_r, which results in an unusually large concentration of thermally activated electrons.

The highly insulating PP matrix blocks these thermally activated electrons when an electric field is applied, creating interfacial polarisation, because of which, the interfacial charges within the BTO nanoparticle may move back and forth resulting in a great loss of conduction. The contact between BTO and PP and bulk BTO are the two potential conduction paths. Out of these two, the primary conduction pathway is the PP/BTO interface [38]. A thin layer of linear ceramic dielectrics is typically employed such as amorphous SiO_2, Al_2O_3, and TiO_3 using techniques like layer-by-layer coating of methyl aluminoxane, sol–gel preparation, or even atomic layer deposition to further reduce the interfacial conduction. Although the amorphous oxide surface coating's highly insulating qualities allowed for a reduction in interface conduction, hysteresis loss persisted.

5.2.2 Dielectric Breakdown Strength

One of the major benefits for films of polymer is their higher dielectric breakdown strength (E_b), and BOPP currently has the highest breakdown strength of all commercially available films (E_b at 63.2% failure probability of ~ 730 MV/m for 2cm^2 sample area) [39]. The E_b value results from the following variables:

a. PP resins that are ultraclean;
b. minimal leakage current;
c. consistent film thickness.

When working at ~ 200 MV/m and 85 °C, this high E_b guarantees a continuous lifetime of > 100,000 h. The energy bandgap (E_g) of the material strongly influences the intrinsic E_b (E_b is proportional to $E_g{}^{\alpha}$) [40, 41] with $\alpha = 3$ for semiconductors and $\alpha = 1$ for insulators. Due to their wide energy band gap and slow carrier movement, polymers have a high E_b value [42, 43]. The highest observed value of E_b for most insulating polymers is only 750 MV/m, which is due to intrinsic/extrinsic flaws and degradation (such as broken bonds) under high-field stresses [44]. Theoretically, the intrinsic E_b value for PP is larger than 2 GV/m.

The four primary processes for breakdown in any polymer thin film are electronic breakdown, thermal breakdown, electromechanical breakdown, and partial discharges [44]. Low temperatures are necessary for electronic breakdown, and a strong electric field is needed to introduce hot metal electrons into the polymer dielectric [31]. Hot electrons that are injected into a polymer will break chemical bonds to produce Auger electrons, which then disrupt neighbouring molecular bonds [45]. Thermal breakdown in polymer dielectrics occurs via joule heating and typically takes place at higher temperatures with a smaller electric field. At the locations of flaws and impurities in polymers, both breakdown mechanisms cause gas-phase production. Precleaning is used in the film capacitor manufacturing process to minimise these flaws. Distribution of electric field in nonuniform way becomes another crucial element in multicomponent systems because of the significant ε_r and conductivity of electrons differences between ceramic fillers and polymer as matrix. Interfacial polarisation was seen in this type of system along the E_o direction at the two poles of a single particle. Because of the compatibility of the POSS ligand and the polypropylene matrix, the high dielectric constant nanoparticles of BTO dispersed well in the polypropylene /BTO@POSS nanocomposite than in the PP/BTO nanocomposites. The Weibull E_b of the PP/BTO@POSS nanocomposites dramatically lowered as compared to the hot-pressed PP, and as a result of the BTO particles clustering, the PP/BTO nanoparticles had an even lower Weibull E_b. According to the results, it is crucial to distribute ceramic nanoparticles evenly throughout the polymer matrix. The affinity of high dielectric constant particles with the polymer as matrix can be improved by functionalising surfaces of nanofillers with tiny molecules like dopamine, organosilane, and organophosphate ligands, or by grafting polymer brushes. Al_2O_3, Ta_2O_5, ZrO_3, and TiO_2 are some examples of ceramic nanoparticles with a moderate ε_r that is utilised to prevent the field concentration caused by increased interfacial polarisation near high-κ particles. Where there is insufficient filler to cause percolation, these high-κ ceramic nanofillers will produce low energy density and a low ε_r. Creating nanoparticles with a core–shell morphology that have a gradient ε_r profile from the polymer matrix to the core and shell of the nanofillers is an alternative strategy [46]. According to a phase-field modelling when the thickness of the shell is roughly equivalent to the diameter of the core, the ideal core–shell structure is produced [47]. In addition, the distinctive 2D nanofillers of plate-like structure offers a stronger capability of blocking charges to impede movement of electrons within the nanocomposite. Low level of filler can minimise filler agglomeration and hence concentration of local fields. When ultrathin boron nitride nanosheets (BNNS) were used as filler in P(VDF-TrFECFE) in trials, the E_b value was recorded as 610 MV/m, which was a 70% increase over the E_b value of the film with a pristine PVDF-TrFECFE solution-cast sample (362 MV/m) [48].

Nanoparticles have large surface area, which cause lot of flaws on their surfaces. These flaws also could be caused by foreign atoms on the surface, irregularities in the surface's geometry, or the molecules close to the particle's surface being non-centrosymmetric (coordinative unsaturation) [49]. These flaws lead to the generation of new localised energy levels in a band that would otherwise be forbidden. The created energy levels could be charge traps that are located in the band gap between the valence and conduction bands (localised states in the band gap). They prefer to fill these traps rather than remain in the conduction or valence band when mobile charges, such as electrons or holes, are present [50]. Once it has filled the trap, the electron or hole finds it difficult to return to the conduction or valence band, especially if the trap is deep enough; however, it can migrate to nearby similar traps if they are close enough [97].

As seen in Fig. 5.5a, they may transition from one particle interphase to another. This could worsen the characteristics of electrical insulation and promote charge mobility in nanocomposites. If the traps are deep enough and far from each other, these phenomenon will not take place. In contrast, if the aforementioned defects produce energy states that are deeper in the conduction/valence band, this may enlarge the band gap in the particle interphase region and act as an energy barrier for the passage of charges. These phenomenon is schematically displayed in Fig. 5.5b. This

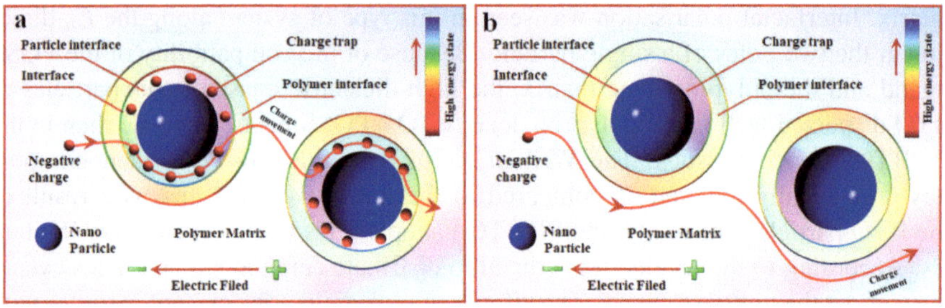

Fig. 5.5 **a** Possible mechanism for charge movement through the charge traps in the particle interphase and **b** the particle interphase represents an energy barrier for charge movement [49]. Adapted with permission from ref. [49] Copyright (2018) (Elsevier)

may indicate a potential mechanism that enhances the nanocomposites' dielectric characteristics.

A novel method of stacked nanodielectrics has been developed by researchers to enhance both E_b and ε_r of nanocomposites [51]. The fundamental idea behind this technique is to produce a dielectric "hard" layer with a high E_b and a dielectric "soft" layer with a high ε_r in a rationally constructed film architecture. A small percentage of insulating nanofillers are present in the high E_b layer, whereas the high-κ layer contains a substantial number of high-κ fillers. To maximise the U_e and enhance the synergy between E_b and ε_r, tuning is required for the stacked nanodielectrics' architecture, volume percentage, and number of layers.

5.2.3 *Discharged (U_e)*

In general, raising either ε_r or electric field or both can result in a high U_e. Since the power frequency has no resonance losses region, improving electric and atomic polarisation is the best way to boost ε_r [42]. When E_g is above 4 eV for insulting polymers, the deformation $\varepsilon_r s$ of the polymer cannot exceed 5. This is because of the molecular bonding of localised electrons and atoms. In order to increase the ε_r for polymers, orientational and interfacial polarizations are applied; however, they result in dielectric losses in the power frequency. It is not possible to simultaneously increase ε_r/U_r and E_b for this reason. Therefore, we are limited to finding an ε_r/U_e and E_b compromise. Therefore, for polymer nanodielectrics, getting high U_e should not be as important as obtaining low loss and high E_b. Low dielectric loss is ensured, and U_e should be evaluated at appropriate temperatures and electric fields to actual application. A trilayer BNNS-BT-BNNS structure and stacked c-BCB nanodielectric demonstrated a U_e of 1.1 J/cm^3 at a η of 93% at 150 °C and 200 MV/m, which was around 2.8 times that of BOPP seen at 70 °C [52].

Dielectric properties estimation of nanodielectric interphase by numerical modelling and dielectric spectroscopy is studied by Andrei et al. [53]. In their studies, to calculate the frequency-dependent dielectric properties of the polymer-nanoparticle interphase, an electro-quasistatic numerical model of a PP-SiO_2 nanodielectric is suggested. It is based on a combination of dielectric spectroscopic results and semi-analytical formulations. The nanodielectric model is made up of spherical SiO_2 nanoparticles with a diameter of 15 nm that are evenly dispersed across a PP matrix. Each nanoparticle is encircled by an interphase that is 10 nm thick.

5.3 Polymer Nanodielectrics Based on Conductive Nanoparticles

Metallic and conductive nanoparticles [54] and carbonaceous nanoparticles (such as carbon black (CB) [55], carbon nanotubes (CNTs) [56], and graphene nanoplates (GNPs) [57]) are utilised as fillers in polymer nanodielectrics. When a polymer serves as the matrix, the conductive particles become polarised when an electric field is applied, and this causes the internal electric field to be zero.

When linear dielectric regime as determined by broadband dielectric spectroscopy (BDS), the great dielectric constant has been reported for some of these nanodielectrics in recent years, e.g. adding Ag nanoparticles as filler to epoxy could result in a relative dielectric constant (ε_r) of more than 300 [54].

Conductive particles in thin polymers function as nanoscale electrodes, and nanocapacitors are created at the interface between the polymer and the nanoparticles. The percolation barrier is nearly reached when the filler material, the connection between nanocapacitors to one another as well as to metal electrodes [57]. A high $\varepsilon_{r,\mathrm{app}}$ (colossal apparent) occurs when the thickness or interparticle distance of the boundary layer becomes very tiny. Since there is no conduction or tunnelling of electrons at low fields (<1 MV/m), the tan δ value for these nanodielectrics likewise stays low.

Large ceramic granules that are semiconductive and thin insulating grains borders form the grain boundary barrier layer (GBBL) in ceramic capacitors, and their polarisation results in a colossal apparent ($\varepsilon_{r,\mathrm{app}}$) resistance of up to 10^5 [58] which is given by,

$$\varepsilon_{r,\mathrm{app}} = \frac{\varepsilon_{gb} d_g}{t}$$

where ε_{gb} is dielectric constant of insulating grain boundary layer, d_g is average grain size of dielectric material, and t is thickness of insulating grain boundary layer.

Local electron tunnelling is provided at the grain boundaries via a Schottky barrier. In order to prevent charges from being injected via the conductive medium, from metal electrodes to bulk composite channel, it is crucial to ensure that there will be no direct contact between the metallic electrodes and semiconducting granules [59]. Researchers used nanosized (18 and 100 nm) surface functionalised aluminium nanoparticles using polymer brushes and a matrix made of isotactic polypropylene (iPP) to study the improvement of dielectric constant by incorporating nanosized conductive particles into polymer/conductive particle nanodielectrics. The ε_r of PP/nAl nanodielectrics was investigated using BDS at various nAl concentrations under an applied electric field of 1MV/m. It was found that the $\varepsilon_{r,\text{app}}$ was only 5.7 at 25 vol% of 100 nm nAl, which was 2.5% less than iPP [60].

Conducive electrons may tunnel into the polypropylene matrix in the presence of strong electric fields, causing dielectric loss. Due to conduction loss, these injected electrons also significantly reduce the E_b (breakdown strength) of polymer/conductive particle nanodielectrics. This was seen in an experiment where E_b was reduced from 620 MV/m for iPP on 5 vol% nAl to 300 MV/m on 25 vol% loading. When metallic nanoparticles are polarised, the charges in the interfacial region cause a strong changes the polymer matrix's local electric field in the direction of the electric field. These close-by areas cause dielectric breakdown and electron tunnelling in the iPP matrix, making them hot spots.

Therefore, under an elevated local field, the metallic particles transform into providing free electrons to the PP matrix by electron sources via electron tunnelling, making nanocomposite conductive above a particular threshold, ultimately resulting in catastrophic dielectric breakdown. The increase of $\varepsilon_r s$ in polymer nanodielectrics is facilitated by these kinds of electrons conducting under a strong electric field. Because the internal electric field for metallic particles in conductive nanoparticle nanodielectrics is laterally zero, they are unable to store any electric energy. As a result, the augmented ε_r and U_r can only be achieved by raising the polymer matrix's local electric field [45].

The metallic particle size must be lower than 2–3 nm at normal temperature to display an efficient Coulomb blockade effect [61], yet it is challenging to disperse these tiny metallic nanoparticles uniformly in the polymer matrix. As a result, high-voltage capacitor applications cannot use composites of polymer and metallic nanoparticles.

Due to their high aspect ratio, CNTs and GNPs can easily aggregate or percolate when used as fillers, which increases the conductivity of polymer/CNT and polymer/GNP nanocomposites. Again, this behaviour led to a situation where conductive pathways transported electrons from the filler to the polymer matrix, increasing the ε_{eff} for those nanodielectrics. Regarding a PVDF/GNP nanocomposite made of polyvinylidene fluoride, researchers obtained more than 200 and 2700 $\varepsilon_r s$ values at 1000 and 100 Hz, respectively [62]. Additionally, in polymer/carbon nanodielectrics, researchers noticed a nonlinear behaviour in the link between current and voltage [63].

Researchers have suggested a variety of additional methods to increase E_b and decrease conductive filler/polymer nanocomposites' dielectric loss, including a third filler, sandwich/multi-layered buildings, as well as surface coating. One research team created polyurethane (PU) nanocomposites based on CNT fillers with GO capsulation, where the function of the Go was to support the CNT dispersion by π-π interaction and serve as a barrier to prevent direct contact between two CNT particles [64]. As a result, this nanocomposite was found to have a resistivity that was almost four times as high as nanocomposite with CNT-only fillers and to have no electric percolation. In that experiment, the E_b value was reported at 120 MV/m, substantially doubling its previous value, and it declined as the filler quantity rose. In a further set of experiments, PI-(PI/NH_2-MWCNT)-PI-three-layered films in which the insulating layers block the conductive paths were made resulting in ultralow dielectric loss and high dielectric constant [65]. There were not colossal ε_r values reported between 0 and 20 wt% filler content; however, this three-layer film reduced E_b values 300 to 75 MV/m at a 20 weight percent loading, which was better than a single-layer composite of PI and CNT.

5.4 Applications of Nanodielectrics

The advantages of nanodielectrics are excellent mechanical properties, high dielectric breakdown strength, low dielectric loss, high power and energy density, and good processibility. Multiscale models and design methodologies are needed to be compliant for nanodielectrics to be applicable in the industry. Not only to control the dispersion of nanodielectrics, understand the interface contributions, and measuring ageing, there should have necessary tools that will enable customization of the constituent properties to meet the needs of the bulk material for a particular application [4]. These properties of nanodielectrics have been successfully used in many areas such as space charge suppression, high energy density storage in capacitors, generators and motors, wire enamels, and in high thermal conductivity. The selected applications of nanodielectrics are shown in Fig. 5.6.

5.4.1 Space Charge Suppression

The superior characteristics of nanodielectrics reduce the internal electric field in them. The breakdown properties and the anti-ageing capabilities of nanodielectrics can be enhanced, especially in direct current (HDVC) cable operating voltage level and high voltage [66]. Improvement of breakdown strength originates from the interfacial zones formed around the nanoparticles and these zones can modify the trap properties, carrier mobility and charge injection in polymer matrices by changing the chain conformation, aggregation structure, or morphology. These modifications can cause the suppression of space charge accumulation as well as improvement of electrical breakdown strength. By the addition of nanoparticles in the polymer matrix, some nanodielectrics exhibit smaller space charge accumulation. The injected charge is captured by the traps which are formed as the result of the introduction of nanoparticles. This is effective only at lower concentration, and higher charge injection causes the degradation of polymer chains. It has been showed by the various research that the addition of inorganic particles into the polymer matrix can prevent the injection and space charge suppression [67]. Zhang et al. addressed the issue by attaching matrix-compatible polymer brushes to spherical colloidal SiO_2 NPs (10–15 nm diameter) to

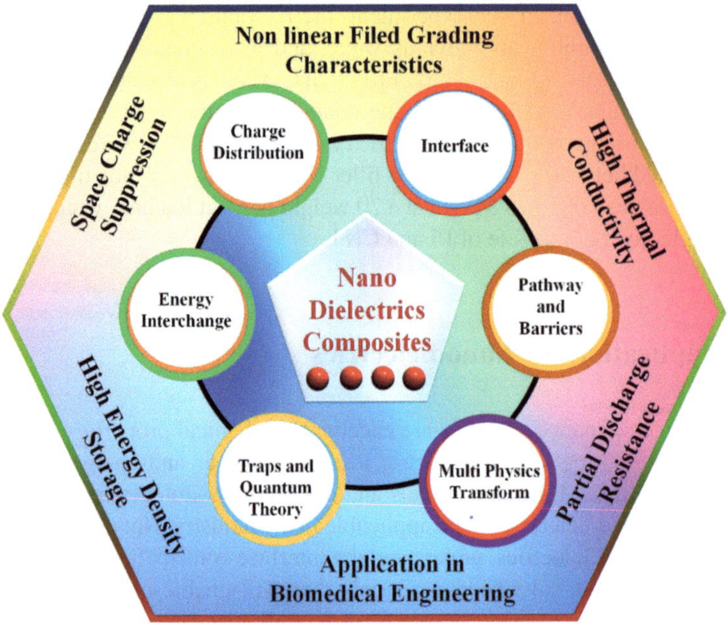

Fig. 5.6 Selected applications of nanodielectrics in various fields

create a uniform NP dispersion in cross-linked polyethylene and were able to achieve improved space charge suppression, improved DC breakdown strength, and limited internal field distortion (10.6%) over a broad range of external DC fields from 30 to 100 kV/mm at room temperature [68]. Zhang et al.in their study showed that the introduction of MgO nanoparticles in low-density polyethylene (LDPE) improved the average trap depth in MgO/low-density polyethylene (LDPE) nanodielectrics. As a result, there was a significant build-up of space charge close to the electrodes. This was measured by using improved pulsed electro-acoustic (PEA) system [69]. Wang et al. evaluated the impact of vinyl silane coupling agent on the electrical characteristics of Al_2O_3/LDPE nanodielectrics and came to the conclusion that the modified Al_2O_3 nanoparticles have a greater space charge suppressing effect than the unmodified ones [70]. In addition to that, Zhou et al. studied the morphology, dielectric properties, DC volume resistivity, and space charge behaviour of surface modified MgO, TiO_2, ZnO, and $A12O_3$ blended with polypropylene (PP). The results show that the space charge suppression and breakdown strength first increase with the increase in nanoparticle content and then decreases [71].

5.4.2 High-Density Energy Storage

Polymer-based nanocomposites with high permittivity, high breakdown strength, flexibility, and lower cost are required for deployment in distributed power production systems and high compact electronic circuits [66]. Development of nanodielectric materials is closely related to the advancement of capacitor technology, and it is shown in Fig. 5.7. Polymers serving as the matrix and the fillers are used to increase the effective dielectric constant. Capacitors that can store a higher amount of energy and capable of instantly releasing it are becoming more and more necessary. These capacitors are used in a variety of military and commercial applications. For such applications, a base polymer with a high permittivity, such as poly(vinylidene fluoride) (PVDF), is a good choice to start. In theory, the fillers aid to increase the effective dielectric constant of the system while maintaining the polymer's built in resistance to disintegration. Many studies are reported on the breakdown strength and composite dielectric permittivity of polymer nanocomposites. Most of the research concentrate on improving the dielectric permittivity by using ferro-metal oxides such as Pb(Zr, Ti)O_3 (PZT), Pb($Mg_{0.33}Nb_{0.77}$)O_3-PbTiO_3 (PMNT), and BaTiO_3 (BT). Studies show that the dielectric permittivity of ceramics can be extended from tens (TiO_2, Al_2O_3) to thousands ($BaTiO_3$), and can even can extend to over 10,000 for $CaCu_3Ti_4O_{12}$ [72]. In conductive polymer nanocomposites for high energy density applications, several studies on percolation phenomenon that occurs when the volume percent of fillers exceeds the percolation threshold have been reported [73]. Metal particles, carbon nanotubes, and graphene sheets are used as the conductive fillers. Their polymer-based nanocomposites indicate an exponential increase in permittivity around the percolation threshold. Additionally, a hierarchical structure for nanocomposites has been devised to concurrently increase the breakdown strength and permittivity [74]. Flexible nanocomposites containing polymer and high dielectric ceramic nanoparticles have gained greater importance for dielectric and energy storage application. A core–shell Ag@polydopamine@$BaTiO_3$ nanocomposites with strawberry-like properties was created by Yang et al. The strawberry-like BT-PDA-Ag show greater enhanced energy density and can be used for energy storage [75]. Wang et al. created a dielectric composite material with extremely high discharge energy density by changing the interface between nanoparticles and poly(vinylidene fluoride-co-hexafluoropropylene) (PVDF-HFP]). After coating SiO_2 outside, barium titanate dielectric concentration and current density inside BT particles decrease. This coating with PDA increases the interface interactions with the nanoparticles, and as a result, ultra-high discharge energy has been achieved. With varied $BaTiO_3$ compositions steadily increasing layer by layer in the nanocomposites, Wang et al. created a new sandwich-like three-layer structure for them [76]. Luo et al. created a 3-D BaTiO3 network in polymer composite with increased permittivity and energy storage density [77].

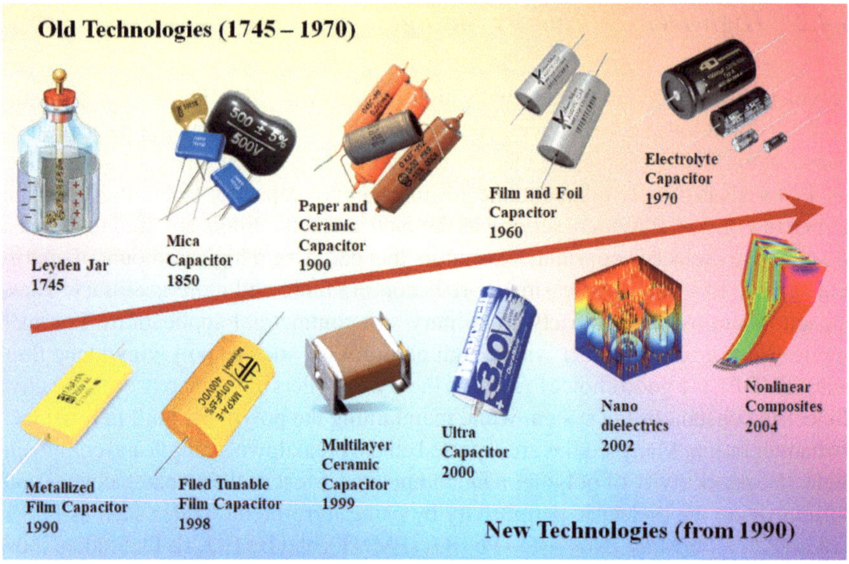

Fig55..7 Development of materials is closely related to the advancement of capacitor technology

5.4.3 Generators and Motors

In rotating machines, it is desirable to have higher capability of withstanding electrical and thermal stress along with improved thermal conductivity [78]. The main advantages of generators are good mechanical strength and environmentally friendly [79]. This was one of the first application of nanodielectric and driving force for improvements in motor systems. Research has been carried out with materials like mica, epoxy resin, and fibre glass. Mica can split in way to form a thin, flat laminate [80], with excellent dielectric qualities, low dielectric losses, heat resistance, and high dielectric strength. High-voltage rotating machines experience the combined effect of thermal, electrical, and mechanical loads. The improvement of voltage endurance is arguably one of the biggest advantages of nanodielectrics [4]. The observed improvement in the short-term AC breakdown strength of roughly 30% opens the door to the possibility of raising working electric stress and/or decreasing ground wall dimensions [81]. The ground wall insulation of large form-wound generators and motors, for which voltage endurance is essential and standards are available, is thus obviously a top target application [82]. Generator design and development is undergoing radical change, not only in the hardware that is being designed but in the way in which they are designed. To meet current and future market demands, improved generator design solutions are required due to the highly flexible peak load regime and the decreasing energy price at the liberalised market. The use of nanoparticles can improve the properties of the dielectrics. A study done on using SiO_2 nanoparticles in epoxy mica has led to a more efficient stator winding insulation system that is customised to the needs of the new, highly flexible grid without sacrificing longevity on the one hand and the specific power output on the other. In the near future, it might be applied to both new-generation designs and the renovation and upgrading of already installed generators [83].

5.4.4 Application in Wire Enamels

Wire enamels also benefit from nanodielectrics' ability to withstand electrical stress for an extended period. This suggest a way to prepare insulating materials that are resistant to coronas, along with the regularly reported improvement in erosion-resistant under stress [4]. A common low-voltage generator was constructed using stator coils with many turns, ranging from one to sixteen turns per coil which can be random wound or form wound. Random-wound coil generators can be made at low cost, and through the application of vacuum-pressure impregnation, their ability to survive harsh weather conditions can be increased. The low-voltage motors coated with polymer nanocomposites have many applications. When the low-voltage motors are random wound with round wire and coated with polyamide-imide, it offers the necessary abrasion resistance for winding at high working temperatures. The middle layers contain different nanosized oxide fillers, including tin oxide, which was thought to be the most prevalent filler, for increased resistance to partial discharge. Normally, production with greater filler loading reduced wire flexibility, which caused enamel cracks, whereas resistance to partial discharge rose with filler loading. The resistance to partial discharges was considerably enhanced at modest filler loading. Strands of magnet wire with a rectangular cross-section coiled around it puncture the coils for medium- to high-voltage motors. Depending on the magnetic wire's application and type, it goes through the coating a number of times. Here, the construction is layer-by-layer coating which results in multiple polymers to polymer interface. These wound coil wire enamels are applicable in transformers, motors, coils, low-current cables, and telephone installations. Random-wound conductors, which use enamelled magnet wire, are particularly common in the design of tiny motors (and other devices like solenoids) [78].

5.4.5 Nanodielectrics with High Thermal Conductivity

The dissipation from the leakage current grows in the high-voltage field as the level of conducting voltage rises. As electronic components are arranged in integrated circuits with ever-increasing compactness, there is a corresponding increase in heat generation per unit volume. The electrical power system and electronic system needs high thermal conductivity and excellent electrical insulation [66]. There are different types of fillers that can increase the thermal conductivity of nanocomposites [84]. These fillers can be metals, inorganic particles, and carbon materials. Wu et al.

studied the improvement of dielectric properties, high dielectric constant, and low dielectric loss and thermal conductivity nanocomposites utilised in film capacitors for energy storage [85]. Seyhan et al. showed that the thermal conductivity of polymer nanocomposites with boron nitride nanosheets (BNNSs) can be improved while maintaining the characteristics of electrical insulation [86]. Zhang et al. created a nanocomposite of uniformly dispersed strongly orientated graphite in a polymer matrix [87]. Sun et al. added a small amount of silver nanoparticles in order to reduce the interfacial thermal resistance between adjacent BNNSs [88]. Hong et al. created a method to construct 3D hexagonal boron nitride network in a polymer composite which is verified with high thermally conductive and mechanically stretchable [89].

5.5 Future Prospects

5.5.1 *Self-Healing Dielectrics*

It is desirable to have electrical insulation that has self-healing capacity, and it is especially attractive if healing is not dependent on an external source. One such approach is to incorporate encapsulated fluids that are discharged owing to damage and locally cross-link to perform mechanical repair [90]. One illustration of this is the well-known clearing phenomenon built into the metallised film capacitors.

Partial discharges, such those that take place in voids, are cause of majority of localised damage in high-voltage insulation. These discharges also involve the release of photon energy at a precise site where healing is required and can serve as the foundation for a focused self-healing approach. A means of introducing active substances at the interior of the polymer matrix is provided by the insertion of nanoparticles. When these active agents are photosensitive, the energy formed by the light-emitting process can be utilised. Despite being an active field, high-voltage dielectric structures hardly ever use it [4]. However, in recent times, efforts are being made to apply the chemistry and theories of self-healing phenomenon to electrically insulating materials, including methods that use plasma energy.

5.5.2 *Increased Stress Applications*

Any practice use of solid dielectrics is to prevent electrical trembling that results from such as in. In cast spacers for gas-insulated equipment, moulded bushings, and cast transformers, one of the problems limiting their durability is electrical treeing caused due to electrical stresses. Electrical treeing is caused due to electrical stresses developed in the material that arises from defects arising in moulding processes like in extrusions, moulding, or casting. Hence, the average stress inside the material is to be kept at low level. Nanomaterials dispersed in polymer dielectric act as barrier that slow the formation of trees and enhances the resistance of materials to electrical treeing in a wide variety of materials [8, 10, 91]. The dielectric strength of HVDC cables is significantly influenced by charge injection, trapping, and space charge accumulation [92]. In XLPE-nanoclay composites, the stress concentration was found to be maximum at the interface of nanoclay and XLPE, while stress concentration inside the nanoclay platelet was lower. The tortuous path created by dispersed nanoclay in XLPE prevented the propagation of both electrical treeing and water tree [10]. Conducting nanosized fillers improve electrical conductivity, which lowers the tendency of materials to accumulate charge [93].

5.5.3 Biomedical Applications

When activated by an outside electric field, flexible materials known as electroactive polymers (EAP) show a change in size or shape. Dielectric elastomers (DEs), a type of EAP, are representative of potential applications ranging from artificial muscle to robotics interaction due to their satisfied mechanical and dielectric properties, such as lightness, softness, flexibility, high deformability, quick response, and low cost [94]. Wang et al. investigated dielectric elastomer nanocomposites with multi-walled CNT and concluded that the electric fields inside the material are concurrently decreased and the deforming performance is improved [95]. A fascinating experiment by Kwak et al. used artificial muscle based on DE to simulate human facial expressions on a robot [96]. In addition to their employment in artificial muscles, these soft dielectric materials' electromechanical capabilities are frequently employed as functional sensors to pick up on environmental changes or human motion. Chhetry et al. stated that the strategy of the microstructuring dielectric material is leveraging the sensor's performance by enhancing the effective permittivity and reducing the stiffness [97].

5.6 Challenges and Future Trends

In this chapter, we have discussed about some of the important features of nanodielectrics and the major research related to the applications. Over the past 15 years, researchers in the field of dielectrics have made significant advancements in the manufacturing of nanodielectric films and materials. The knowledge in dielectric polymer and nanodielectric composites have grown as a result of scientific investigation from the atomic scale to the nanometre and micrometre levels and continues to grow in terms of applications. Adopting antiferroelectric fillers and creating atypical composites may result in a high energy density. A potential and workable method for increasing the energy density without reducing the dielectric strength is to keep fillers at a lower loading fraction. It is possible to increase the dielectric strength of dielectric polymers by molecularly altering the chemistry of the polymer; however, this will call for more research and a scale-up procedure. Though film scale-up might be difficult, the logical design of polymer dielectrics may offer scientific direction for dielectric material synthesis in practice [98]. The most recent advancements in enhancing the dielectric strength of polymer films should be further evaluated on a wide scale and thinner gauges. More research should be done on economical polymers. The current market demands for advanced film capacitors are still uncertain, and the high cost of capacitors makes it difficult to implement new technologies. There should be more research done on polymers with intermediate dielectric constants and polymer composites, like high-temperature polymers that have undergone molecular modification for achieving desired qualities without sacrificing dielectric loss. It is important to continue paying attention to the basic principles of nanodielectrics, polymer chemistry, and computer modelling. In the long run, this will provide superior dielectrics with a high dielectric constant, high dielectric strength, and low dielectric loss [97]. However, overcoming the high cohesive energy of nanoparticles and ensuring homogenous dispersion remains difficult, particularly for production at an industrial scale [66]. The dielectric characteristics of nanoparticles can be effectively improved by surface modification.

References

1. Zhang G, Li Q, Allahyarov E, Li Y, Zhu L (2021) Challenges and opportunities of polymer nanodielectrics for capacitive energy storage. ACS Appl Mater Interf 13(32):37939–37960. https://doi.org/10.1021/acsami.1c04991
2. Ashok N, Prakash K, Selvakumar D, Balachandran M (2020) Synergistic enhancement of mechanical, viscoelastic, transport, thermal and radiation ageing characteristics through chemically bonded interface in nanosilica reinforced EPDM-CIIR blends. J Appl Poly Sci 138(12):50082. https://doi.org/10.1002/app.20201087
3. Schadler LS, Nelson JK (2020) Polymer nanodielectrics—short history and future perspective. J Appl Phys 128:120902. https://doi.org/10.1063/5.0019865
4. Tan DQ (2020) The search for enhanced dielectric strength of polymer-based dielectrics: a focused review on polymer nanocomposites. J Appl Polym Sci 137(33):49379. https://doi.org/10.1002/app.49379
5. Tanaka T, Imai T (2013) Advances in nanodielectric materials over the past 50 years. IEEE Elect Insulat Magazine 29(1):10–23. https://doi.org/10.1109/MEI.2013.6410535
6. Yasoda KY, Baji DS, Kumar MS, Santhanagopalan D, Batabyal SK (2023) Sustainable development of manganese sulfoselenide nanoparticles anchored graphene oxide nanocomposite for high-performance supercapacitor and lithium-ion battery applications. J Alloys Compound 930:Art. no. 167282. https://doi.org/10.1016/j.jallcom.2022.167282
7. Murali B, Baiju KG, Prasad RK, Kumaresan D (2023) Fabrication of barium titanate nanowires-GNP composite bilayer photoanodes for the high-performance dye-sensitized solar cells. Appl Surf Sci 610:Art. no. 155316. https://doi.org/10.1016/j.apsusc.2022.155316
8. Kavitha D, Sindhu TK, Nambiar TNP (2017) Impact of permittivity and concentration of filler nanoparticles on dielectric properties of polymer nanocomposites. IET Sci, Measurement Tech 11(2):179–185. https://doi.org/10.1049/iet-smt.2016.0226
9. Sreehitha PR, Durga B, Balachandran M (2020) Dielectric properties, thermal characteristics and degradation kinetics of PMMA nanodielectrics. Mat Today: Proceed 24(2):772–781, 17
10. Kavitha D, Balachandran M (2019) XLPE—layered silicate nanocomposites for high voltage insulation applications: mechanical and treeing behavior. IET Sci, Measure Tech 13(7):1019–1025. https://doi.org/10.1049/iet-smt.2018.5417

11. Buscaglia V, Randall CA (2020) Size and scaling effects in barium titanate. an overview. J Eur Ceram Soc 40(11):3744–3758. https://doi.org/10.1016/j.jeurceramsoc.2020.01.021
12. Paniagua SA, Kim Y, Henry K, Kumar R, Perry JW, Marder SR (2014) Surface-initiated polymerization from barium titanate nanoparticles for hybrid dielectric capacitors. ACS Appl Mater Interfaces 6:3477–3482. https://doi.org/10.1021/am4056276
13. Teranishi T, Hoshina T, Tsurumi T (2009) Wide range dielectric spectroscopy on perovskite dielectrics. Mater Sci Eng B 161(1–3):55–60. https://doi.org/10.1016/j.mseb.2008.11.039
14. Arlt G, Hennings D, de With G (1985) Dielectric properties of fine-grained barium titanate ceramics. J Appl Phys 58(4):1619–1625. https://doi.org/10.1063/1.336051
15. Allahyarov E, Löwen H, Zhu L (2016) Dipole correlation effects on the local field and the effective dielectric constant in composite dielectrics containing high-k inclusions. Phys Chem Chem Phys 18(28):19103–19117. https://doi.org/10.1039/C6CP03149H
16. Goswami AK (1969) Dielectric properties of unsintered barium titanate. J Appl Phys 40(2):619–624. https://doi.org/10.1063/1.1657443
17. Hsiang H-I, Yen F-S (1993) Dielectric properties and ferroelectric domain of $BaTiO_3$ powders. J Appl Phys 32(Part 1, No. 11A):5029–5035. https://doi.org/10.1143/JJAP.32.5029
18. Guo HZ, Mudryk Y, Ahmad MI, Pang XC, Zhao L, Akinc M, Pecharsky VK, Bowler N, Lin ZQ, Tan X (2012) Structure evolution and dielectric behavior of polystyrene-capped barium titanate nanoparticles. J Mater Chem. https://doi.org/10.1039/c2jm35600g
19. Li Y, Liao Z, Fang F, Wang X, Li L, Zhu J (2014) Significant increase of curie temperature in nano-scale $BaTiO_3$. Appl Phys Lett 105(18):182901. https://doi.org/10.1063/1.4901169
20. Polking MJ, Han M-G, Yourdkhani A, Petkov V, Kisielowski CF, Volkov VV, Zhu Y, Caruntu G, Paul Alivisatos A, Ramesh R (2012) Ferroelectric order in individual nanometre-scale crystals. Nat Mater 11(8):700–709. https://doi.org/10.1038/nmat3371
21. Polking MJ (2016) Deciphering the physics and chemistry of perovskites with transmission electron microscopy. Nanoscale 8(12):6237–6248. https://doi.org/10.1039/C5NR06186E
22. Paniagua SA, Kim Y, Henry K, Kumar R, Perry JW, Marder SR (2014) Surface-initiated polymerization from barium titanate nanoparticles for hybrid dielectric capacitors. ACS Appl Mater Interfaces 6(5):3477–3482. https://doi.org/10.1021/am4056276
23. Kim P, Doss NM, Tillotson JP, Hotchkiss PJ, Pan M-J, Marder SR, Li J, Calame JP, Perry JW (2009) High energy density nanocomposites based on surface-modified $BaTiO_3$ and a ferroelectric polymer. ACS Nano 3(9):2581–2592. https://doi.org/10.1021/nn9006412
24. Zhang H, Marwat MA, Xie B, Ashtar M, Liu K, Zhu Y, Zhang L, Fan P, Samart C, Ye Z (2020) Polymer matrix nanocomposites with 1D ceramic nanofillers for energy storage capacitor applications. ACS Appl Mater Interfaces 12(1):1–37. https://doi.org/10.1021/acsami.9b15005
25. Pan Z, Zhai J, Shen B (2017) Multilayer hierarchical interfaces with high energy density in polymer nanocomposites composed of $BaTiO_3@TiO_2@Al_2O_3$ nanofibers. J Mater Chem A 5(29):15217–15226. https://doi.org/10.1039/C7TA03846A
26. Tang H, Zhou Z, Sodano HA (2014) Relationship between $BaTiO_3$ nanowire aspect ratio and the dielectric permittivity of nanocomposites. ACS Appl Mater Interfaces 6(8):5450–5455. https://doi.org/10.1021/am405038r
27. Joshi UA, Yoon S, Baik S, Lee JS (2006) Surfactant-free hydrothermal synthesis of highly tetragonal barium titanate nanowires: a structural investigation. J Phys Chem B 110(25):12249–12256. https://doi.org/10.1021/jp0600110
28. Calame JP (2006) Finite difference simulations of permittivity and electric field statistics in ceramic-polymer composites for capacitor applications. J Appl Phys 99(8):084101. https://doi.org/10.1063/1.2188032
29. Tang H, Lin Y, Sodano HA (2012) Enhanced energy storage in nanocomposite capacitors through aligned PZT nanowires by uniaxial strain assembly. Adv Energy Mater 2(4):469–476. https://doi.org/10.1002/aenm.201100543
30. Makdessi M, Sari A, Venet P, Bevilacqua P, Joubert C (2015) Accelerated ageing of metallized film capacitors under high ripple currents combined with a DC voltage. IEEE Trans Power Electron 30(5):2435–2444. https://doi.org/10.1109/TPEL.2014.2351274
31. Chi KK (2004) Dielectric phenomena in solids. Elsevier

32. Tsurumi T, Li J, Hoshina T, Kakemoto H, Nakada M, Akedo J (2007) Ultrawide range dielectric spectroscopy of $BaTiO_3$-based perovskite dielectrics. Appl Phys Lett 91(18):182905. https://doi.org/10.1063/1.2804570
33. Zhu L, Wang Q (2012) Novel ferroelectric polymers for high energy density and low loss dielectrics. Macromolecules 45(7):2937–2954. https://doi.org/10.1021/ma2024057
34. Soulestin T, Ladmiral V, Santos FDD, Améduri B (2017) Vinylidene fluoride- and trifluoroethylene-containing fluorinated electroactive copolymers. How does chemistry impact properties? Prog Polym Sci 72:16–60. https://doi.org/10.1016/j.progpolymsci.2017.04.004
35. Li Y, Makita Y, Zhang G, Rui G, Li Z-M, Zhong G-J, Miyoshi T, Huang H-D, Zhu L (2020) Effects of rigid amorphous fraction and lamellar crystal orientation on electrical insulation of poly(ethylene terephthalate) films. Macromolecules 53(10):3967–3977. https://doi.org/10.1021/acs.macromol.0c00646
36. Chen X, Allahyarov E, Li Q, Langhe D, Ponting M, Schuele DE, Baer E, Zhu L (2020) Reducing dielectric loss by nanoconfined impurity ion transport in multilayer films under low electric fields. Compos Part B Eng 190:107908. https://doi.org/10.1016/j.compositesb.2020.107908
37. Chen X, Allahyarov E, Langhe D, Ponting M, Li R, Fukuto M, Schuele DE, Baer E, Zhu L (2020) Reducing dielectric loss and enhancing electrical insulation for multilayer polymer films by nanoconfined ion transport under high poling electric fields. J Mater Chem C 8(18):6102–6117. https://doi.org/10.1039/D0TC00522C
38. Zhang G, Brannum D, Dong D, Tang L, Allahyarov E, Tang S, Kodweis K, Lee J-K, Zhu L (2016) Interfacial polarization-induced loss mechanisms in polypropylene/$BaTiO_3$ nanocomposite dielectrics. Chem Mater 28(13):4646–4660. https://doi.org/10.1021/acs.chemmater.6b01383
39. Ho J, Ramprasad R, Boggs S (2007) Effect of alteration of antioxidant by UV treatment on the dielectric strength of BOPP capacitor film. IEEE Trans Dielectr Electr Insul 14(5):1295–1301. https://doi.org/10.1109/TDEI.2007.4339492
40. Wang L-M (2006) Relationship between intrinsic breakdown field and bandgap of materials. In: 2006 25th International Conference on Microelectronics; IEEE: Belgrade, Serbia and Montenegro, pp 576–579. https://doi.org/10.1109/ICMEL.2006.1651032
41. Sun Y, Bealing C, Boggs S, Ramprasad R (2013) 50+ years of intrinsic breakdown. IEEE Electr Insul Mag 29(2):8–15. https://doi.org/10.1109/MEI.2013.6457595
42. Wang CC, Pilania G, Boggs SA, Kumar S, Breneman C, Ramprasad R (2014) Computational strategies for polymer dielectrics design. Polymer 55(4):979–988. https://doi.org/10.1016/j.polymer.2013.12.069
43. Sun Y, Boggs S, Ramprasad R (2015) The effect of dipole scattering on intrinsic breakdown strength of polymers. IEEE Trans Dielectr Electr Insul 22(1):495–502. https://doi.org/10.1109/TDEI.2014.004690
44. Dissado LA, Fothergill JC, Electrical degradation and breakdown in polymers.
45. Chiu F-C (2014) A review on conduction mechanisms in dielectric films. Adv Mater Sci Eng 2014:1–18. https://doi.org/10.1155/2014/578168
46. Wang J, Guan F, Cui L, Pan J, Wang Q, Zhu L (2014) Achieving high electric energy storage in a polymer nanocomposite at low filling ratios using a highly polarizable phthalocyanine interphase. J Polym Sci Part B Polym Phys 52(24):1669–1680. https://doi.org/10.1002/polb.23554
47. Wang YU, Tan DQ, Krahn J (2011) Computational study of dielectric composites with core-shell filler particles. J Appl Phys 110(4):044103. https://doi.org/10.1063/1.3624660
48. Li Q, Zhang G, Liu F, Han K, Gadinski MR, Xiong C, Wang Q (2015) Solution-processed ferroelectric terpolymer nanocomposites with high breakdown strength and energy density utilizing boron nitride nanosheets. Energy Environ Sci 8(3):922–931. https://doi.org/10.1039/C4EE02962C
49. Alhabill FN, Ayoob R, Andritsch T, Vaughan AS (2018) Introducing particle interphase model for describing the electrical behaviour of nanodielectrics. Mat Design 158:62–73. https://doi.org/10.1016/j.matdes.2018.08.018

50. Anta JA, Marcelli G, Meunier M, Quirke N (2002) Models of electron trapping and transport in polyethylene: current–voltage characteristics. J Appl Phys 92:1002. https://doi.org/10.1063/1.1489714
51. Shen Z, Wang J, Lin Y, Nan C, Chen L, Shen Y (2018) High-throughput phase-field design of high-energy-density polymer nanocomposites. Adv Mater 30(2):1704380. https://doi.org/10.1002/adma.201704380
52. Li Q, Liu F, Yang T, Gadinski MR, Zhang G, Chen L-Q, Wang Q (2016) Sandwich-structured polymer nanocomposites with high energy density and great charge-discharge efficiency at elevated temperatures. Proc Natl Acad Sci 113(36):9995–10000. https://doi.org/10.1073/pnas.1603792113
53. Andrei L, Ciuprina F (2017) Dielectric properties estimation of nanodielectric interphase by numerical modeling and dielectric spectroscopythe. In: 10th International Symposium on Advanced Topics in Electrical Engineering March 23–25. Bucharest, Romania. 519–523. https://doi.org/10.1109/ATEE.2017.7905100
54. Qi L, Lee BI, Chen S, Samuels WD, Exarhos GJ (2005) High-dielectric-constant silver-epoxy composites as embedded dielectrics. Adv Mater 17(14):1777–1781. https://doi.org/10.1002/adma.200401816
55. Xu H-P, Dang Z-M, Jiang M-J, Yao S-H, Bai J (2008) Enhanced dielectric properties and positive temperature coefficient effect in the binary polymer composites with surface modified carbon black. J Mater Chem 18(2):229–234. https://doi.org/10.1039/B713857A
56. Liu H, Shen Y, Song Y, Nan C-W, Lin Y, Yang X (2011) Carbon nanotube array/polymer core/shell structured composites with high dielectric permittivity, low dielectric loss, and large energy density. Adv Mater 23(43):5104–5108. https://doi.org/10.1002/adma.201102079
57. Yousefi N, et al (2014) Highly aligned graphene/polymer nanocomposites with excellent dielectric properties for high-performance electromagnetic interference shielding. Adv Mater 26(31):5480–5487. https://doi.org/10.1002/adma.201305293
58. Chiou B-S, Lin S-T, Duh J-G, Chang P-H (1989) Equivalent circuit model in grain-boundary barrier layer capacitors. J Am Ceram Soc 72(10):1967–1975. https://doi.org/10.1111/j.1151-2916.1989.tb06008.x
59. Simon P, Gogotsi Y (2008) Materials for electrochemical capacitors. Nat Mater 7(11):845–854. https://doi.org/10.1038/nmat2297
60. Zhang G, Li Y, Tang S, Thompson RD, Zhu L (2017) The role of field electron emission in polypropylene/aluminum nanodielectrics under high electric fields. ACS Appl Mater Interfaces 9(11):10106–10119. https://doi.org/10.1021/acsami.7b00095
61. Lu J, Moon K-S, Xu J, Wong CP (2006) Synthesis and dielectric properties of novel high-K polymer composites containing in-situ formed silver nanoparticles for embedded capacitor applications. J Mater Chem 16(16):1543. https://doi.org/10.1039/b514182f
62. He F, Lau S, Chan HL, Fan J (2009) High dielectric permittivity and low percolation threshold in nanocomposites based on poly(vinylidene fluoride) and exfoliated graphite nanoplates. Adv Mater 21(6):710–715. https://doi.org/10.1002/adma.200801758
63. Celzard A, McRae E, Furdin G, Marêché JF (1997) Conduction mechanisms in some graphite - polymer composites: the effect of a direct-current electric field. J Phys Condens Matter 9(10):2225–2237. https://doi.org/10.1088/0953-8984/9/10/011
64. Wu C, Huang X, Wu X, Xie L, Yang K, Jiang P (2013) Graphene oxide-encapsulated carbon nanotube hybrids for high dielectric performance nanocomposites with enhanced energy storage density. Nanoscale 5(9):3847. https://doi.org/10.1039/c3nr00625e
65. Chen Y et al (2014) Enhanced dielectric properties of amino-modified-CNT/polyimide composite films with a sandwich structure. J Mater Chem A 2(34):14118. https://doi.org/10.1039/C4TA01818D
66. Zhong S-L et al (2018) Past and future on nanodielectrics. IET Nanodielect 1(1):41–47. Accessed 19 Sept 2021. https://doi.org/10.1049/iet-nde.2018.0004
67. Chen G, Li ST, Zhong LS (2015) Space charge in nanodielectrics and its impact on electrical performance. In: IEEE 11th International Conference on the Properties and Applications of Dielectric Materials (ICPADM), Sydney, NSW, Australia, July, pp 36–39

68. Zhang L, Khani MM, Krentz TM et al (2017) Suppression of space charge in crosslinked polyethylene filled with poly(stearyl methacrylate)-grafted SiO_2 nanoparticles. Appl Phys Lett 110:130903
69. Zhang L, Zhou YX, Tian JH et al (2014) Experiment and simulation of space charge suppression in LDPE/MgO nanocomposite under external DC electric field. J Electrostat 72:252–260
70. Wang SJ, Zha JW, Wu YH et al (2015) Preparation, microstructure and properties of polyethylene/alumina nanocomposites for HVDC insulation. IEEE Trans Dielectr Electr Insul 22(6):3350–3356
71. Zhou Y, Hu J, Dang B et al (2017) Effect of different nanoparticles on tuning electrical properties of polypropylene nanocomposites. IEEE Trans Dielectr Electr Insul 24(3):1380–1413
72. Barber P, Balasubramanian S, Anguchamy Y et al (2009) Polymer composite and nanocomposite dielectric materials for pulse power energy storage. Materials 2:1697–1733
73. Dang ZM, Zhou T, Yao SH et al (2009) Advanced calcium copper titanate/ polyimide functional hybrid films with high dielectric permittivity. Adv Mater 21:2077–2082
74. Dang ZM, Zheng MS, Zha JW (2016) 1D/2D carbon nanomaterial-polymer dielectric composites with high permittivity for power energy storage applications. Small 12(13):1–14
75. Yang K, Huang XY, He JL et al (2015) Strawberry-like core-shell Ag@polydopamine@$BaTiO_3$ hybrid nanoparticles for high-k polymer nanocomposites with high energy density and low dielectric loss. Adv Mater 2:1500361
76. Wang YF, Wang LX, Yuan QB et al (2017) Ultrahigh electric displacement and energy density in gradient layer-structured BaTiO3/PVDF nanocomposites with an interfacial barrier effect. J. Mater. Chem. A. 5:10849–10855
77. Luo SB, Shen YB, Yu SH et al (2017) Construction of a 3D-BaTiO3 network leading to significantly enhanced dielectric permittivity and energy storage density of polymer composites. Energy Environ Sci 10:137–144
78. Anandraj J, Joshi GM (2017) Fabrication, performance and applications of integrated nanodielectric properties of materials—A review. Compos Interf. https://doi.org/10.1080/09276440.2017.1361717
79. Chauhan RC, Manmohon S, Baljit S (2007) Rotating machine insulation materials and techniques—an overview. Indian J Eng Mater Sci 7:370–374
80. Mitsui H, Yoshida K, Inoue Y et al (1981) Mechanical degradation of high voltage rotating machine insulation. IEEE Trans Elect Insul EI-16:351–359
81. Gott BEB, Electr IEEE (1996) Insul Mag 12(4):28
82. IEEE Standard 1043 (1989) IEEE recommended practice for voltage-endurance testing of form wound bars and coils
83. Hidinger T, Brockschmdt M, Gröppel P, Weil M, Weidner JR (2016) Improved generator performance with a nanocomposite high voltage insulation system for stator windings—a status report. GIGRE Paper A1-109
84. Jiang PK, Chen J, Huang XY (2017) Research status of thermally conductive but electrically insulating polymer nanocomposites (in Chinese). High Voltage Eng 43(9):2791–2799
85. Wu K, Li YW, Huang R et al (2017) Constructing conductive multi-walled carbon nanotubes network inside hexagonal boron nitride network in polymer composites for significantly improved dielectric property and thermal conductivity. Compos Sci Technol 151:193–201
86. Seyhan AT, Goncu Y, Durukan O et al (2017) Silanization of boron nitride nanosheets (BNNSs) through microfluidization and their use for producing thermally conductive and electrically insulating polymer nanocomposites. J Solid State Chem 249:98–107
87. Zhang XM, Zhang JJ, Zhang XL et al (2017) Toward high efficiency thermally conductive and electrically insulating pathways through uniformly dispersed and highly oriented graphites close-packed with SiC. Compos Sci Technol 150:217–226
88. Sun JJ, Yao YM, Zeng XL et al (2017) Preparation of boron nitride nanosheet/nanofibrillated cellulose nanocomposites with ultrahigh thermal conductivity via engineering interfacial thermal resistance. Adv Mater Interfaces 4:1700563
89. Hong HJ, Kwan SM, Lee DS et al (2017) Highly flexible and stretchable thermally conductive composite film by polyurethane supported 3D networks for boron nitride. Compos Sci Technol 152:94–100

90. Toohey T, Sottos NR, Lewis JA, Moore JS, White SR (2007) Nat Mater 6:581
91. Danikas MD, Tanaka T (2009) Nanocomposites—a review of electrical treeing and breakdown. IEEE Electr Insul Mag 25(4):19–25
92. Dissado LA, Mazzanti G, Montanari GC (1997) The role of trapped space charges in the electrical aging of insulating materials. IEEE Trans Dielectr Electr Insul 4(5):496–506
93. Krivda A, Tanaka T, Frechette M, Castellon J, Fabiani D, Montanari GC, Gorur R, Morshuis P, Gubanski S, Kindersberger J, Vaughn A, Pelissou S, Tanaka Y, Schmidt LE, Iyer G, Andritsch T, Seiler J, Anglhuber M (2012) Characterization of epoxy microcomposite and nanocomposite materials for power engineering applications. IEEEElectr. Insul. Mag. 28(2):38–51
94. Liu L, Zhang C, Luo M et al (2017) A biologically inspired artificial muscle based on fiber-reinforced and electropneumatic dielectric elastomers. Smart Mater Struct 26:085018
95. Wang Y, Sun LZ (2017) Development of dielectric elastomer nanocomposites as stretchable actuating materials. Appl Phys Lett 111:161904
96. Kwak JW, Chi HJ, Jung KM et al (2005) A face robot actuated with artificial muscle based on dielectric elastomer. J Mech Sci Technol 19(2):578–588
97. Chhetry A, Yoon H, Park JY (2017) A flexible and highly sensitive capacitive pressure sensor based on conductive fibers with a microporous dielectric for wearable electronics. J. Mater. Chem. C 5:10068–10076
98. Tan DQ (2019) Review of polymer-based nanodielectric exploration and film scale-up for advanced capacitors. Adv Funct Mat 30(18):1808567. https://doi.org/10.1002/adfm.201808567

Chapter 6
Effect of Nanofillers-Reinforced Polymer Blends for Dielectric Applications

Debashish Nayak, Ram Bilash Choudhary, Sanjeev Kumar, Jayanta Bauri, and Sarfaraz Ansari

Abstract The dielectric properties of nanofiller-reinforced polymer blends were discussed in this chapter. The fast development of modern technology means that innovative, eco-friendly, flexible, cost-effective, and light materials must be made for human development. This chapter study looks at the wide range of dielectric materials discovered, from ceramics to polymer composites, and the principles that led to their use in real-world applications. This chapter has outlined the fundamental concepts of dielectric behavior in composite systems. This chapter also looks at how polarization models produce composites to give them suitable dielectric properties and combinations of the different composite parts to solve engineering and scientific problems. Also, recent discoveries of dielectric composite materials have been put into groups, and their properties have been examined. This chapter discusses polymer-ceramic dielectric composites, ceramic-based dielectric composites, natural fiber-reinforced polymer composites (NFRPCs), and bio-based polymer dielectric composites. Finally, the production procedures of all of the dielectric materials discussed, as well as their key applications, have been carefully investigated.

Keywords Polymer dielectrics · Dielectric materials · Ceramic dielectrics · Characterization of materials

6.1 Introduction

The introduction of electricity to the globe ushered in a period of rapid and profound change that was impossible to foresee. The intelligent people on this planet have been eager to use the sciences and theories that electricity has made possible to look into the unknown and spread information about it. Soon after that, in the 1700s and 1800s, much progress was made that led to the basic ideas that make induction an excellent way to produce electricity. Since then, scientists and engineers have kept

D. Nayak (✉) · R. B. Choudhary · S. Kumar · J. Bauri · S. Ansari
Nanostructured Composite Materials Laboratory, Indian Institute of Technology (Indian School of Mines), Dhanbad, Jharkhand 826004, India
e-mail: debashishiitism@gmail.com

S. Moharana et al. (eds.), *Emerging Nanodielectric Materials for Energy Storage*, Nanostructure Science and Technology, https://doi.org/10.1007/978-3-031-40938-7_6

looking into ways to make electricity. This has led to the invention of light bulbs, which have brought light to parts of the world that were previously in the dark [1].

Regarding any aspect of innovation, the material employed is a primary consideration [2]. This is because different materials have different properties that change the need for the invention. Because of this, materials science and engineering research have become an essential part of many of the world's most important discoveries. Over the past 50 years, improvements in technology related to semiconductors, lasers, microscopes, light-emitting diodes, lithium-ion batteries, nanoparticles, fiber-reinforced plastics, and metamaterials have been made possible, thanks to advancements in materials and research in material science [3]. Only because of the progress made in materials science, these researchers have been able to find breakthroughs in their respective domains. Material science is essential in almost every part of modern life, from making a pen's capillary tube to a high-resolution lens for a space mission. In addition to this, they have shown the critical role that materials engineering plays by breaking barriers in the fields of aeronautics, medicine, and information technology.

Dielectrics research is one of the most intriguing fields in materials science and electronics. Michael Faraday, a famous scientist, was the first to use this term to describe the occurrence that occurred when an insulating material was placed in between an electric field and the object being tested. When the insulator underwent charge distribution and storage for a limited duration, this factor was eventually shown to represent a sort of polarization. Capacitors, which utilized insulating materials or a dielectric substance to store charge and, in turn, energy, were invented thanks to Faraday's principles. Since that time, the fields of electronics and materials have advanced to create new dielectric materials that may be employed in applications such as capacitors, pulsed power release, integrated circuits, battery storage, and piezoelectric materials [4]. It is unquestionably anticipated that the nature of materials and their behavior would improve with the fast growth of global markets and technological breakthroughs.

This chapter covers the fundamental theory, fundamental quantities, and fundamental principles of the dielectric. This chapter discussed the basic variables that affect the dielectric as well as the method used to test the dielectric. Additionally, we discussed the high-K and low-K dielectric qualities using several polymer categorization schemes based on permittivity. Dielectric materials based on polymers and their composites are made from a variety of materials, including those with ceramic, carbon, and semiconducting bases. We explained the dielectric qualities and their imitators in the electrochemical application in several instances.

6.2 Dielectric Fundamental Theory

William Whewell created the term dielectric, which derives from the Greek word 'Di' or 'Día' means across and electric. "Dielectric" and "Insulator" are synonyms. Dielectric materials have few free or loosely bound electrons and a high specific resistance, which slows down the flow of electricity. Charge carriers do not move through a dielectric material when an electric field is present. Instead, their equilibrium positions just move a little bit. A positive charge moving along the direction of an applied electric field creates an electric dipole moment. Polarization is the number of dipole moments per unit volume. Due to opposing charge separation, an internal electrical field opposes the material's external field. Random orientation deployments result in zero internal fields without external fields. Direct or alternating current will transfer electric charges via any conductor or insulator. If the material is an insulator or dielectric, the movement of these charges will be hindered, forming a dipole moment between the two substances and polarizing them. To improve material qualities and comprehend dielectric behavior, we must grasp numerous fundamental quantities.

6.2.1 Electric Dipole Moment

An electric dipole is made up of two point charges that are equal and opposing one another, and they are separated by some distance. A dipole moment will be the term used to describe the ensuing occurrence, and polarization will be the term used to describe the accumulation of dipole moments inside a volume. When all of the dipole charges are aligned in the same direction, the polarizability of the material will be determined by the number of dipole moments that are contained inside each unit. Since the dipole moment contains a component that specifies the direction of the dipole moment that it generates, it is a vector quantity.

When two charges, $+q$ and $-q$, are separated by a distance r, the electric dipole moment, denoted by m, is a vector with magnitude $m = qr$ and a direction from the negative charge to the positive charge. When placed in an electric field from the outside, an electric dipole experiences a torque equal to the product of m, E, and the sine of the angle between m and E ($\tau = mE \sin\theta$). Torque has a tendency to align the dipole moment, denoted by m, in the direction of E. The orientation of dipole moments in the presence of an external electric field is shown in Fig. 6.1. If materials have n numbers of an electric dipole, then the total dipole moment is represented by the given equation,

$$M = \sum_{i=1}^{n} q_i r_i \tag{6.1}$$

where i represent the ith number of dipole moments. The polarization (P) of a dielectric material is defined as the dipole moments per unit volume.

$$P = \frac{M}{V} \tag{6.2}$$

where V volume of the dielectric materials.

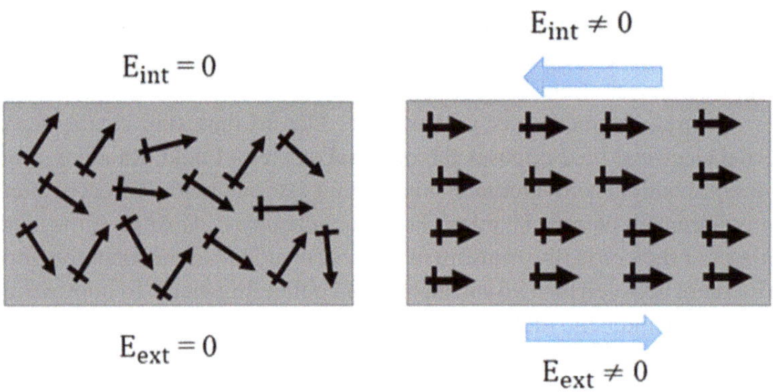

Fig. 6.1 Orientation of dipole moments in the presence of the external electric field

The formula for calculating the potential energy of a dipole is $U_e = --pE$ cos, which may also be written as $U_e = -p \cdot E$ in vector notation. In an electric field that is not homogeneous throughout, the potential energy of an electric dipole will vary depending on its location, and the dipole will be able to experience a force. When p is aligned with E, the direction of decreasing field strength corresponds to the order in which the force acting on the dipole will be. This is because the potential energy U_e will decrease in this direction.

6.2.2 Dielectric Constant

The charge density (*D*), or dielectric displacement caused by an applied electric field *E* in a dielectric medium is proportional to the magnitude of the field. This direct proportionality necessitates the inclusion of a constant, denoted by the symbol $k(\varepsilon)$, which represents the dielectric constant or permittivity of the material that is sandwiched in between the parallel plates.

$$D = \varepsilon E \tag{6.3}$$

An absolute permittivity (ε_0) is discovered to exist in a vacuum, which is the location where the electric field is produced, and this property will be expressed as follows:

$$D = \varepsilon_0 E \tag{6.4}$$

The absolute permittivity is 8.854×10^{-12} F/m because there is no air to interact with. Accordingly, the dielectric constant (also known as the relative permittivity) may be thought of as the density of the electric field created in the medium divided by the density of the electric field generated in a vacuum when the electric field is held at a constant. It is also the ratio between the material's capacitance and the vacuum's capacitance. This section describes the potential difference between the parallel plates under certain situations:

$$V = |E_0|r \tag{6.5}$$

The decrease in field strength and the potential difference that results from the addition of a dielectric material between two parallel plate conductors is indicative of the charge storage capacity of the dielectric material. For all practical purposes, the static dielectric constant may be expressed in terms of capacitances in a vacuum and with a dielectric in the center. Here is the permittivity or dielectric constant equation.

$$E_0 = \frac{C}{C_0} \tag{6.6}$$

In addition, the capacitance of the circuit, while it is empty or under vacuum, is denoted by the letter C, whereas the capacitance when it is filled with the dielectric material is denoted by the letter C_0. Although this word is not often used, permittivity has a link to electrical susceptibility. According to one idea, the permittivity values of materials are determined by their electrical susceptibility.

6.2.3 Dialectics Under the Applied Electric Field

A dielectric material's properties may be studied by applying an electric field. Polarization, dissipation, conduction, and breakdown are important characteristics of materials. Direct current aligns dipoles in a dielectric material's field direction. Furthermore, the charge migrates from one end to the other, ensuring polarization. By neutralizing electrode charges, dielectric substances store charge and act as capacitors. The permittivity of a substance is determined by its capacitance value. There are two types of current in an alternating current electric field between parallel plate capacitors. A charging current (I_C) and a loss current (I_L) determine the material's dielectric constant. The loss current is the parallel material's conductance. Some polarization and conduction currents are the same.

$$I = I_C + I_L \tag{6.7}$$

Dielectric constants may be used to indicate charge storage, where the aforementioned equation may be changed.

$$I = V(i\omega C_0)\left(\in' - i \in''\right) \tag{6.8}$$

$$\varepsilon^* = \in' - i \in'' \tag{6.9}$$

One rotation of the object is equal to $\omega = 2\pi f$ (radians per second), where the AC voltage supply's frequency is f (Eq. 6.8). $\in'$ is the real part of the dielectric constant, and $\in''$ is the imaginary part of the dielectric constant, where ε^* represents the electrical losses in the dielectric material. The dielectric constant may be stated to better explain electric losses, a complicated number in relation to Eq. (6.9). When charging is in progress, the charge is stored in the material because of polarization, and when an alternating current is flowing, the phase of the polarization current is 90 degrees out of phase with the external electric field, contributing to losses or dielectric dissipation. Loss current is a current that runs through resistance and causes energy loss. It is regarded as one of the most significant dielectric losses in the critical elements that engineers, among others, find interesting to determine the suitability of a material for the task at hand. When a capacitor is placed between parallel plate electrodes in a vacuum, the voltage is shown as follows:

$$v = V_m \cos \omega t \tag{6.10}$$

The current through the capacitor is introduced as i_i, where v represents the instantaneous voltage, V_m represents the maximum value of v, and ω represents the angular frequency.

$$i_i = I_m \left(\cos \omega t + \frac{\pi}{2} \right) \tag{6.11}$$

And I_m can be represented by,

$$I_m = \frac{V_m}{z} = \omega C_0 V_m \tag{6.12}$$

The optimum dielectric material has a current–voltage relationship of 90, when the voltage and the current components are not in phase. When a result of the increase in capacitance caused by the dielectric material being positioned between the electrodes, the current is as follows:

$$i_i = I_m Cos\left[\omega t + \left(\frac{\pi}{2} - \delta \right) \right] \tag{6.13}$$

$$I_m = \omega C_0 V_m \in \tag{6.14}$$

In the equations shown above, the dimensionless dielectric constant is introduced. When the current and voltage are out of phase at an angle of 2°, this angle is known as the loss angle in Eq. (6.11). The current between the electrodes may be split into two halves. While the charging current is identical to that of an ideal capacitor, the dielectric loss is located in the 90° trailing component of the applied voltage, and the corresponding equation for the loss angle is,

$$\delta = \tan^{-1}\frac{\varepsilon''}{\varepsilon'} \quad (6.15)$$

As a result, tan is known as the dissipation factor, and ε″ is known as the loss index. As a result, the loss index in a dielectric material is defined as the complex portion of the dielectric constant, and the dielectric loss may be calculated using the equation above. The letter *D* is used to symbolize the dissipation component, tan, commonly known as the loss tangent. The following Eq. (6.18) may also be expressed in terms of the charging current and loss current.

$$\tan\delta = \frac{I_L}{I_C} = \frac{\varepsilon''}{\varepsilon'} \quad (6.16)$$

The loss factor may be expressed in another equation when it comes to resistance and impedance in a capacitor, supposing the resistance is a theoretically infinite quantity.

$$\tan\delta = \frac{1}{\omega RC} \quad (6.17)$$

6.2.3.1 Polarization

Polarization is the spatial organization of a charged particle in response to an applied electric field. In order to counteract the effects of the electric field, the charges will become polarized. This will result in the negative charges moving in opposite directions. The presence of an external field is what leads to the generation of an electric moment over the whole of the dielectric material's volume, and this occurs in each polarizing unit, whether it is an atom, an ion, or a molecule. The polarizable unit will acquire an electric dipole moment, denoted by the symbol *p*, whose magnitude is proportional to the strength of the external field, denoted by the symbol *E*. This direct proportionality is shown by linear dielectrics and is given by the equation $p = \alpha E$, where $\alpha =$ polarizability, which represents the qualities of individual polarizable units and is denoted by a symbol. Linear dielectrics exhibit this direct proportionality.

The degree to which a dielectric substance may be polarized is crucial in determining its electrical characteristics. This parameter is independent of the dielectric volume; therefore, it is very crucial. Because polarization causes charges that may be displaced to collect at physical barriers like grain boundaries, interfacial polarization, and space-charge polarization can both be said to have occurred as a consequence of polarization. Dielectric behavior may be caused by several different dielectric processes, which can be seen at the microscopic level. The dielectric constant is determined by the polarization process, which changes with frequency. At microwave frequencies, for instance, the dipole of water molecules, which spins in response to an alternating electric field, interacts strongly with ionic conduction. This happens because water molecules are polarized in a way that allows them to conduct electricity. Atomic and electrical mechanisms are, on the whole, quite feeble. Each dielectric mechanism has a unique "cutoff frequency" as a defining property.

Dipole moments per unit volume are called the polarization of dielectric materials. If the materials have n numbers of dipole moments, then the total polarization of the materials is defined as,

$$P = \sum_{j=1}^{n} m_j \tag{6.18}$$

In the case of polar materials, there are permanent dipole moments. Due to the random arrangement's dipole moments, the resultant dipole moments of the materials are zero, but if the materials applied a strong electric field, then the materials showed two types of phenomenon, one is intrinsic dipole moments (m_i) which tends to align along the direction of the electric field to provide a net dipole moment, and the second is induced dipole moments (m_{ind}) developed in the material.

$$m_T = m_i + m_{\text{ind}} \tag{6.19}$$

Based on the polarization mechanism, dielectric polarization is divided into five major categories such as.

1. Electronic polarization (P_e),
2. Ionic or atomic polarization (P_a),
3. Orientational polarization (P_{or}),
4. Space-charge/interfacial polarization (P_{sp}),
5. Spontaneous polarization (P_{st}).

$$P = P_e + P_a + P_{\text{or}} + P_{\text{sp}} + P_{\text{st}} \tag{6.20}$$

It takes a certain amount of time for the dipoles' dielectric polarization to reach its maximum value when an external field is applied. The dipoles' dielectric polarization will diminish after some time when the external field is no longer there. The polarizations of the dipoles in the material and the alterations to the electric field occur at different times. An example of relaxation is when polarization (dipole alignment) lags after an applied field. Because of the lag, Gibbs's free energy is always decreasing and can never be restored. Transmission, or relaxation, is a function of frequency. Debye relaxation refers to the phenomenon that occurs in ideal systems when the dipole populations do not interact and may be described by the Debye equation. When all the material's dipoles relax at the same rate (same relaxation time), that is, when the relaxation time is equal to 1, where is the angular frequency of the applied field, we say that the material is undergoing Debye relaxation. In most cases, it is written as the frequency dependence of the medium's complex permittivity.

$$\in(\omega) = \in_{\alpha} + \frac{\Delta \in}{1 + i\omega\tau} \tag{6.21}$$

where $\in_{\alpha}$ is the dielectric constant at the highest frequency and $\Delta\varepsilon = \in_s - \in_{\alpha}$ where $\in_s$ is the dielectric constant at the lowest frequency and τ is the relaxation time. Real and imaginary parts of the complex dielectric permittivity yield

$$\in(\omega) = \in_{\alpha} + \frac{\in_s - \in_{\alpha}}{1 + i\omega\tau} \tag{6.22}$$

$$\in' = \in_{\alpha} + \frac{(\in_s - \in_{\alpha})\omega\tau}{1 + \omega^2\tau^2} \tag{6.23}$$

$$\in'' = \frac{(\in_s - \in_{\alpha})\omega\tau}{1 + \omega^2\tau^2} \tag{6.24}$$

The dielectric loss is given by,

$$\tan(\delta) = \frac{\in''}{\in'} = \frac{(\in_s - \in_{\alpha})\omega\tau}{\in_{\alpha} + \omega^2\tau^2} \tag{6.25}$$

In contrast, many dielectric materials have multiple relaxation times for dipoles. This means that instead of the Debye equation, they adhere to the Cole–Cole equation.

6.2.3.2 Dielectric Relaxation

Response time to a periodic external electric field is a key issue in dielectrics. Dielectric relaxation occurs when polarization lags behind an oscillating electric field. Dielectric relaxation may occur at a typical period for a certain polarization; hence, it may assist to identify the polarization process. It also causes considerable energy loss (or conversion, like in an oscillator, etc.), which is vital for engineering applications. Because time t and frequency f are inverse ($f = 1/t$), dielectric relaxation is explored in the frequency domain through dielectric spectroscopy. Impedance analyzers (LCR meters) typically work from a few Hz to 10 MHz. A vector network analyzer works from MHz to GHz. Higher frequency spectroscopy uses wave-guided systems, THz time-domain spectroscopy, etc. In many circumstances, these techniques are not marketed.

Different electric parameters, such as the complex dielectric constant $\varepsilon^* = \varepsilon' - \varepsilon''$, the complex impedance $Z^* = Z' - iZ$, the complex admittance $Y^* = Y' - iY''$, and the complex electric modulus $M^* = M' - iM''$ may be used to elucidate a relaxation process. A material's electric modulus, which is equal to the inverse of its dielectric permittivity, represents the rate at which its electric field relaxes while the electric displacement remains the same ($M^* = 1/\varepsilon^*$).

Debye Relaxation

The Debye relaxation equation is the most straightforward and tasteful mathematical expression utilized to explain the phenomena of relaxation to date.

$$\varepsilon_r = \varepsilon_\infty + \frac{\varepsilon_s - \varepsilon_\infty}{1 + i\omega\tau} \tag{6.26}$$

The static and high-frequency dielectric permittivity are shown here as ε_s and ε_∞, respectively. The typical time for relaxing is τ. Peter Debye, a scientist, was the first to describe this relaxation for noninteracting dipoles in an alternating external electric field with a distinct relaxation duration. ε'-ε" plots the optimal semicircle for Debye relaxation in the complex plane. The semicircle transforms into an ellipse, known as the Cole–Cole circle, as a result of the reciprocal interactions of dipoles in actual materials. The relaxation is known as Cole–Cole relaxation, and the following equation describes it:

$$\varepsilon_r = \varepsilon_\infty + \frac{\varepsilon_s - \varepsilon_\infty}{1 + (i\omega\tau)^{1-\alpha}} \tag{6.27}$$

Here, the value of the exponent parameter may range from 0 to 1. The Cole–Cole equation becomes the Debye equation when = 0. In various polymers, Cole–Cole relaxation is seen.

Maxwell–Wagner Relaxation

Electric inhomogeneities in materials, such as grain boundaries, the contact between dielectrics and electrodes, and others, generate Maxwell–Wagner relaxation. This is sometimes shown using a two-layer equivalent electric circuit, as in Fig. 6.2.

Maxwell–Wagner relaxation's dielectric permittivity may alternatively be stated as a Debye relaxation style:

$$\varepsilon_r = \varepsilon_\infty + \frac{\varepsilon_s - \varepsilon_\infty}{1 + i\omega\tau} - i\frac{\sigma}{\omega} \tag{6.28}$$

If the following conditions are fulfilled, the result will be: $C_0 = \varepsilon_0 S/d$; $\varepsilon_\infty = C_1\ C_2/[C_0(C_1 + C_2)]$; $\varepsilon_0 = [R^2{}_1 C_1 + R_2^2\ C_2]/[C_0(R_1 + R_2)^2]$; $\delta = 1/[C_0(R_1 + R_2)]$; $\tau = R_1 R_2(C_1 + C_2)/(R_1 + R_2)$. It seems to reason that when there is a large disparity between two resistances, the relaxation would be visible. Each RC unit in

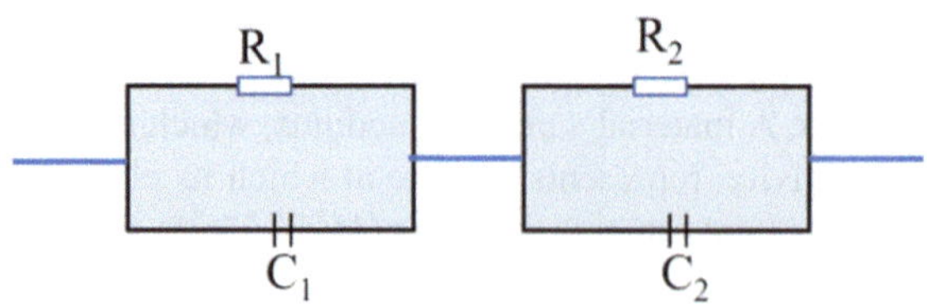

Fig. 6.2 Two-layer Maxwell–Wagner equivalent electric circuit

Fig. 6.2 corresponds to a component of the researched dielectrics, such as a grain, grain border, or electrode interface. It is usual practice to plot data in both impedance and electric modulus to distinguish between these various component relaxations. The circuit's impedance is as follows:

$$Z' = \frac{R_1}{1+(\omega R_1 C_1)^2} + \frac{R_2}{1+(\omega R_2 C_2)^2} \tag{6.29}$$

$$Z'' = R_1 \frac{\omega R_1 C_1}{1+(\omega R_1 C_1)^2} + R_2 \frac{\omega R_2 C_2}{1+(\omega R_2 C_2)^2} \tag{6.30}$$

The electric modulus of the circuit is as follows:

$$M' = \frac{C_0}{C_1} \frac{(\omega R_1 C_1)^2}{1+(\omega R_1 C_1)^2} + \frac{C_0}{C_2} \frac{(\omega R_2 C_2)^2}{1+(\omega R_2 C_2)^2} \tag{6.31}$$

$$M'' = \frac{C_0}{C_1} \frac{\omega R_1 C_1}{1+(\omega R_1 C_1)^2} + \frac{C_0}{C_2} \frac{\omega R_2 C_2}{1+(\omega R_2 C_2)^2} \tag{6.32}$$

From these equations, we may deduce that, whereas normalizing the electric modulus shows just a little amount of relaxation in capacitance, normalizing the impedance reveals a significant amount of relaxation in resistance. In light of the fact that the electric circuit corresponding to Maxwell–Wagner relaxation changes from case to case, a more complex circuit model may be utilized to characterize the dielectric behavior of Maxwell–Wagner relaxation. Here, we just provide the simplest scenario, which calls for the use of only two dielectric constituents.

Universal Dielectric Relaxation

The term universal dielectric response also refers to universal dielectric relaxation. We first present complicated conductivity to help you grasp it better.

$$\sigma = \sigma_0 + \sigma' - i\sigma'' \tag{6.33}$$

In this case, the DC conductivity is $\sigma_{0,}$ and the AC conductivity is σ'. It is unclear if the formula has a value of σ_0. You may look at the link between σ_0 and σ as $\sigma_0 = \lim \sigma$ as $f \rightarrow 0$. Typically, the AC conductivity is determined from dielectric measurement by:

$$\sigma = \varepsilon_0 \varepsilon'' \omega \tag{6.34}$$

Jonscher observed that the frequency-dependent AC conductivity for several dielectrics complies with the following exponent connection.

$$\sigma = A\omega^n (0 < n < 1) \tag{6.35}$$

This is referred to as "universal dielectric relaxation". It is referred to as a "virtually constant loss" if $n = 1$. The ion hopping effect is thought to explain the universal dielectric response. But at high frequencies, electric heterogeneity may also cause similar phenomena. At sufficiently low temperatures or high frequencies, almost continuous loss occurs, and the cause was formerly thought to be vibration relaxation or off-center relaxations that spread out into a wide distribution.

It has so far been unable to pinpoint a particular relaxation. The temperature typically has a considerable impact on the characteristic relaxation time constant, which may be described as an exponent function of temperature:

$$\tau = \tau_0 e^{E_a/\mathrm{k_B}T} \tag{6.36}$$

Here, the temperature is T, the Boltzmann constant is $\mathrm{k_B}$, and E_a is the activation energy. The activation energy is a reflection of the power required for dipoles to jump from one potential well to another. As a result, it provides details on potential relaxation mechanisms.

6.3 Techniques for Dielectric Measurements

When it comes to measuring dielectric properties, there is no one method that can describe all materials throughout the whole frequency range. Because it is difficult to get an accurate measurement for both kinds of materials (lossy and low loss), it means that each band and their respective losses need a distinct procedure. When attempting to characterize materials based on their dielectric measurements, there is always some degree of ambiguity. Some significant factors on that the whole process depends,

- Temperature,
- Nature of the materials,
- Thickness of the materials,
- Size of the materials,
- Frequency,
- Coating and non-coating,
- Destructive and nondestructive,
- Sample cost,
- Sample prepetition process.

Considering the above parameters, testing frequency and test material should be addressed while choosing a measurement procedure. Even the measuring equipment, sample holder, and design are determined by frequency and substance. As for measurement techniques, they may be divided into two primary categories: resonant and non-resonant. Materials may be characterized at single or several discrete frequencies using resonant techniques. This method employs a dielectric material as the resonant element. However, it is applicable only to test samples with minimal loss.

Resonators that are dielectric, planar, and split are all good examples. Perturbation and perturbation-induced resonant frequency shift are two further techniques. This approach suits low- to moderate-loss samples. Non-resonant methods may measure a wide frequency range. It may classify materials by calculating their reflection and transmission coefficients. Here are several prominent and relevant dielectric measuring methods.

6.3.1 Coaxial Probe Method

A coaxial probe, coaxial-line probe, or open-ended coaxial-line technique are all terms used to describe this kind of probe and its associated technology. Measuring lossy materials at high frequencies (such as RF and microwave) may be accomplished using this approach, which is one of the most practical and regularly used methods. In order to measure the complex reflection coefficient, it is used with the VNA.

A metallic probe is used by the coaxial line in order to detect the reflected signal (both phase and magnitude) from the MUT. When testing liquids, the probe is submerged into the sample, but when testing flat surfaces, it contacts the sample. Dielectric measurements may be carried out using this approach across a wide frequency range, anywhere from 0.5 to 110 GHz, and the process itself is relatively straightforward. However, this technique exhibits some deflection for the measurements of low-permittivity materials across a large frequency range of around 0.5–110 GHz [5]. However, some deflection may be seen using this approach for materials with a low permittivity (Fig. 6.3).

6.3.2 Transmission Line Method (Waveguide)

Another common method involves placing a sample of the chemical under examination in the center of a closed transmission line (as shown in Fig. 6.4). When taking measurements, both the reflection and transmission coefficients come in handy. Although it outperforms the coaxial line in terms of precision and sensitivity, the frequency range that it is capable of is much less than that of the coaxial cable. Because a sample has to span the whole cross-sectional area of a line, it needs to be prepared in the shape of a slab or an annular geometry slab. This makes sample preparation very complicated and, as a result, more time consuming.

6.3.3 Free-Space Method

This approach is risk-free since it does not require physical touch. Though often used at a higher frequency, it is also capable of being employed at a lower one; however,

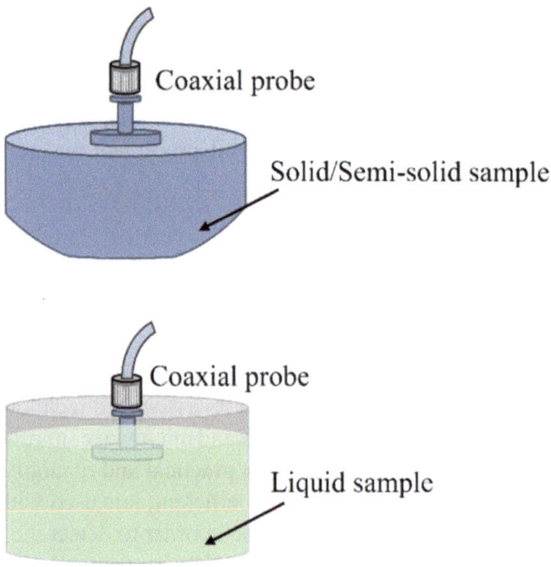

Fig. 6.3 Schematic diagram for coaxial probe method [6]

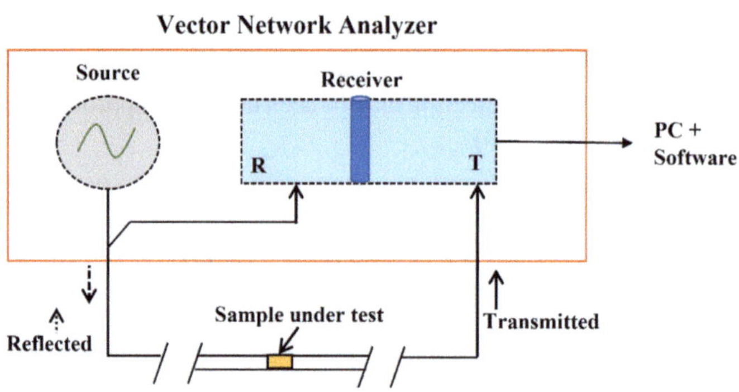

Fig. 6.4 Schematic diagram for transmission (waveguide) line method [7]

the latter application is restricted in terms of the sample sizes it can practically handle. The dielectric properties of solid materials are measured using the free-space approach, which requires the samples to be big, flat, and properly faced on both sides. Homogeneous materials are those that are uniform and free of any imperfections. Material is positioned in between two horn antennas to direct energy onto or through the material in combination with a vector network analyzer. The use of the free-space measurement approach has several benefits, the most important of which is the ability to get reflection and transmission coefficients without making any kind of physical contact with the sample (Fig. 6.5).

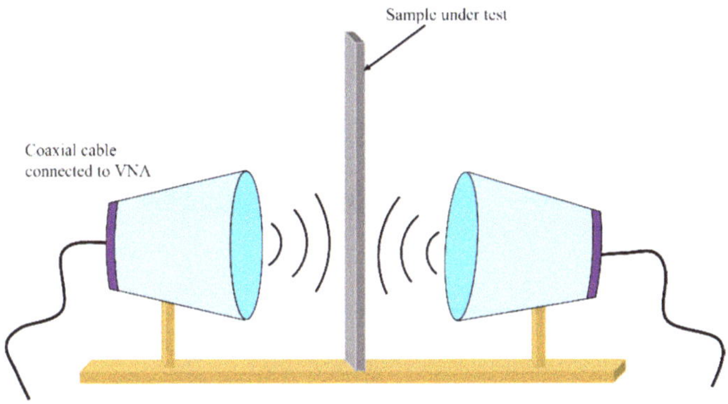

Fig. 6.5 Schematic diagram for free-space method [8]

6.3.4 Resonant Cavity Method

This method does not involve any physical contact and is not harmful. It is most often used at a higher frequency, but it can also be used at a lower frequency; however, the latter application is restricted in terms of the sample sizes it can practically handle. The dielectric properties of solid materials are measured using the free-space approach, which requires the samples to be big, flat, and properly faced on both sides. Homogeneous materials are those that are uniform and free of any imperfections. Material is positioned in between two horn antennas with the intention of directing energy onto or through the material in combination with a vector network analyzer. The use of the free-space measurement approach has a number of benefits, the most important of which is the ability to get reflection and transmission coefficients without making any kind of physical contact with the sample (Fig. 6.6).

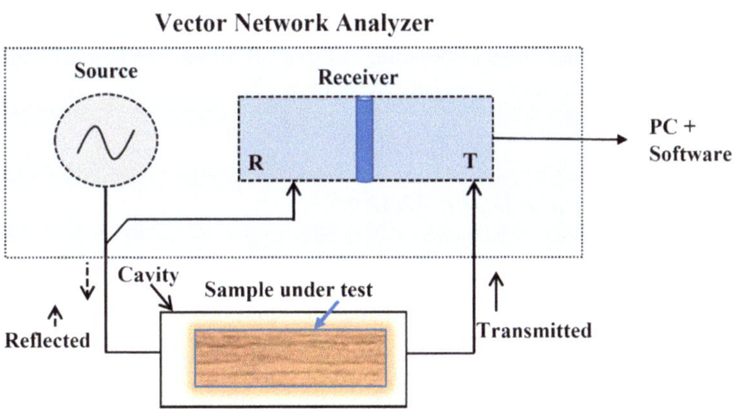

Fig. 6.6 Schematic diagram for cavity resonator method [7]

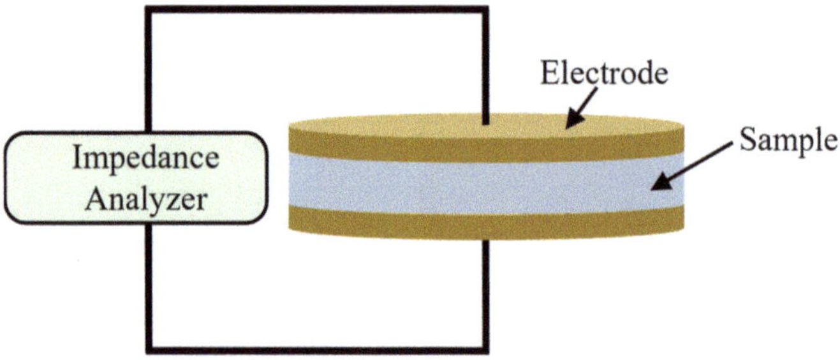

Fig. 6.7 Schematic diagram for parallel plate method [6]

6.3.5 Parallel Plate (Electrode) Method

To create a capacitor using the parallel plate capacitor technique, a dielectric material is placed between two electrodes in the form of a thin sheet sample in order to create a capacitor. The LCR meter, an impedance analyzer, and a dielectric fixture are the three instruments that may be used to collect measurements (see Fig. 6.7). In this technique, generally low frequencies of less than one gigahertz are used. When doing a test using parallel plates, it is necessary to take into account the size of the material as well as measure its capacitance and dissipation factor. In order to determine the dielectric constant of a material, a dielectric sample must first be placed in a sample holder before the capacitance value can be used to determine the permittivity. It includes extremely easy sample preparation and setup, and the frequency range that is generally used for this purpose is from 20 Hz to 1 GHz. It has a high measurement precision (approximately 1% for *r*' and 5% 0.005 for tan), and it has several other benefits as well. If the air gap and its consequences are not taken into consideration and calibrated, however, this may lead to substantial inaccuracy. Poor results are also brought about by the influence of electrode polarization, which leads to false observations. It is possible to lessen its impact by using electrodes with a large microscopic surface area or by operating at higher frequencies; this is because the effect weakens significantly with rising frequency [3].

6.3.6 Planar Transmission Method

These different kinds of transmission lines find widespread use in a variety of RF and microwave components. The microstrip line approach is the most frequent and straightforward method for measuring dielectric materials. It is simple to construct, there are few costs associated with its production, and its small construction makes it particularly well-suited for use in industrial settings. For the purpose of determining the dielectric permittivity, the test sample may function either as a substrate (when the sample is solid) or as a superstrate (when the sample is either solid or liquid).

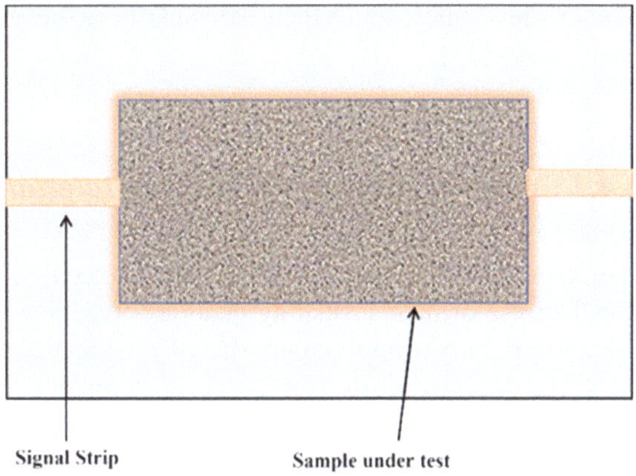

Fig. 6.8 Schematic diagram for planar transmission line method (View from the top side of MSL) [6]

When a dielectric sample is placed over a quasi-TEM transmission line, it is generally known that the effective permittivity of MSL will change, a behavior that is highly dependent on the permittivity of the sample itself. In order to obtain the sample's dielectric characteristics, this will alter MSL's effective dielectric constant (εeff) and characteristic impedance (Z_0). First, you need to get the unloaded effective dielectric constant so you can take a reading. Then, you will need to attach the sample to the signal strip (as shown in Fig. 6.8). This study examines the alterations in the effective dielectric constant (εeff) and the characteristic impedance of a microstrip line (MSL). These changes can be utilized to determine the dielectric properties of a sample through the plane transmission line method. Bernard conducts dielectric measurements using a microstrip ring resonator. The approach demonstrates a change in the quality factor and a variation in the resonant frequency. When a sample is put on top of the ring, its presence alters the substrate or the air barrier, hence shifting the resonance frequency. When used as a substrate, rather than an overlay test material, the influence of the dielectric sample is able to shine through more clearly.

6.4 Dielectric Materials

Studies have revealed that polymers offer appealing qualities such as high flexibility, attractive chemical stability, cost efficiency, and simplicity of fabrication, which has led to their rising use in electronic applications [9]. Because polymers possess this inherent capacity, they may serve as viable substitutes for inorganic and ceramic-based dielectric materials. Polymers provide a number of benefits, but the fact that they have poor thermal stability and display poor dielectric constants restricts the majority of the applications to which they may be put. However, by manufacturing nanocomposites and including inorganic fillers in the composites, it is possible to enhance the dielectric constant values [10]. Research has been done on polymer dielectrics in many different aspects of their applications. One example is in the field of microelectronics, which focuses primarily on the reduction in the size of electronic components [11].

6.4.1 Classification of Dielectric Materials

Dielectric materials are sometimes called "electrically insulating materials" materials because of their insulating qualities. Traditional dielectric capacitors may be categorized into several groups based on the various dielectric materials. First, there are dielectric materials based on polymers. This kind of capacitor mostly uses polymer materials as its dielectric materials. Then, there are dielectric capacitors made of ceramic. Such capacitors' dielectric components may be multi-phase ceramics, glass–ceramics, ceramic films, etc. Graphene and carbon nanotubes (CNT) are two examples of the types of carbon-based materials that are utilized to make dielectric materials. In a variety of contexts and applications, dielectric metal alloys made from oxides, sulfides, and nitrides of various metals are used. Perovskite materials are now being employed in applications that need them to be dielectric as well. In this chapter, we quickly go through every polymeric material and every composite that can be made from it.

6.4.2 Dielectric Polymer and Polymer–Polymer Composite Materials

Improving polymer breakdown strength is a challenge. Ramprasad investigated the process of polymer breakdown strength mathematically and found that dipoles increased dipole-induced scattering, which reduced and constrained hot electrons. Electron scattering by dipoles and phonons accounted for the temperature dependence of the breakdown field. Amorphous polymers have a high-voltage potential due to their randomly dispersed dipoles. Dipole scattering restricted mobility may analyze dipole scattering's effect on the polymer breakdown field. Because of their low cost, high dielectric strength, high energy density, and low dielectric loss, polymer-based dielectrics are preferable over ceramic dielectrics for various applications. Since these capacitors contribute a significant amount of space and weight (> 30%), further improvements in energy and power density of dielectric materials are necessary for miniaturization and better functionality in contemporary electronics and electrical power systems. High electric energy density with minimal dielectric loss is a key difficulty in producing polymer-based dielectrics. High-*k* polymer and low-*k* polymer and its composites play vital roles in the dielectric applications which are discussed below. All the polymer and polymeric composites were synthesized by different polymerization processes which are described in Table 6.1.

Table 6.1 List of the different polymerization process and their importance with examples

Polymerization process	Methods and importance of the polymerization process	Example of polymer and its composites	Refs.
Sol-gel method	Because the synthesis process is done at room temperature and organic ingredients may be incorporated at the early phases of the reaction, the sol-gel approach is frequently employed in the preparation of polymer compounds. Polycondensation and hydrolysis may also occur concurrently to form glass-like polymer matrices. This process may also be used to create polycrystals and organic and inorganic substances	Sol-gel PPy	[12]
Melt blending	Melt blending, in which polymeric components are melted and chilled (cooled), is an efficient method for generating polymer composites	Clay/polymer composites	[13]
Solvent method	The solvent strategy employs a process called polymer intercalation from a solution, which is a solvent system in which the polymer or monomer may swell. Water, toluene, DMSO, DMF, and chloroform are the most common solvents used to dissolve the polymer into the intercalates	POP	[14]
Free-radical polymerization	A wide variety of polymers and material composites may be synthesized using the free-radical polymerization synthesis method. It is one of the most flexible types of polymerization because simple reactions may occur between polymeric free-radical chain ends and other chemicals or substrates due to the generality of free-radical chemical interactions	PMMA, PMMA-ZnS	[2, 15]
In situ polymerization	To initiate polymerization, a liquid monomer solution is absorbed by a modified silicate layer, where it then seeps into the void between the layers	PAni, PPy, PIn, etc	[16–18]
Spin-on technique	Making polymer thin films using spin-on or spin-coating requires dissolving soluble linear or branched polymers in a solvent. To evenly distribute the polymer solution, we place it in a jar and spin it at rates between 1,500 and 500 revolutions per minute	Macroporous polymer	[19]

(continued)

Table 6.1 (continued)

Polymerization process	Methods and importance of the polymerization process	Example of polymer and its composites	Refs.
Chemical vapor deposition	An electrically charged gas begins a reaction with plasma in polymer CVD, while the substrate temperature is adjusted independently. The coating material is heated, and the pressure surrounding the vessel is decreased, allowing the polymer plasma within the chamber to evaporate. The vapor is kept from settling on the substrate, resulting in a homogeneous covering. The thickness of the film deposition and the time required for the vaporization process are determined by temperature and pressure	Polysiloxane	[20]
Solvothermal	In solvothermal synthesis, the desired product is formed in solution using a closed-system technique that employs high temperatures and pressures to trigger a chemical reaction or breakdown of the precursor components	1D zinc coordination polymer	[21]
Electrochemical	Electropolymerization, or the production of polymer films in an electric field, typically uses either a constant potential (potentiostatic) or a linearly scanned potential (potentiodynamic) technique. Electropolymerization, or the production of polymer films using an electric field, typically makes use of either a constant potential (potentiostatic) or a linearly scanned potential (potentiodynamic) technique	PDMS	[22]
Hydrothermal	Hydrothermal synthesis is a common method used to create nanomaterials. It is simply a reaction-solution process. During hydrothermal synthesis, nanomaterials may be created at temperatures ranging from room temperature to thousands of degrees Celsius	Porous organic polymer	[23]

6.4.3 *Low-*k *Polymer and Its Composites*

If the dialectic permittivity of a polymer or polymer composite is less than 4.2, then the polymer or polymer composite may be referred to as having a low-*k* value. To reduce the permittivity of polymers, a number of different modifications can be applied to the polymer chains. Some examples of these adjustments include

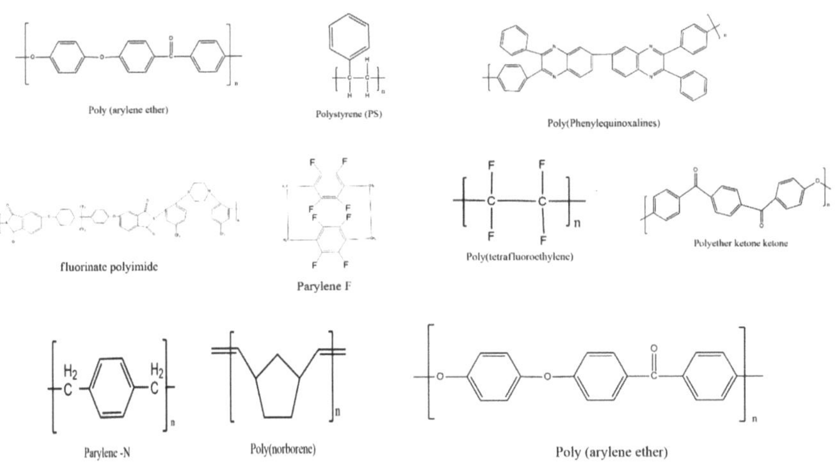

Fig. 6.9 Chemical structure of low-*k* polymer used for dielectric application

decreasing the ionic and orientation contributions and adding fluoride or carbon to the polymer chain.

Polymers, when used in dielectric applications, may be split into two categories: thermoplastic polymers and thermoset polymers. The addition polymerization process is most commonly used to manufacture thermoplastic polymers. Common thermoplastic polymers used in insulation systems include polypropylene (PP), polyethylene (PE), polyvinyl chloride (PVC), polypropylene (PP), and polyamide (PA). Insulation systems in electrical devices benefit greatly from the usage of thermoplastic polymers. In contrast, when heat or chemicals are used to cure thermostable or thermoset polymers, the polymerization process is irreversible and the resulting material cannot be altered. However, the composite's electrical and mechanical characteristics will undergo a cataclysmic shift due to treating these thermosetting polymers. These polymers are capable of taking on the shape of a resin. Epoxy, polyester, phenol, and silicon resins are all examples of these types of resins. The structure of low-*k* polymer is shown in Fig. 6.9.

6.4.4 Analysis of Important Low-k Dielectric Polymer

6.4.4.1 Dielectric Polymers Based on Silicon-Based Materials

These polymers are a hybrid mix with the formula RSiO3, where R may be any number of different things like alkyl, aryl, carboxylic acid, etc. Common silicon-based polymers with low dielectric permittivity and microwave losses include poly(hydrogen silsesquioxane) (PHSSQ) and polymethyl silsesquioxane (PMSSQ).

Table 6.2 At normal temperature, the dielectric permittivity of common traditional polymers

Polymer	Dielectric permittivity	Polymer	Dielectric permittivity
Poly(arylene ether)	2.9	Poly(arylene ether oxazole)	2.6–2.8
Polyquinoline	2.8	Poly(ether ketone ketone)	3.5
Poly(norbornene)	2.4	Silsequioxane	2.8–3.0
Polystyrene	2.6	Fluorinated polyimide	2.6–2.8
Poly(phenyl quinoxaline)	2.8	Poly(tetrafluoroethylene)	1.9
Polynaphthalene	2.2	Non-fluorinated aromatic polyimide	3.2–3.6
Parylene N	2.6	Parylene F	2.4

Because of their outstanding thermal, chemical, mechanical, and electrical capabilities, they have been widely researched for use in microelectronic packaging. The dielectric constants decreased gradually as frequency increased (2.96–2.55 from frequency ranges 100–1000 kHz). Dielectric loss rises with frequency, with a loss of 0.02–0.032 predicted from 100 kHz to 1 MHz [24].

The dielectric properties of nano-hybrid composites containing the silica-based polymer tetraethyl orthosilicate have been investigated. The dielectric properties of samples made by combining polyvinylpyrrolidone and silicon dioxide using the sol-gel technique in the presence of a coupling agent are studied. Dielectric constant and loss were determined using an impedance analyzer on samples with varying weight ratios. The estimated loss tangents demonstrate that when the silica concentration rises from 31.27 to 55.35 wt.%, the losses increase at low frequencies spanning from 1 to 100 kHz [25] (Table 6.2).

6.4.4.2 Low-*k* Epoxy Resins

Low-*k* dielectrics make use of epoxy resins. The properties they possess make them perfect for use in dental products. Their mechanical qualities, great adhesion strength, sensible heat, and electrical resistance make them ideal for application in coatings, adhesives, electronic materials, and polymer matrices for fiber-reinforced composites. Their primary use is as a microscale organic filler in dry distribution transformers. In order to create trifunctional epoxy resins, trimethylolpropane, and epichlorohydrin are combined. Low viscosity, noncrystalline, and plastic resin does not meet the application [26]. At low temperatures (127–348 °C), curing progresses slowly at first, then levels out after a few minutes. Epoxy systems that already include fluorine are modified with fluorinated monomers and oligomers for use in dielectrics. Fluorine boosts the blend's hydrophobicity but reduces its overall surface energy. Between 2 and 10 GHz, dielectric constants ranged from 6.8 to 5.5. Epoxy resins may be manufactured from biodegradable components such as sugars, starches, proteins, lipids, and oils. There is a high demand from business because of their low production cost, small carbon footprint, and biodegradability. Epoxy resins may be made at a low cost by reacting soybean oil and castor oil with glacial acetic acid.

6.4.4.3 Fluorinated Polymers

Fluorocarbons are nonpolar, making them hydrophobic. Fluorinated polymers include at least one fluorine atom in the polymer chain, which limits water absorption. Fluorine may reduce the dielectric constant of polyimides. Fluorine hinders electronic polarization by being nonpolar and low-polarizable. Altering the polymer chain's free volume changes the composite's polarizability. PIs have strong chemical and thermal stability, and fluorine adds performance to electronics. Fluorination is used to enhance PI thin-film dielectric characteristics. Fluorine groups reduce surface free energy and dielectric permittivity of PI films, according to studies. When a reduction in dielectric constant is necessary, the fluorination procedure may be extended to epoxy resins. Epoxy resins have been used as insulation material because of their high electrical resistance and widespread use as molding compounds in electronic packaging. The regulated dielectric constants for fluorinated epoxy resins range from 2.71 to 3.25, and the related loss tangents have been estimated to be between 1.3×10^{-3} and 3.5×10^{-3}, respectively.

6.4.4.4 Porous Polymers

Hydrogen methyl silsesquioxane compounds, which provide low-*k* dielectric and loss properties, are promising porous polymers with low dielectric constants. They have a dielectric permittivity of 2.06 and have pores that are smaller than 10 nm in size because of their strong hydrogen bonding, which inhibits phase separation (on average). They are created by hybridizing the porous polymers with an amphiphilic block copolymer, namely poly[styrene-block-(2-vinylpyridine)] or PS-block-P2VP. The creation of regular nanopores in thin films is facilitated by this hybrid composite [27]. It is shown that the loss at 10 kHz is 0.04 and the dielectric constants are 3.6. The significant spontaneous polarization is what causes the consequences of these features. Additionally, the polymer chain's vast contact area leads to various interfacial polarizations [28]. The benefits of porous poly(vinylidene fluoride), often known as PVDF composites, in terms of mechanical strength, solvent resistance, and thermal stability make them valuable in applications involving multilayer dielectrics. At normal temperatures, PVDF composites have dielectric constants ranging from 7 to 13 [29]. Porous PVDF composites will have two compact surface layers and a dielectric permittivity of 1.5 to 2.5. Studies have shown the conductivity of copper nanowires in porous PVDF's dielectric behavior. When the number of nanowires was increased, low dielectric permittivity between 11.8 and 17.5 was observed. As the frequency was raised, the permittivity gradually decreased. Results from the dielectric permittivity have also shown that the Maxwell–Wagner–Sillars effect and interfacial polarization were identified.

6.4.5 High-k Polymer and Its Composites

In order to maximize performance, polymers with excellent dielectric properties at working frequencies and voltages are desired. These properties include high permittivity, low dielectric loss, and a high breakdown field [30]. High dielectric constant materials have a dielectric constant greater than silicon oxide ($k \sim 4.2$). The intrinsic dielectric constants of polymers have been increased using a variety of methods, including modifying the chain architecture of polymers, mixing polymers, dispersing ceramics, and conducting nanoparticles throughout polymers. There are many high-*k* polymers, which are good for dielectric application, and by the addition of other polymers to it, their dielectric properties increased. In this chapter, we described some important polymers and their composites for high-*k* dielectric properties.

6.4.5.1 Basic Polymer for High-*k* Dielectric Application

Polyvinyl Alcohol

Poly(vinyl alcohol), or PVA for short, is a water-soluble polymer having a high dielectric constant (between 5 and 8) between 10 Hz and 1 MHz, making it useful in low-temperature electronics. When PVA-infused composite substrates are cross-linked with an ammonium dichromate or another cross-linking agent, the resulting materials have a higher dielectric strength.

Polyvinylidene Fluoride

Mechanically flexible, low thermal conductivity, strong chemical and corrosion resistance, a dielectric constant of 8, and considerable dipole moments in the CH_2-CF_2 units all contribute to thermoplastic fluoropolymers' widespread usage as electrical insulators. With a leakage current of (4×10^{-6} A) or less, PVDF composite thin films are ideal for use in microelectronics. They are spin-coated on substrates whose amorphous phases at interfaces exhibit greater temperature relaxation. PVDF composites have a decent dielectric breakdown strength and minimal power dissipation compared to other high-*k* polymeric materials. PVDF composites with high dielectric constants may be employed in "gated" dielectric actuators, electric stress management, and capacitors [29]. There are various dielectric PVDF-based composites materials with dielectric loss and dielectic constant are tabulated in Table 6.3.

Cyanoethyl Polymers

CEP is an organic polymer with hydroxyl groups. A polysaccharide chain with a cyanoethyl side chain has a huge dipole moment. Their dielectric constant might

Table 6.3 Composites based on polyvinylidene fluoride (PVDF) and its dielectric property

PVDF-based composites	Filler materials	Dielectric loss	Dielectric constant	Refs.
PVDF-10 T	TiO_2	3.29	23.06	[31]
PVDF/PVA-MK (30%)	PVA and MK	0.1 (10^3 Hz)	4.27 (10^3 Hz)	[32]
PVDF/GO (20%)	GO	0.28	1.83	[33]
Zn/PVDF	Zn core–shell	3	60	[34]
$Ba_{0.6}Sr_{0.4}TiO_3$/PVDF	BST		8.7	[35]
RuO_2@$n$$BaTiO_3$/PVDF	RuO_2@nBT	1.467	3837.16	[10]

reach 16. Spin-coating CEP composites into thin films provides superior dielectric and semiconductor interfaces, like PVA composites. Cross-linking agents up to 50 wt% of CEP polymers reduce leakage current the greatest, meeting application requirements [36].

6.4.5.2 High-*k* Polymer–Polymer Blends and Composites

Polymer systems' desired qualities may be achieved through blending polymers with the right structure and properties in order to maximize the blend's permittivity. Researchers may create high-permittivity polymer blends by combining high-permittivity polymers with low-permittivity polymers. The improved permittivity comes from polymer–polymer interactions and novel mix topologies. The permittivity of polymer blend system (k_{blend}) is intermediate between components, $k_{\text{polymer-A}} < k_{\text{blend}} < k_{\text{polymer-B}}$. $k_{\text{polymer-A}}$ and $k_{\text{polymer-B}}$ are mix polymers. This blending process is a straightforward and effective path for organic materials, compared to other difficult synthetic procedures, generating excellent dielectric and mechanical characteristics, notably useful for capacitor manufacturing.

Researchers also have concentrated on polymer composites to boost the permittivity of polymer systems to fulfill electronic industry needs. High-permittivity polymer composites use various fillers distributed in the polymer matrix. Polymer composites will have high permittivity, great temperature stability, and frequency properties by mixing high-*k* inorganic materials with organic insulation. Two fillers are often used. High-*k* ceramic particles and electrically conductive additives like metals or carbon are used. These two additives improve polymer composites in various ways. Nanofiller agglomeration and poor compatibility between inorganic fillers and organic polymer matrices are major roadblocks to composite advancement. When fillers are added to composites made from high-*k* polymer systems, the leakage current and dielectric loss of such materials rise; finally, the vast disparity in permittivity causes the electric field within the composites to be distorted when subjected to an external electric field. There are different polymer blenders and polymer composites which showed high-*k* dielectric constant are discussed below.

Polymer–Polymer Blends and Composites for High-*k* Dielectric Properties

Dang investigated a PP/PVDF blend, which stands for polypropylene/polyvinylidene fluoride [37]. Novel high PVDF is combined with PP matrix to increase permittivity. Permittivity reached ~5.4 at 1000 Hz, approximately 3 times pure PP's (~2.2). This blended solution combines the dielectric characteristics of PVDF with the processability of polymers. They might offer a new generation of capacitor dielectric films. Using solution mixing and casting, polymer blend films were made. This film's 100 Hz permittivity is 12. This is higher than PMMA (~2.9) and nears PVDF.

Nan et al. explained the combined ArPTU with PVDF for solution-cast composite films for dielectric application [38]. Permittivity and dielectric loss at 1 kHz are 9.2 and 0.02. Blend films restrict early polarization saturation at low fields, improving breakdown strength and energy density. PVDF/ArPTU (90/10) film has 10.8 J/cm^3 energy density at 700 MV/m and 83% charge–discharge efficiency. At certain volume fractions, the permittivity of polyvinylidene fluoride-polyamide (PA11)/PVDF blends can approach 60, which is twelve times greater than PA11 (~5) alone and five times higher than PVDF (~12). Dielectric permittivity (k_{blend}) surpassed those of the constituent polymers, unlike ordinary polymer blends [39]. These findings show that polymer-blended systems with high-permittivity components are a new, versatile, easy-to-process material. They may be used in flexible electronics.

For low-loss, high-energy-density capacitors, Zhang et al. suggested polar-fluoropolymer blends with tailed nanostructures [40]. Poly(ethylene-chlorotrifluoroethylene) [P(VDF-CTFE)] and poly(vinylidene fluoride-chlorotrifluoroethylene) [P(VDF-CTFE)] are two examples of such mixes (ECTFE). Classical composite theory predicts a dielectric constant of 7 and modest loss (~1%) for blend and cross-linked blend films. Cross-linking blends increased nanoscale mixing, reducing losses in blend films at high fields while preserving energy density.

*6.4.6 Ceramic and Polymer-Ceramic Composites as Dielectric Materials for High-*k *and Low-*k *Dielectric Applications*

Linear or nonlinear dielectric materials exist. An electric field induces polarization in linear dielectrics, and removing it depolarizes the material. There is a subcategory for both polar and nonpolar substances. Polar molecules retain their dipole moments even when exposed to no external electric field. Generally speaking, the dipole moments of molecules in polar dielectrics are larger when they are not symmetric. They experience electronic and ionic polarization, making them molecules without permanent dipole moments. Alkali halides, metal oxides, paraffin, benzene, and carbon chlorides are all examples of ionic crystals. Molecules in nonpolar dielectrics do not exhibit a persistent dipole–dipole moment because of the symmetry between the positive and negative charges. External fields cause them to become polarized.

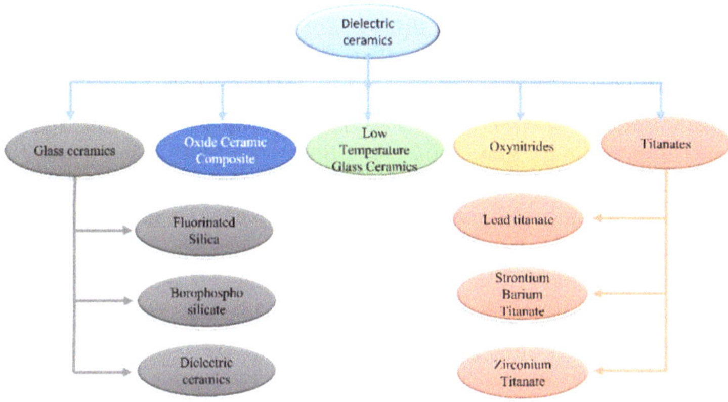

Fig. 6.10 Classification of dielectric ceramic materials

When it comes to polar molecules, atoms or molecules with spherical symmetry have their electron clouds pointing in the direction of the electric field. Positive and negative charges are generated via polarization. These materials include atoms or molecules such as silicon, as well as inert gases, liquids, and solids. These chemicals are polarized by electrons. Ceramics are composed of inorganic, nonmetallic, and nonorganic components, as seen in Fig. 6.10. Heat processing is possible with single or polycrystalline ceramics. Ceramics are corrosion-resistant while remaining stiff and brittle. Because new technologies do not need traditional ceramic compounds, better ceramics have been developed to keep up. Despite its mechanical shortcomings, ceramics have a high stiffness factor, hardness, and abrasion resistance. They are resistant to corrosion and heat. Modern ceramics are employed in the electrical industry due to their material properties. Ceramics are resistant to both weather and electricity. Ceramics are used in the electrical and engineering industries because of their properties. Advanced ceramics that are often used include aluminum oxides and zirconium oxides. Carbides, nitrides, silicides, and borides are examples of ceramics that are not oxides. Single-crystal silicates are used in high-capacity capacitors, whereas steatite tubes are characterized by low permittivity and capacitance. The high permittivity of titanium oxide led to the development of ultracapacitors.

6.4.6.1 Glass–Ceramic

Some of us may be aware that sand from the soil is heated and melted to create glass. The amorphous compounds of silicon and lime make up glasses. Glasses made from soda-lime and silicate are rather common. Despite their names, soda-lime and silicate glass–ceramics do not include any of the aforementioned elements. Pure silicate and borate glasses have a 3.2 to 3.8 MHz dielectric constant. Even though the composites have a higher dielectric permittivity than pure borates and silicates, their values are still below 4.2, making them low-k dielectric ceramics. The dielectric constant might be increased to more than 4.2 with the help of certain 'dopants. Alumina doping helps ceramics attain 5 and 1 MHz permittivity. Lead borosilicate glass (~7.0), magnesium aluminosilicate glass (~5.0), calcium aluminosilicate glass (~6.0), and lithium silicate glass (~6.5) have low permittivity as compared to others. Glass–ceramics have high dielectric constants and thermal expansion coefficients, making them ideal for multilayer ceramics. High-performance packaging and pin grid arrays employ these chemicals [41].

Borosilicate glass has a dielectric constant between 3.9 and 4.2, placing it in the family of low-*k* dielectrics with two other amorphous phases. The matrix was formed at 950 °C, matching silica's thermal expansion coefficient. Extra dopants are needed to prevent silica matrix martensitic transformation. Pure silica has a permittivity of 4 and is the inorganic limit. Porosity and how porous channels increase dielectric characteristics have been studied in this substance [42]. Polymers are included in these silica matrices. Polymer-modified ceramics fulfill low permittivity and losses. Molecular clusters may be seen in microscopic examinations of silica and polymer-modified silica gels. In order to create nanopores, a two-phase technique is required, consisting of a continuous solid phase and a porous air phase.

Connection in ULSI circuits, including computer DRAM chips, MPUs, and application-specific ICs, requires multilayer dielectric thin films. Therefore, the dielectric permittivity of these layers should be kept low to minimize signal latency and cross talk. Fluorinating silica matrices is one approach to reducing dielectric permittivity. Silicon oxide-fluoride films are formed on silicon via chemical vapor deposition (CVD) [43].

Borophosphosilicate (B_2O_3–P_2O_5–SiO_2) ceramic glass is a good electronic packaging material. With their glass–ceramic composition, boron trioxide and phosphorous pentoxide boost the coefficients. The 1:1:2 stoichiometric ratios characterize these glass–ceramics dielectric constant and loss tangent. Glass-derived ceramics internalize these chemicals. Additional sintering and surface crystallization yields external nucleation in these composites, and the use of silver, gold, and copper in metallization indicates the viability of tape casting for fabricating multilayer glass–ceramic sheets. Borophosphosilicate glass–ceramics made by heating fine-grained boron phosphate glass–ceramics. Silica renders the composite acid, basic, and water-resistant. Sol-gel synthesis includes introducing reagents at the correct temperature and pressure in an alkaline environment. Boron (tetramethyl borate), phosphorus (tetramethylphosphite), and modified silica (tetramethylorthosilicate) alkoxides are used. (Tetramethoxysilane will also be utilized as a catalyst) [44].

6.4.6.2 Oxide Composites of Glass–Ceramic

To create samples with low permittivity and optimum thermal expansion coefficients, dielectric interlayers mix cordierite, quartz, and borosilicate substrates. This oxide glass–ceramic is used in high-temperature ceramic filters, structural ceramics, and solid oxide fuel cells. Ceramic fillers include bismuth, silica, alumina, and magnesium oxides. Impedance analyzer dielectric constants are 8.8–9.04 at 13.4 GHz. The calcium titanate powder dielectric resonator is sintered for 4 h at 1450 °C at 6.55 GHz, and dielectric permittivity was 40.5–20.7 [45]. It has been shown that oxides of ceramic-glass composites, such as bismuth, sodium, potassium, and titanates, all contribute to better energy storage. High sintering temperatures and sintering aids maximize microwave ceramics. These aids may boost permittivity and thermal expansion. Discontinuous and increased grain development in oxide ceramics reduce performance. To prevent holes and discontinuous grain development, chemicals and many precautions are utilized.

6.4.7 Dielectric Perovskite and Dielectric Polymer-Perovskite Composite Materials

Due to the fact that electro-ceramics may be made from more than 3000 different compounds, perovskites have drawn particular interest in the field of dielectrics for microwave communication. Perovskites are one of the few ceramics that enable applications demanding optimal dielectric characteristics and low TCF values, making them a promising candidate for use as microwave dielectrics for wireless communication. There are five distinct lattice configurations in which the simple perovskite structures may be found: the cubic, tetragonal, hexagonal, orthorhombic, and rhombohedral structures. The structure of these influences the operating temperatures in microwave devices, and each of them has a distinct transition for providing paraelectricity [46].

Common structures in low-cost perovskite electronics include A-titanate, A-zirconate, and R-BO3. These crystals have a non-perovskite structure with a high packing factor and include a, B, and R as substitutes for barium, calcium, aluminum, strontium, magnesium, and zinc. They are straightforward perovskite ceramics, thanks to the chemicals in their metal oxide composite. In addition, the addition of rare earth metals like neodymium and lanthanum, which are stoichiometrically regulated, results in modified perovskite ceramics with specific uses. In recent years, a barium magnesium and tantalum (BMT) composite has shown to be an effective hexagonal perovskite, thanks to its stable temperature and optimum temperature coefficients. Dopants like niobium and manganese may be used to dope BMTs, which have high dielectric characteristics (permittivity of 25 at 10 GHz) and a TCF of 2.7 ppm/K. It was also proposed that at high sintering temperatures of 1650 °C, this might be difficult for commercial application since defects in the material would be introduced [47].

6.4.8 Dielectric Rare-Earth and Polymer-Rare Earth Composite Materials

Since lead-free ceramics ($BaTiO_3$) are not dangerous to people or the environment, they are particularly popular in the electronics sector. Due to its exceptional electrical characteristics, high permittivity, and use in cutting-edge electrical appliances, rare earth (RA)-doped barium titanate ($BaTiO_3$) has had its dielectric properties examined. Here, research on the dielectric properties of $BaTiO_3$ ceramics doped with RA (lanthanum (La), erbium (Er), neodymium (Nd), samarium (Sm), and cerium (Ce)) was described. The primary goal of this review is to investigate the effects of doping on the described RA's $BaTiO_3$'s dielectric characteristics.

6.4.9 Carbon and Polymer-Carbon-Based Composite Materials for High- and Low-k Dielectric Application

High-k dielectric polymer composites may be made by combining carbon nanostructures with polymers. Despite their low percolation threshold, nanocomposites have high permittivity. Polymers are the first step. Practical applications are limited by the high dielectric loss and low breakdown field. The biggest challenge is resolving conflicts. Carbon-based compounds may improve the dielectric characteristics of polymers. Table 6.4 summarizes some polymer-carbon composite studies for high dielectric applications. Carbon nanotubes and graphene have the potential to improve the dielectric properties of dielectric polymers at low concentrations while maintaining composite flexibility [52]. High electrical conductivity and significant dielectric loss are possible near percolation when conductive particles tunnel. Composites are unable to endure electric fields over the percolation threshold. This flaw hampered the formation of conductive high-k composites. It is critical to raise the dielectric constant while lowering dielectric loss and composite flexibility. Carbon additions improve matrix polymers mechanically and electrically. Graphite, carbon fiber, CNTs, graphene, and reduced graphene oxide are examples of carbon fillers. According to experts, polymer-carbon nanocomposites have a variety of uses. Carbon improves the electrical and mechanical characteristics of polymers. Carbon-containing polymer composites have received a lot of attention. Carbon comes in a variety of shapes, sizes, and characteristics. The enormous surface area of carbon strengthens composites. Under mild stress, carbon nanotubes increase their mechanical and electrical qualities. CNTs benefit from a high aspect ratio.

Filler geometry, size, surface area, physical and chemical properties, interfacial interactions with the matrix-polymer, filler–filler interactions, filler concentration, mixing method, temperature, shear effects, composite morphology, and operating frequency can all have an impact on the dielectric properties of polymer-carbon composites [53]. It is possible to use one-, two-, or three-dimensional composite fillings. Thanks to their excellent functional and structural features,

Table 6.4 Rare-earth-doped ceramics dielectric properties

$Ba_{1-x}La_{2x/3}TiO_3$	Curie temp. (K)	Permittivity	Ref.	$Ba_{1-x}Sm_{2x/3}TiO_3$	Curie temp. (K)	Permittivity	Ref.
$X = 0.00$	390	5900	[48]	$X = 0.00$	390	5800	[49]
$X = 0.02$	353	4600		$X = 0.02$	373	5200	
$X = 0.04$	318	6500		$X = 0.04$	353	7000	
$X = 0.06$	293	9100		$X = 0.06$	343	7400	
$X = 0.08$	230	8600		$Ba_{1-x}Ce_xTiO_3$	Curie temp (K)	Permittivity	Ref.
$Ba_{1-x}Er_xTiO_3$	Curie temp (K)	Permittivity	Ref.	$X = 0.00$	110	1000	[50]
$X = 0.00$	133	5350	[51]	$X = 0.01$	100	2050	
$X = 0.01$	133	19,606		$X = 0.01$	100	5050	

Table 6.5 Polymer-carbon-based dielectric materials and its properties

Composites	Dielectric constant	Dielectric loss	Refs.
PAni/rGO	51.8	0.0093	[54]
PAni/rGO/$BaTiO_2$	54	7.83	[55]
PVDF/CNT	39	87	[56]
PVA/GA	70	0.9	[57]
CNT/GNP	41	0.9	[57]

carbonaceous nanofillers like CNTs and graphene play a crucial role. Table 6.5 showed some polymer-carbon-based dielectric material properties that are highly useful for optoelectronic applications.

6.4.10 Negative Permittivity Polymer Composites

There has been a lot of research on the positive permittivity of dielectric composites over the years; however, certain met composites, because of their unusual electromagnetic properties, may generate a negative permittivity. Negative permittivity from metallic wire arrays and electric fields provide negative permeability in epsilon-negative materials (ENMs). Theoretically, applied voltage and ENM current follow a new dielectric theory basic. They are mostly used in electronic packaging as interference shields or insulators between layers of dielectric composites. This fitting approach improves space-charge polarization owing to two heterogeneous interfaces. New literature reviews ENM-promising polymer-based composites. A nickel-rutile cermet negative permittivity metamaterial was explored in liquid–solid structure compounds. The conduction mechanism shifted from hopping to metallic when nickel concentration increased in percolative cermet. The magnetic behavior of the epsilon-negative cermet was less than 1, confirming that conductive networks caused a high-frequency diamagnetic response of eddy currents [58]. After sintering, indium oxide-doped ceramics were studied for negative permittivity. The negative permittivity seen during monophase indium-tin-oxide sintering was traced back to an increase in the concentration of charge carriers, as predicted by the Drude–Lorentz oscillator model. The dielectric properties were influenced by the presence of free carriers [59].

Negative permittivity metamaterials with "bilayers" have been created to find the frequency zones with the greatest and lowest dielectric permittivity. Wang et al. (2016) discovered that permittivity increased by 40% while dielectric loss remained at 0.06. (considered low at relaxation frequencies). The author also showed thickness ratio and single-layer composites versus dielectric characteristics [60].

For the purpose of studying diamagnetic ENMs, Engheta et al. [61] recommended using a narrow subwavelength cavity dielectric resonator constructed from metamaterials with negative permittivity and permeability. This discovery is important, as metamaterial-based composite production uses mathematical procedures [61]. After this investigation, a laminated composite with a negative permittivity intermediate layer was investigated to increase its permittivity. Wang et al. [62] found that strong interfacial polarization and low-loss tangent owing to the ohmic barrier effect increased permittivity. Table 6.6 showed some examples of negative polymeric composites that showed a negative dielectric constant.

6.5 Summary

This chapter provides a comprehensive examination of dielectric materials, highlighting their enormous variety in terms of both materials and manufacture. The majority of the materials included in this study are ceramics, metals (oxide, sulfide, and nitride), carbon-based materials, and polymers, which provide a clear distinction between their uses for energy storage, insulation, energy storage, and power transmission due to their dielectric properties. Both low-permittivity and high-permittivity materials that may be used in a variety of applications are distinguished by the materials discussed above.

There is great potential in the development of hybrid polymer composites that include natural fiber reinforcement, and it is essential that researchers learn how to optimize their characteristics for use in industry. These composites have the potential to be employed as intrinsically conductive polymers (ICPs) in the semiconductor industry; nevertheless, lowering the material's high contact resistance remains a challenge. When a change in pressure or temperature is given to a functional model, hybrid polymer composites are created to simplify the flow of electricity in ICPs and maximize load transfers. Dielectric constants must be nominal for proper charge

Table 6.6 Dielectric characteristics of epsilon-negative polymeric composites

Composites	Filler materials	Dielectric loss	Dielectric constant	Refs.
Ti_3AlC_2-polyimide	Ti_3AlC_2	0.1–100 (between 10 MHz and 1 GHz)	−1200 toward 0 (between 10 MHz and 1 GHz)	[63]
Graphene/ polyvinylidene fluoride	Single-layer graphene	0.1 lowest and 1 M highest (between 200 kHz and 1 MHz)	−7250 constant (between 200 kHz and 1 MHz)	[64]
Polydimethylsiloxane/ multi-walled carbon nanotubes (PDMS/ MWCNTs)	Membranous metacomposite	0.01–100 (between 10 MHz and 1 GHz)	300 toward −20 @ 933 MHz (between 10 MHz and 1 GHz)	[65]
Flexible graphene/ PDMS metacomposite	Flexible graphene addition	–	−61,250 toward – 40,855 (between 10 kHz and 1 MHz	[66]
MWCNT/TiN/CCTO	Sintered in argon and nitrogen	1000 to 100 (between 93 and 683 MHz)	−200 max (between 93 and 683 MHz)	[67]
$Ag/Al_2O3/3D-BaTiO3/$ epoxy	Silver-conductive particles	0.02 toward 10 (between 10 kHz and 1 MHz)	−4 M toward 0 (between 10 kHz and 1 MHz)	[68]

export, and dielectric losses must be reduced to avoid the leakage of undesired charges that might harm the composite material at the working temperature.

Even though ceramics have excellent dielectric properties, they cannot be used in current applications due to their high production temperature, high cost, and low mechanical strength. The great electrical performance of polymers and their composite combinations make them good material for use in integrated circuits and multilayer dielectric printed circuit boards. In an electric field, polymer composites undergo polarization processes that give them a high dielectric constant while exhibiting little dielectric loss, similar to ceramic composites. Fiber-reinforced polymer composites (FRP) are a kind of polymer composite that may be used for specific purposes that might impact the cost, processing power, green material usage, and waste management of energy storage and insulation systems.

Acknowledgements The authors would like to express their gratitude to the Indian Institute of Technology (Indian School of Mines) Dhanbad, India.

Conflict of Interest The authors confirm that there are no personal or professional conflicts of interest that may be seen as influencing the findings given in this chapter.

References

1. Al-Muntaser AA, Pashameah RA, Sharma K, Alzahrani E, Farea MO, Morsi MA (2022) α-MoO3 nanobelts/CMC-PVA nanocomposites: hybrid materials for optoelectronic and dielectric applications. J Polym Res 29:274
2. Nayak D, Choudhary RB (2019) Augmented optical and electrical properties of PMMA-ZnS nanocomposites as emissive layer for OLED applications. Opt Mater 91:470–481
3. Al-Muntaser AA, Adel Pashameah R, Sharma K, Alzahrani E, Hameed ST, Morsi MA (2022) Boosting of structural, optical, and dielectric properties of PVA/CMC polymer blend using $SrTiO_3$ perovskite nanoparticles for advanced optoelectronic applications. Opt Mater 132:112799
4. Adimule V, Yallur BC, Bhowmik D, Gowda AHJ (2022) Dielectric properties of P_3BT doped $ZrY_2O_3/CoZrY_2O_3$ nanostructures for low cost optoelectronics applications. Trans Electr Electron Mater 23:288–303
5. Jha SN, Narsaiah K, Basediya AL, Sharma R, Jaiswal P, Kumar R, Bhardwaj R (2011) Measurement techniques and application of electrical properties for nondestructive quality evaluation of foods—a review. JFST 48:387–411
6. Jilani MT, Rehman AMK, Khan SMAMT (2012) A brief review of measuring techniques for characterization of dielectric materials. ITEE 1
7. Venkatesh MS, Raghavan GSV (2005) An overview of dielectric properties measuring techniques. CSBE 47:7.15–7.30
8. Saeed K, Shafique MF, Byrne MB, Hunter IC (2012) Planar microwave sensors for complex permittivity characterization of materials and their applications. Applied Measurement Systems. InTech
9. Qiu J, Gu Q, Sha Y, Huang Y, Zhang M, Luo Z (2022) Preparation and application of dielectric polymers with high permittivity and low energy loss: a mini review. J Appl Polym Sci 139:52367
10. Silakaew K, Swatsitang E, Thongbai P (2022) Novel polymer composites of RuO_2@n$BaTiO_3$/PVDF with a high dielectric constant. Ceram Int 48:18925–18932
11. Deng Q, Huang Y, Chen B, Bo M, Feng Y (2022) Conductive V2C MXene and paralelectric $SrTiO_3$ containing polymer composites with high dielectric constant. Colloids Surf A: Physicochem Eng Asp 632:127763
12. Zhang Z, Wu D, Jiang L, Liang F, Rui Y, Tang B (2022) One-step synthesis based on non-aqueous sol-gel conductive polymer-coated SnO2 nanoparticles as advanced anode materials for lithium-ion batteries. J Alloys Compd 899:163274
13. Dlamini DS, Mishra SB, Mishra AK, Mamba BB (2011) Comparative studies of the morphological and thermal properties of clay/polymer nanocomposites synthesized via melt blending and modified solution blending methods. J Compos Mater 45:2211–2216
14. Ashirov T, Song KS, Coskun A (2022) Salt-templated solvothermal synthesis of dioxane-linked three-dimensional nanoporous organic polymers for carbon dioxide and iodine capture. ACS Appl Nano Mater 5:13711–13719
15. Nayak D, Choudhary RB, Kandulna R, Mandal G (2019) Investigation of structural, optical and electrical performance of ZnS sensitized PMMA nanocomposite as an emissive layer for OLED application, p 020017
16. Nayak D, Choudhary RB (2022) Influence of ZnS on the structural, morphological, optical and thermal properties of Polyindole for an emissive layer. Inorg Chem Commun 144:109824
17. Choudhary RB, Nayak D (2021) Tailoring the properties of 2-DrGO-PPy-ZnS nanocomposite as emissive layer for OLEDs. Optik 231:166336
18. Verma A, Choudhary RB, Nayak D, Mandal G (2020) Structural analysis with augmented optoelectronic performance of polyindole–polypyrrole–cadmium sulfide nanohybrid: PIN/PPY/CdS. J Inorg Organomet Polym Mater 30:2683–2691
19. Jiang P, McFarland MJ (2004) Large-scale fabrication of wafer-size colloidal crystals, macroporous polymers and nanocomposites by spin-coating. J Am Chem Soc 126:13778–13786

20. Montméat P, Dechamp J, Enyedi G, Fournel F, Zavvou Z, Jousseaume V (2022) Initiated chemical vapor deposition of polysiloxane as adhesive nanolayer for silicon wafer bonding. Mater Sci Semicond Process 148:106808
21. Wu J, Liu C, Sun P, Tang J, Wang X, Wu T (2022) A 1D zinc coordination polymer built from the in situ generated ligand of bisimidazole-tetrathiolate via solvothermal reaction. J Solid State Chem 312:123180
22. Shi K, Zou H, Sun B, Jiang P, He J, Huang X (2020) Dielectric modulated cellulose paper/PDMS-based triboelectric nanogenerators for wireless transmission and electropolymerization applications. Adv Funct Mater 30:1904536
23. Das S, Chowdhury IH, Chakraborty A, Naskar MK, Sarkar M, Manirul Islam SK (2022) Porous organic polymer (POP) nanosheets: an efficient photo-catalyst for visible-light assisted CO_2 reduction. Adv Mater 3:3165–3173
24. Zhao X-Y, Liu H-J (2010) Review of polymer materials with low dielectric constant. Polym. Int
25. Bahari A, Shahbazi M (2016) Electrical properties of PVP–SiO_2–TMSPM hybrid thin films as OFET gate dielectric. J Electron Mater 45:1201–1209
26. Haque SM, Rey JAA, Masúd AA, Umar Y, Albarracin R (2017) Electrical properties of different polymeric materials and their applications: the influence of electric field. In Properties and applications of polymer dielectrics. InTech
27. Cao M, Li L, Bin HW, Wu SY, Chen XM (2021) Greatly enhanced permittivity in $BaTiO_3$-epoxy dielectric composites with improved connectivity of ceramic phase. J Materiomics 7:1–7
28. Zhu X, Yang J, Dastan D, Garmestani H, Fan R, Shi Z (2019) Fabrication of core-shell structured Ni@$BaTiO_3$ scaffolds for polymer composites with ultrahigh dielectric constant and low loss. Compos Part A Appl Sci 125:105521
29. Li L, Cheng J, Cheng Y, Han T, Liang X, Zhao Y, Zhao G, Dong L (2020) Polymer dielectrics exhibiting an anomalously improved dielectric constant simultaneously achieved high energy density and efficiency enabled by CdSe/Cd 1–x Zn x S quantum dots. J Mater Chem 8:13659–13670
30. Qenawy SA, Nasrat LS, Ismail HM, Asaad JN (2020) Evaluation of dielectric strength of SiR/TiO_2 composites using feed-forward neural network. IET Nanodielectrics 3:74–80
31. Kulkarni ND, Kumari P (2023) Development of highly flexible PVDF-TiO2 nanocomposites for piezoelectric nanogenerator applications. Mater Res Bull 157:112039
32. Alhusaiki-Alghamdi HM (2022) Molecular structural, optical analyses, and dielectric properties of PVDF/PVA-MK composites. Polym Bull
33. Panda M, Sultana N, Singh AK (2022) Structural and optical properties of PVDF/GO nanocomposites. Fullerenes, Nanotubes, Carbon Nanostruct 30:559–570
34. Li T, Zhou W, Li B, Li Y, Cao D, Zhou J, Zuo J, Cai J, Wang G, Liu D, Cai H (2022) Enhanced dielectric and thermal properties of Zn/PVDF composites by tailoring core@double-shell structured Zn particles. Compos Part A Appl Sci 157:106947
35. Guo Y, Wu S, Liu S, Xu J, Pawlikowska E, Szafran M, Rydosz A, Gao F (2022) Enhanced dielectric tunability and energy storage density of sandwich-structured Ba0.6Sr0.4TiO_3/PVDF composites. Mater Lett 306:130910.
36. Wang B, Huang W, Chi L, Al-Hashimi M, Marks TJ, Facchetti A (2018) High- k gate dielectrics for emerging flexible and stretchable electronics. Chem Rev 118:5690–5754
37. Dang Z-M, Yan W-T, Xu H-P (2007) Novel high-dielectric-permittivity poly(vinylidene fluoride)/polypropylene blend composites: the influence of the poly(vinylidene fluoride) concentration and compatibilizer. J Appl Polym Sci 105:3649–3655
38. Li W, Jiang L, Zhang X, Shen Y, Nan CW (2014) High-energy-density dielectric films based on polyvinylidene fluoride and aromatic polythiourea for capacitors. J Mater Chem A 2:15803–15807
39. Li R, Xiong C, Kuang D, Dong L, Lei Y, Yao J, Jiang M, Li L (2008) Polyamide 11/poly(vinylidene fluoride) blends as novel flexible materials for capacitors. Macromol Rapid Commun 29:1449–1454
40. Nan C-W (1993) Physics of inhomogeneous inorganic materials. Prog Mater Sci 37:1–116

41. Zhao H, Chen L, Luan X, Zhang X, Yun J, Xu T (2017) Synthesis, pyrolysis of a novel liquid SiBCN ceramic precursor and its application in ceramic matrix composites. J Eur Ceram Soc 37:1321–1329
42. Rodeghiero E, Moore B, Wolkenberg B, Wuthenow M, Tse O, Giannelis E (1998) Sol-gel synthesis of ceramic matrix composites. Mater Sci Eng A 244:11–21
43. Xu J, Wong CP (2007) Characterization and properties of an organic–inorganic dielectric nanocomposite for embedded decoupling capacitor applications. Compos Part A Appl Sci 38:13–19
44. Yi L, Zhang J, Yang J, Sun F, Zhang H, Zhao L (2018) Effect of annealing induced crystalline evolution on the scratch resistance of polylactide. Tribol Int 128:328–336
45. Reaney IM, Iddles D (2006) Microwave dielectric ceramics for resonators and filters in mobile phone networks. J Am Ceram Soc, 060428035142006
46. Youssef AM, Farag HK, El-Kheshen A, Hammad FF (2018) Synthesis of nano-structured strontium titanate by sol-gel and solid state routes. SILICON 10:1225–1230
47. Imran Z, Rafiq MA, Ahmad M, Rasool K, Batool SS, Hasan MM (2013) Temperature dependent transport and dielectric properties of cadmium titanate nanofiber mats. AIP Adv 3:032146
48. Ganguly M, Rout SK, Sinha TP, Sharma SK, Park HY, Ahn CW, Kim IW (2013) Characterization and rietveld refinement of A-site deficient lanthanum doped barium titanate. J Alloys Compd 579:473–484
49. Ismail FA, Osman RAM, Idris MS (2016) Review on dielectric properties of rare earth doped barium titanate. 090005
50. Yasmm S, Choudhury S, Hakim MA, Bhuiyan AH, Rahman MJ (2011) Effect of cerium doping on microstructure and dielectric properties of $BaTiO_3$ ceramics. JMST 27:759–763
51. Leyet Y, Peña R, Zulueta Y, Guerrero F, Anglada-Rivera J, Romaguera Y, de la Cruz JP (2012) Phase transition and PTCR effect in erbium doped BT ceramics. MSEB 177:832–837
52. Nayak S (2019) Dielectric properties of polymer–carbon composites, 211–234
53. Mittal G, Dhand V, Rhee KY, Park S-J, Lee WR (2015) A review on carbon nanotubes and graphene as fillers in reinforced polymer nanocomposites. J Ind Eng Chem 21:11–25
54. Cho S, Kim M, Lee JS, Jang J (2015) Polypropylene/polyaniline nanofiber/reduced graphene oxide nanocomposite with enhanced electrical, dielectric, and ferroelectric properties for a high energy density capacitor. ACS Appl Mater Interfaces 7:22301–22314
55. Tabhane GH, Giripunje SM, Kondawar SB (2021) Fabrication and dielectric performance of RGO-PANI reinforced PVDF/$BaTiO_3$ composite for energy harvesting. Synth Met 279:116845
56. Al-Saleh MH (2019) Carbon-based polymer nanocomposites as dielectric energy storage materials. Nanotechnology 30:062001
57. Wang Z, Han NM, Wu Y, Liu X, Shen X, Zheng Q, Kim J-K (2017) Ultrahigh dielectric constant and low loss of highly-aligned graphene aerogel/poly(vinyl alcohol) composites with insulating barriers. Carbon 123:385–394
58. Fan G, Wang Z, Wei Z, Liu Y, Fan R (2020) Negative dielectric permittivity and high-frequency diamagnetic responses of percolated nickel/rutile cermets. Compos Part A Appl Sci 139:106132
59. Fan G, Wang Z, Sun K, Liu Y, Fan R (2021) Doped ceramics of indium oxides for negative permittivity materials in MHz-kHz frequency regions. JMST 61:125–131
60. Wang J, Shi Z, Mao F, Chen S, Wang X (2017) Bilayer polymer metacomposites containing negative permittivity layer for new high-k materials. ACS Appl Mater Interfaces 9:1793–1800
61. Engheta N (2002) An idea for thin subwavelength cavity resonators using metamaterials with negative permittivity and permeability. IEEE Antennas Wirel Propag Lett 1:10–13
62. Wang Z, Sun K, Xie P, Hou Q, Liu Y, Gu Q, Fan R (2020) Design and analysis of negative permittivity behaviors in barium titanate/nickel metacomposites. Acta Mater 185:412–419
63. Wu H, Mu Z, Qi G, Zhang Y, Wang X, Xie P, Wu N, Yuan H, Sui K, Fan R, Liu C (2021) Negative permittivity behavior in Ti_3AlC2-polyimide composites and the regulation mechanism. J Mater Sci Mater 32:10388–10397
64. Wang Z, Sun K, Qu Y, Wang Z, Tian J, Li X, Fan R (2021) Negative-k and positive-k layers introduced into graphene/polyvinylidene fluoride composites to achieve high-k and low loss. Mater Des 209:110009

65. Sun K, Xie P, Wang Z, Su T, Shao Q, Ryu J, Zhang X, Guo J, Shankar A, Li J, Fan R, Cao D, Guo Z (2017) Flexible polydimethylsiloxane/multi-walled carbon nanotubes membranous metacomposites with negative permittivity. Polymer 125:50–57
66. Sun K, Dong J, Wang Z, Wang Z, Fan G, Hou Q, An L, Dong M, Fan R, Guo Z (2019) Tunable negative permittivity in flexible graphene/PDMS metacomposites. J Phys Chem C 123:23635–23642
67. Sun K, Yang P, He Q, Tian J, Duan W, Wu X, Qu Y, Du H (2021) Weakly negative permittivity of MWCNT/TiN/CCTO ternary ceramics sintered in argon and nitrogen atmosphere. Ceram Int 47:32297–32302
68. Han M, Shi Z, Zhang W, Zhang K, Wang H, Dastan D, Fan R (2021) Significantly enhanced high permittivity and negative permittivity in Ag/Al2O3/3D-BaTiO3/epoxy metacomposites with unique hierarchical heterogeneous microstructures. Compos Part A Appl Sci 149:106559

Chapter 7
Metal and Metal-Oxide-Based Polymeric Nanodielectrics for Energy Storage: Opportunities and Challenges

Minal Bafna, Nipun Bafna, Farah Deeba, and Ankur Jain

Abstract In recent times, with devices becoming smaller, cheaper, and faster, it has become a dire need to provide devices that can store an abundant amount of energy, which gets delivered also nearly instantaneously. Such devices are not only for finding applications in complex military and medical needs but also for commercial day-to-day applications too. Metal-oxide-based traditional capacitors do have high dielectric constant, but they inherit low breakdown field strength resulting in low energy density and poor mechanical properties to be cast as high energy storage capacitors. Contrary to these, the polymeric materials, which are lightweight and easily amenable to be cast in various sizes and shapes and are cost-effective for mass production, possess relatively high breakdown strengths but very low values of dielectric constant and poor high-temperature performance become unfit for be manufactured as high energy storage devices. Polymer nanodielectrics are a new class of materials that combine the processability, breakdown field strength, and low-loss features of polymers with the high dielectric constant of metal oxide fillers. This chapter examines this class of materials, first shedding light on their current level of development, including those created for both low-temperature and high-temperature applications. A meticulous summary of underlying theories and concepts that construe the correlation between physical property and the energy storage capability of materials has been presented. The contrast in electric behaviors of metallic nanofillers and the host polymer tends to provide real challenges in achieving viable high energy storage materials. The last

M. Bafna (✉)
Department of Physics, Agrawal PG College, Jaipur, India
e-mail: drminalphysics@rediffmail.com

N. Bafna
Mechanical and Industrial Engineering Department, IIT Roorkee, Roorkee, India
e-mail: n_bafna@me.iitr.ac.in

F. Deeba
SS Jain Subodh P G College, Jaipur, India

F. Deeba · A. Jain
School of Applied Sciences, Suresh GyanVihar University, Jaipur, India

A. Jain
Center for Renewable Energy and Storage, Suresh GyanVihar University, Jaipur, India

S. Moharana et al. (eds.), *Emerging Nanodielectric Materials for Energy Storage*,
Nanostructure Science and Technology, https://doi.org/10.1007/978-3-031-40938-7_7

part of the review addresses mitigating such challenges and how and what future research should focus on.

Keywords Energy storage · Composite dielectrics · Polymer nanodielectrics · Metallic nanoparticles dielectric constant · Dielectric breakdown strength · Energy density

7.1 Need for Dielectric Energy Storage

In today's world, electronics have become indispensable and are at the heart of human development as it gets consumed versatility in simple routine devices like mobiles, laptops, and TV screens to complex hybrid/electric vehicles, space machines, medical or defense, and surveillance equipment. For these electronic gadgets to be smarter, they need to have devices that are smaller, lighter, faster, flexible, cheaper, and reliable in a wide range of temperatures and are environmentally sustainable. Hence, there is a growing demand for developing cost-efficient and high-performance energy storage devices like batteries, electrolytic and electrochemical capacitors, fuel cells, etc., that can store an abundant amount of energy that gets delivered also nearly instantaneously [1–4]. The power transmission applications, electromagnetic appliances in the defense sector, and vehicle protection in the automobile industry all are looking forward to devices that perform excellently without withering in a varied range of temperatures (>100 °C) and that possess high-frequency stability and high energy and power density with enhanced efficiency [5, 6]. Among the available energy storage devices, the electrolytic or dielectric capacitors are favored in applications that require short-term charge storage and high-power density. The ultra-high-power density typically in the range of 10^8 w/kg allows the rapid release of large current densities in a very short period, making the charge/discharge speed in μs to ns, cost-effectiveness, wide operating temperature range (−40 to 240 °C), and voltage range in kV, and significant lifetimes compared to other energy storing alternatives are those attributes that make the dielectric capacitors potentially desirable component in an array of electrical applications ranging from personal appliances to complex systems [7, 8]. These capacitors when cast in thin films have added advantage of being lighter and smaller and capable of enduring a high electric field due to possession of low defect density [9, 10]. The dielectric capacitors need to improve upon their energy density performance to compete with the super-capacitors that are reported to possess an excellent energy density of around 30 J cm^{-3} [11]. Although super-capacitors have better energy performance, yet they have the limitations of their being bulky, requiring large installation space, the threat of chemical spills, and lack of suitable constituent materials for electrodes or materials suffering from chemical corrosion [8, 12–14]. The usual mechanism for the storage of electric energy in conventional batteries is according to the chemical interactions taking place between its electrolyte and electrodes, and due to these chemical reactions being slow and irreversible, these batteries take a long time to charge–discharge and have a short lifespan, and the capacitors, on the other hand, is devoid of such limitations as the very mechanism for the storage of electric energy in it is due to the phenomenon of electric field-induced polarization [15–18]. However, the existing dielectric capacitors need to improve upon the issue of possession of high energy and power density when they are made smaller and lighter for portable purposes and the processes need to be economically viable [19, 20].

Here, in this review, we initiate our discussion by explaining the fundamentals related to the factors responsible for enhancing the properties of the dielectric material required for energy storage applications wherein we will explain the various mechanism of polarization and breakdown and then shall present an overview of the present state of the art of these materials and discuss the real challenges of achieving the viable high energy storage materials by synthesizing the metal-oxide–polymer dielectrics made by doping the two materials having contrast in electric behaviors.

7.2 Fundamentals of Dielectric Energy Storage Mechanism

An electrolytic capacitor is an energy storage device that comprises a layer of a dielectric substance kept between two conducting electrodes (shown in Fig. 7.1) and works on the principle of storing electrical energy due to the segregation of equal amounts of charges of opposite polarity on either side of the dielectric substance when an external electric field is applied. The process of storing electrical energy is termed the charging of the capacitor. This stored energy is released in the form of a pulse almost instantaneously and the process is known as discharging of the capacitor. As no electrochemical process is involved in this charge–discharge cycle, the discharge speed is in µseconds rendering the dielectric capacitors a high power density of around 100 W/kg [20, 21]. Apart from charge storage capability, the other key aspects that characterize the performance of a capacitor as an energy storage device are its *charge–discharge profile, energy and power density, potential scalability,* and the *cost incurred* for such storage. All these stated parameters are related directly to electrical phenomena like polarization, conduction, and breakdown within the dielectric material filled inside the capacitor. Thus, the performance of the capacitor is quite detrimental to the thermo-physical characteristics of the dielectric material used within the capacitor. When developing a dielectric capacitor for large-scale energy storage, for ideal performance, the dielectric material within must possess a high value of dielectric constant ε_r, large dielectric breakdown strength E_b, high discharge efficiency η, high resistivity, and must be capable of operating in a broad temperature window [20–23].

The atoms share the arrangement of their electrons to form a molecule. If the charge distribution of positive core and negative electrons is balanced, the molecules are non-polar. But during the rearrangement of electrons, if imbalances in charge distribution occur to develop a slight displacement in the center of mass of positive and negative charge distribution of the atom/molecule, then permanent dipoles get created within the material and the material is rendered to be polar and the process

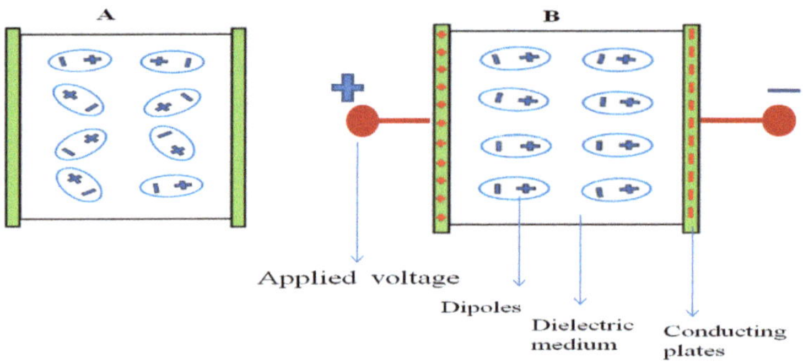

Fig. 7.1 Schematic of a dielectric capacitor

is known as polarization. In absence of an external field, the dipole moments of these are arbitrarily orientated; the resultant polarization may get nullified. Dielectric materials do not possess loosely bound free electrons; they act as electrical insulators in the context of charge conduction. However, they can be polarized by applying an external electric field. {The various mechanisms of polarization have been discussed in Sect. 2.1}. The electric displacement (D) of induced dipoles, under the influence of the external electric field (E_{external}), produces local internal electric fields (E_{induced}) within the dielectric material that counteract to decrease the applied field (E_{external}—E_{induced}) and is expressed about the applied electric field and induced polarization (P) as

$$\mathbf{D} = \varepsilon_0 \mathbf{E} + \mathbf{P}. \tag{7.1}$$

Here, **P** the induced polarization is the measure of dipole moment per unit volume and is given by $P = \varepsilon_0 \chi_e E$, where ε_0 is free space permittivity ($= 8.854 \times 10^{-12}$ F/m) and $\chi_e = \varepsilon_0(\varepsilon_r - 1)$ is the electrical susceptibility of the material. We define the electric field as the gradient of electric potential so when the overall effective electric field within the dielectric material gets reduced due to polarization, the potential (V) also diminishes resulting in an enhancement of the capacitance $C = Q/V$ of the material. The ratio of C to C_0 is termed as dielectric constant (ε_r) or relative permittivity (also mentioned in certain related literature as k), where C and C_0 represent the capacitances when dielectric materials are considered to be taken or not in between the plates of the capacitor.

The more the electric displacement, the higher the value of the dielectric constant (ε_r) or (k) for the material. The dielectric constant or electrical permittivity is a complex quantity expressed as a sum of its real and imaginary parts as $\varepsilon_r = \varepsilon' + \varepsilon''$. . The phase lag ($\delta$) of **D** to **E** is defined as the dielectric loss given by $\tan\delta = \varepsilon''/\varepsilon'$. Here, the real part ε' demonstrates the ease of polarization of the material and determines the size of the surface charge densities on both sides of a dielectric and $\tan\delta$ or the losses develop as a result of bound charges moving in reaction to an applied electrical field. All these dielectric parameters ε', ε'', and $\tan\delta$ are

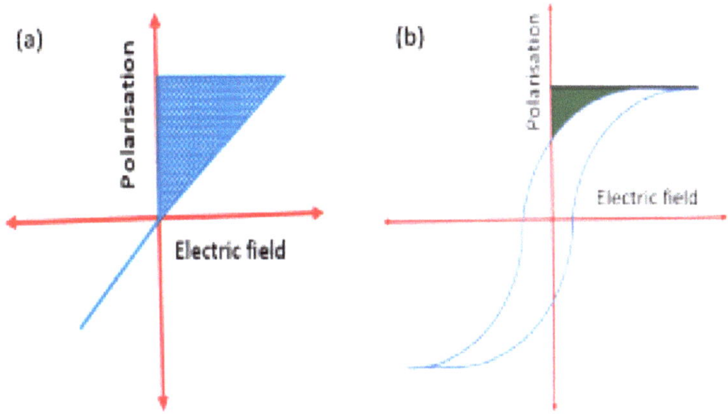

Fig. 7.2 The *P–E* hysteresis loops of **a** linear dielectric and **b** nonlinear dielectric [21]

measured by broadband dielectric spectroscopy or impedance analyzer. The capacity of a dielectric material to store energy is analyzed from its polarization–electric field (*P–E*) hysteresis loop. The materials that do not exhibit hysteresis and their *P–E* relationship is a straight line are referred to as linear dielectrics and are shown in Fig. 7.2a where the shaded region of the graph depicts the recoverable energy storage density rendering these materials high breakdown strength (BDS) and lower energy loss and the reason for preference of these for energy storage applications. The dielectrics that exhibit hysteresis are termed nonlinear dielectric materials as shown in Fig. 7.2b.

If *D* is the induced electric displacement within the dielectric material placed between the electrodes of a capacitor under the influence of an electric field E, then we define the energy density to be the area enclosed by the displacement–electric field curve of Fig. 7.1 and given by Eq. (7.2)

$$U_e = \int E \partial D \tag{7.2}$$

U_e is energy density or electrical energy stored per unit volume in J/cm^3.

For linear dielectrics, when the material is isotropic, that is, the dielectric constant or relative permittivity of the material ε_r does not change with the electric field, the energy density or can be evaluated through Eq. (7.3).

$$U_e = 1/2\varepsilon_0\varepsilon_r E_b^2 = 1/2CV_m^2/Ad \tag{7.3}$$

where E_b is the threshold electric field above which the breakdown of dielectric material occurs. Thus, it seems that having a large dielectric constant ε_r and high breakdown strength, E_b will provide high energy density and will ideally spark the performance of a dielectric material as an excellent capacitor [8, 24].

Once the energy density is analyzed, it is easier to calculate the power density ($\mathbf{P}_e$) for linear dielectrics. For energy density, U_e and loss tangent $\tan\delta$ also defined as the ratio of equivalent series resistance R_s to capacitive reactance $X_c = 1/2\,\pi\,fC$, the power density ($\mathbf{P}_e$) for linear dielectrics can be written using the expression

$$\mathbf{P}_e = U_e/2R_sC = \pi f U_e/\tan\delta \tag{7.4}$$

Thus, the above discussion brings forth that for energy storage applications why it is essential to have materials that exhibit large saturation polarization, small remnant polarization, high dielectric constant (ε_r), and high dielectric breakdown strength. Hence, materials with high capacitance or high dielectric constant are sought after for energy storage applications.

7.2.1 Mechanism of Polarization

The term dielectric constant (ε_r) or dielectric permittivity (k) is the measure of the response of the dipoles (whether permanent or induced) within the dielectric material to the applied external electric field. A material's total permittivity may be influenced by some polarization mechanisms. The various mechanism through which dielectric polarization takes place are broadly classified as (i) electronic polarization, (ii) atomic/ionic polarization, (iii) orientation polarization, and (iv) space charge (i.e., ions and electrons/holes) polarization. Each type of polarization is accompanied by its dielectric loss at a specific frequency (Fig. 7.3). When atoms share the arrangement of their electrons, they combine to form a molecule. If the charge distribution of positive core and negative electrons is balanced, the molecules are non-polar. But during the rearrangement of electrons, if imbalances in charge distribution occur to develop a slight displacement in the center of mass of positive and negative charge distribution of the atom/molecule, then dipoles get created within the material and the material is rendered to be polar. As the dipole moments of these are arbitrarily orientated, the resultant polarization may get nullified. But when an external electric field is applied to such dielectric material, there occurs a slight relative displacement of the negative electron cloud and positive nucleus in opposite directions causing *electronic polarization* of the material. Such effects are observed when an external electric field is applied in the optical frequency range of 10^{15}–10^{18} Hz. When an external field in the infrared frequency range 10^{12}–10^{15} Hz is applied to the dielectric material, there occur relative displacements in the mean position of the atomic nuclei within the molecules rendering the *atomic/ionic polarization* of the material. Atoms exhibit oscillatory behavior with damping effects resembling a mechanical spring and mass system. Any frequency other than the resonant frequency will have a tiny oscillation amplitude. The electronic and atomic polarizations are collectively termed *resonance polarization* and dominate the GHz range observed in power and radio frequency applications. The electronic and atomic systems are almost lossless and only make a minor continuous contribution to $\varepsilon_r\prime$ well below resonance. *Orientation polarization*

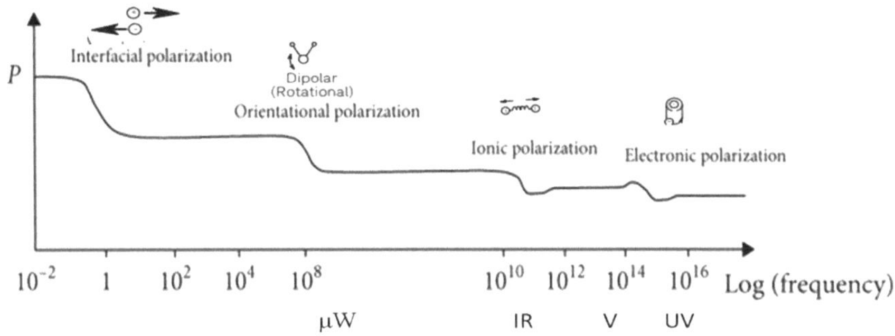

Fig. 7.3 Frequency-dependent polarization process

is a relaxation process that arises when permanent molecular dipoles tend to align along the direction of the applied electric field. The occurrence of such polarization depends on the molecular dynamics of the material and can take place in the frequency range of 1 Hz to 20 GHz. In *space charge polarization* actual physical movement of ionic species (ions/electrons/holes) occurs over nanometers of distance.

In any dielectric material in which process of polarization shall be active depending on the frequency of operation, all sources of polarization are present at low frequencies, giving the materials a high value of *k*. As the frequency is increased, however, the dipoles become unable to orient themselves at such high frequencies, and as a result, only one or two sources of polarization or no sources of polarization are present, which results in a decrease in the *k* value with frequency [26]. The increase in temperature allows the motion of frozen dipoles, and hence, the value of *k* in material increases with the increase in temperature [27]. The dielectric constant in metal oxides is majorly caused due to *ionic polarization* mechanism where a slight displacement of the asymmetrical central cation takes place. The dielectric constant in polymers results due to *orientation, ionic, and electronic polarization* mechanisms depending upon the applied frequency.

The mechanism of polarization in surfaces of multi-component systems is explained using Maxwell–Wagner–Sillars model and is referred to as MWS interfacial polarization in literature [28–32]. Such heterogeneity can be seen in polymeric materials, where different phases coexist when the polymer is reinforced by a filler or plasticizer [25, 26, 30], or in semi-crystalline polymers, where amorphous and crystalline phases coexist [28, 30], or in metal-oxide–polymer dielectrics, where two or more constituent substances, the conducting phase, and non-conducting phase, have distinctly different electrical properties. In metal-oxide–polymer dielectric, the dielectric material can be seen as composed of conducting regions separated by resistive non-conducting boundaries. The application of an electric field at low frequency (i.e., 100 Hz to 1 MHz) compels the unbound charges to move to the boundary of the phases, forming large dipoles such that at such frequencies, the conducting regions' edges have time to accumulate charges, which causes $\varepsilon_r\prime$ to rise. The interfaces within the material tend to trap the motion of the charges and allow them to accumulate within boundaries where a slight distortion brought by the applied field raises the material's overall capacitance. At higher frequencies, however, as the charge displacement is minimal in comparison to the size of the conducting zone, the charges do not have time to concentrate along the material's edges, and polarization does not take place. Thus, it can be understood that the dipoles induced at interfaces require sufficient time to follow the variations of the applied electric field, and hence, the interfacial polarization or the Maxwell–Wagner–Sillars effect is predominant at low frequencies in polymeric materials.

7.2.2 Mechanism of Breakdown

The other parameter in the energy density equation stated in Eq. (7.2) is the breakdown strength E_b that scales quadratically in the energy density function and it appears that manipulating it may be the easiest approach to get the desired dielectric material. The breakdown or dielectric strength is the value of the threshold electric field that can be applied to the material above which the breakdown of dielectric material occurs, i.e., it makes a transition from being insulating to conducting. The breakdown in materials may occur due to any of these three mechanisms—(i) *Free volume breakdown,* (ii) *thermal breakdown, and* (iii) *mechanical breakdown.* When the applied electric field accelerates the free electrons at enough speed to collide and generate more free electrons within the material and disrupt the whole system, the breakdown is referred to as *free volume breakdown* [33]. The presence of defects and voids aids in lowering the electronic breakdown strength. When the energy dissipated due to dielectric losses and conduction losses accumulates enough heat for the catastrophic failure of the device, the breakdown is termed *thermal breakdown* [34]. On application of a high electric field, the induced stress when exceeds the yield stress of the material creates a fracture. Thus, in the process in which electrostatic attraction is far more than the mechanical strength, the material gets deformed and the process is classified as *mechanical breakdown* [35]. Hence, materials may be designed and developed for the high value of breakdown strength taking cognizance of these breakdown modes.

7.2.3 Charge–Discharge Profile

The parameter energy density is significant in applications where the capacitor is desired to serve as an energy reservoir. The discharged energy density per unit of time known as power density is another significant parameter in applications that require high power pulse where it needs to release high energy density in a short period. The charge–discharge profile provides an evaluation of the capacitor's performance and is measured by the parameter charge–discharge efficiency η, which is defined as the ratio of the discharged energy density U_d to storage energy density, U_e i.e., $\eta = U_d/$

U_e. This discharge energy density is significantly controlled by losses. But to have materials with large ε_r or K, materials have large high losses too. The dielectric loss also referred to as loss tangent (tan δ) provides a measure of the electrical energy dissipated during the charging–discharging process. Dielectric loss, conduction loss, or polarization hysteresis in ferroelectric materials although low at room temperature but significant as temperature or electric field increases makes the discharging process inefficient thereby rendering low discharge energy. A low efficiency implies low discharged energy density as good points toward the thermal runaway of the device due to the generation of joule's heat due to these losses and must be avoided.

Apart from room temperature applications, these capacitors are required to perform at elevated temperature conditions like in aircraft, mining fields, or in hybrid electric vehicles where temperatures are above 120 °C. To maintain the peak performance of these capacitors, we are currently employing a cooling system, which increases the overall size and mass of the device.

Consequently, research on capacitive energy storage devices currently focuses on the development of high-performance dielectric materials that possess' high energy density and those that can operate smoothly in wide temperature range with very low dielectric losses. The idea is to increase the degree of polarization throughout the dielectric while reducing losses.

7.3 Current State of Art for Materials for Dielectric Energy Storage Applications

The basic definition of capacitance ($C = \varepsilon_r A/d$) for a parallel plate capacitor suggests that the capacity can be enhanced either by increasing the area (A) of the plates or by reducing the distance (d) between the two plates or by filling a dielectric material of high ε_r or k value. Augmenting the areas of the plate is not a judicious choice as it increases the device size too and restricts device miniaturization [36]. If the distance between the dielectric plates is tried to be reduced for attaining higher C, then after a certain limit of dielectric thickness filled between the plates, there are chances of dielectric breakdown or leakage current to start flowing through it. So for obtaining enhanced capacitance, we are left with only one approach and that is to use materials having a high value of dielectric constant (k) material [37]. Based on the k values or the charge storage capability, the materials are categorized as high k materials and low k materials about the dielectric constant value of silicon nitride whose k value is 7 and less than that of silicon dioxide whose k value is 3.9, respectively. This definition originates from the need for miniaturization of microelectronic components, where we need to replace silicon dioxide or silicon nitride with materials that have a value of dielectric constant less than or more than that for materials having k values of mentioned silicon compounds [5]. Here, it can be recalled that air having dielectric constant $k = 1$ is the material with the lowest k and relaxor ferroelectrics have been reported to have the highest value of k up to ~24,000 at 1 kHz [4]. Low *k* materials are utilized for the fabrication of EMI shields [38–40], ultrafast IC packaging [41, 42], communication devices, etc. [43], but it is the high *k* materials that are sought after for energy storage applications besides their being widely used in gate dielectrics in MOS transistors and TFTs [44, 45].

7.3.1 Dependence of **k** *Value on Structural Attributes*

In an earlier section, we have seen that the mechanism of polarization and hence dielectric constant is highly dependent upon the frequency of applied electric field and temperature around the material. Apart from these, the structural arrangement of the constituent atoms or molecules within the material like *crystallite size, porosity, composition, and defects* also affects the *k* value. For instance, the larger the *crystallite size*, more are the number of dipoles contained within it, resulting in a larger number of the orientation of dipoles on the application of an electric field. So a large crystallite size shall produce material of a higher value of *k* compared to a material of small crystallite size [46, 47]. In contrast to it, a material with higher porosity shall have a low value of permittivity or *k* as air filled inside its random pores has the lowest value of *k* [48]. The dielectric constant of a composite is the resultant of the *k* values of its constituents. The effect of a constituent material of low *k* is to lower the overall effective value of *k* of the composite material. How much it can reduce will depend upon its percentage of concentration. Analogously a constituent material of high *k* will yield an increase in the overall effective value of *k* of the composite material when doped within percolation limits [49, 50]. The defects and grain boundaries present within the material also aid in enhancing the value of *k*. The bonds between atoms in a material are all the same. If any atom has less number of bonds than the bonds other atoms of the remaining bulk material possess, it is characterized as a *defect,* and the loose bonds are called *dangling bonds.* The larger the number of defects present within the material, the larger would be the number of dangling bonds and it implies that the higher would be the density of charge carriers to create dipoles lead to an increased polarization on the application of an external electric field [51]. Sometimes atoms are missing in their regular geometrical distribution within a crystal. Such a missing atom is called *vacancy* which helps to trap the charge carriers. Hence, the presence of vacancies helps generate high polarization resulting in a large value for *k* [52, 53]. Defects are also introduced by introducing an impurity or filler within the host material. The atom substituting the host atom place is referred to as a *substitution defect* and this too helps in trapping the charge carriers to produce greater polarization [54]. In a heterogeneous substance, due to the inhomogeneity of the phases of the constituent material interfaces gets developed, they are also referred to as *grain boundaries.* The grain boundaries aid to trap the charges at the boundary/interface and do not allow them to pass through. Hence, interfacial polarization enhances and grain boundaries contribute to the increased value of *k*. The greater the number of grain boundaries or interfaces larger the value of *k* for the material [55].

7.3.2 Ceramics and Metal Oxides

Metal oxides represent an important class of ceramic materials that have attracted considerable research interest because metals form a large diversity of oxide materials that can adopt varied structural geometries with unique electronic band structures. Among these metal oxide dielectric materials, the perovskites having an ABO_3 structure with A being metal and B a transition group metal is quite interesting; as the application of external electric field on the usual symmetrical cubic perovskite structure distorts the structure creating a small displacement between the positive and negative charge center, and aligning the created dipoles in ferroelectric domains thus culminating in a finite polarization [36]. Due to the possession of very high values of dielectric constant, ceramics have always been potential contenders for dielectric capacitor applications. For a long time, the low-cost titanium dioxide (TiO_2) and barium titanate ($BaTiO_3$)-based inorganic ceramics have been at the core of capacitor development as these possess high dielectric constant as evident in Table 1, and could operate in a wide temperature range. But these ceramic capacitors are heavier and brittle, depend on manufacturing procedures that involve high temperatures, and possess very low dielectric breakdown strength ($E_b < 50$ MV/m). All this limits their use in flexible, compact devices. Aluminum electrolytic is another class of electrolytic capacitors which is easy to fabricate and can operate up to 600 V. But this is also bulkier than BOPP film capacitor and tends to dry up at elevated temperatures and cannot bear fluctuations and ripple currents [24, 26, 36].

In contrast to the nonlinear ferroelectric oxides, the linear metal oxide dielectrics possess a dielectric constant that is independent of the applied electric field, has high breakdown strength, low energy loss, and better energy storage efficiency than other inorganic dielectric systems. Examples of a few linear dielectric metal oxides are the calcium titanium compounds, namely $CaTiO_3$, $CaZrTiO_3$, $CaTiO_3$–$BiScO_3$, $CaSrTiZrO_3$, and $SrTiO_3$. The dielectric properties exhibited by linear dielectric metal oxide $CaTiO_3$ are quite fascinating. It exhibits $k = 175$ and has an exceptionally high breakdown strength of 300 kV/cm and energy storage density of about 1.5 J/ cm^3 and is quite stable in a wide temperature and frequency range. Its performance is hindered in high electric fields by the thermal breakdown and increased electrical conductivity due to electron hopping of multivalent titanium ions and charge migration favored by weak grain boundaries. Most of the other linear dielectric metal oxide do not have a such high dielectric constant (k), and their energy storage density is also quite low, less than 1 J/ cm^3 [6, 7, 23, 24, 26, 36].

Besides having excellent dielectric properties, the metal oxides also exhibit good mechanical and optical properties and hence are potential candidates used for varied thin film electronic systems. To name a few metal oxides Ta_2O_5, Nb_2O_5, Y_2O_3, TiO_2, Al_2O_3, HfO_2, and ZrO_2 that have a high value of dielectric constant ranging between 10^1 and 10^2 [36, 56, 57]. Titanium dioxide (TiO_2) is versatile and used in varied optoelectronic applications due to its high dielectric constant in its rutile form [59]. The other metal oxide with a high dielectric constant is Tin dioxide (SnO_2) which is an n-type semiconductor having a direct band gap energy of 3.37 eV and has a stable chemical, mechanical, and thermal attributes to be used in applications about spintronics, lithium-ion batteries (LIBs), and super-capacitors [60–63]. Another metal oxide ZnO, which is also a wideband gap semiconductor, exhibits not only superb dielectric properties but also exceptional thermo-physical properties. A literature survey reveals that these metal oxides offer greater dielectric constant when doped with some other metals, thereby meeting the requirements of energy storage devices. The oxides compounds of transition metals have undergone a huge development process, and with an efficient volumetric efficiency and an innovative fabrication in multilayers, now the dielectric permittivity has been achieved in thousands as reported for $CaCu_3Ti_4O_{12}$ ($k = 17{,}000$) and La_2SrNiO_4 ($k = 10{,}000$) by Lunkenheimer et al. [57] and colossally a value in order 10^5 for Sr_2TiMnO_6 by Meher et al. [58].

Unfortunately, these metal oxide ceramic capacitors inherit very low breakdown field strength resulting in low energy density and poor mechanical properties to be cast as high energy storage capacitors. The causes of low breakdown strength are flaws in the fabrication process. A lot of literature reports on the various types of research undertaken to mitigate the defects of bulk ceramic material using techniques that involve high-temperature or high-pressure treatment are tedious and incurs huge cost. An alternative way is to prepare the nanoparticles (NP) of these dielectric materials and then form composites of these with materials like glass or polymers that are considered to be "defect-free" intrinsically. This method seems fascinating as the fabrication process can be done at low temperatures in an easy, cost-effective, simple manner with resulting material of improved performance. The only drawback is the formation of an interface between the dielectric NP and high breakdown glass or polymer matrix that creates flaws that reduce the dielectric strength substantially. So the formation of NP of appropriate size, shape, and stoichiometry and their proper distribution in the host matrix is of great significance for the best performance of the formed nanocomposite. The literature survey reveals several synthesis protocols and techniques applied for the production of such nanocomposites as enlisted in Table 1. The conventional method comprises grinding the desired ceramics into fine powders and then heating them together at elevated temperatures (>1000 °C). Such procedures provide superb control of stoichiometry and produce fine crystals but do not provide any regulation on particle size, and the procedures of grinding and heating are elaborate (Fig. 7.4).

7.3.3 Polymeric Dielectrics

Recent developments in the microelectronics industry require materials that facilitate the miniaturization of devices. The low values of dielectric permittivity, dielectric loss, and leakage currents aid the polymeric materials to be drawn in nano-scale layers while maintaining their functionality. Further, their ease of mass manufacturing and flexibility to be drawn in myriad shapes and sizes at atmospheric conditions with minimal heat treatment has gained them a lot of attention to be used for transistor

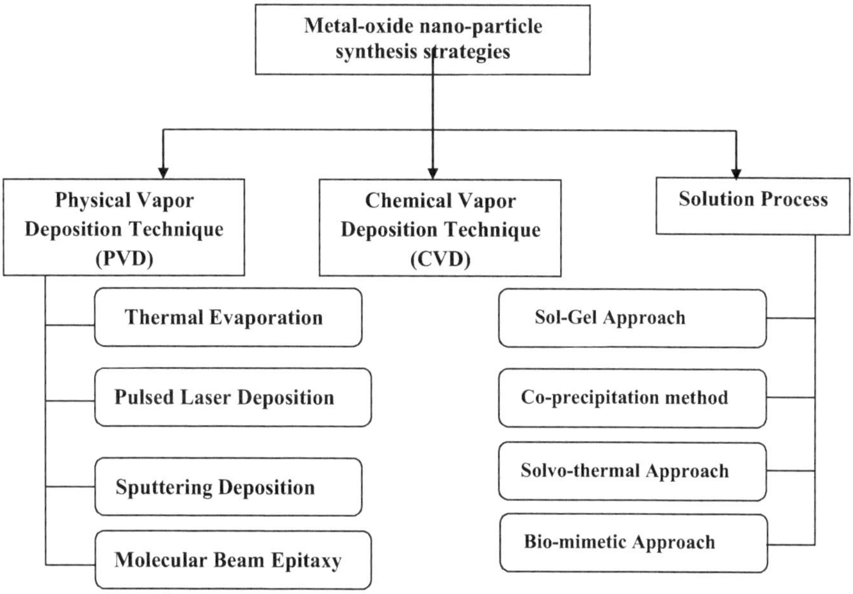

Fig. 7.4 Various techniques for the synthesis of metal oxide nanoparticles

gate dielectrics applications. The low values of dielectric parameters are beneficial in transistor gate dielectric applications, but for energy storage applications, a high value of dielectric permittivity is required as governed by Eq. (7.2), and the ferroelectric polymers that have a considerably large value of dielectric constant have been a subject of research for such applications.

As mentioned earlier, the dielectric constant in polymers results due to *orientation, ionic, and electronic polarization* mechanisms. Functional groups that have highly reactive electron clouds and the aromatic rings in a polymer are observed to cause electronic polarization and are included in the dielectric material to augment the dielectric constant. The dielectric constant due to electronic polarization for the organic polymers is associated with their band gap and has the highest value of 3.6 when the band gap is 4 eV. Generally, the linear dielectric polymers exhibit a dielectric constant in the range of 2–5; as the dielectric constant due to atomic/ionic polarization is less than 25% of the value due to electronic polarization, and hence for the polymers, the total of dielectric constant due to resonance polarization can reach up to a maximum of 4.5. Contrary to this, the atomic/ionic polarization is significantly very high in metal oxide ceramics leading to high values of dielectric constant for them. In the film form of polymers, the mobility of the polymeric chains is constrained and hence the switching of the dipoles is pretty slow generating the dielectric constant due to *orientation polarization* to be less than 10. Here, one should note that the increase in chain mobility may lead to a phase transition of the polymer from glassy to viscous rubbery fluid. One noteworthy example of such polarization is the ferroelectric semi-crystalline polymer polyvinylidene fluoride (PVDF) with a chemical structure $[CF_2\text{-}CH_2\text{-}]_n$, where the dipoles created by negatively charged fluorine and positively charged hydrogen atoms stretch along the applied field. Depending upon the trans and gauche arrangements of the molecules, PVDF is found to exist in four types of conformations (α, β, γ and δ), of which in the beta β phase, where all the molecules align in all-trans manner is most electroactive. The segregation of hydrogen and fluorine atoms gives rise to discrete ferroelectric domains that polarize on the application of electric field and contribute to a high dielectric constant ~15. Thus, processes that form discrete charge-aligned regions are one of the ways to develop materials with high dielectric constant. Electrical poling or mechanical stretching at high temperatures is such a process, and the PVDF molecules that predominantly exist in the alpha α phase are transformed into the β phase by them. Incorporation of bulky chlorinated co-monomers by irradiation in PVDF forms copolymers like P(VDF-TrFE) and P(VDF-TrFE-CTFE) where the chlorinated co-monomer molecules facilitate all-trans conformations thereby increasing the polar phase resulting in a dielectric constant = 50 @ 1 kHz. Burlingame et al. [80] reported that inserting carbonyl atoms in urea to replace a few sulfur atoms increases the dipole interactions between inter-chains of poly-urea, thereby creating a strong local internal field increasing dielectric constant.

The capacitance of electric double layer (EDL) capacitors is observed to be 1000 times that of a typical dielectric polymer film such as BOPP. The contribution to such high values of dielectric constant comes from *space charge polarization* where the actual physical movement of ionic species (ions/electrons/holes) occurs over nanometers of distance. Such polarization is beneficial for super-capacitor applications but becomes a major cause of dielectric loss in dielectric polymers and hence must be avoided.

Although ferroelectric polymers exhibit high dielectric strength, they possess a low value of young's modulus (a few hundred of MPa's) and undergo mechanical breakdown phenomena. Of the various breakdown mechanisms discussed above, the thermal breakdown is of prime concern. In designing capacitive energy storage devices, care must be taken that resistive heating due to leakage currents must not increase the internal temperature of the device to melt and get decomposed. Dissado has reported that the electrical conductivity and thermal conductivity of material have the same degree of effect on breakdown strength, but the variation in electrical conductivity is of several orders of magnitude whereas variation in thermal conductivity is only of one order. Hence, electrical conductivity of the material is significant for designing of the device and can be controlled by manipulating the degree of crystallinity or incorporating dipole traps that confine the accelerated electrons mitigating the prospects of any breakdown [34].

Thus, for developing high-performance dielectric materials, we need to incorporate additives or fillers that reduce the free volume and have high thermal conductivity. Also, polymers with rigid backbones having high Young modulus (> 2 G Pa) must be preferred.

The prime reason for using polymeric matrices is their high breakdown strength. The reported energy densities of neat polymer dielectric capacitors are over 15 J/

cm^3 in comparison to bulk ceramics capacitors that have energy densities over 3 J/cm^3. The energy storage function varies quadratically with breakdown strength and it seems that manipulating it may be the easiest approach to get the desired dielectric material. But literature survey suggests that for polymers E_b of around 1GV/m is already achieved [24, 26] which is quite close to the benchmark of 2 GV/m for diamond [4, 19] thus limiting the scope of its further enhancement. However, the dielectric constant for polymeric materials is often in the range of 2–10, thus there is an opportunity for improvement in achieving greater values, up to perhaps the range for water (about 80).

7.4 Metal or Metal-Oxide-Filled Polymer Composites for Enhanced Energy Storage: Opportunities and Challenges

By far discussion has made it clear that to develop higher energy density capacitors, one needs to improvise on the material's relative dielectric permittivity and electric breakdown strength without compromising on mechanical attributes and its performance temperature window. The metal oxides possess a large value of k but have low breakdown strengths, and in contrast, the polymeric materials exhibit high breakdown strength (>500 MV/m) but with poor permittivity values (<3). So on combining the two electrically contrasting materials, it is intended to combine a highly polarizable dielectric (metal oxide) with a high linear dielectric strength (polymer) to achieve composites that have a balance of attributes of both and results in a material with high energy storage as clearly depicted schematically in Fig. 7.5.

The response of the composite material is largely dependent on the attributes of the dispersed metal oxide and chosen host polymer and is significantly dominated by the interfaces they create and interactions happening at these interfaces. The nano-inclusion of metal oxides into the host polymer can have any of the electric behavior, i.e., it can be conductive, dielectric, or semiconductor, and can be in any shape—spherical-like particles, rods, or needle like or in the form of sheets. The morphology of the polymer within the composite too can be any spherical beads to granular

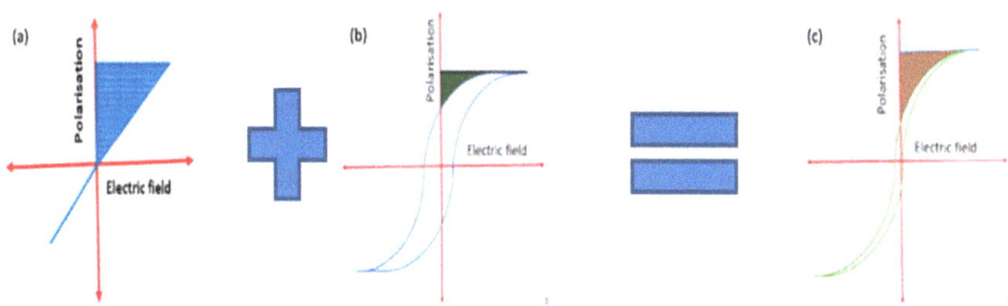

Fig. 7.5 Schematic representation of intent of developing polymer nanocomposite

membranes, to fibers. Based on the internal connectivity, order, and symmetry, there is a total of ten types of structures that can be fabricated for dual-phase composites namely (0–0), (0–1), (0–2), (0–3), (1–1), (2–1), (2–2), (2–3), (1–3), and (3–3). Here, the first number represents the dimension of connectivity of the filler and the second number stands for the dimension of connectivity of the polymer. A general polymer nanodielectric represented by (0–3) nomenclature implies that fillers of spherical shape have been inserted in a three-dimensionally connected polymer matrix [35].

The initial interest in fabricating polymer nanodielectrics was to disperse high *k* metal oxides into a ferroelectric polymer. Li and coworkers [64] found that dispersion of 30% volume fraction of ethylene diamine-modified $BaTiO_3$ nanoparticles into P(VDF-TrFE-CTFE) could enhance the energy density from 3 to 7 J/cm^3. Similarly Dang et al. [65] reported the nickel dispersed PVDF composite to exhibit colossal *k* equal to400. But soon, it was realized that such permittivity contrast enhances the interfacial polarization, which however produces an unbalanced distribution of electric fields such that there is a strong field experienced by the polymer molecule whereas almost negligible field inside the metal filler. So the approach of increasing the dielectric constant of material was coming at the expense of a significant reduction in breakdown strength and was hence not promoted further. To obtain enhanced synergistic effects, metal oxides that had less contrast in dielectric nature were tried upon. Wang et al. [66] found metal oxides ZrO_2 and TiO_2 to have permittivity nearly close to that of PVDF copolymers and they reported that the 9 wt % ZrO_2 in P(VDF-TrFE) produces an energy density of 11.2 J/cm^3 at 270 MV/m and 10 wt % of TiO_2 in PVDF terpolymer produces total energy density of 6.9 J/cm^3 at breakdown strength over 200 MV/m. The results of the approach of using a metal-oxide–polymer pair having comparable dielectric permittivity have been quite motivating, and several researchers have attempted this approach. Choudhary et al. have reported the variation in complex dielectric permittivity, electric modulus, AC electrical conductivity, and the impedance behavior with frequency and temperature for various metal-oxide-doped polymers, and the results of their findings have been shown in Fig. 7.6 as an example [67]. Each polymer composite shows unique dielectric behavior. They have reported such variations for polymer blends too, e.g., poly(vinyl alcohol) (PVA) and poly(ethylene oxide) (PEO) blend matrix (50/50 wt%) dispersed with zinc oxide (ZnO) nanoparticles [68], a comparative study of the effect of metal oxide on the dielectric dispersion and relaxation processes and ionic conductivity in PEO-PMMA blend dispersed with 3% wt of Al_2O_3, SiO_2, SnO_2, and ZnO nanoparticles films of PVDF/PEO blend with Al_2O_3, SnO_2, TiO_2, ZnO nanofillers [69, 70]. Their findings show that the metal oxide nanoparticles produce significant alterations in the polymer spherulitic morphology, create a large number of micro- to nano-sized pores, lower the β-phase content of the PVDF and crystalline phase of PEO, and also the degree of crystallinity of the host matrix. Among the investigated metal oxides, the Al_2O_3 nanoparticles were found to significantly enhance the dielectric permittivity as well the AC electrical conduction. Bafna et al. have reported the variation of dielectric properties on insertion of carbon soot found formed from different oils in PMMA [71] metal salts like $SnCl_2$ [60], $FeCl_3$ [72], and metal oxides of potassium $KMnO_4$ and K_2CrO_4 in PMMA [73, 74] and ZnO doped in polymer PVDF PMMA blend [75, 76]. Deeba et al. have reported a systematic study of variation of dielectric parameters with frequency when a variety of metal oxides have been doped in PMMA and its copolymers or blends [77].

The only limitation of this approach is that composites with extremely high dielectric constants cannot be achieved. To achieve this, the dispersion of core–shell or multi-core metal oxide into polymer has been done [78, 79]. The multi-core metal oxide is made in such a way that the high permittivity metal is kept in the core and material with gradually decreasing permittivity is onto it in such a fashion that permittivity reduces gradually from the core to the shell and finally is comparable to the value of the polymer matrix.

Such an "interfacial engineering" idea was implemented by Marks and coworkers [80] who have used a series of metal oxides viz.$BaTiO_3$, ZrO_2, MgO, $SrTiO_3$, TiO_2, $Ba_{0.5}Sr_{0.5}TiO_3$, and yttria-stabilized zirconia as core and Al_2O_3as shell in polypropylene polymer. They demonstrated that in such materials the shell behaves as a buffer material between the equates the dielectric permittivity of filler and host and high k-valued core aids to raise the overall permittivity of the composite. Another interesting outcome of the study was that the Al_2O_3 shell also aids to reduce the dielectric losses significantly and that the dielectric loss at the strong field was largely proportional to the particle interfacial area, regardless of the constituent nanoparticle type. A similar multi-core strategy has been reported by Shen et al. for conductive Ag nanofillers encapsulated by an organic shell and dispersed in an epoxy matrix resulting in a composite with improved permittivity and increased breakdown strength.

The above study provides certain designing rules for viable metal-oxide–polymer dielectric composites as listed below.

(i) If high permittivity/high conductivity metal oxide nanoparticles such as ABO_3 perovskites are used as filler, the dielectric losses alleviate owing to the internal AC conduction due to space charges. So metal oxides having bulk conductivity and permittivity matching to the host polymer like TiO_2 ($k = 30$), ZrO_2 ($k = 25$), SnO_2, and ZnO should be preferred for dispersion as filler. The bulk conductivity of these metal oxides can reach comparably to the low value of polymer, i.e., below 10–13 S/m. To avoid AC conduction issues, metal-oxide–polymer dielectrics should be preferably used in DC applications. A discovery of materials with high permittivity but poor conductivity is awaited.

(ii) To build a high dielectric constant composite material, we require nano-inclusions of high permittivity metal oxide in an amount before percolation sets in. But this reduces the breakdown strength of the composite. The solution to this is to increase the apparent dielectric constant of nanocomposites at low-volume fractions. The use of 2D nanofillers (rod-like or wire-like) is effective in increasing the dielectric constant considerably by 10% [26, 36]. But one has to cautiously implement such strategies as the use of nano-wire content close to the percolation threshold has been observed to reduce the breakdown strength substantially by 15% due to enhanced current leakage. The addition of a high permittivity interface, i.e., metals or conducting polymers although effectively increase the dielectric constant at a low filler volume fraction but suffer from

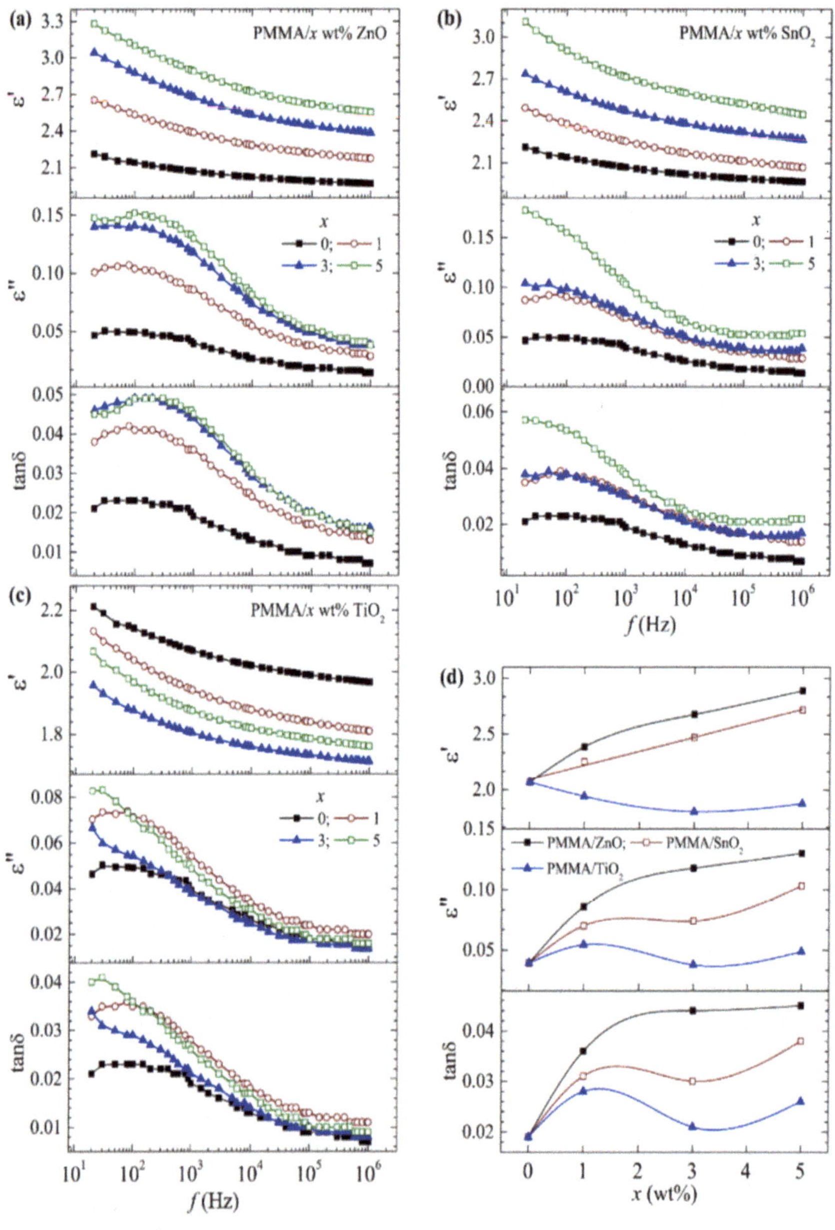

Fig. 7.6 Variation of dielectric parameter with frequency and content of metal oxide dopant in PMMA [67]. Adapted with permission from Ref [67] Copyright (2021) (Elsevier)

high dielectric loss initiated by leakage current through electron tunneling or field electron emission mechanism [10].

(iii) While controlling the electrical behavior of materials through controlling the interactions at the interface, a delicate balance is required to achieve the advantages of getting enhanced MWS interfacial polarization through effective electron trapping and mitigate the demerits of internal AC conduction and increased local electric field in the polymer matrix. The fabrication of materials in layered geometry helps achieve this balance as layered interfacial charges serve as effective charge traps [79]. The multi-layer concept is yet to be explored to achieve the desired results.

7.5 Summary and Perspectives

Polymers are preferred materials in micro- and nano-electronics as they can be easily cast into film form using spin coating, dip coating, or solution cast technique. Owing to the fascinating properties of being thermo-mechanically stable, corrosion resistant, easily and cost-effectively amenable to be cast in varied shapes and forms, and with a high dielectric breakdown strength, the polymers offer to be the most sought-after materials for making composites by doping them with metal oxides that can infuse their characteristic high dielectric permittivity to the composite system. The fabrication of metal-oxide–polymer composites involves the dispersion of metal or metal oxide nanoparticles within the polymer phase via in situ strategies such as solution and melt blending, sol–gel processing, where polymerization occurs around the metal oxide nanoparticle or the ex situ strategies wherein the pre-synthesized metal oxide particle is directly incorporated into the polymer matrix by external mechanical blending, etc. The overall thermo-physical properties of the composite material are largely determined by the (i) morphology of the host polymer: size, surface area, porosity, and shape of the micropores within the polymer determine the size and nature of the dispersed metal and metal oxides nanoparticles [81], (ii) concentration and nature of the functional group in it as it determines the host material's ability to bind with metal complex, and (iii) the method of dispersion of the metal oxide in the host polymer. The dispersion of metal or metal oxides inside the polymer can be achieved through a combination of different mechanisms such as complexation/chelation, electrostatic interactions, precipitation, or reduction reactions, etc. The diverse nature of the functional groups and the different factors controlling the sizes of the dispersed nanoparticles often make it a challenging task to synthesize and impart desired properties in the composite material. Presently the approach to developing metal-oxide-based polymer composite for targeted and tailored electrical properties is by choosing the appropriate combination of host polymer matrix and filler metal oxide, the correct amount of dispersion of metal oxide into the polymer, and controlling their interactions at the interfaces of the two. For these composites to act as favored energy harvesting materials, one needs to consider both the real and imaginary parts of dielectric permittivity, the band gap, dielectric breakdown strength, thermo-mechanical behavior, glass transition temperature, morphology, manufacturing procedure, and cost of the component of the desired composite. The appropriate dispersion of nano-metal/metal oxide particles develops a network of nano-capacitors within the host polymer, which can be charged and discharged as required, thus presenting themselves as energy storage devices. The amount of stored and retrieved energy can be enhanced by tailoring the properties of the interface between the filler and polymer.

The major challenge with these metal oxide polymer nanodielectrics is the uniform dispersion of the metal oxide within the polymer and maintaining the low percolation threshold to achieve both high dielectric permittivity and high breakdown strength at a low-level loading. By increasing the polymer's permittivity, today's metal oxide fillers weaken the material's mechanical properties and decrease its dielectric breakdown strength.

The enormous amount of research findings in the last decade has generated a wealth of data to be used with novel computational methods based on quantum mechanical ab initio methods, classical force-field-based molecular dynamics simulations, and data-driven paradigms, such as quantitative structure–property relationship and machine learning schemes to predict to design and develop novel metal-oxide polymer-based dielectrics for energy harvesting applications.

Metal oxide polymer nanodielectrics offer themselves as a promising class of materials for use in energy storage applications such as capacitors and batteries. They offer several advantages over traditional dielectric materials, including high dielectric constant, low-loss tangent, and good thermal stability. Here in, we summarize the opportunities and challenges in these materials.

Opportunities for metal oxide polymer nanodielectrics include:

- Improving the energy density of capacitors, which could lead to smaller and more powerful electronic devices.
- Increasing the power density of batteries, which could enable faster charging and longer-lasting devices.
- Reducing the cost of energy storage by using less expensive materials.

Challenges for metal oxide polymer nanodielectrics include:

- Developing a synthetic method to produce metal oxide polymer nanodielectrics with controlled size, shape, and composition.
- Improving the stability of the materials under different operating conditions.
- Understanding the mechanism of charge storage and transport in the materials.

Overall, metal oxide polymer nanodielectrics have a lot of potential for energy storage applications, but further research is needed to fully realize their capabilities.

References

1. SarjeantW J, Clelland IW, Price RA (2001) Capacitive components for power electronics. Proc IEEE 89(6):846–855
2. Johnson RW, Evans JL, Jacobsen P, Thompson JR, Christopher M (2004) The changing automotive environment: high temperature electronics. IEEE Trans Electron Packag Manuf 27:164–176
3. Ahmed Z, Al S, Kamaruzzaman S, AmerAH (2020) Review of energy storage services, applications, limitations, and benefits. Energy Rep 6(7):288–306
4. Wentian G, Lu W, Gleb Y (2017) World scientific series in current energy issues energy storage. Capacitive Energy Storage, 167–214. https://doi.org/10.1142/9789813208964_0005
5. Jow TR (2015) Pulsed power capacitor development and outlook. In: IEEE Pulsed Power Conference (PPC), 1–7
6. Hassan YA, Hu H (2020) Current status of polymer nano-composite dielectrics for high-temperature applications. Compos A 138:1060–1064
7. Dang ZM, Yuan JK, Yao SH, Liao RJ (2013) Flexible nanodielectric materials with high permittivity for power energy storage. Adv Mater 25:6334–6365
8. Chauhan A, Patel S, Vaish R, Bowen CR (2015) Anti-ferroelectric ceramics for high energy density capacitors. Materials 8:8009–8031
9. Choi KJ (2004) Enhancement of ferro-electricity in strained $BaTiO_3$ thin films. Science 306:1005–1009
10. Wang Y, Zhou X, Cheng Q, Chu B, Zhang Q (2010) Recent development of high energy density polymers for dielectric capacitors. IEEE Trans Dielectrics, Elect Insulat 17:1036–1042
11. Yao ZH, Song Z, Hao H, Yu ZY, Cao MH, Zhang SJ, Lanagan MT, Liu HX (2017) Homogeneous/inhomogeneous-structured dielectrics and their energy-storage performance. Adv Mater 29:16017–16027
12. Zhang Y, Feng H, Wu X, Wang L, Zhang A, Xia T, Dong H, Li X, Zhang L (2009) Progress of electrochemical capacitor electrode materials: a review. Int J Hydrogen Energy 34:4889–4899
13. Wang G, Zhang L, Zhang J (2012) A review of electrode materials for electrochemical supercapacitors. Chem Soc Rev 41:797–828
14. Sharma P, Bhatti TS (2010) A review on electrochemical double-layer capacitor. Energy Convers Manage 51:2901–2912
15. Chen Q, Shen Y, Zhang S, Zhang QM (2015) Polymer-based dielectrics with high energy storage density. Annu Rev Mater Res 45:433–458
16. Shen Z, Wang X, Luoa B, Lia L (2015) $BaTiO_3$–$BiYbO_3$ perovskite materials for energy storage applications. J Mater Chem A 3:18146–18153
17. Puli VS, Pradhan DK, Chrisey DB, Tomozawa M, Sharma GL, Scott JF, Katiyar RS (2013) Structure, dielectric, ferroelectric, and energy density properties of (1.2×)BZT–xBCT ceramic capacitors for energy storage applications. J Mater Sci 48:2151–2157
18. Zou K, Dan Y, Xu H, Zhang Q, Lu Y, Huang H, He Y (2019) Recent advances in lead-free dielectric materials for energy storage. Mater Res Bull 113:190–201
19. Zhu L (2014) Exploring strategies for high dielectric constant and low loss polymer dielectrics. J PhysChemLett 5(21):3677–3687
20. Sarjeant WJ, Zirnheld J, MacDougall FW (1998) Capacitors. IEEE Trans Plasma Sci 26:1368–1392
21. Balaraman AA, Dutta S (2022) J Phys D: Appl Phys 55:183002–183008. https://doi.org/10.1088/1361-6463/ac46ed
22. Christen T, Carlen MW (2000) Theory of ragone plots. J Power Sources 91:210–216
23. Zhu L, Wang Q (2012) Novel ferroelectric polymers for high energy density and low loss dielectrics. Macromolecules 45:2937–2954
24. Riggs BC, Adireddy S, Rehm CH, PuliSV ER, Chrisey DB (2015) Polymer nanocomposites for energy storage applications. Mat Today: Proceed 2:3853–3863. https://doi.org/10.1016/j.matpr.2015.08.004

25. Siddabattuni S, Schuman TP, Dogan F (2013) Dielectric properties of polymer−particle nanocomposites influenced by electronic nature of filler surfaces. ACS Appl Mater Interfaces 5:1917–1927. https://doi.org/10.1021/am3030239
26. Psarras GC (2018) Dielectric polymer materials for high-density energy storage chapter 2: fundamentals of dielectric theories. https://doi.org/10.1016/B978-0-12-813215-9.00002-6 Elsevier Inc.
27. Psarras GC (2010) Physical properties and applications of polymer nanocomposites chapter 2—conductivity and dielectric characterization of polymer nanocomposites. Woodhead Publishing Series in Composites Science and Engineering, 31–69.
28. SteemanP AM, Turnhout JV (2003) Dielectric properties of inhomogeneous media. In: Kremer F, Schonhals A (eds) Broadband dielectric spectroscopy. Springer, Berlin, pp 4951–5022
29. Boyd RH, Liu F, (1997) Dielectric spectroscopy of semicrystalline polymers. In: Runt JP, Fitzgerald JJ (eds) Dielectric spectroscopy of polymeric materials. American Chemical Society, Washington, DC, pp 1071–1136
30. Hedvig P (1977) Dielectric spectroscopy of polymers. Adam Hilger Ltd., Bristol
31. Korzhenko A, Tabellout M, Emery JR (1999) Influence of a metal-polymer interfacial interaction on dielectric relaxation properties of polyurethane. Polymer 40:7187–7195
32. TsangarisGM PC, Kouloumbi N (1998) Electric modulus and interfacial polarization in composite polymeric systems. J Mater Sci 33:2027–2037
33. Barber P, Balasubramanian S, Anguchamy Y, Gong S, Wibowo A, Gao H, Ploehn HJ, Loye HC (2009) Polymer composite and nanocomposite dielectric materials for pulse power energy storage. Materials 2(4):1697–1733
34. Dissado LA, Fothergill JC (2008) Electrical degradation and breakdown in polymers
35. Zhang G, Allahyarov E, Zhu L (2018) Nano/micro-structured materials for energy and biomedical applications, chapter 1 polymer nanodielectrics: current accomplishments and future challenges for electric energy storage. Springer Nature Singapore Pvt. Ltd. https://doi.org/10.1007/978-981-10-7787-6_1
36. Kaur D, Bharti A, Sharma T, Madhu C (2021) Dielectric properties of ZnO-based nanocomposites and their potential applications. Int J Opt. https://doi.org/10.1155/2021/9950202
37. Guang Z (2008) Handbook of advanced dielectric, piezoelectric and ferroelectric materials synthesis properties and applications. Woodhead Publishing, Sawston, UK
38. Bafna M, Gupta AK, Khanna RK, Vijay YK (2020) Development of potassium permanganate (KMnO4) doped Poly methyl methacrylate (PMMA) composite using layered structure for electromagnetic shielding purpose. Mat Today: Proceed 30(6). https://doi.org/10.1016/j.matpr.2020.03.736
39. Gupta AK, Bafna M, Khanna RK, Shrivasata S, Vijay YK (2021) Study of electromagnetic shielding effectiveness of metal oxide polymer composite in their bulk and layered forms. Environ Sci Pollut Res 24(6):1–810.1007/s11356–020–09322-x
40. Chang J, Zhai H, Hu Z, Li J (2022) Ultra-thin metal composites for electromagnetic interference shielding. Compos B Eng 246:110269. https://doi.org/10.1016/j.compositesb.2022.110269
41. Zadehnazari A (2022) Metal oxide/polymer nanocomposites: a review on recent advances in fabrication and applications. Polymer-Plastics Tech Mat. https://doi.org/10.1080/25740881.2022.2129387
42. Nikolic MV, Vasiljevic ZZ, Auger S, Vidic J (2021) Metal oxide nanoparticles for safe active and intelligent food packaging. Trends Food Sci Technol 116:655–668
43. Won JJ, Hwang HS, Choi D, Ma BC, Jung J, Chang M (2020) Hybrid polymer/metal oxide thin films for high performance. Flexible Transis. Micro-Machines 11(3):264. https://doi.org/10.3390/mi11030264
44. Sun Z, Wong R, Yu M, Li J, Zhang M, Mele L, Hah J, Kathaperumal M, Wong CP (2022) Nanocomposites for future electronics device Packaging: A fundamental study of interfacial connecting mechanisms and optimal conditions of silane coupling agents for Polydopamine-Graphene fillers in epoxy polymers. Chem Eng J 439:135621
45. Held M, Schie SP, Miehler D, Gannott F, Zaumseil J (2015) Polymer/metal oxide hybrid dielectrics for low voltage field-effect transistors with solution-processed, high-mobility semiconductors. Appl Phys Lett 107:083301. https://doi.org/10.1063/1.4929461

46. Mote VD, Purushotham Y, Dole BN (2016) Structural, morphological, physical and dielectric properties of Mn doped ZnO nanocrystals synthesized by sol-gel method. Mater Des 96:99–105
47. Ahmad MP, Rao AV, Babu KS, Rao GN (2019) Particle size effect on the dielectric properties of ZnO nanoparticles. Mater Chem Phys 224:79–84
48. Mangalaraja RV, Manohar P, Gnanam FD, Awano M (2004) Electrical and magnetic properties of Ni0.8Zn0.2Fe_2O_4/silica composite prepared by sol-gel method. J Mat Sci 39(6):2037–2042
49. Silva RLD, Franco A (2020) Effect of porosity on dielectric properties of ZnO ceramics. J Eur Ceram Soc 40(4):1307–1311
50. Bergman DJ (1978) The dielectric constant of a composite material—A problem in classical physics. Phys Rep 43(9):377–407
51. Sareni B, Krähenbühl L, Beroual A, Brosseau C (1997) Effective dielectric constant of random composite materials. J Appl Phys 81(5):2375–2383
52. Mohamed GA, Abd El-Moiz AB, Rashad M (2005) Li-doping effects on the electrical properties of ZnO films prepared by the chemical-bath deposition method. Physica B 370(1–4):158–167
53. Huang D, Liu Z, Li Y, Liu Y (2017) Colossal permittivity and dielectric relaxation of (Li, In) co-doped ZnO ceramics. J Alloys Comp 698:200–206
54. Khan R, Zulfiqar F, Levartoski CI (2018) Influence of oxygen vacancies on the structural, dielectric, and magnetic properties of (Mn, Co) co-doped ZnO nanostructures. J Mater Sci: Mater Electron 29(12):9785–9795
55. Silva R, Banerjee P, Júnior AF (2019) Functional properties of donor- and acceptor-co-doped high dielectric constant zinc oxide ceramics. Phys Chem Chem Phys 21(18):9456–9464
56. Treichel H, Mitwalsky A, Tempel G (1995) Deposition, annealing and characterization of high-dielectric-constant metal oxide films. Adv Mat Optics Elect 5(3):163–175
57. Lunkenheimer P, Krohns S, Riegg S, Ebbinghaus SG, Reller A, Loidl A (2009) Colossal dielectric constants in transition-metal oxides. European Phys J Special Topics 180(1):61–89
58. Meher K, Varma BR (2009) Colossal dielectric behavior of semiconducting Sr_2TiMnO_6 ceramics. J Appl Phys, 105
59. Huang X, PuZ TL, Wang Z, Liu X (2012) Preparation and dielectric properties of surface modified TiO_2/PEN composite films with high thermal stability and flexibility. J Mater Sci: Mater Electron 23(12):2089–2097
60. Bafna M, Garg N, Gupta AK (2018) Dielectric properties of stannous chloride doped PMMA composite films. J Emerg Technol Innov Res 5:494
61. Jahnavi VS, Tripathy SK, Rao AVNR (2019) Structural, optical, magnetic and dielectric studies of SnO_2 nano particles in real time applications. Physica B 565:61–72
62. Pradyumna E, Sreelekha N, Reddy DA, Gunasekhar KR, Subramanyam K (2015) Dopant induced room temperature ferromagnetism in spintronic SnO_2: Co nanoparticles. Int J Adv Eng Nano Tech 2:22
63. Zhao Q, Ma L, Zhang Q, Wang C, Xu X (2015) SnO_2-Based nanomaterials: synthesis and application in lithium-ion batteries and supercapacitors. J Nanomat 15
64. Li J, Seok SI, Chu B, Dogan F, Zhang Q, Wang Q (2009) Nanocomposites of ferroelectric polymers with TiO_2 nanoparticles exhibiting significantly enhanced electrical energy density. Adv Mater 21:217–221. https://doi.org/10.1002/adma.200801106
65. Dang ZM, Yan WT, Xu HP (2007) Novel high-dielectric-permittivity poly(vinylidene fluoride)/polypropylene blend composites: the influence of the poly(vinylidene fluoride) concentration and compatibilizer. J Appl Polym Sci 105:3649–3655
66. Wang H, Xie H, Wang S, Gao Z, Li C, Hu GH, Xiong C (2018) Enhanced dielectric property and energy storage density of PVDF-HFP based dielectric composites by incorporation of silver nanoparticles-decorated exfoliated montmorillonitenanoplatelets. Compos Appl Sci Manuf 108:62–68. https://doi.org/10.1016/j.compositesa.2018.02.020
67. Sengwa RJ, Dhatarwal P (2021) Polymer nanocomposites comprising PMMA matrix and ZnO, SnO_2, and TiO_2 nanofillers: a comparative study of structural, optical, and dielectric properties for multifunctional technological applications. Opt Mater 113:110837
68. Choudhary S (2017) Dielectric dispersion and relaxations in (PVA-PEO)-ZnO polymer nanocomposites. Phys B: Conden Matt 522(1):48–56

69. Choudhary S, Sengwa RJ (2017) Effects of different inorganic nanoparticles on the structural, dielectric and ion transportation properties of polymers blend based nano-composite solid polymer electrolytes. Electrochimica Acta 247:924–941
70. Sengwa RJ, Dhatarwal P, Choudhary S (2020) A comparative study of different metal oxide nanoparticles dispersed PVDF/PEO blend matrix-based advanced multifunctional nanodielectrics for flexible electronic devices. Mater Tod Comm 25:101380–101392
71. Tailor R, Vijay YK, Bafna M (2019) Environmental emissions. IntechOpen. https://doi.org/10.5772/intechopen.92389
72. Bafna M, Gupta AK, Agarwal A, Sain N, Jain V (2021) Dielectric studies and alternating current conductivity studies of polymethylmethacrylate (PMMA) doped with ferric chloride ($FeCl_3$) in varying concentration. Mat Today: Proceed38:1209–1213
73. Bafna M, GuptaA K, Khanna RK (2018) Dielectric properties of potassium permanganate ($KMnO_4$)—PMMA composite films. J Emerg Techno Inno Res 5:433–436
74. Bafna M, Gupta AK (2022) Study of dielectric permittivity and electrical modulus of K_2CrO_4 doped. PMMA 67(6):890–893
75. Deeba F, Gupta AK, Kulshestra V, Bafna M, Jain A (2022) Investigation of dielectric properties of PVDF-PMMA blends. Mat Today Proceed. https://doi.org/10.1016/j.matpr.2022.06.417
76. Deeba F, Gupta AK, Kulshestra V, Bafna M, Jain A (2022) Analysing the dielectric properties of ZnO doped PVDF/PMMA blend composite. J Mat Sci: Mat Elect 33:23703–23713
77. Deeba F, Shrivastav K, Bafna M, Jain A (2022) Tuning of dielectric properties of polymers by composite formation: the effect of inorganic fillers addition. J Compos Sci 6(12):355. https://doi.org/10.3390/jcs6120355
78. Kao KC (2004) Dielectric phenomena in solids: with emphasis on physical concepts of electronic processes, electric polarization and relaxation, dielectric phenomena in solids 10:41–114
79. Vryonis O, Anastassopoulos DL, Vradis AA, Psarras GC (2016) Dielectric response and molecular dynamics in epoxyBaSrTiO$_3$ nanocomposites: effect of nanofiller loading. Polymer 95:82–90
80. Sanida AS, Stavropoulos SG, Speliotis TH, Psarras GC (2017) Development, characterization, energy storage and interface dielectric properties in $SrFe_{12}O_{19}$/epoxy nanocomposites. Polymer 120:73–81
81. Rozenberg BA, Tenne R (2008) Polymer-assisted fabrication of nanoparticles and nanocomposites. Progress Polym Sci 33(1):40–112

Chapter 8
Fabrication and Properties of Dielectric Elastomer-Based Nanocomposites

Tajamal Hussain, Rabia Batool, Khurram Shehzad, Adnan Mujahid, Adeel Afzal, and Muhammad Zahid

Abstract Dielectric elastomers (DEs) are one of the leading electroactive polymer (EAPs) technologies used for actuation application. These elastomers possess the benefits of large strain and stresses, easy processability, fast speed, recovery, efficiency long life, and reliability. Due to their inherently low dielectric constant, nanofillers are introduced into these elastomers to raise their dielectric constant for increased energy storage and large strain. Enhancement in dielectric constant is largely influenced by the surface functionalization of the nanofiller and structural design of the nanocomposites. In this chapter, we will review the recent developments regarding, the fabrication, characterization, and applications of dielectric elastomer-based nanocomposites.

Keywords Dielectric elastomer · Electroactive polymer · Nanocomposites

8.1 Introduction

Dielectric elastomers (DEs) are multi-functional, incompressible, stretchable, highly extensible, and soft poly-active polymer structures [1, 2]. DEs act basically on a transduction mechanism. Being electromechanical transducers, DEs can convert electrical energy to mechanical energy; however, they act as reversible transducers, converting mechanical energy into electrical one as well. Dielectric elastomer actuators (DEAs), undergoing this reversible operation, are said to be working in electromagnetic

T. Hussain (✉) · R. Batool · A. Mujahid · A. Afzal
School of Chemistry, University of the Punjab, Lahore 54590, Pakistan
e-mail: tajamalhussain.chem@pu.edu.pk

K. Shehzad
School of Micro-Nano Electronics, ZJU-Hangzhou Global Scientific and Technological Innovation Center, Zhejiang University, Zhejiang 311200, China

M. Zahid
Department of Chemistry, University of Agriculture Faisalabad, Faisalabad 38040, Pakistan

S. Moharana et al. (eds.), *Emerging Nanodielectric Materials for Energy Storage*, Nanostructure Science and Technology, https://doi.org/10.1007/978-3-031-40938-7_8

transduction-based generator mode [3, 4]. No doubt, DEs are seen to compete for actuators being used in many latest systems such as loudspeakers, cars, washing machines, and computers. DEAs are also named 'artificial muscles' [1, 5] and can be employed in the formation of soft robots [1, 6]. Large actuation strain, high specific energy density, low density, and low modulation make DEs superior over conventional actuation technologies such as electrostrictive, shape memory, piezoelectric, and magnetostrictive materials [4]. Over a decade ago, it was discovered that dielectric elastomers show about 100% strain by the influence of an electric field [1, 7]. As sensors, they are employed to monitor pressure and strain [1, 8–10]. Along with this special characteristic, DE also exhibits low cost, low operation noise, lightweight, fast response, etc.; leading to its application in different fields as a transducer [10, 11]. Thus, DEs find vast applications in the fields of soft robotics [12–14], energy harvesting [15, 16], artificial muscles [17, 18], haptic interfaces [19–21], and tunable lenses [22, 23].

8.2 Working Mechanism of Dielectric Elastomer

The charged particles present in matter move with an applied voltage. As compared to a conductor, charged particles show movements over small distances in dielectric materials; where two coupled processes, i.e., polarization and deformation can be observed. All dielectric materials are electroactive, i.e., in an electric field, they change shape or size [10].

The basic structure of DEA comprises a soft elastomer membrane sandwiched between two compliant electrodes. When a voltage is applied to the electrodes, an electric field is generated. The presence of electrostatic force and induced Maxwell stress causes a decrease in thickness and expansion in the area of the membrane, as reported by Rontgen in 1880 [2]. The pre-strain behavior of DEs plays an important role; (i) thinning of the film occurs in the width direction (ii) creation of large electric field strength by applying a low voltage (iii) orientation of pre-strained elastomer in such directions which improve its resistance toward electric voltage [24]. Actuation mechanisms in DEAs are reported by the SRI International group when they used polyimide and silicone rubber layers to synthesize a bending actuator in 1998 [25] and showed 100% actuation strain in a DEA prepared using a pre-stretched acrylic elastomer [26]. This discovery gathered the attention of researchers to use DEAs in artificial muscle technology [27, 28].

When observed more closely, it was found that when a power source via a conducting wire is connected to the electrodes, a partial leakage of electric current from the conducting wire can be noted, adding charges to the electrodes partially. The presence of opposite charges on both electrodes creates polarizability in the dielectric material. The forces of attraction between these opposite charges expand the membrane but cause a decrease in its thickness; in this way, DE acts as an actuator. However, when the DE membrane is pre-stretched and pre-charged and then undergoes relaxation in the circuit, the voltage between the electrodes is boosted,

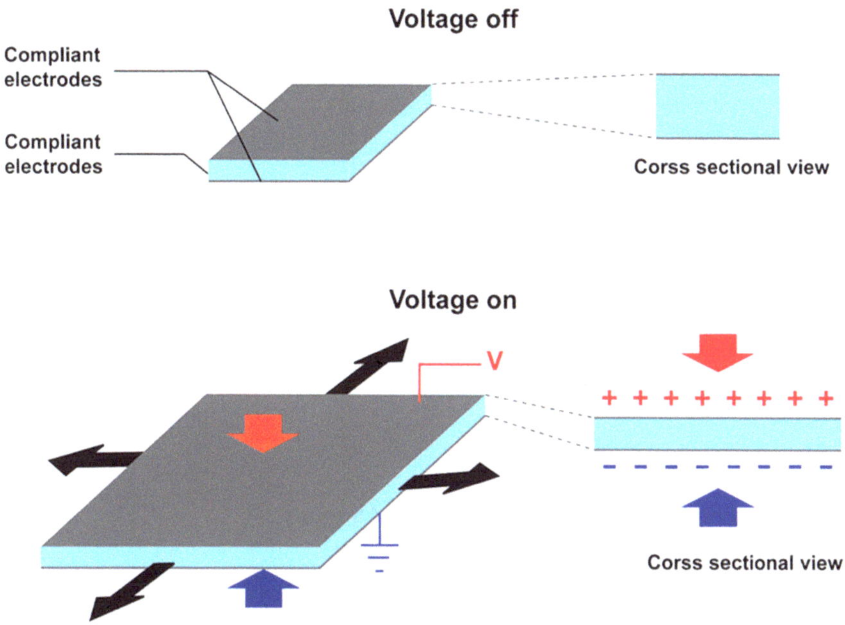

Fig. 8.1 Working mechanism of DEs [28]. Adapted with permission from ref. [28]. Copyright (2019) (American Chemical Society)

and the DE acts as a generator [29, 30]. For the best understanding of DEs, refer to Fig. 8.1 [28].

8.2.1 DE Materials

To achieve better efficiency of DEAs, dielectric elastomers with high electrical breakdown strength, low viscosity, low modulus, and high dielectric constant must be used. Employment of several materials with such properties has been reported by researchers; including fluoroelastomers [31], styrene-ethylene-butylene-styrene [32], natural rubber [33], polyurethanes (PUs) [34], silicones [35], and acrylates [8]. Out of these, the last three materials are the most promising ones [36].

a. ***Polyurethane***

Owing to the presence of polar urethane groups, the higher dielectric constant value (up to 7) of polyurethane (Pus) enables them to work in DEAs at a low electric field. On the contrary, polarity in the groups also leads to a lower breakdown strength [36, 37].

b. ***Silicones***

For the synthesis of DEAs, the most widely used silicone is polydimethylsiloxane (PDMS); which has a low viscosity loss, high conversion efficiency, long life span, fast response rate, weak intermolecular forces, longer bond lengths of the basic unit, i.e., Si–O, and higher flexibility [36, 37].

c. ***Acrylates***

Low price, better adhesive properties, and good performance of commercially available adhesive tapes (VHB 4905/4910) make them excellent candidates to be used in DEAs. Pre-stretched acrylate materials show exceptionally high breakdown strength. However, acrylate-based DEAs have poor adaptability as compared to silicone rubber-based DEAs [8, 36].

8.2.2 Electrodes Used in DE-Based Devices

The functionality of DE-based machines is also affected by the electrodes being used in the formulation of DEAs. The use of stable, highly conductive, strongly adhesive, and good compliancy electrodes results in better working of DEAs. Carbon nanotubes (CNTs), carbon grease, graphite, and carbon powder are the most commonly available choices used for electrodes for this purpose [38–40]. Being cheap, easily accessible, and highly compatible, carbon grease is widely used, but it dries out after a long time. Graphite and carbon black are thin and easily applicable to the membrane's surface, so more suitable to be used in multi-layers stacked DEAs. There is also a drawback in using these electrodes; when a high strain is applied, the contact between the powders is lost, reducing the conductivity values. The high transparency and conductivity of silver nanowires, self-healing, and optical transparency of ionic conductors make them applicable as electrodes in DEAs which are then employed in optical devices [41, 42].

8.2.3 Deformation of Dielectric Elastomer

A thin membrane is unstable and undergoes different morphological patterns such as buckling, crumpling, and wrinkling. If it is compressed by applying strain on boundaries, it undergoes buckling. If uniaxial tensile strength is applied to the membrane, it undergoes wrinkling. Wrinkling and buckling can be seen when the membrane undergoes non-uniform shrinkage. If we stretch the membrane over a liquid drop, symmetrical breakage can be observed leading to crumpling or wrinkling. In the case of DE, where a thin membrane of elastomer is sandwiched between two compliant electrodes; when the applied voltage reaches a particular value, the membrane exhibit instabilities [43].

As DE interconverts electrical and mechanical energies, polymer and electrodes must contract and stretch together [3]. For a considerable deformation of DE, electrodes made up of softer objects with a relatively lower mechanical stiffness, such as carbon grease should be used. By applying a voltage, charged particles begin to flow from one electrode of a transducer to the other. The presence of opposite charges on both electrodes causes the structure to expand in the plane and compress in thickness, leading to deformation in the DE [1, 10].

8.2.4 Configurations of Dielectric Elastomers

Based on the working mechanism of DEs, they are employed to create different configurations of DEAs such as zipping [44], multilayer stacked [45], rolled [46], balloon [47], cone-shaped [48], folded [49], hinge [50], rotary [51], bending [52], and planar [53].

The simplest DEA structure is single-layered. However, deformation in uni-layered structures is too small to be utilized, so multi-layered stacking of DEs can be done to increase the deformation effect in them [54]. As reported by Kovacs and co-workers, increasing the number of DE layers causes difficulty in the synthesis of DEA. This problem was solved by the introduction of an automated process to fabricate DEA with reproducible properties [55]. Fiber-stiffened multi-layered stacked DE sheets can be used to produce a rolled DEA. DE membrane inflated into a balloon can achieve a high area strain and thus can be used in trigger actuators and volume fluid pumps. Rossiter and co-workers [56] presented a cone-shaped DEA; the applied voltage reduces the tension and causes expansion of the DE membrane and it is deformed in the direction of bias force. A double-conical DEA configuration was also formulated by Wang et al. [36]. In the form of an active hinge, DEA was prepared, where a pre-stretched DE membrane is attached on both sides of a hinge support structure [50]. On applying voltage on one side, the hinge structure rotates to the opposite side due to the reduction in the stress of the membrane. Although DEs are nowadays in limelight to be used in the development of soft robots, however, a rigid framework is not much favorable. So, self-organized dielectric elastomer minimum-energy structures (DEMES) are introduced by Kofod et al. [57], where DE is attached to a thin framework of a flexible polymer, releasing a low amount of energy while regaining their original position.

8.2.5 State of Equilibrium in DE

Four different variables affect the state of equilibrium in a DE-based transducer: voltage, force, charge, and displacement. Two of these states are dependent on the properties of the DE, while the other two can be varied according to the requirements. For example, in an electromechanical system of DE, transducer, force, and voltage are variable, and it takes time to gain a state of equilibrium according to its capability of charge polarization and kinematic deformation. The whole process is called 'relaxation'. The time required to achieve this new state of deformation is said to be viscoelastic relaxation time. The process of relaxation can also be explained in terms of the entangled polymer chains responding to the force applied to DE. In elastomers, a polymer network is formed, where flexible polymer chains are interconnected; they may coil or twist or entangle with each other. Applying a force to DE can cause sliding or localized slipping of these entangled chains, resulting in the relaxation of the elastomer. In response to voltage, polarity is created in the elastomers; the molecular dipoles re-orient themselves, causing dielectric relaxation in DE material. This process involves rotation and localized translation of molecules [29, 31, 58].

When a power source is connected to the dielectric material, current leakage takes place, i.e., transport of ions, electrons, or both causes a small current to flow through the material. The density of this current leakage depends upon the applied field; it increases with the increasing applied field [58, 59].

8.2.6 Theories for Dielectric Elastomer-Based Actuators

The shrinkage in the size of the elastomer by applying a voltage across its electrodes is the result of Maxwell stress (interaction between quasistatic electric charges). The Maxwellian formula is derived based on a basic model of materials, assuming: (i) electrode materials have an infinity value of conductivity and zero elastic modulus, (ii) there is ignorable current leakage and dielectric loss, (iii) the elastomer used is perfectly elastic and exhibit very low loss modulus, (iv) the polymer used is isotropic, (v) both elastomer and electrodes are incompressible, and (vi) elastomer shows a constant value of elastic modulus [59, 60].

The application of DEs in actuators is limited by the low dielectric permittivity of the elastomer being used in the actuator. So, Maxwell stress can be improved by increasing its permittivity values [61].

8.3 Applications of Dielectric Elastomers

8.3.1 Actuators

High flexibility and capability to show large deformations allow DEs to be employed in the formulation of actuators such as tapered drivers, diamond drivers, folding drivers, and cylindrical drivers [36]. A bionic robot is made by Li et al. [12] using acrylic (3 M VHB) as the dielectric elastomer. The two flexible electrodes used are (i) surrounding water and (ii) conductive ion gel. Water drawing was achieved by using mechanical sections of DE material. Silicone rubber is used as a water-impermeable encapsulant. Another kind of DEA was fabricated by Jun et al. [62] to control surface wettability. VHB 4905 was the DE material used; carbon black was used as the flexible electrode. Electromechanical instability (EMI) phenomenon was employed in making this actuator. Wrinkling was observed in DE after relaxation time. Deformation of the material can be controlled by changing voltage and thus wettability can be adjusted accordingly [60].

8.3.2 Generators

Dielectric elastomers are known to interconvert electrical and mechanical energy; hence, they can also be used to generate electricity. Energies from different sources such as blowing wind, waves of an ocean, walking of humans, and flowing eaters can be harvested using DEs [36, 63]. A generator to harvest energy from the motion of human bodies that can run low-consumption systems was formulated by Jean-Mistral and co-workers [64]. Energy from the human body, being cost-free, stable, and green energy, is advantageous over the other forms of energy. A heel generator consisting of a multi-layered DE membrane was developed by Stanford Institute (SRI) in the United States; this generator uses the effect of a heel to generate electric power. 0.8 J energy can be generated per step while walking. Its better illustration is given in Fig. 8.2 [63].

Kornbluh et al. [66] formulated a wave energy harvesting device that can be actuated by water wave fluctuations; DE units employed in this device are disc-shaped, floating on the sea and converting wave energy into electricity.

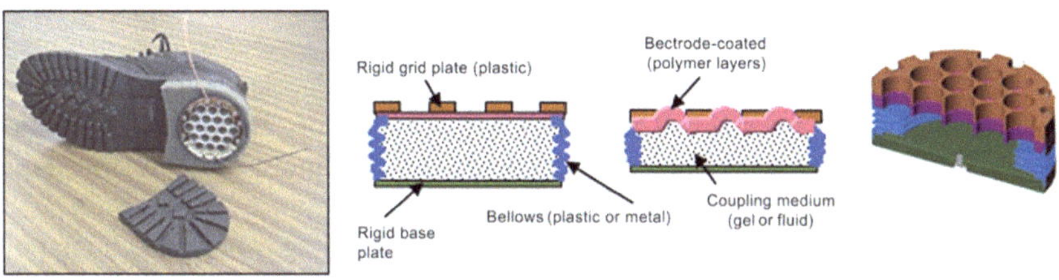

Fig. 8.2 Heel strike generator based on dielectric elastomer fitted in the shoe and its cross-section view. *Source* SRI International [65]. Adapted with permission from ref. [65] Copyright (2012) (Springer Nature)

8.3.3 Sensors

DEs-based sensors find their applications in the field of biomedicine, helping in the detection of prosthetic pressure, thus improving its comfort level [67]. DEs-based pressure sensors, bending sensors, and tension sensors have been developed by New Zealand Stretch Sense Company [36].

8.3.4 Soft Robots

Gripping objects of different types, sizes, and shapes is one of the biggest challenges in the formulation of soft robots; this problem is being resolved by the use of DEAs. Light weighted and greater capability to deform can be achieved by self-organization of the flat state of DEMES into an out-of-plane complex structure. DEMES-based tulip-shaped soft gripper; based on a push–pull configuration; that opens to hold an object when subjected to an electric field was fabricated [68]. A multi-segmented gripper is also formulated by Araromi and co-workers; the multiple segments allow the gripper to envelop objects of different shapes and sizes [69]. Moreover, Shian et al. showed that the integration of stiff fibers can induce bending and wrapping, providing aid in gripping objects [70]. DE balloons can also be utilized as a hydraulic source in the making of grippers in soft robots.

The first biomimetic walking robot was prepared by Eckerle and co-workers [8]; DEA used in this robot named FLEX has six legs with two degrees of freedom at each leg. However, the walking speed of this robot was too slow. Spring roll actuators were used to make a six-legged robot. Inspired by crawling natural creatures, crawling robots have also been made along with legged robots [71]. Electronic components are integrated into the feet of these robots, whose actions move them stably in forward and backward directions [72]. A tethered robot that can climb a wall is also made [73]. In the soft crawling robot formulated by Li et al. [74], oriented plastic fibers in the form of a bundle are attached at both ends of the actuators; these bundles provide a fraction to the robot. Later on, the artificial nervous system was also integrated into a caterpillar-like crawling soft robot [75].

Hopping robots, jumping robots, flying robots, swimming robots, and humanoid robots are also being designed using DEAs.

8.4 Dielectric Elastomer-Based Nano Composites

No doubt, dielectric polymer materials are wonderful materials because of their ease in processibilities and high flexibilities but their use is limited because of low dielectric constant value and low thermal stabilities. No single component possesses the multidimensional properties required for functional materials employed in the recent day's technology. Nanoparticles, ceramics, Mxene, and carbon-based materials (carbon nanotubes, graphene oxide, reduced graphene oxide, etc.) are famous for their excellent electrical, mechanical, and thermal properties and are incorporated as filler in polymer matrix for the synthesis of polymer nanocomposites, a system, with excellent multidimensional features. Such dielectric elastomer-based nanocomposites (DENC) have extensive applications because of excellent conductivity and remarkable dielectric constant values with ultra-low tangent loss and high flexibility on the addition of a minor quantity of fillers.

8.4.1 Fabrication of Dielectric Elastomer-Based Nanocomposites

Synthesis and designing of DENC because of their extensive applications have been the subject of a great number of research works reported in the last two decades. Solution mixing method, melt blending method, in situ polymerization, and injection molding are often used because of their unique features and advantages.

a. ***Solution mixing method***

The solution mixing method is one of the simple, cost-effective, and less time-consuming methods. In this approach, two main components of the composites (filler and matrix) are mixed or dispersed in the same or two miscible solvents in two separate containers followed by their mixing with the help of some external force like sonication, mechanical mixing, etc. For the preparation of dielectric polymer composites, in most cases, the solvent selection is based on the ability of the dissolving polymer matrix to disperse fillers and its non-reactive nature, and ease in removal of the solvent after the formation of the composites. This method is one of the commonly used methods since no sophisticated instruments or apparatus is required for this. The solution mixing method is not recommended for polymers that do have not good solubility in a solvent with low boiling points. In that case, the removal of the solvent becomes tedious. Usually, ultrasonication is considered a very effective tool for the good dispersion of MWCNTs and graphene-related materials in water and other common organic solvents. It helps in breaking the agglomerates into separate entities. It is reported that extensive or powerful sonication sometimes leaves adverse effects which could result in a lowering of the aspect ratio of fillers like multi-walled carbon nanotubes. Synthesis of the polymer composites is not possible for polymers that are not well soluble in solvents with low boiling points. The solution mixing method can be used with slight modification to have better results depending upon circumstances. AmronsakChankul and co-workers [76] prepared Zn^{+2}/ZnO-based polydiacetylene composites [Zn^{+2}/ZnO/PDA)] through the solution mixing method with a slight alteration. After mixing all the precursors, they kept the mixture for incubation at room temperature for different time intervals ranging from 0.5 to 24 h and from 4 to 70 °C overnight. After incubation, each sample was irradiated by UV light to evaporate the solvent. They studied the effect of incubation temperature and time on the size of the prepared nanocomposites.

b. ***Melt blending method***

It is also a mixing method as discussed in the previous section. In this method, the filler is forced to mix with the help of some external force in the already melted polymer, or a mixture of fillers and polymer matrix is fed into the heating chamber for melting. Melt blending is a solvent-free method and an environment-friendly one. This method is considered one of the best methods for polymers that have solubility issues and low melting points. The melt blending method is strongly recommended for large-scale synthesis of the composites. On the other hand, this method is not considered as economical one as the solution mixing method. The use of this method is not suitable for the micro-level synthesis of the composites. A major limitation of this method is the synthesis of the composite of thermoset polymer as a matrix. At different places in the literature, it is observed that the melt bending method was used with slight modification for the synthesis of DENC with desired features. The melting process is carried out in a sealed environment at ultra-low pressure followed by quenching in ice-cold water. Such melt blending method is also known as the melt quenching method [77]. In the recent years, the use of two immiscible polymers as a matrix to obtain carbon-related materials based on DENC with improved and unique dielectric properties is becoming a hot trend [78–80]. The mechanism behind the achievement of high dielectric constant and low tangent loss is the controlled migration of the MWCNTs in the interface of the immiscible polymers. For this purpose, various combinations of immiscible polymers like polystyrene-polymethylmethacrylate, polyvinylidene fluoride-poly acetic acid, etc., can be used to control the migration of fillers across the interface of their unfavorable and favorable polymers. For the synthesis of DECN using two immiscible polymers, the melt blending method is considered the most convenient approach.

c. **In situ** ***polymerization method***

In situ polymerization method for the synthesis of the dielectric polymer composites is the best choice for polymers with high melting points and have solubility issues in commonly used low boiling point solvents [81]. In this mode of preparation of composites, polymerization of the polymer matrix has been carried out in the presence of the filler. This method is very commonly used because of its excellent outcome in terms of highly uniform dispersion of the filler in the matrix. It is one of the economical methods like the solution mixing method for the preparation of composites as no sophisticated instrumentation is required. When a series of composites are needed to be prepared with different concentrations of fillers, to study the effect of the contents of filler, this method is not considered as reliable as the solution mixing method and melt blending method. Polymerization is very much sensitive to the experimental conditions as far as the molecular weight and structure of the polymer are concerned. Any difference in molecular weight and structure of polymer for different composites with different filler contents may influence the features of composites which leads to a wrong conclusion regarding the study of the effect of filler contents. This method is also not considered the best available choice when a composite of two immiscible polymers is required to be prepared.

d. ***Injection molding method***

The injection molding method is considered one of the best methods to prepare DENC in any shape using a mold of a particular shape. The basic strategy of this method is analogous to the melt mixing method. A polymer containing the filler is heated above its melting point and the filler gets dispersed in this molten form with less viscosity. Finally, this melt is injected mechanically into a mold, where it gets cool down to give composites of the desired shape. The injection molding method is the most commonly used one for the industrial-scale production of composites. This method is only limited to thermoplastic polymer for the synthesis of the composites. Through control of the alignment of MWCNTs in composites prepared by injection molding method, tuning the dielectric properties of composites is a very reliable approach [82]. The injection molding method can be used in a versatile way and with the combination of the other techniques to have desired results in polymer composites. Physical foaming combined with the injection molding method for the synthesis of the composite will facilitate the generation of interesting microstructures which are desired to have required features in the composites [83]. A brief data of the synthesis method of the DENC with their claimed values of dielectric constant and dielectric loss is given in Table 8.1.

8.4.2 Types of Fillers Used

Different kinds of fillers are being used in the formulation of DENC, mainly divided into two types:

a. Ceramics
b. Conductive fillers

The above-mentioned fillers are of great importance, due to an effective polarization formed between filler layers and polymer matrix [92, 93].

a. ***Ceramics***

Extremely high permittivity values of ceramics such as $BaTiO_3$ [35, 94] and TiO_2 [35, 95] as compared to the polymer matrix, in which they are incorporated, make them an excellent candidate to be used in DENC. The incorporation of ceramics in DE composites improves their electrical properties to a great extent. The requirement of a relatively greater amount of filler in this case endangers the value of breakdown voltage and flexibility of the composite, making it less flexible. On the other hand, conductive fillers are more appealing to be used in DENC as the product composite is easily processible and exhibits high flexibility. Moreover, these fillers are required in a lesser amount as compared to ceramic-based composites. Although, the use of conductive fillers results in a huge increase in the dissipation of electrical and

Table 8.1 Dielectric properties values of dielectric elastomer-based nanocomposites

Polymer matrix	Nano fillers	Dielectric permittivity	Dielectric loss	Synthesis method	References
Polypropylene	0.34% MWCNT	30	0.06	Melt mixing and batch foaming	[84]
Polyimide	1 vol% graphene + 16 vol%$BaTiO_3$	31	0.03	Solution mixing	[85]
Polypyrrole	8 wt% MWCNT	44	0.07	Inverse micro-emulsion polymerization	[86]
Cyanoethyl pullulan polymer	0.06 wt% CNT + Reduced graphene oxide	32	0.05	Solution mixing	[87]
Polydimethyl-Siloxane	2 wt% graphene nanoplates	89	1.5	Solution mixing and vulcanization	[88]
Polysulfone	25 vol% MWCNT	55	0.05	Solution mixing and electrospinning	[89]
Polypropylene	1.25 vol% MWCNT	57.2	0.05	Melt mixing and injection molding	[90]
Polyvinylidene Fluoride	30 vol% $BaTiO_3$ nanoparticles + 3 vol% $BaTiO_3$ nanofibers	27	0.06	Solution mixing	[91]
Polypropylene	1.25 wt% MWCNT	14.1	0.39	Compression molding	[90]

mechanical energy; affecting the stability and lifetime of the composites in which they are used [59].

b. ***Conductive fillers***

The mostly employed conductive fillers include carbon black [96], carbon nanotubes [97], graphene [59], metal nanoparticles [98], etc. It is observed that metal nanoparticles used as fillers are less effective as compared to carbon-based nanofillers; based on the fact that polymer matrix has more affinity with carbon-based nanofillers.

Recently, 1D (one-dimensional) and 2D (two-dimensional) carbon-based nanofillers have grabbed much attention from researchers since they fabricate composites with high dielectric permittivity. These fillers form a percolative network through the polymer matrix and thus improve the electric properties of the composites drastically. The percolation threshold value for the composite is greatly affected by the geometric factors of filler particles and their affinity with the matrix at the interface. Graphene nanosheets (2D) have a unique structure, i.e., a highly stable planar honeycomb made by single-atom thick sp^2 C-atoms imparting remarkable electron mobility, a very high surface area, excellent electrical conductivity, and the sheets can easily contact with each other as compared to CNTs, thereby graphene nanosheets are known to have a lower percolation threshold value [99]. A relatively lower breakdown strength, high dielectric loss, and vast mechanical hysteresis caused due to the leakage of current between the conductive channels of graphene sheets formed in the polymer matrix are the main drawbacks to using graphenes, chemically different graphene-related materials such as graphene oxide and reduced graphene oxide for the fabrication of graphene/DEA composites. Thus, the synthesis of a stable and improved graphene-based DENC is one of the most notable fields that is being increasingly reported over time.

The actuation behavior of the graphene-based polymer composite can be improved by seeking a solution to the above-mentioned problems, e.g., functionalization of graphene components or limiting the dielectric loss by decorating them with an insulating layer, hence, improving the compatibility between the conductive fillers and matrix components [59].

8.4.3 Factors Affecting the Functionalization of Composite

a. ***Route used to synthesize the filler***

The synthesis technique used to fabricate fillers (e.g., graphene nanosheets) has a dramatic influence on the properties of the final product; top-down and bottom-up being the most-preferable ones [100, 101]. On the other hand, bottom-up technique, synthesis on SiC surface [102], and mechanical exfoliation [103] are not suitable methods to synthesize graphene which is to be used in composites due to high cost and partial scalability [104]. Low yield, comparatively lower yield, and high cost of graphene nanosheets synthesized through chemical vapor deposition of hydrocarbons are not much recommended for this purpose [105]. Owing to the low-cost processability and high scalability of the liquid phase, exfoliation is a better suitable process [104]. Despite the lower quality of graphene obtained through the reduction of graphene oxide (GO), controlled costs and higher yield make this route a good choice for the production of graphene nanosheets which are to be incorporated into composites [106]. Hence, the most effective routes to be used for the synthesis of commercial graphene nanosheets are liquid phase exfoliation and reduction of GO.

b. ***Functionalization of the filler***

Modification of the fillers being employed, e.g., uniform dispersion or deagglomeration, surface modification, ways to enhance the strength of matrix-filler interaction, synthesis of core–shell structures, use of complex fillers, etc., plays an important role in the functioning of the composite [107, 108]. Several chemicals (covalent) [109] and physical (non-covalent) [110] techniques have been employed to achieve these functionalized fillers. Covalent functionalization includes the organization of π-cloud for the conversion of sp^2 carbon network into sp^3 hybridized carbon [111]. On the other hand, non-covalent functionalization includes the use of electrostatic forces, Van der Waals forces, π-π interactions, physisorption, etc.; resulting in the formation of electron–hole puddles and causing changes in the doping density [112]. Preferably, functionalization methods mentioned later (non-covalent) are capable to stop the leakage of current, thus increasing the dielectric permittivity and allowing controllable energy dissipation. Hence, non-covalent modification is more favorable to synthesizing highly efficient nanofillers incorporated composites [113].

c. ***Content of filler in the composite***

Increasing the carbon to oxygen ratio of the filler results in a better elastic modulus and improved dielectric permittivity. The growth rate of elastic modulus is lower as compared to dielectric permittivity. However, higher contents of filler cause a decrease in electrical breakdown strength [59].

d. ***Interactions between matrix and filler***

The electromechanical properties and microstructure of a composite are greatly affected by interactions between matrix and filler particles. For instance, coating of hydrophilic, less toxic, and stable TiO_2 on hydrophobic reduced graphene oxide (rGO) improves its surface wettability, (TiO_2 acts as an insulator controlling the current leakage inside the composite) causing better interactions between the filler and matrix, resulting in greater storage capability and loss modulus as compared to neat polymers [25, 59].

e. ***Use of complex filler***

As reported, increasing the intensity of the electric field reduces the thickness of the composite. However, a composite in which complex filler is incorporated exhibits a much better actuation behavior. This was observed in an experiment performed, where the actuation behavior of PU/rGO and PU/1G-1 T was compared.

The composite with a complex filler showed many results better than the other one [114].

8.4.4 Design and Structure of the Composite

As far as the efficiency and sensitivity of the composites are concerned, the structure and design of the composites are equally important as the selection of matrix and fillers is vital. A lot of studies have been focused on the designing of composites with a specific structure that significantly regulates the efficiency of the composites. Agglomeration of the filler is one of the serious issues which limit the homogenous dispersion of the filler in the matrix. Development of the elastomeric composites is often accomplished with cross-linking phenomena to obtain desired features in the composites. The vulcanization strategy, basically a high-temperature chemical cross-linking method, is commonly used to obtain the cross-linking network in the composite [78, 82, 83, 115]. Electron beam radiation is also considered the best choice for the development of a cross-linking network to have highly uniform dispersion of the filler in the matrix. Cross-linking development by irradiation of electron beam is accompanied at ambient temperature which may lead toward cross-linking without thermal degradation in the polymer matrix and with highly homogenous dispersion of the filler [115]. Filler orientation in DENC is one of the vital factors which influence the dielectric permittivity values. It has been observed that the mean orientation angle demonstrates the resultant dielectric features of the electroactive composites. Theoretical models have been purposed to correlate the experimental data on dielectric properties and numerical results obtained through proposed models and good agreement has been found between them [116]. Selective localization of the filler in the matrix influences morphology which intern demonstrates the rheology and dielectric properties of the composites. Selective localization of the functional filler in the polymer matrix is achieved by the addition of another immiscible polymer as filler. It may result in multiple times improvements in the features of the composites. The addition of 3 wt% of GO in polyurethane-based polylactic acid results in about a 400% increase in the value of dielectric constant as compared to that of polyurethane-based polylactic acid composites without GO [26]. Similarly, the use of the third component as an additive to increase non-covalent interaction between the filler and matrix is also proved very effective and it results in an improvement in the dielectric constant values of ternary composites [117]. In such polymer composites, uniform dispersion of the fillers is obtained, because the third component helps in getting the maximum features of the filler in the composite. The addition of the filler in terms of masterbatch is also a very common strategy for preparing dielectric nanocomposites. Usually, the filler is added in high concentration in a material called carrier materials and this resultant stuff is known as a masterbatch. Finally, this masterbatch is added to the main matrix. The compatibility of the carrier materials with the filler and main matrix is taken into consideration while selecting the carrier material. Composites prepared with this method result high dielectric constant with ultra-low tangent value

[118]. The addition of a surfactant along the filler in the polymer matrix is one of the commonly used strategies for the synthesis of composites with enhanced homogenous dispersion. Control of agglomeration by using surfactants like DBSA, APTES, and CTAB helps fillers in imparting maximum features to the polymer composites [24, 119]. Surfactant makes a spread network in the polymer matrix or gets attached to the surface of the nanofillers and prevents or reduces the chances of interaction filler and thus controls the agglomeration of the nanofiller. Uniform dispersion of the filler in the nanocomposites helps in attaining significant improvement in dielectric constant values. Multilayer polymer composites are well known because of their structure which allows tuning of the features by slight modification in it. Dielectric constant and other electrical properties can be tuned in a very broad range just by increasing the number of layers or thickness of the layer of the same combination of filler and matrix, and the preparation method of the composites. Layer assembly is not only helpful in achieving quantitatively good results in DENC but it is also considered a facile approach to attain out of the plane deformation. In many cases, shape reconfiguration is essential in the working of soft robots [34, 120].

8.4.5 Applications

a. ***Tunable lenses***

Lifelike robots have a very important feature, i.e., eyes that can act exactly like a human eye, that can respond quickly and precisely by tuning the focal length [121, 122]. The eye is connected to the ciliary body with the help of a ligament, i.e., fibrous tissue ring attached at its equator. The relaxation and contraction processes of the ciliary body help in the tuning of the lens. Although most of the robots have digital camera-based eyes that move the components mechanically, thus tuning the focus, this process takes a lot of energy. Several soft actuators are employed to overcome this drawback; which tune the focal length by controlling the liquid droplet curvature; including stimuli-response [22], electromagnetic wave [23], electro-wetting[123], dielectrophoretic forces [17], electrochemistry [124], etc. The pre-stretching of a lens is carried out before using it in the robots. A Maxwell stress is created in the actuating system by inducing an electric field; a shorter focal length is created by the lens if its height increases and diameter decreases, causing it to squeeze. The lens gets back to its actual dimensions, exhibiting a greater focal length and the dielectric elastomer actuator re-attains its shape when the applied voltage is turned off. Unfortunately, only a few DE materials including VHB 4910 and silicone are known to fabricate lenses exhibiting high performance. It has been reported by researchers that DE with incorporated dielectric fillers [94] or DE blended with polymers [28] exhibit better electromechanical properties. Recently, a new model has been proposed, that consists of a hybrid structure where a ring made of DENC acts as an actuator and is connected to the lens made of a sphere made up of elastomer [125].

b. ***Artificial muscles***

Some materials undergo deformations in their structures when subjected to an external stimulus (pressure, electricity, light, humidity, heat, etc.); these materials are often referred to as 'artificial muscles'. They find their applications in different interdisciplinary fields [126–128]. Energy density, response time, output force, energy conversion efficiency, actuated strain, and fatigue life are the key features of an artificial muscle [129, 130]. Many types of artificial muscles exhibiting unique properties and advantages have been reported so far, such as natural muscles [131] and humidity/solvent actuation yarn [132, 133]. At some points, artificial muscles are capable to perform even better than naturally occurring human muscles. Recently, the deformable electrode is used by Shea's group to formulate a stretchable EHD pump; this electrode consists of a dielectric elastomer/carbon black composite which is embedded with a tube made up of monolithic elastomer. Extraordinary mechanical outputs are obtained from this pump, assuring a high mechanical strength. It is reported to be used to create a self-contained fluidic muscle by integration with an elastomeric chamber; eliminating the use of the external compressing system for reversible bending actuation [129]. Miriyev et al. also fabricated a soft actuating composite by combining a polymer matrix like PDMS (poly(dimethylsiloxane) elastomer) with a phase-changing material like ethanol [134]. The deformation of such a composite to maximum volume is done by transitioning the liquid phase of ethanol to the gas phase upon heating.

Embedding magnetic particles into an elastomer matrix is an appealing route to develop soft magnetic actuators. For instance, a cylindrical glass rod was wrapped by a soft magnetic composite by Hu and co-workers [135]; the composite was magnetized by the application of a strong and uniform magnetic field. The uniform external magnetic field interacted with a spatially varying field of magnetic composites; enabling the soft robot to change its shape by generating internal torque. This special method can be applied to fabricate robots out of artificial muscles that enable them to jump, crawl, walk, roll, transport objects, etc.

c. ***Smart skin***

Taking inspiration from nature, i.c., natural skin, researchers are making attempts to create smart electronic skins that can practice LMC (learning-mimicking-creating) cycle; being quite helpful in health monitoring technologies, robotics, prosthetics, etc. The discovery of mechanoreceptors naturally integrated into human skin has opened up a gateway to smart flexible electronics. The tactile sensation of human skin is mimicked by the sensors used inside the smart skin that can detect and respond to haptic stimuli. The idea to create unique features in smart skin such as changing color or shape has been taken from mussels, chameleons, octopuses, etc. Internal organs of the body are also designed in such a way as to derive the LMC cycle along with the external organs. Synaptic, implantable sensors, artificial neurons, energy harvesters, etc., are the major biocompatible electronic devices being used in smart skin [136].

Dielectric elastomer composites are widely employed for the fabrication of smart skin devices. Generally, pre-cured elastomer and conductive filler suspension are blended to form a composite. However, the conductive paths formed inside the matrix should be dispersed homogeneously to avoid variations in the conductivity from one location to another [137]. Several other methods such as the direct incorporation of a conductive network in the elastomer matrix are also being employed in the making of smart skin [138]. A carbonized plain weave cotton fabric was taken by Zhang et al.; it was then encapsulated in an elastomer matrix; resulting in the formulation of a strain sensor [139]; stretchability and sensitivity of such a sensor can be customized according to the requirements. High gauge factors were also attained by an anisotropic conductive material while controlling its flexibility to respond to pressing, bending, and twisting [140].

Implantable devices need bioresorbable batteries for the effective management of energy sources; whose working depends upon mechanical compliance, desorption rates, device dimensions, etc. [141]. The biochemical energy produced by internal body organs such as cardiac movements or breathing is scavenged by the energy devices used. A BD-TENG (biodegradable triboelectric nanogenerator) has been fabricated that can power inter-digit electrodes on which orientation of nerve cell growth was carried out; the BD-TENG consisted of electrode layers, encapsulation structure (biodegradable), and friction layers [142]. An energy harvester was implanted in a smart skin device that can record breathing and heartbeat by Ma et al. [143]. However, large hysteresis and low response time are the main challenges being faced in the field of smart skin sensors.

d. ***Biomedical applications***

Neurostimulation, cardiac pacing, tissue regeneration, therapeutic delivery, biochemical sensing, physiological monitoring, drug delivery, etc., are some of the most common applications of implantable biomedical devices (IMDs) [144–147]. Some of the main drawbacks being faced while using IMDs are the removal of the device when its life ends, and the risk of infection [144, 148]. To overcome these problems, devices have been formulated that can be degraded in different environments safely, named 'biodegradable electronics', or the devices which dissolve automatically in the biofluids without causing toxicity, thus named 'bioresorbable electronics' [73, 149]. These electronics can reduce the risks that can be caused by the removal surgery of conventional devices, and increase the human life span by up to three years [150]. IMDs consist of four elements; conductor, semi-conductor, encapsulant, and dielectric. The elastic components in IMDs provide comparatively better integration with the tissues, in which the device is implanted. Sim et al. synthesized a bioelectronic patch, where ion gel (dielectric), polydimethylsiloxane (PDMS) (encapsulant), and PDMS composite (semi-conductor) were interconnected; this patch provides electric simulation by harvesting energy from heartbeat [151]. Similarly, Schiavone and co-workers developed e-dura, an IMD where the electrode was coated with a PDMS composite and PDMS was used as encapsulate; which can apply stimulations to vertebral column electrically [152].

Ling Yu and co-workers proposed the synthesis of an artificial material, where polycaprolactone (3D skeletons) was incorporated into a dielectric elastomer; the

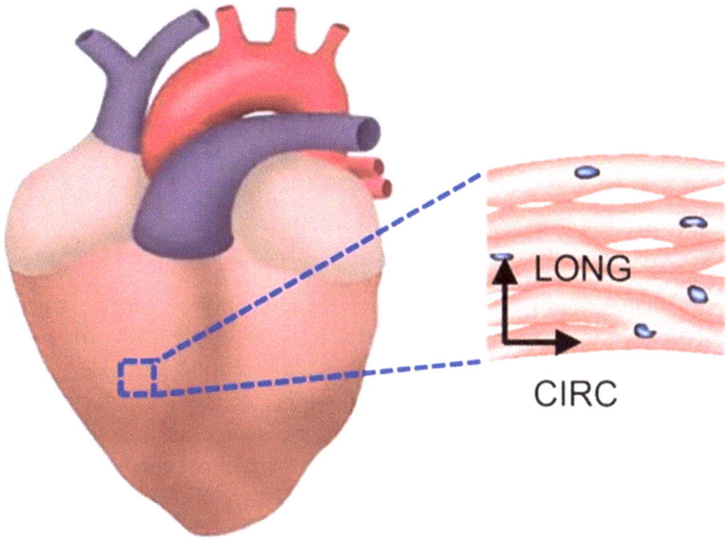

Fig. 8.3 Schematic diagram of the right ventricular myocardium with undulated, preferentially oriented cardiac muscle fibers, showing anisotropic mechanical and electrical properties. Arrows indicate anatomically defined circumferential and longitudinal axes [153]. Adapted with permission from ref. [153]. Open access under a CC BY 4.0 license. Copyright © 2022 Yun Ling et al.

as-prepared composite whose contractions and relaxations were similar to that of heart muscles (Fig. 8.3 [153]).

e. ***Micro-air vehicle***

Naturally occurring flyers commonly show maneuvering and hovering flights. Birds fly in a vertical plane by flapping their wings, and they can change pitch angle at a very small amount. On the other hand, insects fly by flapping their wings in the horizontal plane, they can change pitch angle at a very large amount [154]. Muscles are distributed evenly in the bird's wings causing active wing actuation, however, the wings of insects have no muscles and their actuation is controlled by the thorax (passive wing actuation). Former is a relatively more complex form of actuation [155]. Muscles used in passive wing actuation are divided into two categories; neuro-controlled signals synchronize the actuation process (synchronous flight muscles) and neuro-controlled signals do not synchronize the actuation process (asynchronous flight muscles) [156]. Based on these muscles, a passive flight is classified as direct (each wing can be directly controlled by the muscle) or indirect flight (simultaneous flapping of wings does not allow the muscle to control individual wings) [155]. Thus, indirect flight mechanism can be employed to create flapping wings artificially [157].

Smart material actuators (SMAs) such as IPMC actuators, SMA actuators, DE actuators, and piezoelectric actuators have a great potential to design flapping-wing mechanism-based devices as they provide high scalability, high damage tolerance, multi-functioning, compactness, high integrability, and high customization facility

[158]. Lau et al. combined a rolled DEA with a cross-ply laminate of carbon fiber-reinforced polymer (CFRP) to verify the feasibility of the insect wing mechanism [159]. Dielectric elastomer devices have been reported to be employed in the fabrication of micro-air vehicles [59, 157].

8.5 Conclusion

Highly stretchable, multi-functional, extensible, incompressible, poly-active, and soft dielectric elastomers have gained the attention of researchers as promising candidates to be used in vast fields of energy harvesters, haptic interfaces, soft robots, artificial muscles, and tunable lenses as generators, actuators, and sensors. They are being incorporated into polymer matrices to form a superior material that exhibits exceptional dielectric constant values coupled with excellent conductivity, and lower values of dielectric loss. Different kinds of fillers (such as ceramics and conductive fillers) can also be added to the DE-based polymer composites, thus improving their electrical properties, lifetime, and stability. Dielectric/polymer composites are being used mainly in artificial muscles, biomedical devices, soft robots, smart skin, micro-air vehicles, etc.

Despite all the achievements made in the field of dielectric nanocomposites based on elastomers, researches are facing different challenges including the need for a detailed understanding of mechanisms related to the dielectric behavior of nanocomposites, synthesis of cost-effective and scalable elastomer-based polymer composites at large scale, development of models for life-prediction and performance evaluation of dielectric materials, etc. Thus, further research for improvements in these highly applicable materials is being focused on by scientists.

References

1. Ciarella L, Richter A, Henke E-FM (2021) Digital electronics using dielectric elastomer structures as transistors. Appl Phys Lett 119:261901. https://doi.org/10.1063/5.0074821
2. Röntgen WC (1880) Ueber die durch Electricität bewirkten Form- und Volumenänderungen von dielectrischen Körpern. Ann Phys 247:771–786. https://doi.org/10.1002/andp.18802471304
3. Pelrine R (2008) Electromechanical transduction effects in dielectric elastomers: actuation, sensing, stiffness modulation and electric energy generation. In: Dielectric elastomers as electromechanical transducers, pp 1–12
4. Lu T, Ma C, Wang T (2020) Mechanics of dielectric elastomer structures: a review. Extreme Mech Lett 38:100752
5. Bar-Cohen Y (2002) Electroactive polymers as artificial muscles: a review. J Spacecr Rocket 39(6):822–827
6. Gupta U, Qin L, Wang Y, Godaba H, Zhu J (2019) Soft robots based on dielectric elastomer actuators: a review. Smart Mat Struct 28(10):103002
7. Pelrine R, Kornbluh R, Pei Q, Joseph J (2000) High-speed electrically actuated elastomers with strain greater than 100%. Science 287(5454):836–839

8. Kofod G, Kornbluh RD, Pelrine R, Sommer-Larsen P (2001) Actuation response of polyacrylate dielectric elastomers. In: Smart structures and materials 2001: electroactive polymer actuators and devices. SPIE, pp 141–147
9. O'Brien B, Gisby T, Anderson IA (2014) Stretch sensors for human body motion 9056:905618. https://doi.org/10.1117/12.2046143
10. Suo Z (2010) Theory of dielectric elastomers. Acta Mech Solida Sin 23:549–578. https://doi.org/10.1016/S0894-9166(11)60004-9
11. Carpi F (2010) Electromechanically active polymers. Polym Int 59(3):277–278
12. Li T, Li G, Liang Y, Cheng T, Dai J, Yang X, Liu B, Zeng Z, Huang Z, Luo Y, Xie T, Yang W (2017) Fast-moving soft electronic fish. Sci Adv 3:e1602045. https://doi.org/10.1126/sciadv.1602045
13. Gu G, Zou J, Zhao R, Zhao X, Zhu X (2018) Soft wall-climbing robots. Sci Robot 3(25):eaat2874
14. Christianson C, Goldberg NN, Deheyn DD, Cai S, Tolley MT (2018) Translucent soft robots driven by frameless fluid electrode dielectric elastomer actuators. Sci Robot 3(17):eaat1893
15. Huang J, Shian S, Suo Z, Clarke DR (2013) Maximizing the energy density of dielectric elastomer generators using equi-biaxial loading. Adv Funct Mater 23:5056–5061. https://doi.org/10.1002/adfm.201300402
16. Ellingford C, Zhang R, Wemyss AM, Zhang Y, Brown OB, Zhou H, Wan C et al (2020) Self-healing dielectric elastomers for damage-tolerant actuation and energy harvesting. ACS Appl Mater Interfaces 12(6):7595–7604
17. Ren H, Xianyu H, Xu S, Wu S-T (2008) Adaptive dielectric liquid lens. Opt Express 16:14954–14960. https://doi.org/10.1364/OE.16.014954
18. Chen Y, Zhao H, Mao J, Chirarattananon P, Helbling EF, Hyun NSP, Wood RJ et al (2019) Controlled flight of a microrobot powered by soft artificial muscles. Nature 575(7782):324–329
19. Gao Y, Fang X, Tran D, Ju K, Qian B, Li J, Dielectric elastomer actuators based on stretchable and self-healable hydrogel electrodes. R Soc Open Sci 6:182145. https://doi.org/10.1098/rsos.182145
20. Ji X, Liu X, Cacucciolo V, Civet Y, El Haitami A, Cantin S, Perriard Y, Shea H (2021) Untethered feel-through haptics using 18-μm thick dielectric elastomer actuators. Adv Funct Mater 31:2006639. https://doi.org/10.1002/adfm.202006639
21. Ho TYK, Nirmal A, Kulkarni MR, Accoto D, Mathews N (2022) Soft actuator materials for electrically driven haptic interfaces. Adv Intell Syst 4(2):2100061
22. Carpi F, Frediani G, Turco S, De Rossi D (2011) Bioinspired tunable lens with muscle-like electroactive elastomers. Adv Funct Mater 21:4152–4158. https://doi.org/10.1002/adfm.201101253
23. Pieroni M, Lagomarsini C, Rossi DD, Carpi F (2016) Electrically tunable soft solid lens inspired by reptile and bird accommodation. Bioinspir Biomim 11:065003. https://doi.org/10.1088/1748-3190/11/6/065003
24. Bashir F, Hussain T, Mujahid A, Shehzad K, Raza MA, Zahid M, Athar MM (2018) Tailoring electrical and thermal properties of polymethyl methacrylate-carbon nanotubes composites through polyaniline and dodecyl benzene sulphonic acid impregnation. Polym Compos 39:E1052–E1059. https://doi.org/10.1002/pc.24485
25. Dahl M, Liu Y, Yin Y (2014) Composite titanium dioxide nanomaterials. Chem Rev 114:9853–9889. https://doi.org/10.1021/cr400634p
26. Azadi F, Khonakdar HA, Jafari SH, Arjmand M, Wagenknecht U, Ruckdäschel H, Altstädt V (2022) The effect of tailoring morphology on rheology and dielectric properties of poly (lactic acid)/thermoplastic polyurethane/graphene oxide nanocomposites. Mater Today Commun 33:104497. https://doi.org/10.1016/j.mtcomm.2022.104497
27. Duduta M, Hajiesmaili E, Zhao H, Wood RJ, Clarke DR (2019) Realizing the potential of dielectric elastomer artificial muscles. Proc Natl Acad Sci 116:2476–2481. https://doi.org/10.1073/pnas.1815053116

28. Qiu Y, Zhang E, Plamthottam R, Pei Q (2019) Dielectric elastomer artificial muscle: materials innovations and device explorations. Acc Chem Res 52:316–325. https://doi.org/10.1021/acs.accounts.8b00516
29. Fan P, Zhu Z, Hu Q (2022) Investigation on free relaxation process influencing energy harvesting performance of dielectric elastomer generators in intermittent motion. Sens Actuators Phys 347:113944. https://doi.org/10.1016/j.sna.2022.113944
30. Gupta U, Qin L, Wang Y, Godaba H, Zhu J (2019) Soft robots based on dielectric elastomer actuators: a review. Smart Mater Struct 28:103002. https://doi.org/10.1088/1361-665X/ab3a77
31. Zhao X, Li C, Qi R, Guo H, Peng G (2022) Study on relaxation behavior of all-polymer PVDF-based films containing dielectric fluoroelastomer. Polymer 252:124933. https://doi.org/10.1016/j.polymer.2022.124933
32. Zhu J, Zhang L, Zhao Y, Yin L-J, Zha J-W, Dang Z-M (2022) Advanced dielectric elastomer based on optimized thermoplastic polyurethane–styrene ethylene butylene styrene blend: experiment and simulation. J Appl Polym Sci 139:51595. https://doi.org/10.1002/app.51595
33. Yu L, Hu T, Yang D, Wei Q (2022) Enhanced electromechanical performance of natural rubber dielectric elastomers achieved by in situ synthesis of silver nanoparticles on TiO_2 nanoparticles. IET Nanodielectrics 5:39–49. https://doi.org/10.1049/nde2.12030
34. Ren C, Su Z, Su Y, Wang L (2022) Polyurethane elastomer layered nanocomposite material for sports grounds and the preparation method thereof. BioMed Res Int 2022:e5152911. https://doi.org/10.1155/2022/5152911
35. Ziegmann A, Schubert DW (2018) Influence of the particle size and the filling degree of barium titanate filled silicone elastomers used as potential dielectric elastomers on the mechanical properties and the crosslinking density. Mater Today Commun 14:90–98. https://doi.org/10.1016/j.mtcomm.2017.12.013
36. Zhao Y, Yin L-J, Zhong S-L, Zha J-W, Dang Z-M (2020) Review of dielectric elastomers for actuators, generators and sensors. IET Nanodielectrics 3:99–106. https://doi.org/10.1049/iet-nde.2019.0045
37. Bartlett MD, Fassler A, Kazem N, Markvicka EJ, Mandal P, Majidi C (2016) Stretchable, high-k dielectric elastomers through liquid-metal inclusions. Adv Mater 28:3726–3731. https://doi.org/10.1002/adma.201506243
38. Bhuyan P, Cho D, Choe M, Lee S, Park S (2022) Liquid metal patterned stretchable and soft capacitive sensor with enhanced dielectric property enabled by graphite nanofiber fillers. Polymers 14:710. https://doi.org/10.3390/polym14040710
39. Brochu P, Pei Q (2010) Advances in dielectric elastomers for actuators and artificial muscles. Macromol Rapid Commun 31:10–36. https://doi.org/10.1002/marc.200900425
40. Low S-H, Lau GK (2014) Bi-axially crumpled silver thin-film electrodes for dielectric elastomer actuators. Smart Mater Struct 23:125021. https://doi.org/10.1088/0964-1726/23/12/125021
41. Haghiashtiani G, Habtour E, Park SH, Gardea F, McAlpine MC (2018) 3D printed electrically-driven soft actuators. Extreme Mech Lett 21:1–8. https://doi.org/10.1016/j.eml.2018.02.002
42. Le Floch P, Molinari N, Nan K, Zhang S, Kozinsky B, Suo Z, Liu J (2020) Fundamental limits to the electrochemical impedance stability of dielectric elastomers in bioelectronics. Nano Lett 20:224–233. https://doi.org/10.1021/acs.nanolett.9b03705
43. Godaba H, Zhang Z-Q, Gupta U, Foo CC, Zhu J (2019) Instabilities in dielectric elastomers: buckling, wrinkling, and crumpling. Soft Matter 15:7137–7144. https://doi.org/10.1039/C9SM01145E
44. Maffli L, Rosset S, Shea HR (2013) Zipping dielectric elastomer actuators: characterization, design and modeling. Smart Mater Struct 22:104013. https://doi.org/10.1088/0964-1726/22/10/104013
45. Shintake J, Ichige D, Kanno R, Nagai T, Shimizu K (2021) Monolithic stacked dielectric elastomer actuators. Front Robot AI 8
46. Kunz A (2006) Spring roll dielectric elastomer actuators for a portable force feedback glove. Smart Struct Mater 2006 Electroact Polym Actuators Devices EAPAD

47. Ko UH, Kumar V, Rosen B, Varghese S (2022) Characterization of bending balloon actuators. Front Robot AI 9
48. Yunguang L, Wang H, Zhu Y (2010) Design and implementation of cone dielectric elastomer actuator with double-slider mechanism. J Bionic Eng 7. https://doi.org/10.1016/S1672-6529(09)60237-7
49. Carpi F, Salaris C, Rossi DD (2007) Folded dielectric elastomer actuators. Smart Mater Struct 16:S300. https://doi.org/10.1088/0964-1726/16/2/S15
50. Lochmatter P, Kovacs G (2008) Design and characterization of an active hinge segment based on soft dielectric EAPs. Sens Actuators Phys 141:577–587. https://doi.org/10.1016/j.sna.2007.10.029
51. Wang NF, Guo H, Chen B, Cui C, Zhang X (2018) Design of a rotary dielectric elastomer actuator using topology optimization method based on pairs of curves. Smart Mater Struct 27. https://doi.org/10.1088/1361-665X/aab991
52. Su Y, Ogden RW, Destrade M (2021) Bending control and stability of functionally graded dielectric elastomers. Extreme Mech Lett 43:101162. https://doi.org/10.1016/j.eml.2020.101162
53. Carpi F, Chiarelli P, Mazzoldi A, de rossi D (2003) Dielectric elastomer planar actuators for small-scale applications. Proc SPIE—Int Soc Opt Eng. https://doi.org/10.1117/12.508694
54. Palmić TB, Slavič J (2022) Single-process 3D-printed stacked dielectric actuator. Int J Mech Sci 230:107555. https://doi.org/10.1016/j.ijmecsci.2022.107555
55. Maas J, Tepel D, Hoffstadt T (2015) Actuator design and automated manufacturing process for DEAP-based multilayer stack-actuators. Meccanica 50:2839–2854. https://doi.org/10.1007/s11012-015-0273-2
56. Rossiter J, Walters P, Stoimenov B (2009) Printing 3D dielectric elastomer actuators for soft robotics. In: Electroactive polymer actuators and devices (EAPAD) 2009. SPIE, pp 149–158
57. Kofod G, Paajanen M, Bauer S (2006) Self-organized minimum-energy structures for dielectric elastomer actuators. Appl Phys A 85:141–143. https://doi.org/10.1007/s00339-006-3680-3
58. Chiang Foo C, Cai S, Jin Adrian Koh S, Bauer S, Suo Z (2012) Model of dissipative dielectric elastomers. J Appl Phys 111:034102. https://doi.org/10.1063/1.3680878
59. Panahi-Sarmad M, Zahiri B, Noroozi M (2019) Graphene-based composite for dielectric elastomer actuator: a comprehensive review. Sens Actuators Phys 293. https://doi.org/10.1016/j.sna.2019.05.003
60. Liu L, Han Y, Xing Z, Yong H (2022) Nonlinear deformation and instability of a dielectric elastomer tube actuator. Int J Non-Linear Mech 147:104235. https://doi.org/10.1016/j.ijnonlinmec.2022.104235
61. Boldini A, Porfiri M (2022) On Maxwell stress and its relationship with the dielectric constant in the actuation of ionic polymer metal composites. J Mech Phys Solids 164:104875. https://doi.org/10.1016/j.jmps.2022.104875
62. Jun K, Kim D, Ryu S, Oh I-K (2017) Surface modification of anisotropic dielectric elastomer actuators with Uni-A nd bi-axially wrinkled carbon electrodes for wettability control. Sci Rep 7. https://doi.org/10.1038/s41598-017-06274-0
63. Kornbluh RD, Pelrine R, Prahlad H, Wong-Foy A, McCoy B, Kim S, Eckerle J, Low T (2011) From boots to buoys: promises and challenges of dielectric elastomer energy harvesting 7976:797605. https://doi.org/10.1117/12.882367
64. Jean-Mistral C, Basrour S, Chaillout J-J (2008) Dielectric polymer: scavenging energy from human motion. In: Electroactive polymer actuators and devices (EAPAD) 2008. SPIE, pp 369–378
65. Rasmussen L (2012) Electroactivity in polymeric materials. Springer, US, Boston, MA
66. Kornbluh R, Pelrine R, Prahlad H, Wong-Foy A, McCoy B, Kim S, Eckerle J, Low T (2012) Dielectric elastomers: stretching the capabilities of energy harvesting. MRS Bull 37. https://doi.org/10.1557/mrs.2012.41
67. Lee YR, Kwon H, Lee DH, Lee BY (2017) Highly flexible and transparent dielectric elastomer actuators using silver nanowire and carbon nanotube hybrid electrodes. Soft Matter 13:6390–6395. https://doi.org/10.1039/C7SM01329A

68. Kofod G, Wirges W, Paajanen M, Bauer S (2007) Energy minimization for self-organized structure formation and actuation. Appl Phys Lett 90:081916. https://doi.org/10.1063/1.2695785
69. Araromi OA, Gavrilovich I, Shintake J, Rosset S, Richard M, Gass V, Shea HR (2015) Rollable multisegment dielectric elastomer minimum energy structures for a deployable microsatellite gripper. IEEEASME Trans Mechatron 20:438–446. https://doi.org/10.1109/TMECH.2014.2329367
70. Shian S, Bertoldi K, Clarke DR (2015) Dielectric elastomer based "grippers" for soft robotics. Adv Mater 27:6814–6819. https://doi.org/10.1002/adma.201503078
71. Guo Y, Liu L, Liu Y, Leng J (2021) Review of dielectric elastomer actuators and their applications in soft robots. Adv Intell Syst 3:2000282. https://doi.org/10.1002/aisy.202000282
72. Cao J, Qin L, Liu J, Ren Q, Foo CC, Wang H, Lee HP, Zhu J (2018) Untethered soft robot capable of stable locomotion using soft electrostatic actuators. Extreme Mech Lett 21:9–16. https://doi.org/10.1016/j.eml.2018.02.004
73. Kuzma M, Gerhard E, Shan D, Yang J (2020) Advances in bioresorbable electronics and uses in biomedical sensing, pp 29–72
74. Li J, Liu L, Liu Y, Leng J (2019) Dielectric elastomer spring-roll bending actuators: applications in soft robotics and design. Soft Robot 6:69–81. https://doi.org/10.1089/soro.2018.0037
75. Ishige M, Umedachi T, Taniguchi T, Kawahara Y (2019) Exploring behaviors of caterpillar-like soft robots with a central pattern generator-based controller and reinforcement learning. Soft Robot 6:579–594. https://doi.org/10.1089/soro.2018.0126
76. Chanakul A, Saymung R, Seetha S, Traiphol R, Traiphol N (2021) Solution-mixing method for large-scale production of reversible thermochromic and acid/base-colorimetric sensors. Colloids Surf Physicochem Eng Asp 615:126241. https://doi.org/10.1016/j.colsurfa.2021.126241
77. Nedelcu N, Chiroiu V, Rugină C, Munteanu L, Ioan R, Girip I, Dragne C (2020) Dielectric properties of GeSbSe glasses prepared by the conventional melt-quenching method. Results Phys 16:102856. https://doi.org/10.1016/j.rinp.2019.102856
78. Hussain T, Malik T, Mujahid A, Mustafa G (2018) Polystyrene adsorbed multi-walled carbon nanotubes incorporated polymethylmethacrylate composites with modified percolation phenomena. MRS Adv 3:25–30. https://doi.org/10.1557/adv.2017.625
79. Zhang H, Jiang H, Huang Z-X, Qu J-P (2022) Toward high dielectric constant and low dielectric loss nanocomposite via kinetical migration. Compos Sci Technol 221:109310. https://doi.org/10.1016/j.compscitech.2022.109310
80. Zhou L, Tian Y, Xu P, Wei H, Li Y, Peng H-X, Qin F (2021) Effect of the selective localization of carbon nanotubes and phase domain in immiscible blends on tunable microwave dielectric properties. Compos Sci Technol 213:108919. https://doi.org/10.1016/j.compscitech.2021.108919
81. Danner PM, Iacob M, Sasso G, Burda I, Rieger B, Nüesch F, Opris DM (2022) Solvent-free synthesis and processing of conductive elastomer composites for green dielectric elastomer transducers. Macromol Rapid Commun 43:2100823. https://doi.org/10.1002/marc.202100823
82. Arjmand M, Mahmoodi M, Park S, Sundararaj U (2013) An innovative method to reduce the energy loss of conductive filler/polymer composites for charge storage applications. Compos Sci Technol 78:24–29. https://doi.org/10.1016/j.compscitech.2013.01.019
83. Thomassin J-M, Pagnoulle C, Bednarz L, Huynen I, Jerome R, Detrembleur C (2008) Foams of polycaprolactone/MWNT nanocomposites for efficient EMI reduction. J Mater Chem 18:792–796. https://doi.org/10.1039/B709864B
84. Ameli A, Nofar M, Park CB, Pötschke P, Rizvi G (2014) Polypropylene/carbon nanotube nano/microcellular structures with high dielectric permittivity, low dielectric loss, and low percolation threshold. Carbon 71:206–217. https://doi.org/10.1016/j.carbon.2014.01.031
85. Liu J, Tian G, Qi S, Wu Z, Wu D (2014) Enhanced dielectric permittivity of a flexible three-phase polyimide–graphene–BaTiO3 composite material. Mater Lett 124:117–119. https://doi.org/10.1016/j.matlet.2014.02.105

86. Martinelli NG, Savini M, Muccioli L, Olivier Y, Castet F, Zannoni C, Beljonne D, Cornil J (2009) Modeling polymer dielectric/pentacene interfaces: on the role of electrostatic energy disorder on charge carrier mobility. Adv Funct Mater 19:3254–3261. https://doi.org/10.1002/adfm.200901077
87. Kim J-Y, Kim T, Suk JW, Chou H, Jang J-H, Lee JH, Kholmanov IN, Akinwande D, Ruoff RS (2014) Enhanced dielectric performance in polymer composite films with carbon nanotube-reduced graphene oxide hybrid filler. Small 10:3405–3411. https://doi.org/10.1002/smll.201400363
88. Tian M, Wei Z, Zan X, Zhang L, Zhang J, Ma Q, Ning N, Nishi T (2014) Thermally expanded graphene nanoplates/polydimethylsiloxane composites with high dielectric constant, low dielectric loss and improved actuated strain. Compos Sci Technol 99:37–44. https://doi.org/10.1016/j.compscitech.2014.05.004
89. Gao J, Asadi K, Xu JB, An J (2009) Controlling of the surface energy of the gate dielectric in organic field-effect transistors by polymer blend. Appl Phys Lett 94:093302. https://doi.org/10.1063/1.3086894
90. Ameli A, Wang S, Kazemi Y, Park CB, Pötschke P (2015) A facile method to increase the charge storage capability of polymer nanocomposites. Nano Energy 15:54–65. https://doi.org/10.1016/j.nanoen.2015.04.004
91. Hu P, Shen Y, Guan Y, Zhang X, Lin Y, Zhang Q, Nan C-W (2014) Topological-structure modulated polymer nanocomposites exhibiting highly enhanced dielectric strength and energy density. Adv Funct Mater 24:3172–3178. https://doi.org/10.1002/adfm.201303684
92. Zhang G, Brannum D, Dong D, Tang L, Allahyarov E, Tang S, Kodweis K, Lee J-K, Zhu L (2016) Interfacial polarization-induced loss mechanisms in polypropylene/$BaTiO_3$ nanocomposite dielectrics. Chem Mater 28:4646–4660. https://doi.org/10.1021/acs.chemmater.6b01383
93. Zhu L (2014) Exploring strategies for high dielectric constant and low loss polymer dielectrics. J Phys Chem Lett 5:3677–3687. https://doi.org/10.1021/jz501831q
94. Yang D, Ni Y, Kong X, Wang Y, Zhang L (2019) A mussel-like inspired modification of $BaTiO_3$ nanopartciles using catechol/polyamine co-deposition and silane grafting for high-performance dielectric elastomer composites. Compos Part B Eng 172:621–627. https://doi.org/10.1016/j.compositesb.2019.05.101
95. Maiolino P, Galantini F, Mastrogiovanni F, Gallone G, Cannata G, Carpi F (2015) Soft dielectrics for capacitive sensing in robot skins: Performance of different elastomer types. Sens Actuators Phys 226:37–47. https://doi.org/10.1016/j.sna.2015.02.010
96. Szadkowski B, Marzec A, Zaborski M (2020) Use of carbon black as a reinforcing nano-filler in conductivity-reversible elastomer composites. Polym Test 81:106222. https://doi.org/10.1016/j.polymertesting.2019.106222
97. Kim WJ, Cho S, Hong J, Hong JP (2022) Geometrically versatile triboelectric yarn-based harvesters via carbon nanotubes-elastomer composites. Compos Sci Technol 219:109247. https://doi.org/10.1016/j.compscitech.2021.109247
98. Fredin LA, Li Z, Lanagan MT, Ratner MA, Marks TJ (2013) Sustainable high capacitance at high frequencies: metallic aluminum-polypropylene nanocomposites. ACS Nano 7:396–407. https://doi.org/10.1021/nn3044148
99. Dang Z-M, Zheng M-S, Zha J-W (2016) 1D/2D carbon nanomaterial-polymer dielectric composites with high permittivity for power energy storage applications. Small 12:1688–1701. https://doi.org/10.1002/smll.201503193
100. Kim H, Abdala AA, Macosko CW (2010) Graphene/polymer nanocomposites. Macromolecules 43:6515–6530. https://doi.org/10.1021/ma100572e
101. Kuilla T, Bhadra S, Yao D, Kim NH, Bose S, Lee JH (2010) Recent advances in graphene based polymer composites. Prog Polym Sci 35:1350–1375. https://doi.org/10.1016/j.progpolymsci.2010.07.005
102. Wang Y, Gu P, Zhao Z, Zhou P, Li Z, Li Y (2020) Synthesis of graphene-like carbon film on SiC substrate. J Cryst Growth 531:125356. https://doi.org/10.1016/j.jcrysgro.2019.125356

103. Warner JH, Schaffel F, Rummeli M, Bachmatiuk A (2012) Graphene: Ffndamentals and emergent applications. Newnes
104. Tour JM (2014) Scaling up exfoliation. Nat Mater 13:545–546. https://doi.org/10.1038/nmat3961
105. Wang X, You H, Liu F, Li M, Wan L, Li S, Li Q, Xu Y, Tian R, Yu Z, Xiang D, Cheng J (2009) Large-scale synthesis of few-layered graphene using CVD. Chem Vap Depos 15:53–56. https://doi.org/10.1002/cvde.200806737
106. Acik M, Lee G, Mattevi C, Pirkle A, Wallace RM, Chhowalla M, Cho K, Chabal Y (2011) The role of oxygen during thermal reduction of graphene oxide studied by infrared absorption spectroscopy. J Phys Chem C 115:19761–19781. https://doi.org/10.1021/jp2052618
107. Huang X, Jiang P (2015) Core-shell structured high-k polymer nanocomposites for energy storage and dielectric applications. Adv Mater 27:546–554. https://doi.org/10.1002/adma.201401310
108. Punetha VD, Rana S, Yoo HJ, Chaurasia A, McLeskey JT, Ramasamy MS, Sahoo NG, Cho JW (2017) Functionalization of carbon nanomaterials for advanced polymer nanocomposites: a comparison study between CNT and graphene. Prog Polym Sci 67:1–47. https://doi.org/10.1016/j.progpolymsci.2016.12.010
109. Liao L, Xie Q, Guo X, Liu Z (2015) Fabrication of chemical graphene nanoribbons via edge-selective covalent modification. Adv Mater 27:4093–4096. https://doi.org/10.1002/adma.201501788
110. Wang H, Bi S-G, Ye Y-S, Xue Y, Xie X-L, Mai Y-W (2015) An effective non-covalent grafting approach to functionalize individually dispersed reduced graphene oxide sheets with high grafting density, solubility and electrical conductivity. Nanoscale 7:3548–3557. https://doi.org/10.1039/C4NR06710J
111. Kumar S, Raj S, Kolanthai E, Sood AK, Sampath S, Chatterjee K (2015) Chemical functionalization of graphene to augment stem cell osteogenesis and inhibit biofilm formation on polymer composites for orthopedic applications. ACS Appl Mater Interfaces 7:3237–3252. https://doi.org/10.1021/am5079732
112. Sreeprasad TS, Berry V (2013) How do the electrical properties of graphene change with its functionalization? Small 9:341–350. https://doi.org/10.1002/smll.201202196
113. Vasileiou AA, Kontopoulou M, Docoslis A (2014) A noncovalent compatibilization approach to improve the filler dispersion and properties of polyethylene/graphene composites. ACS Appl Mater Interfaces 6:1916–1925. https://doi.org/10.1021/am404979g
114. Chen T, Qiu J, Zhu K, Li J (2016) Electro-mechanical performance of polyurethane dielectric elastomer flexible micro-actuator composite modified with titanium dioxide-graphene hybrid fillers. Mater Des 90:1069–1076. https://doi.org/10.1016/j.matdes.2015.11.068
115. Prasad Sahoo B (2021) Fabrication of radiation crosslinked and MWCNT reinforced ethylene acrylic elastomer nanocomposites: Evaluation of mechanical, dynamic mechanical, thermal and dielectric properties. Mater Today Proc 41:203–210. https://doi.org/10.1016/j.matpr.2020.08.692
116. Wang Y, Kim S, Li GP, Sun LZ (2015) Filler orientation effect on relative permittivity of dielectric elastomer nanocomposites filled with carbon nanotubes. Comput Mater Sci 104:69–75. https://doi.org/10.1016/j.commatsci.2015.03.022
117. Hussain T, Jabeen S, Shehzad K, Mujahid A, Ahmad MN, Farooqi ZH, Raza MH (2018) Polyaniline/silver decorated-MWCNT composites with enhanced electrical and thermal properties. Polym Compos 39:E1346–E1353. https://doi.org/10.1002/pc.24219
118. Song B (2022) Study on dielectric and thermal properties of ABS/multilayer graphene composites. J Phys Conf Ser 2247:012011. https://doi.org/10.1088/1742-6596/2247/1/012011
119. Chonat A, Palatty S (2020) Enhanced electrochemical performance of a hybrid supercapacitive material based on ternary doped polyaniline/activated carbon composite. Energy Fuels 34:10148–10159. https://doi.org/10.1021/acs.energyfuels.0c01555
120. Guin T, Settle MJ, Kowalski BA, Auguste AD, Beblo RV, Reich GW, White TJ (2018) Layered liquid crystal elastomer actuators. Nat Commun 9:2531. https://doi.org/10.1038/s41467-018-04911-4

121. Li J, Wang Y, Liu L, Xu S, Liu Y, Leng J, Cai S (2019) A biomimetic soft lens controlled by electrooculographic signal. Adv Funct Mater 29:1903762. https://doi.org/10.1002/adfm.201903762
122. Xiao Y, Mao J, Shan Y, Yang T, Chen Z, Zhou F, He J, Shen Y, Zhao J, Li T, Luo Y (2020) Anisotropic electroactive elastomer for highly maneuverable soft robotics. Nanoscale 12:7514–7521. https://doi.org/10.1039/D0NR00924E
123. Krogmann F, Monch W, Zappe H (2008) Electrowetting for tunable microoptics. J Microelectromechanical Syst 17:1501–1512. https://doi.org/10.1109/JMEMS.2008.2007256
124. Hwang T, Kwon H-Y, Oh J-S, Hong J-P, Hong S-C, Lee Y, Ryeol Choi H, Jin Kim K, Hossain Bhuiya M, Nam J-D (2013) Transparent actuator made with few layer graphene electrode and dielectric elastomer, for variable focus lens. Appl Phys Lett 103:023106. https://doi.org/10.1063/1.4812982
125. Jiang L, Wang Y, Wang X, Ning F, Wen S, Zhou Y, Chen S, Betts A, Jerrams S, Zhou F-L (2021) Electrohydrodynamic printing of a dielectric elastomer actuator and its application in tunable lenses. Compos Part Appl Sci Manuf 147:106461. https://doi.org/10.1016/j.compositesa.2021.106461
126. Chen Y, Chen C, Rehman HU, Zheng X, Li H, Liu H, Hedenqvist MS (2020) Shape-memory polymeric artificial muscles: mechanisms, applications and challenges. Mol Basel Switz 25:E4246. https://doi.org/10.3390/molecules25184246
127. Mirvakili SM, Hunter IW (2018) Artificial muscles: mechanisms, applications, and challenges. Adv Mater 30:1704407. https://doi.org/10.1002/adma.201704407
128. Yang J, Wang S, Yao J, Yang X, Yu T, Wei K (2022) Mechanochemical characteristics and influence mechanisms of a biomass hydrogel artificial muscle based on different parameters of the sodium alginate adjustment. Smart Mater Struct 31:055002. https://doi.org/10.1088/1361-665X/ac5aa6
129. Wang J, Gao D, Lee PS (2021) Recent progress in artificial muscles for interactive soft robotics. Adv Mater 33:2003088. https://doi.org/10.1002/adma.202003088
130. Zhang J, Sheng J, O'Neill CT, Walsh CJ, Wood RJ, Ryu J-H, Desai JP, Yip MC (2019) Robotic artificial muscles: current progress and future perspectives. IEEE Trans Robot 35:761–781. https://doi.org/10.1109/TRO.2019.2894371
131. Liang W, Liu H, Wang K, Qian Z, Ren L, Ren L (2020) Comparative study of robotic artificial actuators and biological muscle. Adv Mech Eng 12:1687814020933409. https://doi.org/10.1177/1687814020933409
132. Chen P, Xu Y, He S, Sun X, Pan S, Deng J, Chen D, Peng H (2015) Hierarchically arranged helical fibre actuators driven by solvents and vapours. Nat Nanotechnol 10:1077–1083. https://doi.org/10.1038/nnano.2015.198
133. Gu X, Fan Q, Yang F, Cai L, Zhang N, Zhou W, Zhou W, Xie S (2016) Hydro-actuation of hybrid carbon nanotube yarn muscles. Nanoscale 8:17881–17886. https://doi.org/10.1039/C6NR06185K
134. Miriyev A, Stack K, Lipson H (2017) Soft material for soft actuators. Nat Commun 8:596. https://doi.org/10.1038/s41467-017-00685-3
135. Hu W, Lum GZ, Mastrangeli M, Sitti M (2018) Small-scale soft-bodied robot with multimodal locomotion. Nature 554:81–85. https://doi.org/10.1038/nature25443
136. Nie B, Liu S, Qu Q, Zhang Y, Zhao M, Liu J (2022) Bio-inspired flexible electronics for smart E-skin. Acta Biomater 139:280–295. https://doi.org/10.1016/j.actbio.2021.06.018
137. Yang Y, Zhao G, Cheng X, Deng H, Fu Q (2021) Stretchable and healable conductive elastomer based on PEDOT: PSS/natural rubber for self-powered temperature and strain sensing. ACS Appl Mater Interfaces 13:14599–14611. https://doi.org/10.1021/acsami.1c00879
138. Rentals J, Wang C, Zhang X, Carey T, Chen K, Yin Y, Torrisi F (2016) Environmentally-friendly conductive cotton fabric as flexible strain sensor based on hot press reduced graphene oxide. Carbon 111. https://doi.org/10.1016/j.carbon.2016.10.045
139. Zhang M, Wang C, Wang H, Jian M, Hao X, Zhang Y (2017) Carbonized cotton fabric for high-performance wearable strain sensors. Adv Funct Mater 27:1604795. https://doi.org/10.1002/adfm.201604795

140. Araromi OA, Graule MA, Dorsey KL, Castellanos S, Foster JR, Hsu W-H, Passy AE, Vlassak JJ, Weaver JC, Walsh CJ, Wood RJ (2020) Ultra-sensitive and resilient compliant strain gauges for soft machines. Nature 587:219–224. https://doi.org/10.1038/s41586-020-2892-6
141. Huang X, Wang D, Yuan Z, Xie W, Wu Y, Li R, Zhao Y, Luo D, Cen L, Chen B, Wu H, Xu H, Sheng X, Zhang M, Zhao L, Yin L (2018) A fully biodegradable battery for self-powered transient implants. Small Weinh Bergstr Ger 14:e1800994. https://doi.org/10.1002/smll.201800994
142. Zheng Q, Zou Y, Zhang Y, Liu Z, Shi B, Wang X, Jin Y, Ouyang H, Li Z, Wang ZL (2016) Biodegradable triboelectric nanogenerator as a life-time designed implantable power source. Sci Adv 2:e1501478. https://doi.org/10.1126/sciadv.1501478
143. Ma Y, Zheng Q, Liu Y, Shi B, Xue X, Ji W, Liu Z, Jin Y, Zou Y, An Z, Zhang W, Wang X, Jiang W, Xu Z, Wang ZL, Li Z, Zhang H (2016) Self-powered, one-stop, and multifunctional implantable triboelectric active sensor for real-time biomedical monitoring. Nano Lett 16:6042–6051. https://doi.org/10.1021/acs.nanolett.6b01968
144. Chatterjee S, Saxena M, Padmanabhan D, Jayachandra M, Pandya HJ (2019) Futuristic medical implants using bioresorbable materials and devices. Biosens Bioelectron 142:111489. https://doi.org/10.1016/j.bios.2019.111489
145. Ramesh S, Lungaro L, Tsikritsis D, Weflen E, Rivero IV, Elfick APD (2018) Fabrication and evaluation of poly(lactic acid), chitosan, and tricalcium phosphate biocomposites for guided bone regeneration. J Appl Polym Sci 135:46692. https://doi.org/10.1002/app.46692
146. Turner B, Ramesh S, Menegatti S, Daniele M (2022) Resorbable elastomers for implantable medical devices: highlights and applications. Polym Int 71:552–561. https://doi.org/10.1002/pi.6349
147. Turner BL, Kilgour KM, Stine SJ, Daniele M, Menegatti S (2020) Dual-affinity ratiometric quenching (DARQ) assay for the quantification of therapeutic antibodies in CHO-S cell culture fluids. Anal Chem 92:16274–16283. https://doi.org/10.1021/acs.analchem.0c04269
148. Singh R, Bathaei MJ, Istif E, Beker L (2020) A review of bioresorbable implantable medical devices: materials, fabrication, and implementation. Adv Healthc Mater 9:2000790. https://doi.org/10.1002/adhm.202000790
149. Huang X (2018) Materials and applications of bioresorbable electronics. J Semicond 39:011003. https://doi.org/10.1088/1674-4926/39/1/011003
150. Lau EW (2017) Technologies for prolonging cardiac implantable electronic device longevity. Pacing Clin Electrophysiol 40:75–96. https://doi.org/10.1111/pace.12989
151. Sim K, Ershad F, Zhang Y, Yang P, Shim H, Rao Z, Lu Y, Thukral A, Elgalad A, Xi Y, Tian B, Taylor DA, Yu C (2020) An epicardial bioelectronic patch made from soft rubbery materials and capable of spatiotemporal mapping of electrophysiological activity. Nat Electron 3:775–784. https://doi.org/10.1038/s41928-020-00493-6
152. Schiavone G, Fallegger F, Kang X, Barra B, Vachicouras N, Roussinova E, Furfaro I, Jiguet S, Seáñez I, Borgognon S, Rowald A, Li Q, Qin C, Bézard E, Bloch J, Courtine G, Capogrosso M, Lacour SP (2020) Soft, implantable bioelectronic interfaces for translational research. Adv Mater 32:1906512. https://doi.org/10.1002/adma.201906512
153. Ling Y, Pang W, Liu J, Page M, Xu Y, Zhao G, Stalla D, Xie J, Zhang Y, Yan Z (2022) Bioinspired elastomer composites with programmed mechanical and electrical anisotropies. Nat Commun 13:524. https://doi.org/10.1038/s41467-022-28185-z
154. Shyy W, Lian Y, Tang J, Viieru D, Liu H (2007) Aerodynamics of low Reynolds Number flyers. Cambridge University Press, Cambridge
155. Grimaldi DA, Engel MS (2005) Evolution of the insects. Cambridge University Press, Cambridge, UK and New York
156. Karásek M (2014) Robotic hummingbird: Design of a control mechanism for a hovering flapping wing micro air vehicle
157. Chattaraj N, Ganguli R (2022) Mechatronic approaches to synthesize biomimetic flapping-wing mechanisms: a review. Int J Aeronaut Space Sci. https://doi.org/10.1007/s42405-022-00527-7

158. Gomis-Bellmunt O, Campanile LF (2009). Design rules for actuators in active mechanical systems. Springer Science & Business Media
159. Xiao S, Hu K, Huang B, Deng H, Ding X (2021) A review of research on the mechanical design of hoverable flapping wing micro-air vehicles. J Bionic Eng 18:1235–1254. https://doi.org/10.1007/s42235-021-00118-4

Chapter 9
Physical and Chemical Properties of Inorganic-Polymer Nanodielectrics and Their Applications

Kavya Pulagam Srinivasa Babu, K. P. Chaithra, Arunkumar Chandrasekhar, and T. P. Vinod

Abstract Organic polymers have been widely used as matrices in composite materials for electrical power equipment because of their reliability, ease of fabrication, availability, and low cost. Nevertheless, they have comparatively poor electrical, thermal, and mechanical properties compared to metals and ceramics. Most of the polymer systems, such as copolymers, modified polymers, homopolymers, and blended polymers, are considered to have limitations in delivering the demands of the aforementioned applications. This challenge can be efficiently addressed by blending the polymers with inorganic particles such as metal chalcogenides, nanoclays, silica, transition metal oxides, nanometals, and metallic phosphates. Nanodielectrics, the dielectrics made of nanocomposites, can be prepared by combining inorganic polymer matrices with nanoscale fillers. Inorganic polymer nanodielectrics thus formed show significant improvements in their physical and chemical properties. The properties and applications of inorganic polymer nanodielectrics are discussed in this chapter. Advanced properties of inorganic polymer nanodielectrics are related to their intermolecular interactions, chain flexibility, stability, crosslinking, tactility, dielectric constant, thermal property, processability, texture structure, flexibility, etc. This chapter also summarizes current and future scope of research in inorganic polymer nanodielectrics for applications such as partial discharge (PD) resistance, electric energy storage, and space charge suppression.

Keywords Nanocomposite · Inorganic polymers · Nanodielectrics · Energy storage · Space charge suppression

K. P. S. Babu · K. P. Chaithra · T. P. Vinod
Department of Chemistry, CHRIST (Deemed to be University), Hosur Road, Bengaluru 560029, India

A. Chandrasekhar (✉)
Nanosensors and Nanoenergy Lab, Department of Sensors and Biomedical Technology, School of Electronics Engineering, Vellore Institute of Technology, Tamil Nadu, Vellore 632014, India
e-mail: arunkumar.c@vit.ac.in

S. Moharana et al. (eds.), *Emerging Nanodielectric Materials for Energy Storage*, Nanostructure Science and Technology, https://doi.org/10.1007/978-3-031-40938-7_9

9.1 Introduction

Polymers are one of the most commonly used classes of materials in various fields of technology and engineering. Common organic polymers like polyolefins, nylons, polyesters, and polyurethanes are extensively used because of their salient features, which include excellent processability, durability, lightweight, ease of fabrication, and comparatively low cost [1, 2]. A major problem in polymer research is to extend the application window of such materials while preserving their useful properties while improving modulus, heat resistance, fire performance, and strength. This can be achieved by reinforcing the polymer matrix with nanoscopic inorganic particles (silica, metallic phosphates, nanoclays, nanometals, transition metal oxides, and metal chalcogenides) which can be zero-dimensional (0D) nanoparticles, one-dimensional (1D) nanotubes or nanofibres, and two-dimensional (2D) layered silicates or graphene (Fig. 9.1). Incorporation of these inorganic particles can lead to superior physical and mechanical properties compared to that of polymer composites filled with micro-sized additives.

The combination of one or more additives, typically of the size of 10–100 Å with a polymer matrix, is known as polymer nanocomposites [4]. Inorganic particles of different morphologies and dimensionalities such as spherical, layered natural or artificial clays or micas, either intercalated or exfoliated are dispersed in various polymers. Three important features that distinguish polymer nanocomposites from conventional polymer micro composites are: (i) filler size, (ii) filler quantity, and (iii) large specific surface area. The inorganic particles used as additives can implement or improve optical properties, mechanical behavior, thermal properties, electronic response, optoelectronic properties, density, refractive index, etc. [5].

The characteristic properties of polymer nanocomposites make them useful for energy-related applications. One such usage of them is seen as dielectric materials, owing to the enhanced dielectric properties compared to conventional polymeric materials. A nanodielectric is a solid composite material composed of dielectric matrix (mostly polymer) with filler particles, with sizes less than 100 nm. These

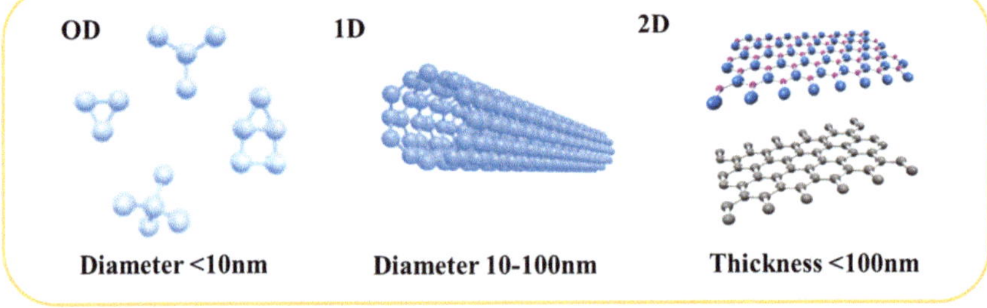

Fig. 9.1 Illustration of nanofillers in different dimensions: 0D nanoparticles, 1D nanotubes, and 2D layered silicates. Recreated with permission from Nature Communications, Copyright © 2021 [3]

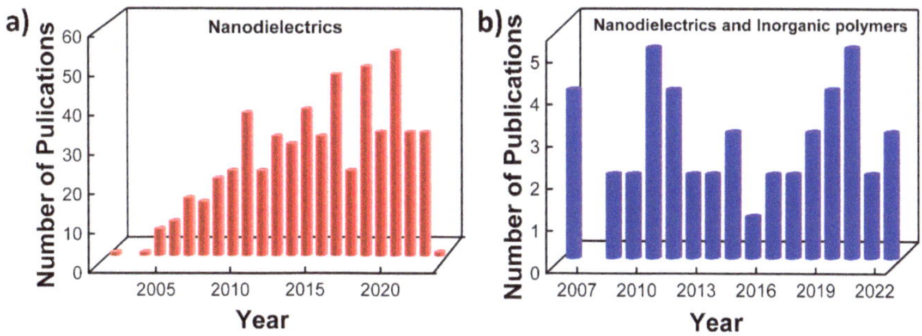

Fig. 9.2 Number of publications with years graph obtained with the search for keywords nanodielectrics (**a**) and nanodielectrics and inorganic polymers (**b**). Data from Scopus October 2022

nanodielectrics exhibit notable improvements in dielectric strength and voltage endurance are so appealing, both commercially and in terms of basic science research. According to the statistics taken from Scopus in October 2022, it is evident that the number of publications in this area (nanodielectrics) are increasing every year owing to their applications in various fields (Fig. 9.2a). Inorganic polymer nanodielectrics are also gaining importance in the field of research as seen in Fig. 9.2b.

Preparation of inorganic polymer nanodielectrics can be done by various methods such as the sol–gel method, in situ polymerization, solution mixing, melt mixing, and in situ intercalative polymerization [1, 4–6]. The preparation process and type of inorganic nanomaterials used influence the physical and chemical properties of dielectric. This chapter will discuss the physical and chemical properties of inorganic polymer nanodielectrics and their applications (Fig. 9.3).

9.2 Physical and Chemical Properties of Inorganic Polymer Nanodielectrics

9.2.1 Electrical Properties

The electrical charge transfer, dielectric breakdown, and endurance characteristics of the materials are significantly impacted by incorporating nano-sized fillers into the insulating polymers. Understanding these requires consideration of how the particles alter the morphology of the polymer as well as the physical structure of the polymer and the local chemistry. It is also important to take into account how electrical characteristics are altered by electrostatic forces that form an interaction zone surrounding the particle surfaces [7]. Their dielectric behavior and conductivity are dependent on the electrical responses of the micro- or nanocomposites polymer matrix. As insulating is the primary electrical character of polymers due to their lower concentration

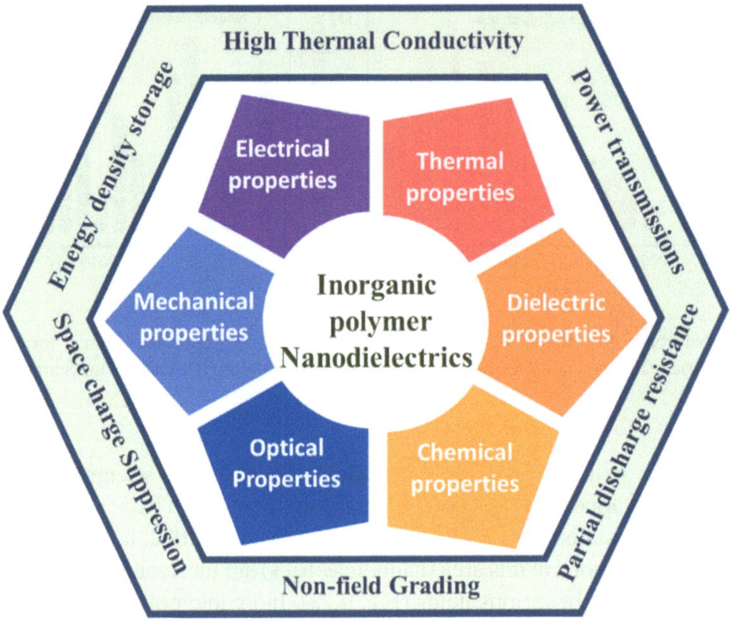

Fig. 9.3 Inorganic polymer nanodielectrics: Its physical and chemical properties align with their applications

of free charge carriers, polymer composites appear to be dielectrics. Classification of polymer nanocomposites includes two main groups: (i) the insulating matrix-dielectric reinforcing phase, and (ii) the insulating matrix-conductive reinforcing phase [8]. The electrical behavior of nanocomposites can be experimentally investigated by means of broadband dielectric spectroscopy (BDS) and direct current (DC) conductivity measurements.

Inorganic nanofillers added to polymers in order to enhance electrical properties could be metallic, ceramic, or any allotropic form of carbon. An interesting group of ceramic inclusions includes the active or functional dielectrics, such as piezoelectric, ferroelectric, and pyroelectric elements. Electrical characteristics include increase in the conductivity, dielectric permittivity, and dielectric strength at low nanofiller content. General electrical properties of nanodielectrics of nanocomposites are discussed below.

(a) **Breakdown strength**: The capacity to withstand the induced electric field is a basic requirement of electrical insulation. The applied electric field is gradually increased until breakdown occurs in order to determine the material's breakdown strength, which is dependent on the electrode geometry, immersion medium, sample thickness, and other factors. Weibull statistical distribution can be used to examine breakdown strength using shape and scale parameters [9, 10]. Shape parameters are analogous to the standard deviation in the normal distribution and scale parameters are analogous to the mean in the normal distribution [10].

DC breakdown data forms the basis for the current interest in employing nanocomposites as high-performance dielectric materials [11, 12]. A few examples from the literature showing the incorporation of inorganic materials in polymers as a dielectric material with improved breakdown strength are: (i) the addition of differently sized 90 wt% boron nitride to the epoxy-based systems increased the breakdown strengthBreakdown strength with decrease in size of the fillerFiller particle [13, 14]. (ii) the DC breakdown strength of polypropylene (PP) increased by roughly one-third when nanomagnesium oxide (MgO) was added and when PP is incorporated with 5% hydrophobic fumed nanosilica, it resulted in an increase from 511 to 778 kV/mm [15] and (iv) the addition of 5wt% silica (SiO_2) to silica/cross-linked polyethylene (XLPE) followed by surface treatment using ethenylsilane gives an increased breakdown strength when compared to polymer [16, 17] (Table 9.1).

Any insulation material is subjected to dielectric breakdown when the applied electric field reaches a certain threshold value, which is influenced by temperature, system geometry, and physical or chemical flaws in the dielectric. Breakdown strength of various insulating polymers is listed in Table 9.2. Breakdown also occurs when the number of mobile charge carriers is suddenly and irreversibly multiplied under the effect of a high field. It has been demonstrated that adding typical (micron-sized) fillers to polymer dielectrics decreases the field electric breakdown (E_{bd}) [23, 24]. The breakdown strength decreases with addition of nanofiller beyond certain point because of the nanofiller aggregation and it behaves like micro-sized inclusions. In other words, presence of sizable particles degrades breakdown performance. A few examples from the literature are: (i) addition of nanoscopic zinc oxide (ZnO) to low-density polyethylene (LDPE) decreased strength with an increase in filler amount, (ii) addition of titania (TiO_2) nanoparticles to LDPE decreased DC breakdown strength with an increase in filler amount [19].

Table 9.1 E_{bd} values of inorganic polymer nanodielectrics

Material	E_{bd} (kV/mm)	References
Bisphenol epoxy + TiO_2 (10 wt % and 1.5 μm)	200	
Bisphenol epoxy + TiO_2 (10 wt % and 38 nm)	280	[18]
Bisphenol epoxy	332	
Bisphenol epoxy + TiO_2 (10 wt % and 15 μm)	305	
Bisphenol epoxy + TiO_2 (10 wt % and 23 nm)	391	[19]
Bisphenol epoxy	347	
Bisphenol epoxy + MMT nanoclay (4–1/2 wt %)	531	[20]
Silica/cross-linked polyethylene (XLPE)	269	
XLPE + SiO_2 (5 wt % and 12 nm)	314	[21]
Low-density polyethylene (LDPE)	630	
LDPE + MgO_2 (2 wt % and 50 nm)	890	[22]

Reused with permission from [21] Copyright © 2007, Springer

Table 9.2 Breakdown strength of various insulating polymers

Polymer	E_{bd} (kV/mm)
LDPE	200
High-density polyethylene (HDPE)	200
XLPE	220
Polypropylene (Biaxially oriented)	200
Polystyrene	200
Polytetrafluoroethylene	88–176
Poly(vinylidene fluoride)	10.2
Polycarbonate	252
Polyester	300
Polyimide	280
Epoxy resin	25–45

(b) **Permittivity**

The number and nature of polarizable species present in the system, as well as their dynamics, are known as permittivity. Dielectric response is affected by its surrounding. Structural information about this can be obtained by dielectric spectroscopy. Permittivity of a material is given by Eq. 9.1 [26, 27].

$$\varepsilon_r = \varepsilon_r\prime + i\varepsilon_r^{''} \tag{9.1}$$

where, in relative permittivity, $\varepsilon_r\prime$ is the real part, it is related to the energy stored in the system, $\varepsilon_r^{''}$ is the imaginary part.

The frequency dependence of the permittivity can be used to describe a material's dielectric response. It is exemplified by the Claussius–Mossotti Eq. 9.2 [28].

$$\alpha = \frac{3\varepsilon_0 M}{N_A \rho} \times \frac{(\varepsilon_r - 1)}{(\varepsilon_r + 2)} \tag{9.2}$$

M stands for molecular mass, ε_r for relative permittivity/dielectric constant, ε_0 for free space permittivity, N_A for avagadro number, ρ for density, and α for molecular polarizability. For multi-component systems like nanodielectrics, permittivity is given by Lichtenecker–Rother Eq. 9.3 [29].

$$\log \varepsilon_c^{'} = \emptyset_p \log \varepsilon_p^{'} + \emptyset_m \log \varepsilon_m^{'} \tag{9.3}$$

where $\emptyset_p$ stands for the volume fraction of particles and $\emptyset_m$ stands for the volume fraction of matrix, $\varepsilon_c^{'}$ stands for the real permittivity of the composite, $\varepsilon_p^{'}$ stands for the real permittivity of particles, and $\varepsilon_m^{'}$. stands for the real permittivity of matrix.

A few examples from the literature showing variation in permittivity due to the addition of inorganic materials are: (i) when excess amount of montmorillonite nanofiller is added into the epoxy matrix, variation in the permittivity from 4.6 to 4.3 was observed [30]. (ii) The addition of SiO2 to LDPE-based nanocomposites-Nanocomposites, the permittivity varied [31, 32]. When the amount of SiO_2 reaches 5%, the permittivity value reduces to that of the base polymer. (iii) When the wt% of silica nanofiller varied from 0.06 to 5.58, the permittivity values varied from 3.26 to 2.27 [26].

It is concluded that adding nanofillers "*The surface performance was significantly improved, and it was able to withstand the detrimental effects of partial discharges.*" [33]. Two distinct processes, one physical and one chemical, are attributed to the advantageous effects of adding a nanofiller to a polymer. The physical consequence is that the surface barrier layer that forms as a result of localized nanofiller accumulation following polymer ablation is thermally stable because the layer abundant in inorganic particles whereas the chemical consequence is that the nanofiller promotes the development of solid over gaseous decomposition products (char).

Improvement in electric properties of nanodielectrics is useful for reducing the local electrical field enhancements that is in filed grading. Field grading is important in the construction of electrical assemblies, where the breakdown occurs. In this case, a dielectric with a higher breakdown strength must be used to level the stress and reduce the electric field in the area of concern. Main application of polymeric dielectrics is to serve as outdoor insulation that is on overhead lines or as outdoor bushings. This reduces the corona activity and surface discharges which degrade the material. Addition of small quantity of nanofiller can be used to avoid these damages [34].

9.2.2 Mechanical Properties

High surface volume ratio of nanofiller makes nanodielectrics possess high reinforcing efficiency even at low concentrations. Many nanocomposites, where nanofiller addition is done to the polymer, change the morphology of the matrix leading to increase in the stiffness, toughness, elastic modulus, and mechanical strength. This increase occurs through modifying polymeric nanocomposite materials' macroscopic mechanical characteristics. Using nanofillers with matching coefficient of thermal expansion results in improved discharge resistance, improved abrasion, long-term mechanical behavior, and wear resistance [35]. When the polymer is reinforced with nanomaterials, it is also discovered that the long-term mechanical properties of nanocomposite, such as wear and abrasion resistance, fatigue, and creep, are greatly enhanced. Knowing the characteristics of the matrix, fillers, and other constituents present allows it to predict the general mechanical properties of the composites as a whole. Numerous theories can be used to predict the individual effects of the filler aspect ratio, filler/matrix volume ratio, filler stiffness, etc., on the mechanical response of these composites [36]. The Mori–Tanaka method was developed to determine the elastic stress distribution within and around an ellipsoidal particle in an infinite matrix. It was based on the principles of Eshelby's particle inclusion model. Analysis of nanocomposites and a range of reinforcing geometries have both been successfully applied to unidirectional composites' stiffness can be predicted by Halpin and Tsai's theory. (i.e., nano fiber, spherical nanoparticles) [37, 38]. It is given by Eq. 9.4.

$$E = \emptyset_f E_f + \left(1 - \emptyset_f\right) E_m \tag{9.4}$$

where and $\emptyset_f$. is the filler's volume fraction, E_f for Young's modulus of the filler, and E_m stands for Young's modulus of the matrix.

These theories demonstrate that, particularly for fillers with high surface to volume aspect ratios, the fraction of the interfacial region increases significantly as filler size is reduced (to nanoscale). The region of the interphase containing nanotube (NT) and nanoplatelet (NP) in polymer matrix is graphically depicted in Fig. 9.4. In Fig. 9.5, the longitudinal modulus results of nanotube and nanoplatelet reinforced nanocomposites with same filler level are shown [39]. It was demonstrated that nanotubes display more notable interphase effects on the elastic modulus than nanoplatelet fillers for high aspect ratio nanofillers. Addition of excess leads to insufficient interfacial contact, interfacial slippage, and insufficient dispersion in nanotubes. This results in decreased modulus [36, 40].

Mechanical properties of nanodielectrics of nanocomposites include:

(a) **Tensile strength**

The nanoencapsulated polymer composites will have improved mechanical properties but when more amount of fillers are added, it leads to aggregation and irregular interfacial bonding [42]. Latest advancement of nanofillers and their processing and chemical functionalization has helped to solve this issue. Liu et al. studied about single-walled carbon nanotube-polyvinyl alcohol (CNT-PVA) nanocomposites for dielectric applications, where it is observed that the strength and Young's

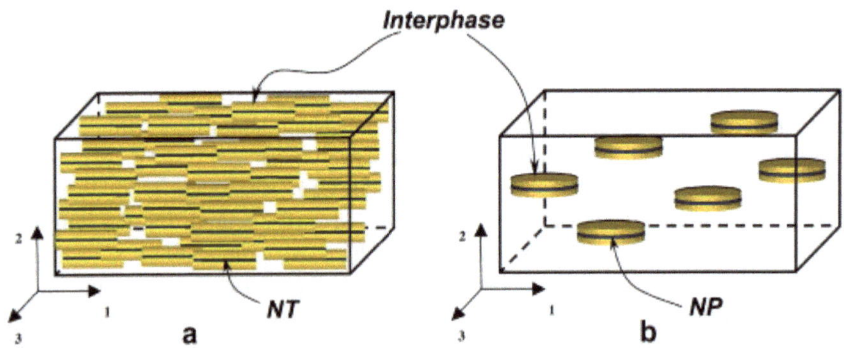

Fig. 9.4 Illustration of aligned **a** nanotube and **b** nanoplatelet nanocomposites with interphase region. Reused with permission from [41] Copyright © 2008 Composites Science and Technology

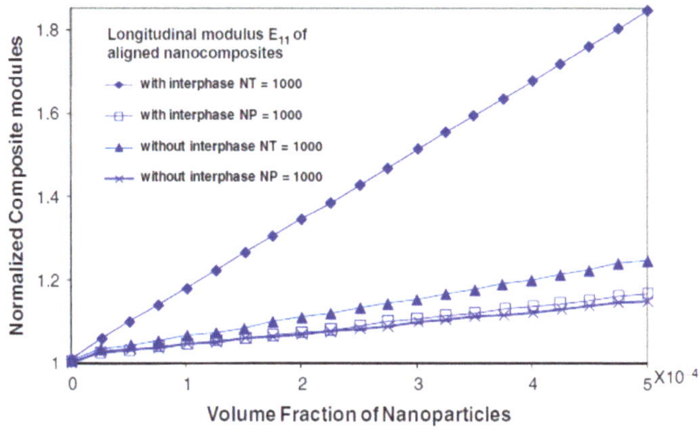

Fig. 9.5 Effects of interphase. Reused with permission from [41] Copyright © 2008 Composites Science and Technology

modulus are improved because of the hydrogen bonding between the polymer matrix and nanotubes [40]. Halpin–Tsai equations' modulus prediction and experimental results have excellent agreement. Ramanathan et al. found that by loading an amide-functionalized single-walled nanotube (SWNT) in a poly (methyl methacrylate) PMMA matrix, drastic improvement in the mechanical properties can be achieved [39, 43].

Addition of nanoparticles increases the stiffness and hence tensile modulus of nanocomposite. This can be explained using an example from literature, where addition of nanosilica to elastosil (E) [EX, x is weight %] increased the tensile strength. But with excess of nanosilica, tensile strength decreased [25, 29] (Fig. 9.6). This is due to the agglomeration and increased roughness of the surface of nanocomposites with increase in the amount of nanosilica [39, 44].

(b) **Elastic Modulus**

The elastic modulus is the essence of stiffness of any material and it can be obtained using dynamic mechanical analysis (DMA). DMA results for the addition of nanosilica to elastosil at various wt% and temperature are showed in Fig. 9.7. These results shows that with increase in wt% of silica and decrease in temperature, the storage modulus (E') and loss modulus (E'') increase. The variation in E' and E'' with temperature is due to the increased mobility of the polymer chains [39]. With DMA, the optimum temperature and weight percentage to get the maximum stiffness can be determined.

Table 9.3 gives an overview of relation between E' and E'' with temperature and concentration [45]. Tan δ the dielectric loss tangent, ($\tan \delta = \varepsilon''/\varepsilon'$) is widely used to measure the amount of electrical energy dissipated when a material undergoes electrical conduction, dielectric resonance, etc. This is indicative of increased flexibility or little interaction between the polymer and the filler and is assessed at the glass transition temperature. A decrease in crosslinking density was correlated with

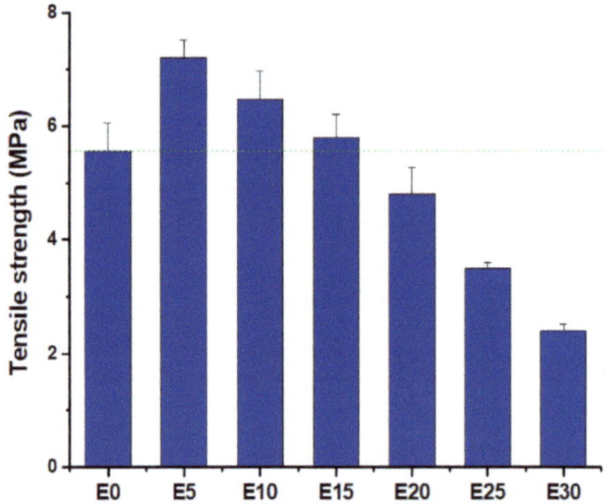

Fig. 9.6 Tensile strength of nanocomposites. Reused with permission from [39] Copyright © 2021 Nanomaterials

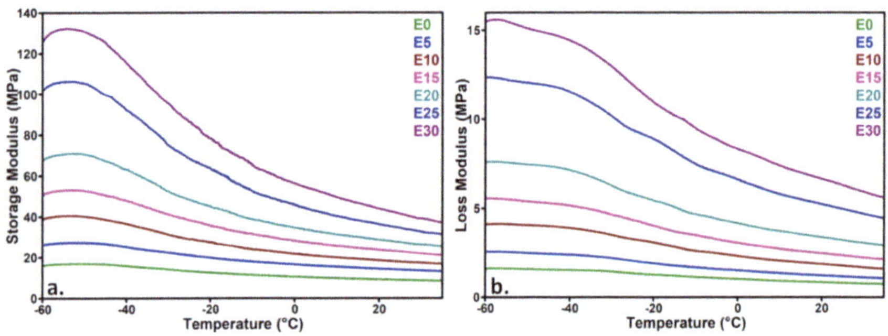

Fig. 9.7 DMA curves of nanocomposites: **a** storage modulus (E') and **b** loss modulus (E'') with change in wt% and temperature. Reused with permission from [39] Copyright © 2021 Nanomaterials

an increase in the tan δ value in polysiloxane nanocomposites. The development of novel dielectric materials for high-frequency, high-voltage, and high-temperature applications depends on improving the glass temperature of polymeric insulation through appropriate nanocomposite engineering.

Inorganic filler particles often boost stiffness while occasionally sacrificing toughness in polymer matrixes [9]. The mechanical strength and elastic modulus of many nanocomposites appear to be significantly improved with large aspect ratio nanofillers. Also, by altering the lengths of the polymeric nanocomposites, the macroscopic mechanical properties can be changed [46].

Table 9.3 Storage modulus (E') and loss modulus (E'') of nanocomposites with various temperatures

Nanocomposites	$\tan\delta_{(-30\,°C)}$	$E'_{-30\,°C}$ (MPa)	$E'_{30\,°C}$ (MPa)	$E'_{-0\,°C}$ (MPa)	$E''_{30\,°C}$ (MPa)
E0	0.101	14.2	8.7	10.5	0.8
E5	0.099	22.3	13.6	16.5	1.1
E10	0.111	31.2	17.4	21.7	1.7
E15	0.112	41.2	22.0	28.2	2.3
E20	0.119	52.5	26.4	34.6	3.1
E25	0.134	75.0	32.7	45.8	4.7
E30	0.135	95.0	39.1	56.3	5.9

Reused with permission [39] Copyright © 2021 Nanomaterials

9.2.3 *Interface Properties*

Polymer composites are reinforced with several inorganic particles of nanometric size. As a size of fillerFiller particle decreases, the total interphase volume gradually increases as the particular surface area increases [47]. This enhances the applications in the field of dielectrics. A multi-core model presented out by Tanaka et al. gives the detailed hypothesis of the process underlying the improved performance due to the addition of inorganic materials. As per the model, the internal interfaces are composed of (a) bonded layer, (b) bound layer, (c) loose coupling layer, and (d) electrical double layer [48–50] (Fig. 9.8). Bonded layer is the thin layer [1] which is tightly bonded to organic and inorganic components with the help of suitable coupling agent. Bound layer is the layer (thickness < 10 nm) that is strongly interacting with the surface of the nanoparticle and/or bounded layer [51]. In this layer, chain mobility and crystallinity are restricted. The glass transition temperature is directly correlated with it. Loose coupling layer is loosely coupled to the first and second layer. Chain mobility, chain conformations, crystallinity, etc., will be differing from undisturbed matrix. The next fourth layer which overlaps all the above three layers called electric double layer. The nanoparticles are charged and polymer contains sufficient mobile charge carriers and it forms a long distance dipole. Some of the polymers like polypropylene (PP), polyethylene (PE), and ethylene vinyl acetate (EVA) will have tendency to gain positively charge. Epoxy, polyamide, and silicone elastomers have a tendency to become negatively charged. This was found out from the triboelectric series (Fig. 9.9) [48, 52]. This mobile charge carriers in polymers effect electrical conductivity and dielectric properties.

The presence of interface will affect the properties of a system which are useful in dielectric applications. Epoxy composites (epoxy resins with layered silicates having large surface area, and large aspect ratio) are common dielectric substances. They have been developed as advanced, high-performing insulating materials.

Models such as, the multi-core model (also known as Wilkes model), and the water shell model are developed to learn more about how nanofillers added to polymers

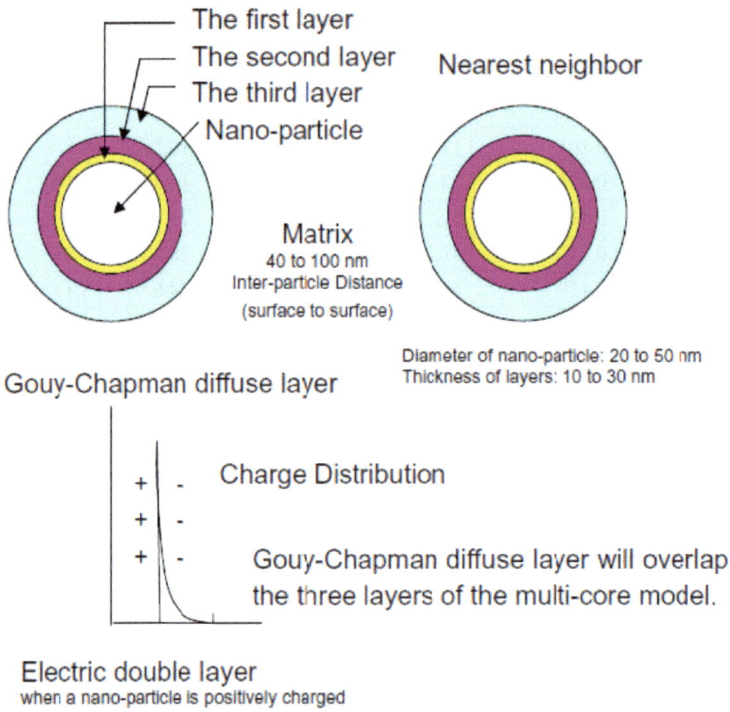

Fig. 9.8 Multicore model for nanocomposite. Reused with permission from [34] Copyright © 2016 Polymers

will interact with one another. Experimental evidence supports the great effectiveness of coupling agents. Also, this statistic emphasizes the significance of "the interaction zone". When silica-based nanofillers are added to the organic polymers, silane coupling occurs. Silane coupling is a type of rather powerful bonding between organic polymers and inorganic materials (mostly, nanofillers). Figure 9.10 illustrates how hydrogen bonds can be used to join two types of immiscible substances [34, 53].

9.2.4 Optical Properties

The optical properties of polymers embedded with nanofillers in a dielectric have long attracted the interest of researchers. As the market for materials for optical applications expands, demand for novel materials with functionality and transparency increases. Addition of inorganic nanoparticle nanocomposites to polymers such as polyetherketone (PEK-c), poly (methylmethacrylate) (PMMA), bisphenol-A polycarbonate (PC), polystyrene (PS), and epoxy resin [54] has considerable promises since they can supply the essential and interesting optical properties. These properties

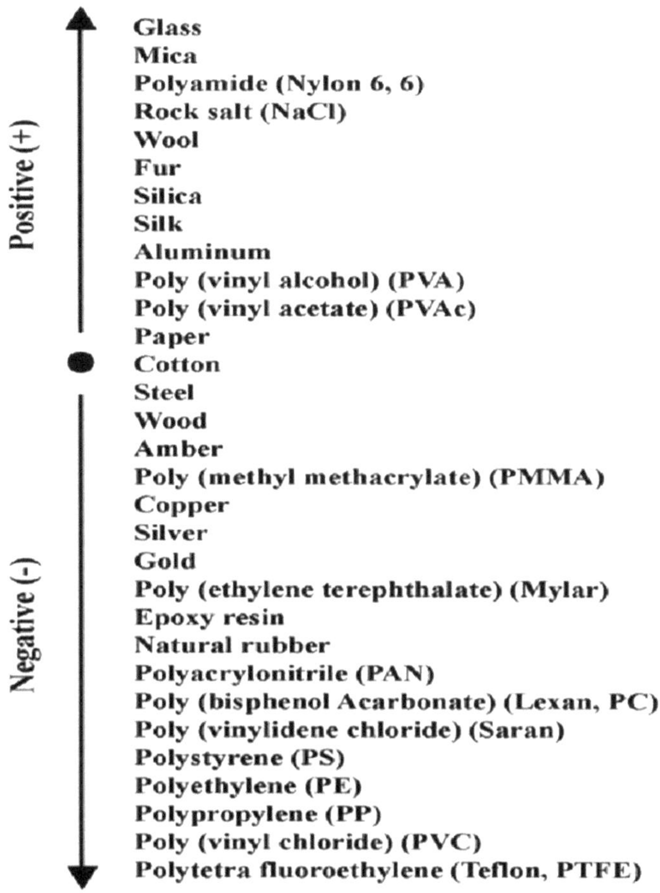

Fig. 9.9 Triboelectricity series of various kinds of materials. Reused with permission from [52] Copyright © 2015, Textile Research Journal

include high thermal stability, low optical loss, and good transparency at particular wavelength, refractive index, stability, and easy processability. Some of them are mentioned in Table 9.4 [55].

Transparency of the polymer nanocomposites relied on the spatial distribution and size of the fillers that are added to the polymer matrix. Sarwar et al. [56] studied about the transparency of pure polyamide and polyamide with different amount of silica particles. The maximum was found for 5 wt% silica and it gradually decreased by increasing the wt % of the filler. Another study by Chen et al. showed that polymer PDMS composites with different amount of nanosilica (MQ silicone resin) shows different transparency. As the loading of the filler increases, transparency decreases and it is found maximum in (83% filler addition) among all others with 64, 59, 56, and 21% addition of MQ silicone resin (Fig. 9.11). This decreasing transparency is

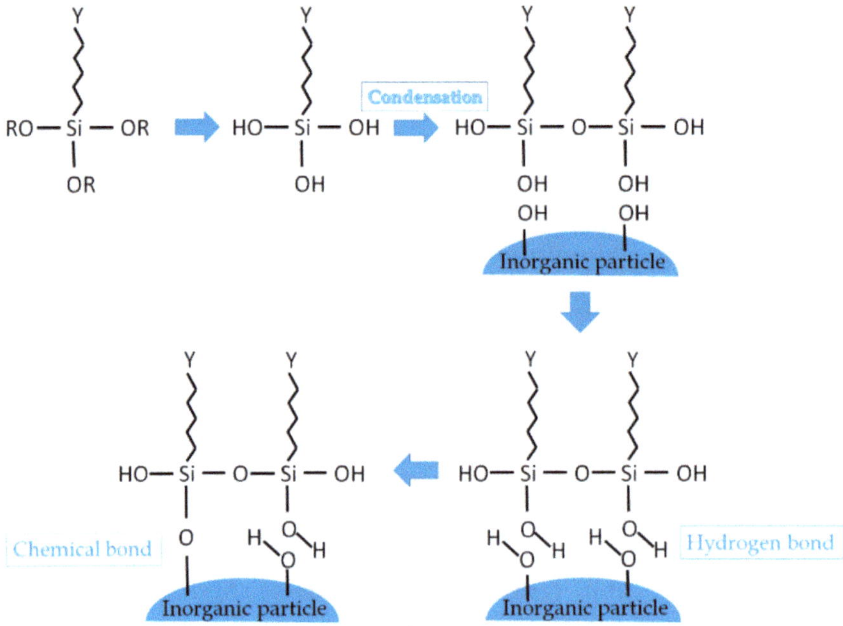

Fig. 9.10 Hydrogen bonding of silane coupling of polymer. Reused with permission from [34] Copyright © 2016, Polymers

Table 9.4 Physical properties of some of the optical polymers.

Materials	Refractive index	Transparency (% at 400–800 nm)	T_g (°C)	Optical loss (dβ/cm)	Dielectric constant
Poly(methylmethacrylate)	1.49	93	105	0.2 (at 850 nm)	2.2–2.3
Polystyrene	1.59	90	100	–	2.7
Polycarbonate	1.58	89	150	–	2.7–3.0
Polyurethane	1.56	–	–	0.8 (at 633 and 1064 nm)	–
Epoxy resin	1.58	–	–	0.3 (at 633 nm) and 0.8 (at 10,640 nm)	3.8
Polyetherketone-c	1.51	90	228	<0.5 (at 633 nm)	3.2

Reused with permission from [55] Copyright © 2010 Woodhead Publishing Limited

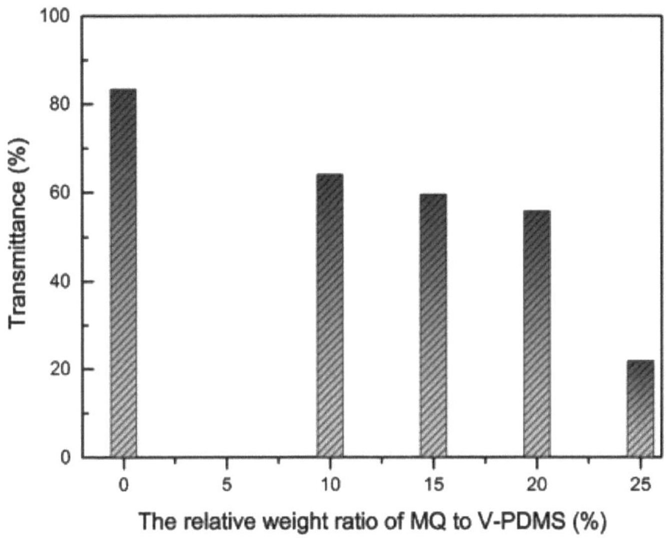

Fig. 9.11 Transmittance of the MQ silicone nanofiller and PDMS matrix. Reused with permission from [57] Copyright © 2015, Elsevier

because of aggregations MQ silicone resin in PDMS matrix. It is seen that light is scattered and leads to opacity, which is verified by SEM and DLS [57].

Chai and coworkers reported that PMMA/$La_{0.45}$ $Ce_{0.45}$ $Tb_{0.1}$ PO_4 showed high transparency for visible range and intense green emission of Tb^{3+} (at 543 nm) on UV irradiation [54]. PMMA/SiO_2 and PMMA/ZrO_2 nanocomposites prepared by Wang et al. showed the transmittance in the visible region persist above 95% and it was increasing with the amount of the filler. As the load increased, the overall light transmittance in polycarbonate (PC)/alumina nanocomposites containing nanoparticles decreased.

9.3 Applications

9.3.1 Space Charge Suppression

High-voltage direct current (HVDC) is the ideal environment for the injection from the electrode and the accumulation of space charge in the dielectrics. The local electric field distribution will be distorted by the space charge, which will also gradually accelerates the degradation of polymer chains. According to several published works, adding inorganic nanoparticles to polymer matrix is a feasible method for reducing the injection and accumulation of space charge [58]. Crosslinked polyethylene (XLPE) was inserted to poly(stearyl methacrylate)-grafted SiO_2 by Zhang et al.

and measurements show that the space charge was effectively suppressed, leading to a limited internal field distortion of 10.6% over a broad range of external DC fields from 30 to 100 kV/mm at room temperature. The charge dynamics in nanodielectrics has been explained using the creation of deep trap when nanoparticles are introduced into the material [58]. According to Zhang et al., nano-sized MgO will improve the trap depth in low-density polyethylene (LDPE) nanodielectrics. Hence, a considerable amount of space charge accumulated near the electrodes. Wang et al. investigated the impact of a coupling agent (vinyl silane) on the electrical characteristics of Al_2O_3/LDPE nanodielectrics and found that the modified Al_2O_3 nanoparticles have a stronger space charge suppressing effect than the unmodified ones. Additionally, Zhou et al. thoroughly examined the space charge characteristic of polypropylene (PP)-based nanodielectrics, which have shown significant potential as a recyclable HVDC cable insulation material [59].

9.3.2 Electric Energy Storage

Polymer-based dielectrics have substantial applications in field-effect transistors (FETs), dielectric-based capacitors, actuators, communication devices, etc., which play a crucial role in electronic device and advanced intelligence equipment's. Low permittivity values (usually $k < 3.5$, except PVDF) are the main disadvantage of the polymers, such as polyethylene, polypropylene, polyethylene terephthalate, poly (dimethylsiloxane), epoxy, and poly (vinylidene fluoride) [60–66]. It is challenging to increase the permittivity (capability of a material to polarize in an alternating current electrical field—according to dielectric theory) of a neat polymer because C-H bonds in polymers can only generate a small amount of polarization. It eventually limits the significant improvement of the practical devices. In order to arouse stronger polarization, high fraction of nano-sized inorganic fillers like, semiconducting oxides, ceramics, giant dielectric oxide particles, carbon, or metals that are mixed with the material.

These dielectrics are widely used as an insulating layer between electrically conducting surfaces when the need for electrical separation of charged materials arises. They have fast charging and discharging capability, hence, dielectric capacitors possessing electrical energy storage have been helping in development of advanced electrical power engineering. However, the main obstacle to their potential industrial uses is they have lower energy storage density on comparison with electrochemical energy storage devices [67, 68]. A material can have a high energy density if it has a number of polarizable domains because these domains will generate a lot of induced dipole moments on application of electric field. Dielectric capacitors helps create large number of such dipoles which induces an electric field opposing the applied electric field in turn, the net electric field will be lowered and it results in conduction pathways for dielectric loss and dielectric breakdown [69].

Energy storage density for linear dielectrics is given by Eq. (9.5) [70].

$$U_{\text{storage}} = \int E\mathrm{d}D = \frac{1}{2}\varepsilon_0\varepsilon_r E^2 \tag{9.5}$$

E-Applied electric field and D-Electric displacement.

For linear dielectric materials such as polyethylene, biaxially oriented polypropylene, and polystyrene, the polarization and depolarization follow linear path. The energy density stored and recovered is same (area ABDA, Fig. 9.12). But practically, the majority of dielectric materials (such as polar polymers, polymers containing impurity ions, and immiscible polymer blends with poor interfaces) react nonlinearly to an applied external electric field. The polarization and depolarization curves generally take different routes resulting in hysteresis loop depicted by region ABCA [70]. This is due to greater coupling interaction in-between the dipole of charging and discharging materials.

The applied electric field and energy storage density are proportional in linear dielectrics (Eq. 9.5). However, for nonlinear dielectrics such as anti-ferroelectrics, ferroelectrics and, relaxor ferroelectrics, the relationship is not proportional due to residual polarization, P_r (the intercept D-E loop on the Y-axis) [53]. The polarization–electric field and polarization axis hysteresis loop should be integrated to get discharge and charge energy densities, Eq. (9.6).

The stored and released energy density (Eq. 9.6), for nonlinear [70, 71].

$$U_{\text{storage}} = \int_{0}^{P_{\max}} E\mathrm{d}P \text{ and } U_{\text{released}} = \int_{P_r}^{P_{\max}} E\mathrm{d}P \tag{9.6}$$

where P_r is the residual polarization and $P_{\max}$ is the maximum polarization.

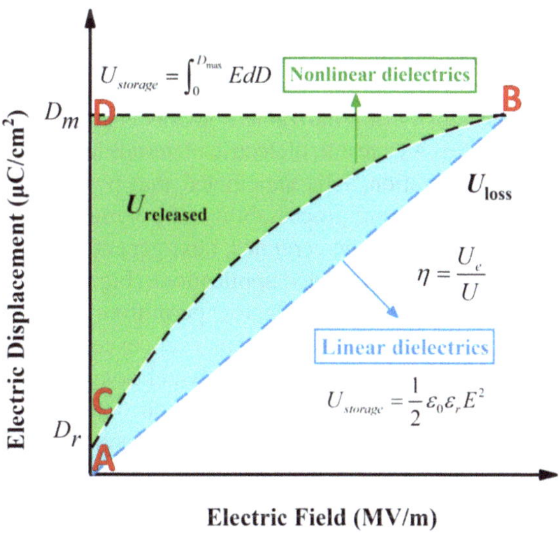

Fig. 9.12 D-E hysterisis loops for energy storage, energy released, and energy loss calculation under high-filed switching. Reused with permission from [70] Copyright © 2022, Polymer

The above equations show that the dielectric can achieve higher energy storage density by increasing the breakdown strength, polarization strength, and dielectric constant of the material. The material should have a low conductivity and low dielectric loss to ensure practical applications under an external electric field.

Polymer-based nanocomposites with high breakdown strength and high permittivity are anticipated with the advancement of distributed power production systems and the high compactness of the electronic circuits due to their adaptability and low cost. However, it is out of the scope of the present chapter to discuss detailed experimental information. Analytical descriptions of the experimental techniques can be found elsewhere (Kremer and Arndt, 1997; Kremer and Schönhals, 2003; Riande and Díaz-Calleja, 2004; Vassilikou-Dova and Kalogeras, 2009) [69, 72, 73].

Polyvinylidene fluoride (PVDF), a common ferroelectric polymer, it has been widely utilized as the host material to produce high energy storage nanodielectrics. As a result of PVDF's high breakdown strength, a terpolymer of vinylidene fluride, chlorotrifluoroethylene (CTFE), and trifluoroethylene (TrFE), known as PVDF-CTFE-TrFE, was created to create poly [(vinylidene fluoride)-r-(chlorotrifluoroethylene)]. These are referred to as single phase polymer dielectrics. These prepared polymers have high energy density but low dielectric permittivity and breakdown strength. This will make it very difficult to attain high energy density along with high breakdown strength, E_b with minimal dielectric loss.

This can be explained using moss rule, Eq. 9.7

$$\varepsilon = \varepsilon_0 \varepsilon_r = 1 + \left(\frac{\hbar \omega p}{E_b} \right)^2 \tag{9.7}$$

Polymer nanocomposites are multiphase materials that provide another method for producing heterogeneous dielectric materials without compromising on the Moss rule. Polymer nanocomposite design consists of nanoscale phase-segregated fillers with high electronic conductivity (yellow) and insulating polymer (green) that prevent inter-domain conduction and percolation. "Nanodipoles" are created when space charge accumulates at phase boundaries with an applied electric field. The number density of polarizable domains is significantly increased due to nanoscale size dipoles. As a result, dielectric materials have energy storage dominated by interfacial polarization. It is anticipated that by combining the high breakdown field strength values and processability of polymers with the high relative permittivity, ε_r values of inorganic ceramics, next-generation polymer dielectrics ca–n meet the demands for power related applications (Fig. 9.13) [74].

Ceramic materials' dielectric permittivity ranges from tens (TiO_2, Al_2O_3) to thousands ($BaTiO_3$), and it can even exceed 10,000 for $CaCu_3Ti_4O_{12}$. Due to which, numerous nanoparticle/polymer composites have been thoroughly studied. Owing to the percolation phenomenon that occurs when volume percent of the fillers reaches the percolation threshold, conductive particle/polymer nanocomposites for high energy density are of significant interest. When utilized as conductive fillers, graphene nanosheets, metal particles, and carbon nanotubes (CNT) show

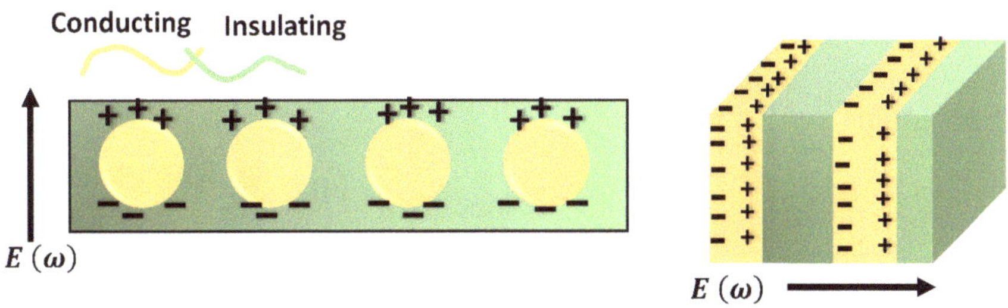

Fig. 9.13 Block copolymers with nano- and micro-phase separation undergo interfacial polarization. Recreated with permission from [74] Copyright © 2022, Polymer

an exponential rise in permittivity near the percolation threshold. Additionally, a hierarchy structure for nanocomposites is built to concurrently increase breakdown strength and permittivity [63]. The incorporation of inorganic nanofillers into polymers showed improved energy-related applications. A strawberry-like core–shell Ag/polydopamine/$BaTiO_3$ nanocomposites with high dielectric performance was created by Yang et al. The Coulomb blockade effect of super-small nano-Ag explained the improvement in the performance of the nanocomposites [75]. Wang and his coworkers fabricated three-layer sandwich-like $BaTiO_3$/PVDF nanocomposites with varying $BaTiO_3$ concentrations layer by layer. Luo et al. created a three-dimensional $BaTiO_3$ network in polymer composites and achieved an enhancement of the dielectric permittivity of polymer composites [76, 77].

9.3.3 Nanodielectrics with Partial Discharge (PD) Resistance

In order to assess the efficiency and durability of nanodielectrics for high-voltage applications such as the wires of randomly wound motors and the stator end windings of rotating machines, partial discharge resistance of an insulating material is a critical component. Partial discharge will erode the insulating materials which causes breakdown and degradation of polymer chain. It can be reduced by adding ceramic nanoparticles as they strengthen the bonds between the nano-sized fillers and polymer matrices [78]. Ceramic nanoparticles operate as a barrier in the direction of high energy charges, shielding those charges from directly attacking the polymer [79].

Surface roughness, PD current, and the erosion depth are the main parameters for the evaluation of PD resistance that can be done using many configurations of electrode system like (a) Rod-to-plane electrode, which allows evaluation of surface roughness and erosion depth that can be helpful for the characterization of nanocomposite materials [this work was done by Krivda et al.] and (b) International Electrotechnical Commission (IEC) electrode gives surface roughness (Fig. 9.14) [53, 77, 79].

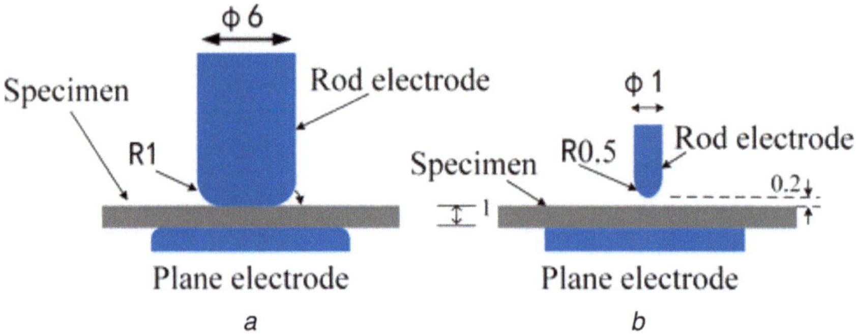

Fig. 9.14 Electrode system for measuring PD resistance of **a** IEC(b) electrode, and **b** Rod-to-plane electrode. Reused with permission from [77] Copyright © 2018 IET Nanodielectrics

It is proved from the results obtained by Krivda et al. that better protection against PD erosion is seen when combination of micro- and nano-sized fillers is added in comparison with the base resin having either solely microfillers or solely nanofillers. Demonstrations have been provided to show that PD resistance notably developed on nanofiller addition. Iizuka et al. analyzed and measured PD resistance of two different forms of epoxy/SiO_2 nanocomposites such as aerosil (prepared by addition of nano SiO_2) and nanopox (prepared by directly curing epoxy and nano SiO_2 mixture). Aerosil had better PD resistance than nanopox among the nanocomposite studied [80]. Tanaka et al. also investigated that addition of fumed silica with XLPE nanocomposites increases PD resistance. Kozako et al. showed that 2 wt% incorporation of exfoliated layered silicate improves the PD resistance of polyamide nanocomposites.

It is difficult to achieve an evenly distributed electric field in a high-voltage setting. There must be some locations where the local electric field strength is greater than the breakdown strength of air domains, such as the endpoint of a high-voltage generator winding and cable insulation, where the local electric field strength is greater than the air domains' 30 kV/mm breakdown strength.

9.3.4 Nonlinear Field Grading

The purpose of using field grading material is to avoid the electrical breakdown (electrical stress) and thermal problems that occurs in high-voltage and medium-voltage applications including generators, motors, cable accessories, and end windings. Field grading applications are used under alternating current (AC) and direct current (DC) conditions.

The field grading materials used in DC-applications have nonlinear I-V (current–voltage) properties which can be explained by this Eq. 9.8.

$$I = I_0\left(\frac{V}{V_0}\right)^{\alpha} \tag{9.8}$$

where I_0 and V_0 are material constants and α is the nonlinearity exponent of the material.

Numerous AC and DC high-voltage applications, such as surface flashover protection of outdoor silicon rubber polymer insulators and corona protection of motor stator bars and cable accessories, are prevented from stress concentration by nonlinear field grading, which homogenizes the electric field distribution [81, 82].

There is a recent set of literature, in which some representative semiconducting nanofillers are added to the polymer matrix to create nonlinear characteristics. Onneby et al. and Donzel et al. investigated the influence of semiconducting nanofillers such as silicon carbide (SiC) and ZnO in polymer to nonlinear field grading characterization, respectively. Other carbon-based materials, such as carbon nanotubes and graphene oxide, are used to examine the field grading properties of nanocomposites in order to develop percolation systems, where percolation concept helps in understanding the non-filed grading properties of nanocomposites with conducting/semiconducting nanoparticles and insulating matrix. It has been found that the percolation threshold of the composite gives the minimum volume percentage required to create a continuous network of filler and the matrix. Percolation threshold of the composite increases with the volume fraction of fillers added which depends on the geometry of the system [83]. Critically, the percolation threshold is affected by a variety of factors, including nanoparticle morphology, filler distribution homogeneity, and the type of conduction between adjacent particles (e.g., tunneling or Schottky barriers), as illustrated in Fig. 9.15 [77].

9.4 Conclusions

Polymers play a major role in most of the technology related applications. However, they have relatively poor mechanical, optical, thermal, and electrical properties compared to metals and ceramics. Nanocomposites with polymer as matrix and nanoparticle as fillers can be used for improved physical and chemical properties. These properties of polymer nanocomposites make them useful for energy-related applications. One such usage of nanocomposites is seen as nanodielectrics, owing to the enhanced dielectric properties compared to conventional polymeric materials.

This chapter describes about the improvements seen in the physical and chemical properties including electrical, mechanical, optical, interfacial, and thermal properties. It briefly discusses how addition of nano-filler to polymers varies the electrical properties in terms of breakdown strength, and permittivity of the polymer. Breakdown strength of nanocomposite using Weibull statistical distribution with respect to shape and size parameters and the substantial improvements seen in the permittivity of the material due to the local accumulation of the nanofiller is explained. Morphological changes and increase in mechanical properties such as stiffness, toughness,

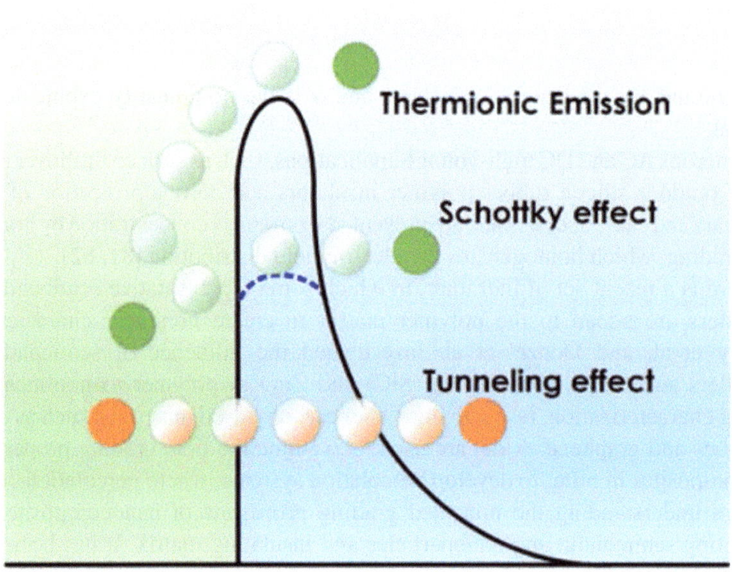

Fig. 9.15 Conduction mechanism of various nanodielectric system. Reused with permission from [77] Copyright © 2018 IET Nanodielectrics

and elastic modulus, and mechanical strength using theories such as Halpin and Tsai, Mori–Tanaka are discussed. The impact of the size of fillers to polymer matrix and their interactions studied by its interface properties using Wilkes multi-core model and their effects on refractive index, transparency, and dielectric constant are also explained.

The nanocomposites with improved physical and chemical properties open the possibilities for nanodielectrics in energy-related applications. Various applications including energy storage, space charge suppression, partial discharge resistance, nonlinear field grading using these composites are discussed in this chapter.

Challenges in the field of inorganic polymer nanodielectrics

Nanodielectrics are a combination of polymer matrix and inorganic nanoparticles that have recently received extensive attention. Polymer matrix has a high breakdown strength and low loss, and inorganic nanoparticles have a high dielectric constant. However, due to non-uniformity in the electric field distribution, the high permittivity contrast between fillers causes these properties to be compromised. Current nanoparticle/polymer interface models have limited applicability and are lacking in fine details, which must be developed in the future research. In the recent developments, applications of nanodielectrics in energy-related applications such as space charge suppression, electrical energy storage, partial discharge resistance, and nonlinear field grading have been demonstrated. In the future, research works are expected to be carried out on biomedical applications and electrical power industries of these materials.

References

1. Jordan J, Jacob KI, Tannenbaum R, et al (2005) Experimental trends in polymer nanocomposites—a review. Mater Sci Eng A 393:1–11
2. Berta M, Lindsay C, Pans G et al (2006) Effect of chemical structure on combustion and thermal behaviour of polyurethane elastomer layered silicate nanocomposites. Polym Degrad Stab 91:1179–1191
3. Shaw ZL, Kuriakose S, Cheeseman S et al (2021) Antipathogenic properties and applications of low-dimensional materials. Nat Commun 12:3897
4. Dantas de Oliveira A, Augusto Gonçalves Beatrice C (2019) Polymer nanocomposites with different types of nanofiller. In: Nanocomposites—recent evolutions. IntechOpen
5. Jeon I-Y, Baek J-B (2010) Nanocomposites derived from polymers and inorganic nanoparticles. Materials 3:3654–3674
6. Murr LE, Martinez E, Hernandez J et al (2012) Microstructures and properties of 17-4 PH stainless steel fabricated by selective laser melting. J Mater Res Technol 1:167–177
7. Nelson JK (ed) (2010) Dielectric polymer nanocomposites. Springer, US, Boston, MA
8. Psarras GC (2010) Conductivity and dielectric characterization of polymer nanocomposites, In: Physical properties and applications of polymer nanocomposites. Elsevier, pp 31–69
9. Fothergill JC (2010) Electrical properties. Dielectric polymer nanocomposites. Springer, US, Boston, MA, pp 197–228
10. Weibull W (1951) A statistical distribution function of wide applicability. J Appl Mech
11. Li Y, Zhou Y, Cheng S et al (2021) Polymer nanocomposites with high energy density utilizing oriented nanosheets and high-dielectric-constant nanoparticles. Materials 14:4780
12. Tuncer E, Sauers I, James DR et al (2007) Enhancement of dielectric strength in nanocomposites. Nanotechnology 18:325704
13. Donnay M, Tzavalas S, Logakis E (2015) Boron nitride filled epoxy with improved thermal conductivity and dielectric breakdown strength. Compos Sci Technol 110:152–158
14. Andritsch T (2010) Epoxy based nanodielectrics for high voltage DC applications: synthesis, dielectric properties and space charge dynamics
15. Vaughan AS (2016) Nanodielectrics: the role of structure in determining electrical properties. Controlling the morphology of polymers. Springer International Publishing, Cham, pp 237–262
16. John VL, Gomathi N, Joseph K et al (2022) Plasma functionalized CNT/Cyanate ester nanocomposites for aerospace structural applications. ChemistrySelect 7:e202201260
17. Han Y, Xue Z, Liu D et al (2021) Research developments in XLPE nanocomposites and their blends: published papers, patents, and production, Presented at the.
18. Smith RC, Nelson JK, Schadler LS (2010) Electrical behavior of particle-filled polymer nanocomposites, In: Physical properties and applications of polymer nanocomposites. Elsevier, pp 70–107
19. Hu Y, Smith R, Nelson J et al (2006) Some mechanistic understanding of the impulse strength of nanocomposites, In: 2006 IEEE conference on electrical insulation and dielectric phenomena. IEEE, pp 31–34
20. Sarathi R, Sahu RK, Kumar PR et al (2006) Understanding the performance of epoxy nano composites-a physico-chemical approach. IEEJ Trans Fundam Mater 126:1112–1120
21. Roy M, Nelson JK, MacCrone RK et al (2007) Candidate mechanisms controlling the electrical characteristics of silica/XLPE nanodielectrics. J Mater Sci 42:3789–3799
22. Murakami Y, Nemoto M, Okuzumi S et al (2008) DC conduction and electrical breakdown of MgO/LDPE nanocomposite. IEEE Trans Dielectr Electr Insul 15:33–39

23. Okuzumi S, Murakami Y, Nagao M et al (2008) DC breakdown strength and conduction current of MgO/LDPE composite influenced by filler size. In: 2008 Annual report conference on electrical insulation and dielectric phenomena. IEEE, pp 722–725
24. Takala M, Ranta H, Nevalainen P et al (2010) Dielectric properties and partial discharge endurance of polypropylene-silica nanocomposite. IEEE Trans Dielectr Electr Insul 17:1259–1267
25. Barber P, Balasubramanian S, Anguchamy Y et al (2009) Polymer composite and nanocomposite dielectric materials for pulse power energy storage. Materials 2:1697–1733
26. Woodward WHH (2021) Broadband dielectric spectroscopy-a practical guide
27. Jonscher AK (1999) Dielectric relaxation in solids. J Phys D Appl Phys 32:R57
28. Zhu T, Yu Q, Zheng W et al (2021) Intrinsic high-k–low-loss dielectric polyimides containing ortho -position aromatic nitrile moieties: reconsideration on Clausius-Mossotti equation. Polym Chem 12:2481–2489
29. Simpkin R (2010) Derivation of Lichtenecker's logarithmic mixture formula from Maxwell's equations. IEEE Trans Microw Theory Tech 58:545–550
30. Tanaka T, Montanari GC, Mulhaupt R (2004) Polymer nanocomposites as dielectrics and electrical insulation-perspectives for processing technologies, material characterization and future applications. IEEE Trans Dielectr Electr Insul 11:763–784
31. Khaled U, Alzahrani S, Khan Y (2016) Dielectric properties improvement of LDPE based on nano fillers. In: 2016 IEEE international conference on high voltage engineering and application (ICHVE) IEEE, pp 1–4
32. Hui S, Chaki TK, Chattopadhyay S (2010) Dielectric properties of EVA/LDPE TPE system: effect of nanosilica and controlled irradiation. Polym Eng Sci 50:730–738
33. Frechette M, Larocque RY, Trudeau M et al (2008) Nanostructured polymer microcomposites: a distinct class of insulating materials. IEEE Trans Dielectr Electr Insul 15:90–105
34. Pleşa I, Noţingher P, Schlögl S et al (2016) Properties of polymer composites used in high-voltage applications. Polymers (Basel) 8:173
35. Griffith G, Tucker S, Milsom J et al (2000) Problems with modern air-cooled generator stator winding insulation. IEEE Electr Insul Mag 16:6–10
36. Irwin P, Zhang W, Cao Y et al (2010) Mechanical and thermal properties. Dielectric polymer nanocomposites. Springer, US, Boston, MA, pp 163–196
37. Affdl JCH, Kardos JL (1976) The Halpin-Tsai equations: a review. Polym Eng Sci 16:344–352
38. Halpin JC, Pagano NJ (1969) The laminate approximation for randomly oriented fibrous composites. J Compos Mater 3:720–724
39. Radu ER, Panaitescu DM, Andrei L et al (2021) Properties of polysiloxane/nanosilica nanodielectrics for wearable electronic devices. Nanomaterials 12:95
40. Liu L, Barber AH, Nuriel S et al (2005) Mechanical properties of functionalized single-walled carbon-nanotube/poly(vinyl alcohol) nanocomposites. Adv Funct Mater 15:975–980
41. Liu H, Brinson LC (2008) Reinforcing efficiency of nanoparticles: a simple comparison for polymer nanocomposites. Compos Sci Technol 68(6):1502–1512. https://doi.org/10.1016/j.compscitech.2007.10.033
42. Podsiadlo P, Kaushik AK, Arruda EM et al (2007) Ultrastrong and stiff layered polymer nanocomposites. Science (1979) 318:80–83
43. Ramanathan T, Liu H, Brinson LC (2005) Functionalized SWNT/polymer nanocomposites for dramatic property improvement. J Polym Sci B Polym Phys 43:2269–2279
44. Liu W, Hoa S, Pugh M (2005) Fracture toughness and water uptake of high-performance epoxy/nanoclay nanocomposites. Compos Sci Technol 65:2364–2373
45. Roy N, Bhowmick AK (2012) Novel in situ silica/polydimethylsiloxane nanocomposites: facile one-pot synthesis and characterization. Rubber Chem Technol 85:92–107
46. Chiulan I, Panaitescu DM, Radu E-R et al (2020) Comprehensive characterization of silica-modified silicon rubbers. J Mech Behav Biomed Mater 101:103427
47. Roy M, Nelson JK, MacCrone RK et al (2005) Polymer nanocomposite dielectrics—the role of the interface. IEEE Trans Dielectr Electr Insul 12:629–643

48. Tanaka T, Kozako M, Fuse N et al (2005) Proposal of a multi-core model for polymer nanocomposite dielectrics. IEEE Trans Dielectr Electr Insul 12:669–681
49. Frechette MF, Trudeau ML, Alamdari HD et al (2004) Introductory remarks on nanodielectrics. IEEE Trans Dielectr Electr Insul 11:808–818
50. Ajayan PM, Schadler LS, Braun PV (2003) Nanocomposite science and technology
51. Lewis TJ (1994) Nanometric dielectrics. IEEE Trans Dielectr Electr Insul 1:812–825
52. Liu S, Hua T, Luo X et al (2015) A novel approach to improving the quality of chitosan blended yarns using static theory. Text Res J 85:1022–1034
53. Tanaka T (2005) Dielectric nanocomposites with insulating properties. IEEE Trans Dielectr Electr Insul 12:914–928
54. Chai R, Lian H, Yang P et al (2009) In situ preparation and luminescent properties of $LaPO_4:Ce^{3+}$, Tb^{3+} nanoparticles and transparent $LaPO_4:Ce^{3+}$, Tb^{3+}/PMMA nanocomposite. J Colloid Interface Sci 336:46–50
55. Li AD, Liu WC (2010) Optical properties of ferroelectric nanocrystal/polymer composites, In: Physical properties and applications of polymer nanocomposites. Elsevier, pp 108–158
56. Sarwar MI, Zulfiqar S, Ahmad Z (2008) Polyamide–silica nanocomposites: mechanical, morphological and thermomechanical investigations. Polym Int 57:292–296
57. Chen D, Chen F, Hu X et al (2015) Thermal stability, mechanical and optical properties of novel addition cured PDMS composites with nano-silica sol and MQ silicone resin. Compos Sci Technol 117:307–314
58. Zhang L, Zhou Y, Tian J et al (2014) Experiment and simulation of space charge suppression in LDPE/MgO nanocomposite under external DC electric field. J Electrostat 72:252–260
59. Zhou Y, Hu J, Dang B et al (2017) Effect of different nanoparticles on tuning electrical properties of polypropylene nanocomposites. IEEE Trans Dielectr Electr Insul 24:1380–1389
60. Carpi F, Bauer S, De RD (2010) Stretching dielectric elastomer performance. Science (1979) 330:1759–1761
61. McKay TG, O'Brien BM, Calius EP et al (2011) Soft generators using dielectric elastomers. Appl Phys Lett 98:142903
62. Molberg M, Crespy D, Rupper P et al (2010) High breakdown field dielectric elastomer actuators using encapsulated polyaniline as high dielectric constant filler. Adv Funct Mater 20:3280–3291
63. Dang Z-M, Zheng M-S, and Zha J-W (2016) 1D/2D carbon nanomaterial-polymer dielectric composites with high permittivity for power energy storage applications. Small 12:1688–1701
64. Guo Y, Liu L, Liu Y et al (2021) Review of dielectric elastomer actuators and their applications in soft robots. Adv Intell Syst 3:2000282
65. Chen Q, Shen Y, Zhang S et al (2015) Polymer-based dielectrics with high energy storage density. Annu Rev Mater Res 45:433–458
66. Zhao Y, Yin LJ, Zhong SL et al (2020) Review of dielectric elastomers for actuators, generators and sensors. Iet Nanodielectrics 3(4):99–106
67. Zha J-W, Zheng M-S, Fan B-H et al (2021) Polymer-based dielectrics with high permittivity for electric energy storage: a review. Nano Energy 89:106438
68. McKay T, O'Brien B, Calius E et al (2010) Self-priming dielectric elastomer generators. Smart Mater Struct 19:055025
69. Rabuffi M, Picci G (2002) Status quo and future prospects for metallized polypropylene energy storage capacitors. IEEE Trans Plasma Sci 30(5):1939–1942
70. Yang Z, Yue D, Yao Y et al (2022) Energy storage application of all-organic polymer dielectrics: a review. Polymers (Basel) 14:1160
71. Zhu L, Wang Q (2012) Novel ferroelectric polymers for high energy density and low loss dielectrics. Macromolecules 45:2937–2954
72. Barshaw EJ, White J, Chait MJ et al (2007) High energy density (HED) biaxially-oriented poly-propylene (BOPP) capacitors for pulse power applications. IEEE Trans Magn 43:223–225
73. Chung TM (2012) Functionalization of polypropylene with high dielectric properties: applications in electric energy storage. Green Sustain Chem 2(02):29

74. Hardy CG, Islam MdS, Gonzalez-Delozier D et al (2012) Oligoaniline-containing supramolecular block copolymer nanodielectric materials. Macromol Rapid Commun 33:791–797
75. Yang K, Huang X, He J et al (2015) Strawberry-like core-shell Ag@Polydopamine@$BaTiO_3$ hybrid nanoparticles for high-*k* polymer nanocomposites with high energy density and low dielectric loss. Adv Mater Interfaces 2:1500361
76. Luo S, Shen Y, Yu S et al (2017) Construction of a 3D-BaTiO 3 network leading to significantly enhanced dielectric permittivity and energy storage density of polymer composites. Energy Environ Sci 10:137–144
77. Zhong S, Dang Z, Zhou W et al (2018) Past and future on nanodielectrics. IET Nanodielectrics 1:41–47
78. Xiao-hong Z, Jun-guo G, Jin-mei Z, et al Characteristics of electrical breakdown and partial discharge of polyethylene/montmorillonite nano-composites
79. Tanaka T, Bulinski A, Castellon J et al (2011) Dielectric properties of XLPE/Sio2 nanocomposites based on CIGRE WG D1.24 cooperative test results. IEEE Trans Dielectr Electr Insul :1482–1517
80. Iizuka T, Uchida K, Tanaka T (2011) Voltage endurance characteristics of epoxy/silica nanocomposites. IEEJ Trans Fundamentals Mater 94:65–73
81. Pradhan M, Greijer H, Eriksson G et al (2016) Functional behaviors of electric field grading composite materials. IEEE Trans Dielectr Electr Insul 23:768–778
82. Onneby C, Martensson E, Gafvert U et al (2001) Electrical properties of field grading materials influenced by the silicon carbide grain size, In: ICSD'01. Proceedings of the 2001 IEEE 7th international conference on solid dielectrics (Cat. No.01CH37117). IEEE, pp 43–45
83. Cherney EA (2005) Silicone rubber dielectrics modified by inorganic fillers for outdoor high voltage insulation applications, pp 1108–1115

Chapter 10
Surface Engineering of Graphene-Based Polymeric Composites for Energy Storage Devices

Debajani Tripathy, Ankita Subhrasmita Gadtya, Bibhuti B. Sahu, and Srikanta Moharana

Abstract Graphene is a promising nanocarbon material with exceptional features such as a large surface area and outstanding electrical and thermal properties. It has the potential to be a new creation of reinforcing material for polymer composites owing to its low mass density, large specific surface area, excellent compatibility, the inexpensive cost to create compared to carbon nanotubes, and attractive flexibility. The several approaches for synthesizing graphene and distributing it in polymer matrices are explored. This chapter focuses on a summary of the surface alternation of graphene with various synthetic techniques and the preparation and properties of graphene-based different polymer nanocomposites. This chapter provides a broad overview of the nanocomposite synthesis, properties, and finally prospective application of polymer-graphene nanocomposites in various energy storage sectors.

Keywords Graphene · Graphene surface alternation · Polymer composites · Energy storage

10.1 Introduction

Nanoscience is a new field that is growing very quickly right now. It can be beneficial to several different sectors, including computing, sensors, biology, and technology, among others. Nanoparticles of different sizes, shapes, and properties have been found, which has led to a lot of interesting progress in this field. Researchers have found that nanoscale fillers in polymer nanocomposite composites make them

D. Tripathy · A. S. Gadtya · S. Moharana (✉)
Department of Chemistry, School of Applied Sciences, Centurion University of Technology and Management, R.Sitapur, Paralakhemundi, Odisha 761211, India
e-mail: srikantanit@gmail.com; srikanta.moharana@cutm.ac.in

B. B. Sahu
Basic Science and Humanities Department, Nalanda Institute of Technology (NIT), Bhubaneswar, Odisha 751024, India

S. Moharana et al. (eds.), *Emerging Nanodielectric Materials for Energy Storage*, Nanostructure Science and Technology, https://doi.org/10.1007/978-3-031-40938-7_10

better at mechanical, electrical, thermal, and optical performances [1–3]. There are various types of multifunctional nanocomposite materials are made with the use of nanofillers like carbon nanotubes, nano-clay, graphene, and metal or ceramic nanoparticles. Recent years have seen extensive research into conductive nanofillers-reinforced polymer composites due to their exceptional multifunctional capabilities when compared to those of traditional conductive polymer composites [4–7]. However, with the incorporation of electrically conductive nanoparticles into non-conducting polymers, electron conduction is obtained [8]. Antistatic materials, electromagnetic interference (EMI) shielding, sensors, and conductors are just a few of the many uses for conducting polymer composites, which are advantageous over intrinsic conducting polymers due to their ease of production, cost-effectiveness, and adaptability in electrical characteristics. Because of the low filler levels possible with these conductive fillers, the composites either keep or improve the electrical and mechanical recital of the matrix. Thus, the composite system with nanomaterial reinforcements ought to have enhanced thermal conductivity [9–11]. The various carbon-based allotropes including carbon nanotubes (CNTs) and graphene-based materials are used to generate superior-performing electrically directing composites [12]. These composites are among the most promising nanofiller materials owing to their outstanding mechanical and thermal capabilities, remarkable chemical inertness, and the ability to have their electrical properties modified [13]. There are different parameters including complex processing, poor surface chemistry control, weak interfacial contact with the polymeric matrix, and agglomeration are some of the CNT's drawbacks [14]. The high cost of producing carbon nanotubes (CNTs) is another issue with their use as nanofillers [15]. Thus, it is difficult to mass-produce useful composite materials based on CNTs. When carbon fibers will not cut it and nanotubes are out of reach due to cost, where does a frugal materials scientist turn for a workable conductive composite? [16]. Graphene is the name of the material in the issue, and it exists exclusively in a two-dimensional form. Graphene-based materials have recently emerged as a viable choice for application in state-of-the-art high-performance nanocomposites [12].

The growing number of research papers on graphene shows that academics and businesses are becoming more interested in this material. Over the past few times, the number of graphene patents has steadily grown. These patents have been filed for a wide range of uses, such as characterization, polymer composites, transparent displays, transistors, capacitors, solar cells, biosensors, conductive inks, windows, saturable absorbers, and photodetectors [17]. The total number of newly found fullerenes climbed substantially over the decade from 2005 to 2014. Yet, graphene is the most recently formed nanocarbon and also shows the greatest rate of growth. Although just a few hundred papers were published on graphene in its first three years, interest in the material grew rapidly. Graphene studies have increased rapidly in popularity. As compared to articles on carbon nanotubes (CNTs), those on graphene were on par in 2013, but in 2014, they substantially exceeded their CNT counterparts. In terms of where nanocarbon research has been published, the United States and China stand out as the most prolific authors, responsible for well over half of all such studies. This is especially true when it comes to the two hottest nanomaterials of the moment, carbon nanotubes (CNTs), and graphene [18]. Significant progress in the investigation of graphene polymer nanocomposites is being driven by the revolutionary improvement in polymer properties, such as mechanical, electrical, and thermal conductivity, and chemical, optical, and gas impermeability at a low-level concentration of filler [19]. There are numerous excellent studies based on unaltered graphene-reinforced polymer nanocomposites available. Moreover, surface functionalization plays an important impact in the enhancement of the electrical and mechanical characteristics of the polymer nanocomposite. This typical chapter focuses on graphene and its properties, as well as its production process. The methods utilized to covalently and non-covalently change surfaces are analyzed in depth and brought up to date in this work. We conclude with a discussion of the varied electronic uses for graphene-based polymer nanocomposites and the consequence that surface functionalization has on the electrical conductivity of these materials.

10.2 Graphene-Based Two-Dimensional Nanostructured Materials

Graphene is a two-dimensional carbon allotrope with a thickness of just one atom. It is composed of a honeycomb arrangement of hexagonal crystalline structure with sp^2 carbon atoms in a conjugated system. Although graphene was theoretically conceived in the 1940s, it lacked the thermodynamic stability required for reliable operation in everyday environments [20–22]. In a 2004 tabletop experiment [21], Geim and colleagues at Manchester University successfully distinguished individual graphene layers. The thinnest known substance has been discovered thanks to this groundbreaking discovery, which has opened up new avenues of investigation in the domains of physics, chemistry, biology, and materials science [22, 23]. Graphene has a low coefficient of thermal expansion (CTE), thermal conductivity, young's modulus (1 TPa), and superior optical transmittance along with other desirable characteristics [20]. This graphene-based material shows more perspective than other nanostructured carbon allotropes, like 0D fullerenes and 1D nanotubes.

10.2.1 Preparation of Graphene-Based Materials

There are two different techniques for synthesizing graphene: either from the top-down (a) or from the bottom-up (b) approach. Top-down methods that aim to lessen the van der Waals contacts between graphene layers include mechanical (Scotch tape), chemical (solution-based exfoliation, graphite oxide exfoliation or reduction), and electrochemical (oxidation or reduction and exfoliation) (as shown in Fig. 10.1) [24]. On the other hand, graphene is made via bottom-up techniques, which entail the assembly of minuscule molecular building blocks into sole or many-layered graphene assemblies through catalytic (e.g., CVD), thermal (e.g., SiC breakdown), or chemical (e.g., organic synthesizing) processes [25]. Graphene is a 2D carbon allotrope, where every single carbon atom connects with every other carbon atom via sp^2. Carbon atoms arrange themselves in honeycomb crystal symmetry, with a bond length of just 0.141 nm. A typical density of graphene is 0.77 g/cm^3. Graphene's surface area per unit mass is thought to be around 2600 m^2/gm^2 [20, 26]. Graphene has amazing electrical, thermal, mechanical, electronic, and optical characteristics, making it the most appealing nanomaterial today. However, it is the reediest and toughest composite ever examined. Tensile strength tests have shown that graphene is comparable to or even exceeds that of carbon nanotubes (CNT), and significantly exceeds that of steel, Kevlar, and additional polymers [27]. To determine the modulus of elasticity of single-layered graphene sheets created by chemically reducing graphene oxide, academic and industrial researchers used the tip of an atomic force microscope to make a small indentation in the core of a modified sheet of graphene. Moreover, an elastic modulus (1TPa) and an inherent strength (130 GPa) have also been calculated for a defect-free monolayer graphene sheet [28, 29]. However, the thermal conductivity of graphene [5000 W/(m^{3}K)] by its electrical conductivity is close to 6000 S/cm [30, 31]. Due to its semi-metallic nature, graphene exhibits several fascinating electrical characteristics. The extraordinary thermodynamic and electrical conductivity of graphene is owing to its exceptionally elevated charge carrier mobility (10 cm^{2}V s^{-1}) [32]. The edge states and any adsorbed or intercalated species have a significant impact on graphene's magnetic properties [33, 34].

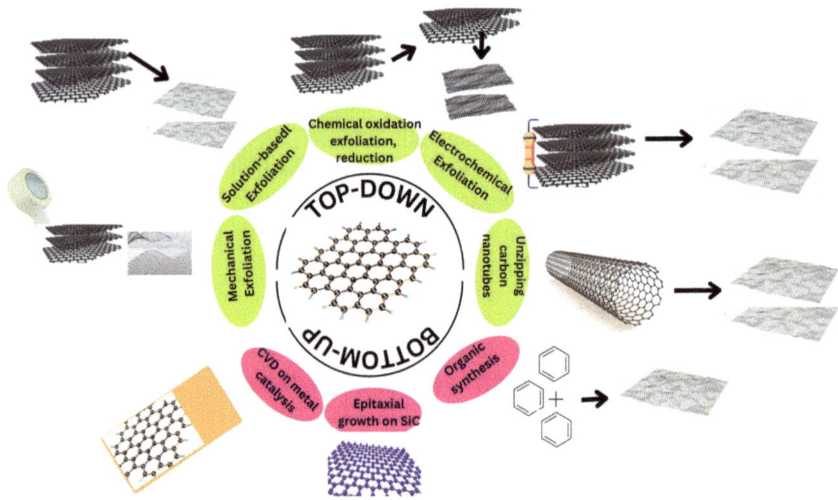

Fig. 10.1 Schematic illustration for preparation of materials based on graphene

10.2.2 Surface Alternation of Graphene

Graphene-based materials tend to stick together more in a polymer matrix, which makes them less friendly to large molecules like polymeric chains [35]. To modify their physiochemical characteristics, graphene and its derivatives are often chemically functionalized. The potential benefits of graphene sheets are immense, but only if the material can be made operational and widely disseminated. Covalent functionalization involves the formation of chemical bonds, while non-covalent functionalization relies only on van der Waals forces.

(a) **Covalent Functionalization**

Covalent superficial alteration of graphene causes hybridization of one or additional sp^2 carbon atoms of the aromatic edifice, leading to forfeiture of electronic conjugation and a strain in the plane. As the hybridization has changed, the carbon atoms (sp^2hybridized) which are ordinarily in a flat plane must now adopt the sp^3 tetrahedral geometry. Specifically, this is because carbon atoms at the edge of a molecule are more likely to undergo covalent addition processes [36]. Graphene's dispersibility in organic solvents and water can be improved by covalent organic functionalization, and its characteristics can be combined with those of extra-functional materials like chromophores or polymers [37]. Functionalization can transpire on the interface, the boundaries, and the defects of materials. The chemical techniques used to create graphene are not without problems, including damage to the carbon lattice, structural errors, and the random adsorption of solvent molecules [38, 39]. There are various classes of covalent functionalization have been reported in recent times, which include free radical adding, atomic radical addition, nucleophilic addition, cycloaddition, and electrophilic substitution.

(i) **Cycloaddition Reactions**

In cycloaddition reactions, bond breaking and bond creating occur simultaneously deprived of the creation of anions or cations as intermediates, resulting in a cyclical motion of electrons. The graphene molecule has undergone cycloadditions [2 + 1], [2 + 2], [3 + 2], and [4 + 2]. The most well-known kind of cycloaddition is the 1,3-dipolar and [3 + 2] cycloaddition reaction [40, 41]. Other cycloaddition reactions including (i) [2 + 1] cycloaddition similar to the Bingel reaction, interactions with nitrenes and leading to the creation of cyclopropane or aziridine adducts [42], (ii) [2 + 2] cycloaddition with an aryne or benzyne [43] and (iii) [4 + 2] cycloaddition are two ways to create a cycle with four carbon atoms. An electron-rich diene and an electron-poor dienophile can undergo a cycloaddition via the Diels–Alder mechanism (electron-deficient species) [44]. Graphene has a wide variety of potential uses owing to its capability of functioning as either the diene or dienophile in the Diels–Alder process. Graphene has the potential to induce unanticipated consequences, and it is notoriously difficult to regulate surface modifications. Nitrides and carbenes, in particular, are capable of reacting with solvents as well as one another [45].

(ii) **Free radical Reactions**

Covalent bonds are formed between the aromatic structure of graphene and free radicals, which are extremely reactive organic molecules. These modifications have been accomplished through thermodynamic or photochemical processes. Though, by using superior-energy reactants, unwanted side reactions or lack of homogeneity might occur in the functionalization reaction [46]. The use of aryl diazonium salts in radical addition reactions is widespread. Peroxide addition, Bergman cyclization [47], and Kolbe electro-synthesis [48] are a few such processes. The reactivity of aryl diazonium salts is the subject of this subsection. Tour and his co-workers [49] suggested first successfully putting aryl diazonium salts into graphene. In this step, nitrogen removal, the aryl diazonium ion transforms into an aryl radical. It is hypothesized that the radical aryl moiety contributes an electron to the sp^2-hybridized graphene [50]. This process, however, can generate organic radicals that are not only capable of covalently reacting with graphene but also of self-polymerizing [51]. The amount of graphene sheets has a significant impact on this reaction. Moreover, the bi or multilayer graphene, the reactivity of a solitary graphene slip is found to be 10 times higher, as reported by Strano and his co-workers [52].

(iii) **Nucleophilic Reactions**

Graphene is the material of choice when it comes to making a nucleophilic addition that accepts electrons. The chemical reaction between graphene and poly-9,9′-dihexyfluorene carbazole Nitrogen anions are created on carbazole when it is subjected to the action of a base, which results in the formation of the anionic moiety. Once the two substances react with one another, a covalent connection is created between the nitrogen anion and the surface of the graphene [53, 54].

(iv) **Reaction with atomic Radii**

By using hydrogen, fluorine, and oxygen from the periodic table instead of organic free radicals, the number of unwanted byproducts is kept to a minimum. As a result, the functionalized graphene becomes even as well as consistent. Hydrogenation makes it facile to procedure the secondary C-H bonding because it changes the shape of the lattice when the first hydrogen atom is added. Graphene in its fully hydrogenated form is known as graphene. Graphene can be hydrogenated when subjected to hydrogen-based plasmas [54]. When you fluorinate graphene, you make a single bond between fluorine and carbon, which is similar to hydrogenation but much stronger. There is a lot of functionalization after fluorination, much like there is after hydrogenation [55]. It has been hypothesized that fluorographene will act as an insulator. Typically, one of three techniques is used: The three methods for exfoliating bulk graphite fluoride are: (i) by exposing it to XeF_2 [56], (ii) by etching it with a fluorinated reagent [57], and (iii) by using a liquid phase.

(v) **Electrophilic Reactions**

Graphene's abundance of free electrons makes it amenable to this class of processes. The Friedel–Crafts acylation [58] and the hydrogen–lithium exchange are two types of substitution processes. Metal-functionalized aromatics are more reactive than their simpler counterparts. Before the hydrogen–lithium exchange, BuLi deprotonates and carbometallizes graphene. An electrophile then interacts with the lithium graphene derivative. The lithium graphene derivative undergoes an instantaneous reaction upon electrophile addition, leading to the conception of covalent bonding [59].

(b) **Noncovalent functionalization**

Non-covalent interactions, such as hydrophobic, van der Waals, and electrostatic interactions, can cause molecules to stick to the surface of graphene [60]. This method of functionalization is noteworthy because it allows molecules to be immobilized straightforwardly and reversibly without disrupting the electrical network [61]. Functionalized non-covalent is essential for the immobilization of proteins, enzymes, medicines, and DNA, particularly in the background of devices, because even a small change in the electrical properties of the system could result in an entire alteration in the structure and characteristics of the system [62, 63]. There has been extensive research into complexes based on graphene, including interactions between non-polar gases and H^+, cations, and anions. Three parts are responsible for its attractive strength: electrostatic, dispersive, and inductive interactions. To make sodium cholate, cetyltrimethylammonium bromide, polyvinylpyrrolidone, triphenylene, and pyrene derivatives are only some of the surfactants that may be employed in an exfoliation approach to creating functionalized graphene from graphite without the use of covalent bonds [64]. On the other hand, along with the single-layer variety, a sizeable amount of few-layered graphene or scattered graphite is likewise produced. Coronene carboxylate produces layers around 100 nm thick [65, 66]. For instance, Bai and his co-workers have developed larger graphene flakes that can be manufactured using perylene derivatives [67, 68].

10.3 Synthesis Process of Graphene-Based Polymer Composites

Graphene-based materials have high strength and high electrical and thermal conductivity making them a potential contender for usage in an extensive range of commercial products, like polymer composites, conductive coatings, fuel cell batteries, and ultra-capacitors. Chemical vapor deposition (CVD), exfoliation, liquid-phase exfoliation, etc., are all examples of top-down ways to make graphene, which can be further divided into bottom-up ways. Owing to their outstanding material characteristics, including yield métier, robustness, electrical and thermal conductivity, and optical properties, polymer nanocomposites (PNC) research has advanced greatly over the last decade, and their applications have grown substantially as a result [69–73]. Graphene possesses excellent dimensional stability, high thermal stability, high gas impermeability, elevated strength and elastic modulus, and high electrical and thermal conductivity [72]. Adding graphene to polymers at a tiny volume percentage may dramatically enhance their characteristics. Furthermore, graphene may be employed at a lower volume percentage since it has a better surface-to-volume ratio than carbon nanotubes [73]. It may enhance a wide variety of polymer matrix characteristics. As an example, graphene-based nanocomposites are in high demand for use in photovoltaic devices likc solar cells because of their low resistivity and high carrier mobility, and because (i) used in lithium-ion batteries (LIBs) such as power density, energy density, and speed of charging in hydrogen fuel cells, (ii) they are used in thermoelectric materials, (iii) used in photovoltaic devices including solar cells owing to their low resistivity and high carrier mobility, and (iv) used as an electrode to raise the electro-catalytic performances [71].

One definition of nanocomposite is a material, which is having any one dimension in the nanoscale. It has made strides in the areas of light, electricity, and magnetism and has desirable features like thermal stability and mechanical strength. Polymer, metal, or ceramic matrices are used to create nanocomposites, which are then filled with nanoparticles (graphene, nanotubes, and clays) [74]. The material's mechanical, thermal, and electrical characteristics are all improved by adding fillers. Polymeric nanocomposites are the most adaptable of all nanomaterials, finding use in several fields such as energy, electronics, health care, and more [71, 72]. Micro-composites, intercalated nanocomposites, and exfoliated nanocomposites are the three main categories of polymer composites based on the distribution of nanosized layers. In the micro-composites' structure, the graphene sheets are disseminated as particles inside the polymer matrix, yet the graphene platelets themselves are unharmed [75]. When distinct polymer chains are inserted between graphene layers, this creates an introduced structure. The exfoliated amalgams' graphene films are evenly distributed throughout the polymer backbone. Since exfoliation maximizes the surface area of contact between the polymer and the filler, it is the preferred morphology for nanocomposites. This leads to improved bonding and superior mechanical characteristics [76]. Graphene nanocomposites' characteristics are affected by a number of factors, including the filler's chemical compatibility with the matrix, the filler's volume fraction, and processing conditions, including dispersion and exfoliation. For optimal performance, it is necessary to use suitable manufacturing techniques. Additionally, the recital eminence of nanomaterials is closely correlated with the rate of dispersity. In-situ polymerization, melt intercalation, and exfoliation adsorption are the three main ways to create polymer nanocomposites [75–77].

Heat or radiation can be used to start polymerization in place by mixing adapted graphene through a monomer or pre-polymer to make a steady chemical bonding and then adding the opposite initiator. It improves stress transfer by making sure that the filler particles are evenly spread throughout the polymer milieu and by swelling the contact between the filler and polymer in the finished composite [46]. These are the primary benefits of in-situ polymerization: (a) It is easier to control the filler's distribution and interaction with the polymer's surface. (b) High-filled composites may be cast-off as a master batch and melt-mixed with other polymers to generate comparable composites, and (c) they might be utilized to manufacture high-fill and ultra-superior relative molecular mass nanocomposites. As it does not involve using lots of organic solvents, it is better for the planet than solution intercalation [46, 78]. However, graphene-based polymer nanocomposites may have their mechanical, electrical, and thermal characteristics vastly improved by in-situ polymerization. By reacting with the polymer, the functionalized graphene forms a chemical connection at the interface that is both robust and long-lasting. The surface of graphene oxide is abundant in oxygen-containing functional groups, so polymers are typically selected whose monomers or pre-polymers can form stable chemical bonds with these oxygen-containing functional groups, like polyimide, polyurethane, epoxy resin, polyaniline, and polyvinyl alcohol. Polyaniline (PANI) and graphene oxide (GO) films remained made via in-situ polymerization by Zhu and his co-workers [79]. With impressive water dispersibility qualities and a beneficial synergistic impact on anti-corrosion performance, the produced PANI-GO nanocomposite material impressed. In addition, the coating layer's ability to prevent water penetration increases, which may serve as a kind of electro-active defense. Zhu et al. [80] have cast-off an in-situ polymerization method to manufacture a GO-polypyrrole (PPy) nanocomposite flick. Waterborne epoxy (WEP) coatings were shown to be more corrosion-resistant after being coated with this layer. Passivation by conductive PPy and the impermeable property of graphene oxide sheets are both present in the GO-PPy nanocomposites that were made. As the first step of in-situ inter-calative polymerization, graphene or altered graphene is stretched in the liquid monomer. Typically, polymerization instigates with the addition of an appropriate originator, followed by the application of heat or radiation [81, 82]. The various composites including polystyrene (PS)–graphene [83], polymethylmethacrylate (PMMA)-based expanded graphite (EG) [84], polystyrene sulfonate (PSS), polyimide (PI), and polyethylene terephthalate (PET)-based layer double hydroxide (LDH) [85–87] have been successfully reported by using this technique.

In this chapter, we have reported many strategies used in the literature to produce graphene and characterize its properties. Camphor, graphitic oxide, nanodiamonds, and silicon carbide are only a few of the precursors that have been inspected aimed at their possible usage in the production of graphene. Graphene aqueous dispersals may be manufactured using these methods. A new route for producing graphene using the hydrothermal method, cryo-milling, and lyophilization has been described. The exfoliation of graphene sheets into individual layers has been achieved by a variety of methods, including heat exfoliation and ultrasonication. In addition, the numerous characterization techniques used to verify the filler's nanostructure have been discussed. As a result of Geim and his co-worker's (2004) successful isolation of graphene, many methods for its production have been uncovered [88]. Scotch tape was proposed as a viable option for physically exfoliating graphite to directly create graphene. A piece of graphite was taped together using this method. Repeated folding and peeling of the tape produced progressively thinner layers of graphite. Finally, a single sheet of carbon called graphene is created by this process. CVD, chemical exfoliation, carbon nanotubes (slicing), and the direct sonication of carbon sheets are only a few of the methods shown in the many research studies to produce graphene. Tkalya et al. [89] created dispersions (aqueous) of graphene by oxidation and exfoliation of graphite, trailed by lessening in the occurrence of a surfactant. Graphite oxide was manufactured using a method similar to that developed by Hummer and used for many years. The Sonic Vibracell VC750, which has a horn with a cylindrical tip, was used to exfoliate the graphene oxide (GO). The reduction procedure, which used hydrazine, took 72 h, and was conducted at a temperature of 120 °C. GO platelets were characterized via atomic force microscopy (AFM). The majority of the analyzed platelets had thicknesses of less than 1 nm or two or three atomic layers. The results confirmed the efficacy of sonic exfoliation and another case of graphene production using a more conventional hydrothermal method. Hummer's method successfully produced graphite oxide, an intermediate in the manufacture of graphene. Graphite oxide went through a series of chemical reactions to become GO. Sonicating an uneven mixture of GO and water, then reducing it with hydrazine hydrate, produced reduced GO. There are two techniques were employed to investigate the structure of graphene: XRD and FTIR spectroscopy. Reducing GO led to the formation of graphene (GR) sheets, as revealed by the presence of characteristic FTIR signals indicating the formation of the sp^2 carbon structure [90]. Pure graphite oxide was synthesized from natural flake graphite powder using Hummer's method. Boiling tetra-ethylene glycol (TEG) was used to dissolve the resultant graphite oxide.

Microwave-assisted solvothermal (MW-ST) reduction was utilized at 300°C to lessen the yellowish-brown colloidal suspension. Indicating that graphite oxide was successfully reduced to graphene nanosheets (GNS) and the solution generated during the reduction process had a dark black hue [91]. X-ray photoelectron spectroscopy (XPS) measurements showed that GO had a higher oxygen concentration than graphene did, at 30.2% against 8.3%. These findings agreed with those obtained using the hydrazine reduction technique [92, 93]. To add insult to injury, the process was conducted in an alkaline medium at a relatively low temperature of around 180 °C. The process caused a quick change in color from yellow to brown then black which indicates that the −ve charged oxygen functional groups of graphite oxide were swiftly deoxygenated. In order to make graphite oxide, a variant of Hummer's method was utilized. Ultrasonic exfoliation was used to create GO. To produce aqueous graphene dispersion, water, hydrazine, and ammonia solutions were mixed chemically together. SEM and TEM pictures were used to examine the micrograph of GR sheets. They exhibited that 2D structures can be made thermodynamically stable by bending, and they did it by scrolling and entwining GR sheets. The success of the successive exfoliation and reduction of graphite to create graphene is shown here. According to the XRD form, the inter-planar spacing of graphene (0.36 nm) is greater than that of graphite (0.34 nm) [94]. It has been shown that the random orientation (turbostratic) of the GR sheets caused by exfoliation is related to the size of the inter-planar spacing and the resulting widening trend [95]. The results of X-ray diffraction (XRD) [96] on graphene samples made this way are the same as those made by chemical reduction. GNS was made by applying heat exfoliation and chemical reduction to graphite oxide made using the Staudenmaier process [97]. McAllister et al. [98] have previously shown that a temperature of roughly 1050 °C is enough for thermal exfoliation and an in-situ decrease in GO during this method. The XRD peaks of GO, pure graphite, and graphene were compared to prove that graphene was synthesized. Ripples in the graphene layer plane were perceived in the TEM image, a signature of the material due to its thermal variability [99]. GO was synthesized from graphite by a variation of Hummer's method. Using hydrazine hydrate as a reductant, graphene was manufactured out of GO [100]. To create the exfoliated GO nanosheets, graphene oxide was either subjected to a solvent exfoliation or a heat exfoliation method. The hydrophilicity of graphite oxide makes solvent exfoliation possible, since water molecules may be introduced between the stacked layers. Mechanical exfoliation, accomplished by ultrasonication or constant stirring, was used to fabricate GO sheets. The resulting graphene oxide (GO) nanosheets were reduced using reducing chemicals like hydrazine hydrate to create reduced graphene oxide (RGO) or graphene [101]. For instance, Patole and his co-workers [102] have suggested the process of creating thermally expanded graphene (EG) from expandable graphite, as seen in Fig. 10.2. The procedure was supported in a chemical vapor deposition apparatus. The size, composition, and structure of EG and GR have been investigated using X-ray diffraction, SEM, and TEM.

In the SEM micrographs, there are several graphene layers were seen in the EG. After being subjected to greater temperatures, the EG samples were shown to have a reduced number of GR layers. Raman spectroscopy findings reveal that graphene has both G and 2D bands. Most graphene 2D bands have a prominent peak at roughly 2670 cm^{-1} [103]. When graphene sheet thickness declines, so does the intensity of the D band. Raman spectroscopy's G, D, and 2D bands provide striking visual representations of the graphene structure's sp^2 carbon atoms. We adapted Hummer's method, using EG as a starting point. Lyophilization and freeze-drying were used to create GNS. Many researchers have employed lyophilization procedures to create

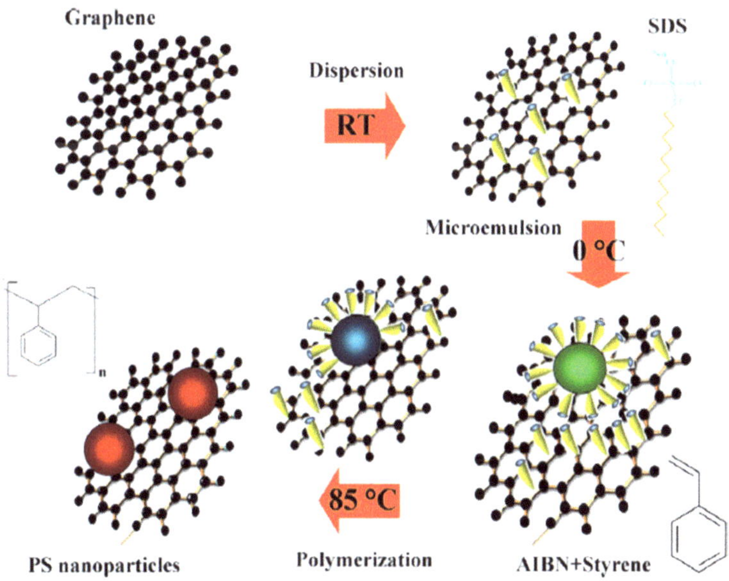

Fig. 10.2 Graphene sheets are gradually modified using PS nanoparticles [102]. Adapted with permission from Ref. [102] Copyright (2010) (Elsevier)

cellulose nano-whiskers. These nano-whisker powders may be easily redistributed in organic environments due to their low van der Waals interactions and loose packing [104, 105]. Using the same method, GNS powders were produced and found to be easily dispersed in solvents for further functionalization. It was found that the resulting GNS granules in this situation were airy and light [106–109]. Core-level-shifts X-ray photoelectron spectroscopy analysis confirmed that graphite oxide and GNS had been decreased. The GNS XRD pattern did not exhibit any peaks, in contrast to graphite oxide's pattern, which included a significant peak at 12 °C. With the use of AFM measurements, it was found that the GNS had fully exfoliated [110]. Chaturvedi et al. [111] used a powerful oxidizing chemical to treat graphite powder before ultrasonic processing in a volatile solvent. Thermal shockwave therapy was used to lessen according to the FTIR spectral task of GR, which exhibited a peak at 1628 cm^{-1}, consistent with the unoxidized graphite realms, GR was not reduced to zero after being subjected to thermal shock treatment. Analogous findings were also reported by Che et al. [112]. Mechanical exfoliation of graphite on an SU-8 exterior yielded graphene with tailor-made atomic layer thicknesses and orientations. Raman spectroscopy was used to pinpoint the monolayer in graphene [113]. Graphite oxide was created by the response of graphite with H_2SO_4, HNO_3, and potassium chlorate. For this procedure, we used a quartz tube with a rubber stopper and a sealed end. After a quick flush by argon gas via the elastic plug, the tube was proximately heated to roughly 1050 °C in a conduit incinerator. To make GO, graphite was imperiled to an adapted Staudenmaier procedure. The resultant graphite oxide was exfoliated into GO nanosheets using ultrasonics. The resulting nanosheets were then treated with ammonia and glucose in the appropriate proportions, therefore lowering GO. Ultraviolet–visible spectroscopy (UV–vis) was measured to confirm the lowering of GO with glucose. The disappearance of the C 1/4 O peak indicates a decrease in GO. Once GO was reduced, the conjugated C–C bond could once again exist, and this change influenced a redistribution of the 14-C bond's peak. According to Liu et al. [114], RGO was synthesized from GO.

In this case, hydrazine monohydrate was included in the last stages of preparation for the RGO-polyaniline (PANI) nanomaterial. GO production has been improved by revising Hummer's methodology [115]. With this method, $NaNO_3$ was removed, while $KMnO_4$ concentration was raised, a change inspired by Hummer's method. This reaction, which produced graphite oxide, was agreed out using a solution of sulfuric acid and hydrogen phosphate. The graphite oxide produced by this process was more oxidized than that produced by Hummer's technique. Thermal reduction at 200 °C for 30 min was used to transform the GO into RGO. The width of the RGO sheets was found to be 1.5 nm using AFM. Additionally, 2D thin sheets of RGO could be seen in SEM images [116]. Zeng et al. [117, 118] reported the synthesis of graphite oxide using a variation of Hummer's technique. Resin-grafted graphite oxide (RGO) was made by further sonicating graphite oxide and then reducing it with hydrazine monohydrate.

There was a significant difference in the BET surface area between camphor, graphite, and diamond samples. It was found that the samples could take in up to 3% by weight of hydrogen. Graphene with a specific surface area of 156 m^2 g^{-1} was found to absorb just 0.4% of the hydrogen; it was subjected to in a separate study. In a simple test, the amount that a commercial graphite intercalation compound grew when it was heated was measured. The EG intercalation compound was mixed with ultrasonic waves in tetrahydrofuran (THF) for 60 min to help it spread out. Sonication resulted in the formation of GnPs with a thickness of 3.57 and 0.50 nm, respectively. Researchers have seen a potential trend toward using GnPs instead of CNTs for making polymer nanocomposite materials [119]. Table 10.1 summarizes the methods used to create graphene-based polymer nanocomposites as well as their benefits and limitations.

Table 10.1 Graphene-polymer nanocomposites: various preparative methods

Different techniques	Techniques used for the preparation of graphene-polymer nanocomposites	Benefits	Limitations
Solution mixing technique	Mechanical mixing, magnetic agitation, or large energy sonication is used to combine the graphene-solvent mixture with the polymer solution, and in the end, evaporating the solvent yields a composite	Dispersed graphene oxide composites	Eliminating solvents is a major problem
In-situ polymerization technique	By reacting monomers or pre-polymers with graphene and graphene-based polymeric composites can be made	Consistently distributed composites with enhanced characteristics	The viscosity of polymerized materials tends to increase during the process, making them difficult to manipulate and reducing their load fraction. Evaporating away the solvent is a challenge
Melt blending	Graphene and polymer mixing facilitated by elevated temperature	Sustainable, cheap, and amenable to industrial-scale production	The large shear stress may diminish the graphene aspect ratios, which would prohibit the composites from achieving their ideal minimum percolation threshold and good electrical conductivity

10.4 Graphene-Filled Different Polymer Composites

Several studies on polymer nanocomposites using various nanofillers have been published. Nonetheless, additional study into graphene-based nanocomposites for high-performance polymers is required. Graphene's usefulness as a nanofiller is demonstrated through the use of a wide variety of polymeric systems. They include epoxy, polystyrene (PS), polyaniline (PANI), polyurethane (PU), polyvinylidene fluoride (PVDF), Nafion, polycarbonate (PC), polyethylene terephthalate (PET), etc. Several academic and industrial researchers have been developing novel polymer-graphene nanocomposites for a variety of applications and may benefit from the subsequent discussion and details on graphene-based polymer nanomaterials.

10.4.1 Epoxy-Graphene Nanocomposites

Epoxy composites containing graphene oxide sheets have their thermal expansion measured using a thermo-mechanical analyzer. The heat conductivity of the epoxy resin was further reduced by the addition of graphene sheets. The embellishments of only 1% GO improved the heat conductivity of epoxy resins as much as filling them with 1% SWNT. The heat conductivity of an epoxy resin encumbered with GO at a 5 wt% was 1 W/mK, which is four-fold greater in contrast with the conductivity of raw epoxy resin. The values stated in the literature are strengthened by these findings. However, 20 wt% of GO, according to the research, might boost thermal conductivity to 6.44 W/mK. According to our findings, the graphene composite shows promise as a heat-conducting thermal interface material. Below the glass transition temperature, thermal extension related to experiments of graphene composites showed a similar influence of the SWNTs on the bulk CTEs (T_g). As compared to normal epoxy resin, which has a CTE of around 8.2 105 C1, epoxy composites containing 5 wt% graphite exhibit a decrease in CTE below T_g of 31.7%. The epoxy/graphene compounds have been fabricated on-site, and EMI shielding studies were conducted on them [120]. The critical phenomena of the percolation threshold in epoxy-graphene composites are described by the DC conductivity,

$$m = h(c)\, t\ (1), \tag{10.1}$$

where the parameters are interchangeable good agreement is found between the conductivity of polymer-graphene compounds and the percolation behavior perceived by Eq. (10.1). The data-fitting procedure yielded a percolation threshold of $c = 0.52$ vol%. Other two-dimensional fillers and an isocyanate-modified PS-based GO nanocomposite also achieved a low percolation threshold [121]. This occurred because the graphene-based sheets were evenly distributed inside the epoxy matrix and had a high aspect ratio. When a consistent graphene-based sheet network was constructed conducting in nature in the sequestering epoxy matrix, the

EMI-defending efficacy improved by enhancing graphene loading over the whole frequency assortment. However, the epoxy-based graphene compounds with a 15% graphene load attained the desired value (20 dB) for an EMI-defending composite. The results showed that polymer-graphene composite material could be used in manufacturing as light, operative composites to block electromagnetic radiation. Epoxy-graphene composites have been made using in-situ polymerization, and their ability to block EMI is evaluated [120, 122]. Graphene loading improved EMI shielding performance across the board. Thus, epoxy-graphene composite materials might be utilized as efficient, lightweight materials for protecting electronic equipment from electromagnetic radiation. We utilized a thermo-mechanical analyzer to determine the thermal expansion of epoxy composites reinforced with graphene oxide sheets [123]. The graphene slips were able to greatly upsurge the heat conductivity of the epoxy resin, which was previously quite weak. An epoxy resin filled with 5% GO composites has four times the heat conductivity of an unfilled epoxy resin [124]. Heat may be dissipated via the use of graphene composites and other thermal interface materials.

10.4.2 Polystyrene (PS)–Graphene Nanocomposites

The fabrication of polystyrene (PS)-based-isocyanate-modified graphene composites in N, N dimethylformamide (DMF) as solvent through a solution mixing approach The composites were reduced with dimethyl hydrazine for 24 h at 80°C. To precipitate the composites, a DMF solution was added very slowly to a very huge capacity of violently stirring methanol (10:1 regarding the bulk of DMF cast-off). It seemed that almost all of the composites were filled with graphene sheets, despite only having a filler loading of 2.4% vol (found in SEM images). The electrical conductivity percolation threshold in PS with 0.1 vol.% GO was achieved. As graphene is so uniformly distributed and has such an enormously superior aspect ratio, it exhibits a percolation, i.e., three times less compared to that seen for any other two-dimensional filler [125]. Thin films loaded at around 0.15 vol.% with the composites met the antistatic criterion for conductivity (106 S m^{-1}), which is rather high. The value increased significantly between 0.4 and 1% of the load. The material's electric conduction varied from 0.1 to 1 Sm^{-1} at 2.5 vol. PS/GNPIL materials are developed in a manner analogous to the development of PS/graphene adapted with isocyanate. Compression-molded thin sheets were used to test the composite sample for electrical conductivity and thermal stability (2 mm).

Pure PS has a conductivity of around 1014 S/m when measured using a four-probe system. Adding 0.38 vol.% GNPIL significantly raised the electrical conductivity of the PS matrix to 5.77 S m1. In this article, the thermal stability of pure polystyrene is contrasted with that of the PS/GNPC8P composite material. The 2nd-step deprivation temperature of the PS-GNPC8P composite was almost 50 °C higher compared to pure PS. These point to a robust connection among the polymer matrix and GNPC8P at the boundary, which might slow down the movement of polymer shackles in the proximity of contact which upsurge the nanocomposites' resistance to heat. Eda et al. [126] used a solution mixing method to produce PS-FGS thin films composite exhibit semiconducting behavior. The thin film displayed an ambipolar field effect. The composite thin films' 1–24 Sm^{-1} electrical conductivity was in close pact with the known standards for bulk composites. During further cooling, the conduction of the reedy coatings increased somewhat once the temperature had fallen to about 50 K. Thin films of reduced graphene oxide [128] and graphitic flakes [127] exhibited the same peculiar behavior. Below 50 K, electron–phonon scattering may limit carrier mobility. Nevertheless, the composite material shifts from n-type to p-type as the temperature increases. Each FGS and sheet-to-sheet connection has a flaw that makes it hard for carriers to move freely. The performance gap between organic and inorganic devices is expected to close as material parameters and reduction conditions are optimized. Hu and his co-workers [129] recently developed in-situ emulsion polymerization techniques used in the production of PS-graphene nanosheet (GNS) nanocomposites. The creation of PS-GNS nanocomposites is shown in simplified form in the diagram below. It was discovered that polystyrene microspheres ranging in size from 90 to 150 nm were embedded in the graphene superficially, predominantly in the crevices between the slanted nanofilms. This demonstrates the compatibility between PS microspheres and GNS, which allows for nanosized dispersion to be produced without further surface treatment. When pure GNS (0.37 nm) was put next to PS/GNS nanocomposites (0.41 nm), the distance between the layers grew. To reduce the strength of the van der Waals forces and increase the inter-planar distance between the slanted nanosheets, PS microspheres may be attached to the edges of the nanosheets. Raman spectra of pure GNS nanocomposites and PS/GNS hybrids are shown. It was determined that pure GNS has two distinct properties, the D band at 1331 cm^{-1} and the G band at 1594 cm^{-1}. The peak intensity ratio, I(D)/I(G), for the PS-GNS nanocomposites, was 1.156. As a result, it is possible that the chemical grafting of polymers to the GNS surface is facilitated by the presence of localized areas of sp^3 blemishes within the sp^2 network of carbon [130]. Nanocomposites have 100 °C higher thermal stability than pure PS. Increased T_g may be traced back to a strong interaction between the polymer matrix PS and GNS, as seen below. At a GNS content of 2.0 wt%, the electrical conductivity of the PS/GNS nanocomposites was found to be 2.9 102 S m1. It was, however, mentioned that pure PS had a conductivity of around 1.0 1010 S m1 [131]. It was recently reported that a nanostructured poly(styrene–isoprene-styrene) block copolymer had PS domains containing functionalized graphene sheets (FGS) (SIS). Analysis of AFM images reveals that poly(styrene-b-isoprene-b-styrene) is made up of well-ordered, spherical PS block domains that are perpendicular to the free surface (SIS). Cylindrical PS domains have been found dispersed perpendicular and corresponding to the free exterior in SIS and FGS composites. Mean diameters of 20 nm (and their long-range order) were found in both the clean matrix and the cylindrical domains of the nanocomposites. T_g for the PS block increased from 78 °C for pure SIS to 84 degrees Celsius for the SIS/FGS composite, whereas the T_g for the PI block stayed relatively constant at around 58 degrees Celsius for both materials. T_g increases because chain mobility in PS domains is impeded. The researchers claim that by incorporating 2D particles into BC, BC is endowed with robust electrical properties that are essential for optoelectronic uses. The solution blending method using DMF as the solvent was used to produce modified graphene composites, including PS and isocyanate [132]. The compounds were coagulated by gradually adding DMF to a large volume of CH_3OH. These composites seemed to be nearly filled using a graphene sheet, although they contained just 2.4% filler by volume. The conductivity of a composite material may be understood as a function of the volumetric proportion of filler loading. By using a similar method, Liu et al. [133] were able to produce a PS-ionic liquid-functionalized graphene composite that displayed enhanced electrical conductivity compared to unmodified PS. Functionalized graphene sheet (FGS) materials have been developed by solution amalgamation [126]. Thin layers of semiconducting composites exhibited the interfacial field effect. The thin coatings' conductivity decreased at temperatures as high as 50 K and then slightly increased with subsequent cooling. These results were also seen in thin films of diminished GO and graphitic flakes [128, 134]. For their nanocomposites, Hu and his research team [135] used in-situ emulsion polymerization to make graphene nanosheets (GNS) materials. Including PS microspheres with GNS may lead to non-sized dispersion without the need for any further surface preparation. The nanocomposite was 100 °C more stable at high temperatures than pristine PS. The electric conduction process of the PS-GNS nanomaterial is greatly more sophisticated than that of pristine PS.

10.4.3 Polyaniline (PANI)-Graphene Nanocomposites

To synthesize PANI-based graphene composite paper, aniline was electro-polymerized in-situ on graphene paper (GPCP) [136]. A three-electrode anodic electro-polymerization cell was used to conduct the polymerization. Graphite paper was used as the active electrode, while a Pt plate was used as the counter electrode and a standard calomel electrode (SCE) as the reference electrode. The electrolyte was composed of aniline and sulphuric acid (at a concentration of 0.5 m) (0.05 M). During a long length of time, PANI was electro-polymerized in situ on graphene paper using a constant potential of 0.75 V against the SCE (60, 300, and 900 s). However, the GPCP's morphology was studied with electron energy loss spectroscopy (EELS). The presence of homogeneous PANI coatings on a single graphene sheet was demonstrated by the identification of nitrogen, carbon, and oxygen that were uniformly distributed throughout the whole surface of the PANI-graphene sheet. In contrast to the regular arrangement of PANI on each 2D graphene, the 3D structure of GPCP is inhomogeneous. The results from cyclic voltammetry (CV) on graphene paper (GP) and graphene-enhanced carbon nanotubes (GPCP). It is because of the transition between quinone and hydroquinone classes, GP has a single redox peak. It is observed that the two groups of redox peaks were seen in the GPCP, indicating the presence of pseudo-capacitive PANI in the composites [136].

The redox characteristics of GP and GPCP are examined in a Nyquist plot (Fig. 10.3a). The higher resistance is owing to the increased concentration of GPCP, and this may be owing to the build-up of poorer conducting PANI. On the other hand, GPCP demonstrated much higher cycle stability than GP (Fig. 10.3b). An excellent composite of PANI-graphene electrodes was developed using a spin coating method [136, 137]. However, a pure aqueous dispersion of GO sheets was deposited via a deep coating method on a quartz substrate and then thermally reduced to create a graphene layer. The graphene sheets were then spin-coated with a dark blue PANI solution in NMP. According to morphological tests, the PANI-graphene electrodes had far smoother surfaces than that of the indium tin oxide (ITO) or PANI-ITO electrodes. The current density of the CV at the PANI-ITO electrodes reduced dramatically after being subjected to 100 cycles at 20 mV s^{-1} in 1 mol L^{-1} H_2SO_4, and the potential change among the oxidation and reduction increased from 87 to 106 mV. When applied to a PANI-graphene electrode, the same process showed negligible impact on the material's overall qualities. So, when making electrochromic devices, the PANI-graphene electrode is better than an ITO electrode. After a few cycles of potential switching, the ITO electrode's performance significantly decreased, despite its initial high optical contrast and short switching time. However, when the voltage was altered, the electrochromic devices using graphene electrodes showed only a slight decrease in performance. The PANI-GNS-CNT composite was fabricated using in-situ polymerization [136, 138]. Electrodes were made by dissolving electroactive materials like carbon black and polytetrafluoroethylene (PTFE) in ethanol. The resulting slurry was spread with a spatula on a nickel substrate before being baked at 100 °C for 12 h under a vacuum. The specific capacitance of the synthesized PANI composites and pure PANI is presented in Fig. 10.4 as a function of scan rates. The PANI-GNS-CNT composites greatly outperformed pure PANI and PANI-CNT composites in terms of specific capacitance. An increased specific capacitance was attributed to a synergistic effect between GNS and PANI [136, 139].

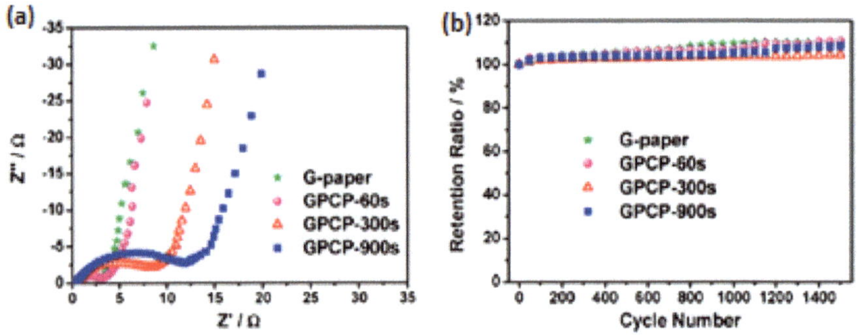

Fig. 10.3 Comparison of cyclic voltammograms captured in 1 M H_2SO_4 at rates ranging from 2 to 20 mV/s (**a**). The Nyquist plots of the G-paper and GPCP (60 s, 300 s, and 900 s) (**b**), with stability at 50 mV/s cycling [136]. Adapted with permission from Ref. [136] Copyright (2010) (Elsevier)

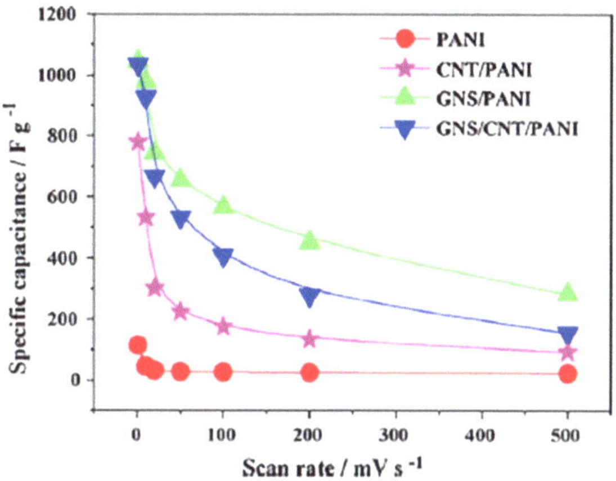

Fig. 10.4 Capacitance data at varying scan rates for polyaniline, carbon nanotube polyaniline, graphene nanosheets polyaniline, and graphene nanosheet carbon nanotube polyaniline. The GNS-PANI electrode material was found to be superior since it had the largest specific capacitance [136]. Adapted with permission from Ref. [136] Copyright (2010) (Elsevier)

10.4.4 Nafion-Graphene Nanocomposites

Graphene and Nafion were combined in a solution to create an electrode modified with tris(2,2-bipyridyl) ruthenium (II) ($Ru(bpy)_3{}^{2+}$) [140]. The electrode was soaked for 30 min in a 1 M $Ru(bpy)_3{}^{2+}$ solution to produce a $Ru(bpy)^{3+}$ modified electrode. The cyclic voltammetry (CV) of the modified Nafion-graphene electrode suggests that graphene's conductivity improves in electron transport. Electrochemical catalysis of TPA oxidation by the $Ru(bpy)_3{}^{2+}$)-Nafion-graphene composite film is demonstrated by an increase in the anodic peak current upon addition of TPA. With the modified electrode, oxalate ions were detected with high sensitivity, selectivity, and stability. Graphene-Nafion-tris(2,21-bipyridyl)ruthenium(II) ($Ru(bpy)_3]^{2+}$) Graphene and Nafion were combined in a solvent to create specialized electrodes [141, 142]. The synthesized electrode was subsequently exposed to 1 M $[Ru(bpy)_3]^{2+}$ solution. The responsiveness, selectivity, and stability of the modified electrode significantly increased.

10.4.5 Poly(Vinyl Alcohol) (PVA)-Graphene Nanocomposites

Liang and his co-workers [143] have reported that the poly(vinyl alcohol) (PVA)-based graphene nanocomposites comprise GO as fillers, PVA as a polymer matrix, and water as the processing solvent. The synthesized GO and PVA-based GO nanocomposites were analyzed by X-ray diffraction studies (as shown in Fig. 10.5). It was found that the peak at 20.9° that crystallized matched the GO and suggested that the GO had exfoliated into individual sheets.

It is reported that the PVA-based graphene nanocomposites performed better mechanically than pure PVA. However, the tensile strength was observed to increase by 76% (from 49.9 to 87.6 MPa) and Young's modulus by 62% (from 2.13 to

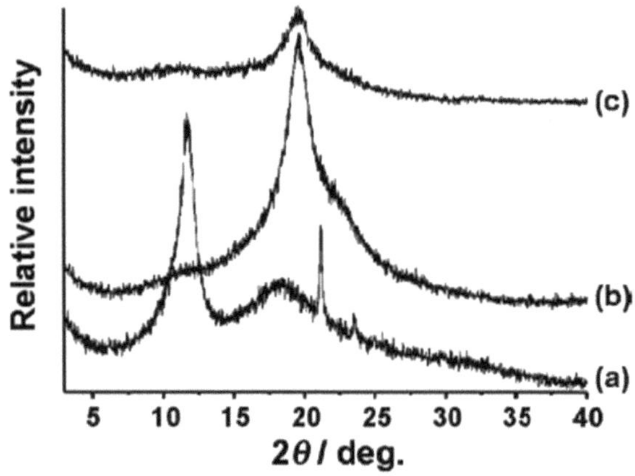

Fig. 10.5 X-ray diffraction (XRD) patterns of **a** graphene oxide (GO), **b** graphene-polyvinyl alcohol (PVA) nanocomposite with 0.5 wt% of GO, and **c** neat PVA [143]. Adapted with permission from Ref. [143] Copyright (2009) (Willey)

3.45 GPa), when the GO loading was increased from 0% to 0.7 wt% (0.41 vol.%). There are various factors for the molecular-level dispersion of graphene sheets in the PVA matrix and their strong interfacial adhesion owing to their H-bonding with the PVA matrix. The incorporation of 0.7 wt% of GO into PVA-graphene nanocomposites corresponded with a rise in glass transition temperature (T_g) from 37.5 to 40.8 °C. This raise in T_g may be due to the H-bonding of graphene and PVA matrix. It has been found that DSC can be used to detect the hydrogen bonding of graphene and PVA. These nanocomposites have better crystallographic and thermal stability than neat PVA. Further, these graphene nanosheets and extensively exfoliated PVA-based nanocomposites were made using a facial aqueous solution [149]. When graphene nanosheets were mixed into the PVA matrix, the resulting composites showed increased strength. A tensile strength of 42 MPa was demonstrated by the composite containing 1.8% graphene by volume, an increase of 150% over the corresponding pure PVA sample. The addition of more graphene relatively significantly raised the tensile strength from 42 to 43 MPa. When graphene loading was increased, the composites' elongation at break reduced dramatically and unexpectedly. The presence of 1.8 vol. % graphene reduced the elongation at break from 220% in the neat sample to 98% in the composite. The study found that graphene-based nanocomposites have a high aspect ratio and well contact with the polymer matrix slowed down the chain's mobility.

10.4.6 Polyurethane (PU)-Graphene Nanocomposites

Nanocomposites made from FGS and water-based polyurethane were produced using an in-situ approach (WPU). The morphological analysis of the nanocomposite showed that the FGS particles were spread out evenly over the WPU matrix. These nanocomposites have superior electrical conductivity owing to their uniform distribution of FGS particles throughout the WPU matrix. However, the percolation threshold was reached at a loading of only 2% FGS. Once an electrically conducting channel formed in the polymer matrix, the electrical conductivity of the polymer changed. Lee and his co-workers have made FGS-WPU nanocomposites that have a higher conductivity [144]. A DSC study found that adding FGS to nanocomposites raises both the melting temperature and the heat of fusion (H_m) of the soft segment of WPU. In contrast, hard segment crystallinity decreased with increasing FGS loading in the nanocomposites.

Besides, three different nanocomposites were developed by Liang et al. through the use of solution mixtures. They used graphene that had been treated with isocyanate, sulfonate, or reduction as the nanofiller and thermoplastic polyurethane (TPU) as the matrix polymer [145]. According to TGA testing, the thermal breakdown rate of TPU nanocomposites made from reduced and sulfonated graphene was much lower than that of isocyanate-modified graphene. Graphene sheets that have been sulfonated have fewer functional groups attached than graphene that has been changed with isocyanate. The infrared-triggered actuation performance of TPU-graphene nanocomposites comprising 1 wt.% sulfonated graphene was fascinating and durable. When exposed to infrared light, this infrared-responsive nanocomposite decreased and rapidly raised a 21.6 g weight of 3.1 cm (0.21 N). There are several cycling tests yielded a maximum energy density of 0.40 Jg^{-1}. Furthermore, the isocyanate-modified graphene-TPU nanocomposites exhibited a poor shape recovery rate and IR-triggered actuation (Fig. 10.6), whereas TPU-sulfonated graphene nanocomposites had improved electrical and mechanical performances. Nanocomposites made with 1% TPU-sulfonated graphene had their tensile strength (100% strain) rise by 75%, and their Young's modulus goes up by 120%. This improvement in mechanical characteristics may be due to the uniform dispersion of graphene in the polymer matrix.

10.4.7 Poly(Vinylidene Fluoride) (PVDF)-Graphene Nanocomposites

Graphene oxide and expanded graphite (EG) were used in a solution processing and compression molding process to create a PVDF based on functionalized graphene sheets (FGS) [146]. The structure and thermal stability of the composites were analyzed through X-ray diffraction and dynamic scanning calorimetric technique. However, it is observed that modified composites have superior mechanical performance than those of pristine polymer and unmodified ones. It is noticed that at 25 °C, pure PVDF had a storage modulus of 1275 MPa, but adding 2 wt% FGS or EG raised it to 1859 or 1739 MPa. Further, with the incorporation of 4 wt% filler into the PVDF matrix, the storage modulus rises to 2460 and 2695 MPa, respectively. The DMA results demonstrated that the nanofiller's reinforcing impact contributed to the

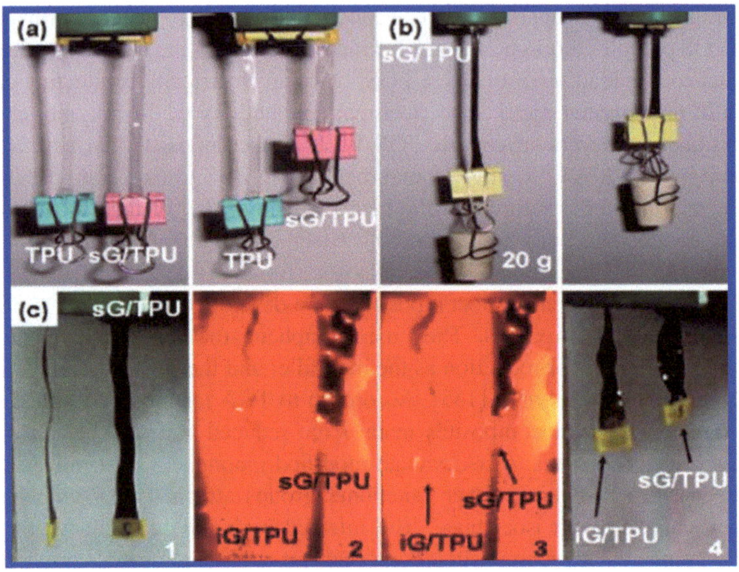

Fig. 10.6 Infrared optical pictures of nanocomposites using graphene as the active material [145]. Adapted with permission from Ref. [145] Copyright (2009) (American Chemical Society)

nanocomposites' improved Tg. At a loading of 5% EG in PVDF-FGS composites and 2% FGS in PVDF-EG composites, the percolation of electrical conductivity was shown to be very significant, owing to their larger aspect ratio, FGS has a lower percolation threshold and a better-conducting network than EG. The influence of temperature on the electrical conductivity of the PVDF-EG nanocomposites was also studied at temperatures ranging from 20 to 170 °C. This nanocomposite resistivity increased steadily with increasing temperature, increasing rapidly after the polymer melt. As the filler particles are so much farther apart at melting, they offer considerably more resistance. It was thought that nanocomposite materials could be used as positive temperature coefficient (PTC) materials because the resistivity changes by three orders of magnitude as the amount of EG changes.

10.4.8 Poly(3,4-Ethyldioxythiophene)-Graphene Nanocomposites

Using in-situ polymerization, a composite of poly (3,4-ethyl dioxythiophene) (PEDOT) and sulfonated graphene was developed [147]. This novel hybrid material has excellent transparency, electrical conductivity, outstanding flexibility, and high thermal stability, and it can be easily treated in both aqueous and organic solvents. Optical transmittances at 550 nm were found to be 96%, 76%, 51%, and

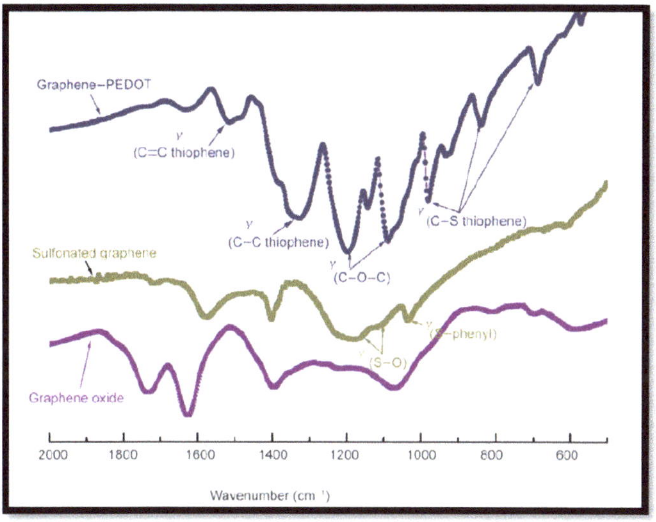

Fig. 10.7 Graphene oxide, sulfonated graphene, and graphene-PEDOT FTIR spectra [147]. Adapted with permission from Ref. [147] Copyright (2009) (Springer Nature)

36% for PEDOT-graphene films of thicknesses of 33, 58, 76, and 103 nm, respectively (Fig. 10.7). The conductivity of the composite films produced on quartz and PMMA substrates was 7 and 10.8 Sm^{-1}, respectively, regardless of the film thickness. PEDOT-graphene composites are favorable owing to their superior thermal stability. Moreover, the temperature of 297 and 325 °C, the composite material exhibits mass loss, and below 325 °C, mass loss reached roughly 19%. The PEDOT-graphene composite also showed improved thermal stability over PEDOT-PSS composites.

10.4.9 Polyethylene Terephthalate and Polycarbonate-Based Graphene Nanocomposites

Melt compounding was used to create graphene nanocomposites based on polyethylene terephthalate (PET) [148]. The morphological analysis of the nanocomposites revealed that the graphene network is composed of several thin stacks of a few sheets of monolayer graphene. The wrinkly, overlapping graphene sheets that come out of this process may have a high electrical conductivity because they connect the individual graphene sheets well and move a lot of current [149]. Melt compounding was used to make composites of polycarbonate (PC) and functionalized graphene sheets (FGS) or graphite. The structural, morphological, and rheological study of the PC-FGS composites revealed that the FGS layers were significantly exfoliated and had viscoelastic properties. After being annealed for 10,000 s, the PC-FGS composites show stiffness percolation at 1wt% to 1.5t5 of FGS loading into the PC matrix. In contrast, a graphite loading of 3–5 wt% was sufficient to achieve this percolation. These composites with a weight percentage of (0.5 wt %) of FGS contents have interesting reversibility among solid and liquid performance, and this is impacted by prior processing. However, the electrical properties of the resultant composites with lower FGS loadings might still result in electrical conductivity percolation, in comparison with utilizing graphite filler. It has been found that PC-FGS nanocomposites have a greater tensile modulus than pure PC. Because of the FGS loading, the CTEof the composites was also greatly diminished. Graphite- and FGS-reinforced polycarbonate films were subjected to N2 and He penetration tests at 35 °C. Both enhancers have the potential to make nitrogen and helium in the air less harmful. However, FGS composites had significantly lower permeability than graphite composites. Gas molecules diffuse slower across membranes because a uniform distribution of FGS with a large aspect ratio may produce complex routes, lowering the cross-sectional area available for penetration [150].

10.5 Application of Polymer-Graphene Composites for Energy Storage Devices

In recent times, one of the most promising methods of energy storage is the super capacitor since it has a high power density, is quick to charge and discharge, and has a long cycle life. The electrodes in super capacitors would be made from a 3D graphene-based conductive structure owing to its elevated specific charge, low density, and superior surface area. The optimum electrode materials for pseudocapacitors are typically mixtures of conducting polymers (PANI and PPy) and graphene that have a high theoretical capacitance owing to their unique redox characteristics [151]. However, the conducting polymer to 3D graphene for supercapacitors stabilizes the polymer chain in the network, increases electrolyte ions' surface accessibility, minimizes their diffusion distance, and improves the composite performance [152]. The 3D graphene-based polymer nanocomposites can be created by cross-linking organic molecules with 3D graphene using a self-assembled method followed by in-situ polymerization. Liu et al. [153] also made a graphene supercapacitor that can withstand a lot of pressure. They also came up with a way to make 3D PPy-graphene (PPy-G) foam in place by cross-linking organic molecules. The good porosity, conductivity, and mechanical durability of PPy-G foam electrodes kept their CV curves and capacitive even after 1000 recorded cycles with a 50% strain. Moreover, the compact PPy sheets and pure 3D graphene have specific capacitances that are substantially lower than the discharge slope prediction of 350 F/g^{-1}. Figure 10.8 shows different application analyses of graphene, from conductive ink to chemical sensors to LEDs to composites to energy to touch screens and high-frequency electronics.

Kulkarni et al. [154] used an in-situ polymerization technique to fabricate PANI nanofibers with a diameter of 20–100 nm using a 3D graphene template. With a current density of 4 mA/cm^2, it was thought that the battery would keep 86.5%

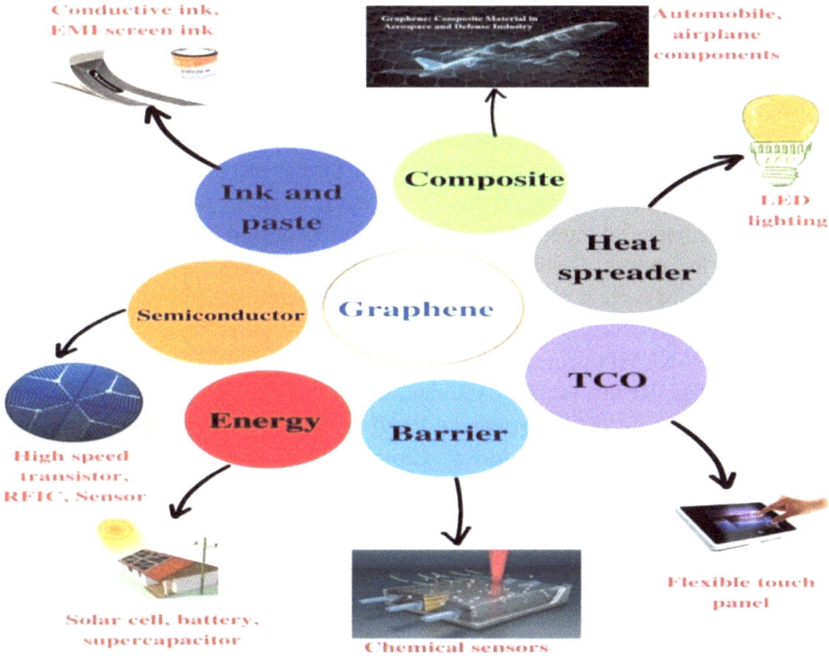

Fig. 10.8 Use of graphene-based materials in a variety of disciplines

of its original capacity after 5000 cycles of charging and discharging. The active PANI nanofibers have huge specific surface area and the graphene backbone's efficient conducting routes enhanced cycle stability and rate capability, resulting in a surprising synergistic efficiency for better super capacitive performance. When scanned from 150 to 800 mV at 10 mV s^{-1}, composite electrodes had a maximum specific capacitance of 1024 F g^{-1} in 1 M H_2SO_4. Another study that used sonication or reduction to make a 3D graphene-poly(anthraquinonyl sulfide) (PAQS) composite was conducted by Zhang et al. [155]. This composite could be used as a bendable cathode in LIBs and $NaIO_2$ batteries (SIBs) if it was put under mechanical pressure. As a result of the stacking relationship between the graphene sheets, the polymer particles (100–200 nm) were evenly dispersed across the graphene surface and enclosed within the graphene sheets. The improved composite has very good cycling stability (71.4% capacity retention after 1000 cycles at 0.5 °C) and excellent rate performance (102 mAh g^{-1} for LIBs at 20 °C and 72 mAh g^{-1} for SIBs at 5 °C). To make it easier for active polymers to react electrochemically, PAQS particles were spread out evenly and encased in 3D graphene with a porous structure that was built up in layers [156]. This allowed for reliable ion transport channels and a viable electron transport path. The microbial fuel cell (MFC) used as a bio-oxidation method is another way to turn chemical energy from organic substrates like sugar, fatty acids, and proteins into electricity [157, 158]. However, the high porosity in the

3D graphene structures allows the bacteria and a growth medium access to the inside of the electrode. Yang et al. [159] have reported that the hydrophilic conducting polymer PANI is on 3D graphene by polymerizing it. This made it easier for bacteria to stick to the surface and form biofilms [160]. Bacteria colonized the PANI-coated graphene surface and established an extensive biofilm. The middle PANI polymers acted as conductive nanowires to increase extracellular electron transport between the electrode and the microorganisms by engaging directly with the redox-active proteins on the bacterial membrane [161, 162]. It has been reported that after 150 h (Fig. 10.9a), the power density of the graphene-PANI MFC as still significantly higher than that of carbon cloth. Furthermore, the charge-transfer resistance of the graphene-PANI MFC was only around 100 (as shown in Fig. 10.9a). Moreover, the 3D graphene-PANI MFC composites reached a maximum power density of about 768 mW/m^2 (Fig. 10.9b). Figure 10.9c demonstrates that the high bacterial loading and efficient EET gave the 3D graphene-PANI anode a naturally high output power density (extracellular electron transfer). Chitosan was used as a backbone for 3D-GPNCs along with conducting polymers because it is biocompatible, can absorb, and is good for the environment. Chen et al. [162] made a 3D chitosan-vacuum-stripped graphene scaffold using a self-assembly method. The hierarchically porous structure with layered, branched macropores, mesopores, and micropores increased bacterial colonization and electron transfer from redox-active compounds. The highest power density in MFC came from a 50% graphene scaffold at 1530 mWm^{-2}, and this is 78 times more powerful than a carbon cloth anode.

Xia and his co-workers [163] have made a composite material by decorating functionalized graphene sheets with Fe_2O_3 quantum dots (QDs, 2 nm) (FGS) through a facile technique and used it as a supercapacitor electrode (as shown in Fig. 10.10). The specific capacitance of these Fe_2O_3 QDs-FGS composites in 1 M Na_2SO_4

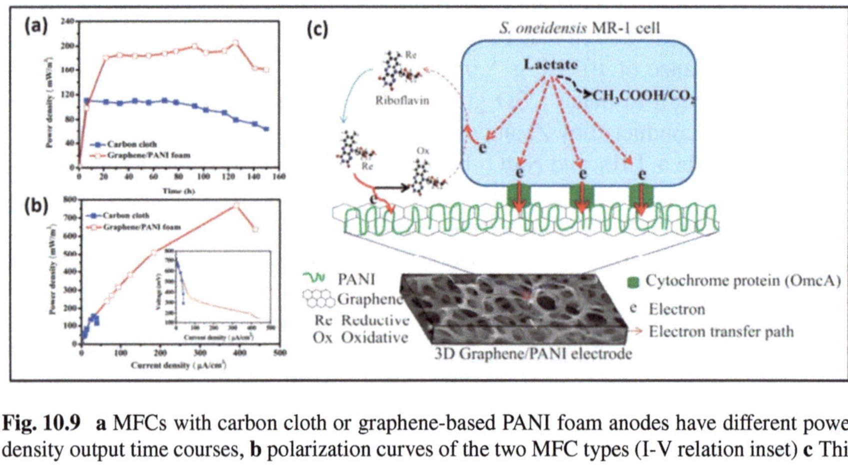

Fig. 10.9 **a** MFCs with carbon cloth or graphene-based PANI foam anodes have different power density output time courses, **b** polarization curves of the two MFC types (I-V relation inset) **c** This picture shows S. oneidensis MR-1 bacteria and a 3D graphene-PANI monolith electrode interacting with each other [162]. Adapted with permission from Ref. [162] Copyright (2012) (American Chemical Society)

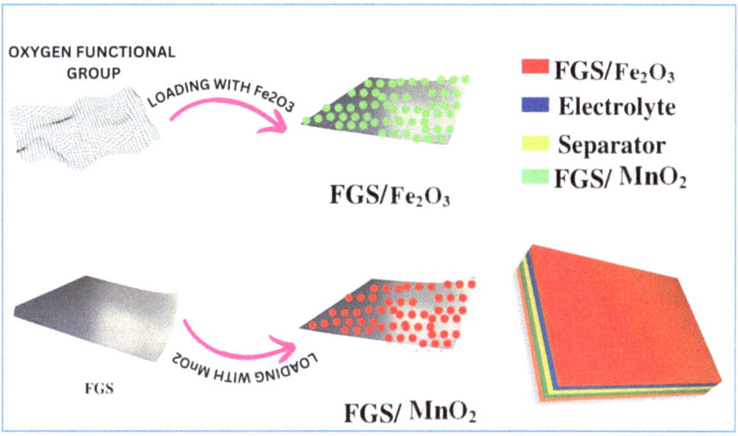

Fig. 10.10 A schematic representation of the creation of an asymmetric supercapacitor and design of the composite electrodes

between -1 and 0 V vs Ag–AgCl can reach up to 347 F/g^{-1}. The Fe_2O_3-FGS-MnO_2-FGSasymmetric supercapacitor has a power density of 100 W kg^{-1} and an energy density of 50.7 Wh kg^{-1}, and it has outstanding cycling stability.

10.6 Conclusions

Graphene research and development is progressing at a breakneck pace right now, with huge breakthroughs being made every day. Recent progress in our understanding of graphene polymer nanocomposites can be traced to the dramatic improvements in polymer properties, such as mechanical, electrical, thermal conductivity, chemical, optical, and gas impermeability, which can be made with low concentrations of filler. In this chapter, we focused on the precise incorporation of surface-functionalized graphene into different polymer matrices to realize their potential applications in fields of energy storage sectors including electronics and composite materials. The enhanced characteristics are a result of the uniform distribution of changed graphene and improved polymer-graphene compatibility via a variety of reactive functional groups on modified graphene. Graphene's emergence as viable nanofillers has ushered in a new era in the development of low-cost, high-performance composites for a wide variety of uses.

Acknowledgements The authors gratefully acknowledge the support provided by Centurion University of Technology and Management, Odisha, India, for carrying out the present research work.

Conflicts of Interest The authors declare no conflict of interest.

References

1. Okada A, Kawasumi M, Usuki A, Kojima Y, Kurauchi T, Kamigaito O (1990) Synthesis and properties of nylon-6/clay hybrids. Polymer-based molecular composites. In: MRS symposium proceedings, Pittsburgh, vol 171, pp 45–50
2. Nurazzi NM, Sabaruddin FA, Harussani MM, Kamarudin SH, Rayung M, Asyraf MRM, Khalina A (2021) Mechanical performance and applications of CNTs reinforced polymer composites-a review. Nanomater 11(9):2186
3. Wazalwar R, Sahu M, Raichur AM (2021) Mechanical properties of aerospace epoxy composites reinforced with 2D nano-fillers: current status and road to industrialization. Nanoscale Adv 3(10):2741–2776
4. Essabir H, Raji M, Bouhfid R, Qaiss AEK (2021) Hybrid nanocomposites based on graphene and nano-clay: preparation, characterization, and synergistic effect. In: Graphene and nanoparticles hybrid nanocomposites: from preparation to applications, pp 153–181
5. Bhanushali H, Amrutkar S, Mestry S, Mhaske ST (2022) Shape memory polymer nanocomposite: a review on structure–property relationship. Polym Bull 79(6):3437–3493
6. Karki S, Gohain MB, Yadav D, Ingole PG (2021) Nanocomposite and bio-nanocomposite polymeric materials/membranes development in energy and medical sector: a review. Int J Biol Macromol 193:2121–2139
7. Siwal SS, Zhang Q, Devi N, Thakur VK (2020) Carbon-based polymer nanocomposite for high-performance energy storage applications. Polymer 12(3):505
8. Benny Mattam L, Bijoy A, Abraham Thadathil D, George L, Varghese A (2022) Conducting polymers: a versatile material for biomedical applications. ChemistrySelect 7(42):e202201765
9. Kruželák J, Kvasničáková A, Hložeková K, Hudec I (2021) Progress in polymers and polymer composites used as efficient materials for EMI shielding. Nanoscale Adv 3(1):123–172
10. Shahapurkar K, Gelaw M, Tirth V, Soudagar MEM, Shahapurkar P, Mujtaba MA, Ahmed GMS (2022) Comprehensive review on polymer composites as electromagnetic interference shielding materials. Polym Polym Compos 30:1–17
11. Samsudin SS, Abdul Majid MS, MohdJamir MR, Osman AF, Jaafar M, Alshahrani HA (2022) Physical, thermal transport, and compressive properties of epoxy composite filled with graphitic-and ceramic-based thermally conductive nanofillers. Polym 14(5):1014
12. Minisha S, Vedhi C, Rajakani P (2022) Methods of graphene synthesis and graphene-based electrode material for supercapacitor applications. ECS J Solid State Sci Technol 11(11):111002
13. Sethulekshmi AS, Jayan JS, Appukutta S, Joseph K (2021) MoS_2: Advanced nanofiller for reinforcing polymer matrix. Phys E Low Dimens Syst Nanostruct 132:114716
14. Lian C, Gu Z, Zhang Y, Ma Z, Qiu H, Gu J (2021) Structural design strategies of polymer matrix composites for electromagnetic interference shielding: a review. Nano-Micro Lett 13(1):181
15. Li Y, Huang X, Zeng L, Li R, Tian H, Fu X, Zhong WH (2019) A review of the electrical and mechanical properties of carbon nanofiller-reinforced polymer composites. J Mater Sci 54:1036–1076
16. Ansari MNM, Sayem MA (2023) Microwave-assisted activated carbon: a promising class of materials for a wide range of applications. In: Radiation technologies and applications in materials science, pp 331–368
17. Hecht DS, Hu L, Irvin G (2011) Emerging transparent electrodes based on thin films of carbon nanotubes, graphene, and metallic nanostructures. Adv Mater 23(13):1482–1513
18. El-Kady MF, Shao Y, Kaner RB (2016) Graphene for batteries, supercapacitors and beyond. Nat Rev Mater 1(7):1–14
19. Kumar SSA, Bashir S, Ramesh K, Ramesh S (2021) New perspectives on graphene/graphene oxide based polymer nanocomposites for corrosion applications: the relevance of the graphene/polymer barrier coatings. Prog Org Coat 154:106215

20. Mohan VB, Lau KT, Hui D, Bhattacharyya D (2018) Graphene-based materials and their composites: a review on production, applications and product limitations. Compos B Eng 142:200–220
21. Shmavonyan G, Cheshev D, Averkiev A, Tran TH, Sheremet E (2022) Nanospectroscopy of graphene and two-dimensional atomic materials and hybrid structures. Opt Nanospectroscopy 401
22. Moharana S, Mahaling RN (2017) Silver (Ag)-Graphene oxide (GO)-Poly (vinylidene fluoride-co-hexafluoropropylene)(PVDF-HFP) nanostructured composites with high dielectric constant and low dielectric loss. Chem Phys Lett 680:31–36
23. Novoselov KS, Geim AK, Morozov SV, Jiang D, Zhang Y, Dubonos SV (2004) Electric field effect in atomically thin carbon films. Sci 306:666–669
24. Asghar F, Shakoor B, Fatima S, Munir S, Razzaq H, Naheed S, Butler IS (2022) Fabrication and prospective applications of graphene oxide-modified nanocomposites for wastewater remediation. RSC Adv 12(19):11750–11768
25. Ashraf G, Aziz A, Iftikhar T, Zhong ZT, Asif M, Chen W (2022) The roadmap of graphene-based sensors: electrochemical methods for bioanalytical applications. Biosensors 12(12):1183
26. Moharana S, Mahaling RN (2021) Enhancement investigations on dielectric and electrical properties of niobium pentoxide (Nb_2O_5) reinforced poly (vinylidene fluoride)(PVDF)-graphene oxide (GO) nanocomposite films. J Asian Ceram Soc 9(3):1183–1193
27. Nitin MS, Suresh Kumar S (2022) Ballistic performance of synergistically toughened Kevlar/epoxy composite targets reinforced with multiwalled carbon nanotubes/graphene nanofillers. Polym Compos 43(2):782–797
28. Dubey PK, Hong J, Lee K, Singh P (2023) Graphene-based materials: synthesis and applications. In: Nanomater, pp 59–84
29. Patra L, Pandey R (2022) Mechanical properties of 2D materials: a review on molecular dynamics based nanoindentation simulations. Mater Today Commun 31:103623
30. Balandin AA (2011) Thermal properties of graphene and nanostructured carbon materials. Nat Mater 10(8):569–581
31. Jung I, Dikin DA, Piner RD, Ruoff RS (2008) Tunable electrical conductivity of individual graphene oxide sheets reduced at “low” temperatures. Nano Lett 8(12):4283–4287
32. Chauhan AK, Gupta SK, Taguchi D, Manaka T, Jha P, Veerender P, Iwamoto M (2017) Enhancement of the carrier mobility of conducting polymers by formation of their graphene composites. RSC Adv 7(20):11913–11920
33. Matte HR, Subrahmanyam KS, Rao CNR (2009) Novel magnetic properties of graphene: presence of both ferromagnetic and antiferromagnetic features and other aspects. J Phys Chem C 113(23):9982–9985
34. Guo Q, Dedkov Y, Voloshina E (2020) Intercalation of Mn in graphene/Cu (111) interface: insights to the electronic and magnetic properties from theory. Sci Rep 10(1):21684
35. Geng Y, Wang SJ, Kim JK (2009) Preparation of graphite nanoplatelets and graphene sheets. J Colloid Interface Sci 336(2):592–598
36. Vacchi IA, Ménard-Moyon C, Bianco A (2017) Chemical functionalization of graphene family members. Phys Sci Rev 2(1)
37. Ioniţă M, Vlăsceanu GM, Watzlawek AA, Voicu SI, Burns JS, Iovu H (2017) Graphene and functionalized graphene: extraordinary prospects for nanobiocomposite materials. Compos B Eng 121:34–57
38. Acik M, Chabal YJ (2011) Nature of graphene edges: a review. Jpn J Appl Phys 50(7R):070101
39. NebolSin VA, Galstyan V, Silina YE (2020) Graphene oxide and its chemical nature: multi-stage interactions between the oxygen and graphene. Surf Interfaces 21:100763
40. Shi L, Xia W (2012) Photoredox functionalization of C-H bonds adjacent to a nitrogen atom. Chem Soc Rev 41(23):7687–7697
41. Cao Y, Osuna S, Liang Y, Haddon RC, Houk KN (2013) Diels-Alder reactions of graphene: computational predictions of products and sites of reaction. J Am Chem Soc 135(46):17643–17649

42. Guday G (2019) Surface chemistry of low-dimensional carbon materials: synthesis and functionalization of graphene. Freie Universitaet Berlin (Germany)
43. He J, Qiu D, Li Y (2020) Strategies toward aryne multifunctionalization via 1, 2-benzdiyne and benzyne. Acc Chem Res 53(2):508–519
44. Rkein B, Bigot A, Birbaum L, Manneveau M, De Paolis M, Legros J, Chataigner I (2021) Reactivity of 3-nitroindoles with electron-rich species. Chem Commun 57(1):27–44
45. Ratwani CR, Abdelkader A(2022) Self-healing by Diels-Alder cycloaddition in advanced functional polymers: a review. Prog Mater Sci, p 101001
46. Ashfaq A, Clochard MC, Coqueret X, Dispenza C, Driscoll MS, Ulański P, Al-Sheikhly M (2020) Polymerization reactions and modifications of polymers by ionizing radiation. Polymer 12(12):2877
47. Shabbir M, Raza ZA, Shah TH, Tariq MR (2022) Recent progress in graphenes: synthesis, covalent functionalization and environmental applications. J Nanostructure Chem 12(6):1033–1051
48. Majumder M, Thakur AK (2019) Graphene and its modifications for supercapacitor applications. Surf Eng, pp 113–138
49. Lomeda JR, Doyle CD, Kosynkin DV, Hwang WF, Tour JM (2008) Diazonium functionalization of surfactant-wrapped chemically converted graphene sheets. J Am Chem Soc 130(48):16201–16206
50. Chua CK, Pumera M (2013) Covalent chemistry on graphene. Chem Soc Rev 42(8):3222–3233
51. Johns JE, Hersam MC (2013) Atomic covalent functionalization of graphene. Acc Chem Res 46(1):77–86
52. Sharma R, Baik JH, Perera CJ, Strano MS (2010) Anomalously large reactivity of single graphene layers and edges toward electron transfer chemistries. Nano Lett 10(2):398–405
53. Xu X, Chen J, Luo X, Lu J, Zhou H, Wu W, Li Z (2012) Poly (9, 9′-diheylfluorene carbazole) functionalized with reduced graphene oxide: convenient synthesis using nitrogen-based nucleophiles and potential applications in optical limiting. Chem Eur J 18(45):14384–14391
54. Li PP, Chen Y, Zhu J, Feng M, Zhuang X, Lin Y, Zhan H (2011) Charm-bracelet-type Poly (N-Vinylcarbazole) functionalized with reduced graphene oxide for broadband optical limiting. Chem Eur J 17(3):780–785
55. Wojtaszek M, Tombros N, Caretta A, Van Loosdrecht PHM, Van Wees BJ (2011) A road to hydrogenating graphene by a reactive ion etching plasma. J Appl Phys 110(6):063715
56. Medeiros PV, Mascarenhas AJ, de Brito MF, de Castilho CM (2010) A DFT study of halogen atoms adsorbed on graphene layers. Nanotechnology 21(48):485701
57. Robinson Jeremy T, James S, Chad Junkermeier E, Stefan Badescu C, Thomas L, Keith Perkins F, Maxim ZK (2010) Properties of fluorinated graphene films. Nano Lett 10(8):3001–3005
58. Chua CK, Pumera M (2012) Friedel-crafts acylation on graphene. Chem Asian J 7(5):1009–1012
59. Yuan C, Chen W, Yan L (2012) Amino-grafted graphene as a stable and metal-free solid basic catalyst. J Mater Chem 22(15):7456–7460
60. Zhan J, Lei Z, Zhang Y (2022) Non-covalent interactions of graphene surface: mechanisms and applications. Chem 8(4):947–979
61. Mao HY, Lu YH, Lin JD, Zhong S, Wee ATS, Chen W (2013) Manipulating the electronic and chemical properties of graphene via molecular functionalization. Prog Surf Sci 88(2):132–159
62. Guo X (2013) Single-molecule electrical biosensors based on single-walled carbon nanotubes. Adv Mater 25(25):3397–3408
63. Lin Y, Taylor S, Li H, Fernando KS, Qu L, Wang W, Sun YP (2004) Advances toward bioapplications of carbon nanotubes. J Mater Chem 14(4):527–541
64. Ma H, Shen Z (2020) Exfoliation of graphene nanosheets in aqueous media. Ceram Int 46(14):21873–21887
65. Brisebois PP, Siaj M (2020) Harvesting graphene oxide–years 1859 to 2019: a review of its structure, synthesis, properties and exfoliation. J Mater Chem C 8(5):1517–1547

66. Ghosh A, Rao KV, George SJ, Rao CNR (2010) Noncovalent functionalization, exfoliation, and solubilization of graphene in water by employing a fluorescent coronene carboxylate. Chem Eur J 16(9):2700–2704
67. Englert JM, Röhrl J, Schmidt CD, Graupner R, Hundhausen M, Hauke F, Hirsch A (2009) Soluble graphene: generation of aqueous graphene solutions aided by a perylenebisimide-based bolaamphiphile. Adv Mater 21(42):4265–4269
68. Bai H, Xu Y, Zhao L, Li C, Shi G (2009) Non-covalent functionalization of graphene sheets by sulfonated polyaniline. Chem Commun 13:1667–1669
69. Ibrahim A, Klopocinska A, Horvat K, Abdel Hamid Z (2021) Graphene-based nanocomposites: synthesis, mechanical properties, and characterizations. Polym 13(17):2869
70. Lawal AT (2020) Recent progress in graphene based polymer nanocomposites. Cogent Chem. 6(1):1833476
71. Zhou T, Zheng Y, Gao H, Min S, Li S, Liu HK, Guo Z (2015) Surface engineering and design strategy for surface-amorphized TiO_2@graphene hybrids for high power Li-ion battery electrodes. Adv Sci 2(9):1500027
72. Shi S, Chen F, Ehlerding EB, Cai W (2014) Surface engineering of graphene-based nanomaterials for biomedical applications. Bioconjug Chem 25(9):1609–1619
73. Azizi-Lalabadi M, Hashemi H, Feng J, Jafari SM (2020) Carbon nanomaterials against pathogens; the antimicrobial activity of carbon nanotubes, graphene/graphene oxide, fullerenes, and their nanocomposites. Adv Colloid Interface Sci 284:102250
74. Bhattacharya M (2016) Polymer nanocomposites—a comparison between carbon nanotubes, graphene, and clay as nanofillers. Mater 9(4):262
75. Nandee R, Chowdhury MA, Ahmed MU, Shuvho BA, Debnath UK (2019) Performance and characterization of two-dimensional material graphene conductivity-a review. Mater Perform Charact 8(1):183–196
76. Clancy AJ, Bayazit MK, Hodge SA, Skipper NT, Howard CA, Shaffer MS (2018) Charged carbon nanomaterials: redox chemistries of fullerenes, carbon nanotubes, and graphenes. Chem Rev 118(16):7363–7408
77. Cui Y, Kundalwal SI, Kumar S (2016) Gas barrier performance of graphene/polymer nanocomposites. Carbon 98:313–333
78. Zhao HY, Yu MY, Liu J, Li X, Min P, Yu ZZ (2022) Efficient preconstruction of three-dimensional graphene networks for thermally conductive polymer composites. Nano-Micro Lett 14(1):129
79. Zhu X, Ni Z, Dong L, Yang Z, Cheng L, Zhou X, Chen M (2019) In-situ modulation of interactions between polyaniline and graphene oxide films to develop waterborne epoxy anticorrosion coatings. Prog Org Coat 133:106–116
80. Zhu Q, Li EN, Liu X, Song W, Li Y, Wang X, Liu C (2020) Epoxy coating with in-situ synthesis of polypyrrole functionalized graphene oxide for enhanced anticorrosive performance. Prog Org Coat 140:105488
81. Zheng W, Lu X, Wong SC (2004) Electrical and mechanical properties of expanded graphite-reinforced high-density polyethylene. J Appl Polym Sci 91:2781–2788
82. Hu HT, Wang JC, Wan L, Liu FM, Zheng H (2010) Preparation and properties of graphene nanosheets—polystyrene nanocomposites via in-situ emulsion polymerization. Chem Phy Lett 484:247–253
83. Ye L, Meng XY, Ji X, Li ZM, Tang JH (2009) Synthesis and characterization of expandable graphite-poly(methyl methacrylate) composite particles and their application to flame retardation of rigid polyurethane foams. Polym Degrad Stab 94:971–979
84. Chen G, Wu D, Weng W, Wu C (2003) Exfoliation of graphite flakes and its nanocomposites. Carbon 41:619–621
85. Kornmann X (2001) Synthesis and characterization of thermoset-layered silicate nanocomposites. Ph.D. thesis. Sweden, Lulea Tekniska Universitet
86. Moujahid EM, Besse JP, Leroux F (2003) Poly(styrene sulfonate) layered double hydroxide nanocomposites Stability and subsequent structural transformation with changes in temperature. J Mater Chem 13:258–264

87. Hsueh HB, Chen CY (2003) Preparation and properties of LDHs= polyimide nanocomposites. Polym 44:1151–1161
88. Novoselov KS, Geim AK, Morozov SV (2004) Electric field effect in atomically thin carbon films. Science 306:666–669
89. Tkalya E, Ghislandi M, Alekseev A (2010) Latex-based concept for the preparation of graphene-based polymer nanocomposites. J Mater Chem 20:3035–3039
90. Latif I, Alwan TB, Al-Dujaili AH (2012) Low frequency dielectric study of PAPA-PVA-GR nanocomposites. Nanosci Nanotechnol 2:190–200
91. Murugan AV, Muraliganth T, Manthiram A (2009) Rapid, facile microwave-solvothermal synthesis of graphene nanosheets and their polyaniline nanocomposites for energy strorage. Chem Mater 21:5004–5006
92. Fan X, Peng W, Li Y (2008) Deoxygenation of exfoliated graphite oxide under alkaline conditions: a green route to graphene preparation. Adv Mater 2:4490–4493
93. Park S, An J, Jung I (2009) Colloidal suspensions of highly reduced graphene oxide in a wide variety of organic solvents. Nano Lett 9:1593–1597
94. Jiang L, Shen XP, Wu JL (2010) Preparation and characterization of graphene/poly (vinyl alcohol) nanocomposites. J Appl Polym Sci 118:275–279
95. Bourlinos AB, Gournis D, Petridis D (2003) Graphite oxide: chemical reduction to graphite and surface modification with primary aliphatic amines and amino acids. Langmuir 19:6050–6055
96. Wang G, Shen X, Yao J (2009) Graphene nanosheets for enhanced lithium storage in lithium ion batteries. Carbon 47:2049–2053
97. Staudenmaier L (1898) Verfahren zur darstellung der graphitsa¨ure. Ber Dtsch Chem Ges 31:1481–1499
98. Michael J, McAllister J-L (2007) Single sheet functionalized graphene by oxidation and thermal expansion of graphite. Chem Mater 19:4396–4404
99. Zhang HB, Zheng WG, Yan Q (2010) Electrically conductive polyethylene terephthalate/ graphene nanocomposites prepared by melt compounding. Polym 51:1191–1196
100. Potts JR, Dreyer DR, Bielawski CW (2011) Graphene based polymer nanocomposites. Polymer 52:5–25
101. Hu H, Wang X, Wang J (2010) Preparation and properties of graphene nanosheets–polystyrene nanocomposites via in situ emulsion polymerization. Chem Phys Lett 484:247–253
102. Patole AS, Patole SP, Kang H (2010) A facile approach to the fabrication of graphene/ polystyrene nanocomposite by in situ microemulsion polymerization. J Colloid Interface Sci 350:530–537
103. Ni Z, Wang Y, Yu T (2008) Raman spectroscopy and imaging of graphene. Nano Res 1:273–291
104. Gawryla MD, van den Berg O, Weder C (2009) Clay aerogel/cellulose whisker nanocomposites: a nanoscale wattle and daub. J Mater Chem 19:2118–2124
105. Van den Berg O, Capadona JR, Weder C (2007) Preparation of homogeneous dispersions of tunicate cellulose whiskers in organic solvents. Biomacromol 8:1353–1357
106. Chen W, Yan L, Bangal PR (2010) Preparation of graphene by the rapid and mild thermal reduction of graphene oxide induced by microwaves. Carbon 48:1146–1152
107. Chen W, Yan L (2010) Preparation of graphene by a low temperature thermal reduction at atmosphere pressure. Nanoscale 2:559–563
108. Zhou X, Liu Z (2010) A scalable, solution-phase processing route to graphene oxide and graphene ultra large sheets. Chem Commun 46:2611–2613
109. Vickery JL, Patil AJ, Mann S (2009) Fabrication of graphene–polymer nanocomposites with higher-order three dimensional architectures. Adv Mater 21:2180–2184
110. Cao Y, Feng J, Wu P (2010) Preparation of organically dispersible graphene nanosheet powders through a lyophilization method and their poly (lactic acid) composites. Carbon 48:3834–3839
111. Chaturvedi A, Tiwari A, Tiwari A (2013) Spectroscopic and morphological analysis of graphene vinylester nanocomposites. Adv Mater Lett 4:656–661

112. Che J, Shen L, Xiao Y (2010) A new approach to fabricate graphene nanosheets in organic medium: combination of reduction and dispersion. J Mater Chem 20:1722–1727
113. Gong L, Kinloch IA, Young RJ (2010) Interfacial stress transfer in a graphene monolayer nanocomposite. Adv Mater 22:2694–2697
114. Liu P, Huang Y, Wang L (2013) Preparation and excellent microwave absorption property of three component nanocomposites: Polyaniline-reduced graphene oxide-CO_3O_4 nanoparticles. Synth Met 177:89–93
115. Marcano DC, Kosynkin DV, Berlin JM (2010) Improved synthesis of graphene oxide. ACS Nano 4:4806–4814
116. Kumar SK, Castro M, Saiter A (2013) Development of poly (isobutylene-coisoprene)/reduced graphene oxide nanocomposites for barrier, dielectric and sensing applications. Mater Lett 96:109–112
117. Zeng X, Yang J, Yuan W (2012) Preparation of a poly (methyl methacrylate)-reduced graphene oxide composite with enhanced properties by a solution blending method. Eur Polym J 48:1674–1682
118. Subrahmanyam K, Vivekchand S, Govindaraj A (2008) A study of graphenes prepared by different methods: characterization, properties and solubilization. J Mater Chem 18:1517–1523
119. Zaman I, Kuan HC, Dai J (2012) From carbon nanotubes and silicate layers to graphene platelets for polymer nanocomposites. Nanoscale 4: 4578–4586
120. Yu J, Lu K, Sourty E, Grossiord N, Koning CE, Loos J (2007) Characterization of conductive multiwall carbon nanotube/polystyrene composites prepared by latex technology. Carbon 45:2897–2903
121. Li J, Vaisman L, Marom G, Kim JK (2007) Br treated graphite nanoplatelets for improved electrical conductivity of polymer composites. Carbon 45:744–750
122. Wang S, Tambraparni M, Qiu J, Tipton J, Dean D (2009) Thermal expansion of graphene composites. Macromolecules 42:5251–5255
123. Kuilla T, Srivastava SK, Bhowmick AK (2009) Rubber/LDH nanocomposites by solution blending. J Appl Polym Sci 111:635–641
124. Yu A, Ramesh P, Itkis ME, Elena B, Haddon RC (2007) Graphite nanoplatelet-epoxy composite thermal interface materials. J Phys Chem C 111:7565–7569
125. Liu N, Luo F, Wu H, Liu Y, Zhang C, Chen J (2008) One step ionic-liquid assisted electrochemical synthesis of ionic-liquid-functionalized graphene sheets directly from graphene. Adv Funct Mater 18:1518–1525
126. Eda G, Chhowalla M (2009) Graphene-based composite thin films for electronics. Nano Lett 9:814–818
127. Mu Q, Feng S (2007) Thermal conductivity of graphite/silicone rubber prepared by solution intercalation. Thermochim Acta 462:70–75
128. Zhang HB, Zheng WG, Yan Q, Yang Y, Wang J, Lu ZH (2010) Electrically conductive polyethylene terephthalate/graphene nanocomposites prepared by melt compounding. Polymer 51:1191–1196
129. Hu H, Wang X, Wanga J, Wana L, Liu F, Zheng H (2010) Preparation and properties of graphene nanosheets-polystyrene nanocomposites via in situ emulsion polymerization. Chem Phys Lett 484:247–253
130. Stankovich S, Dikin DA, Piner RD, Kohlhaas KA, Kleinhammes A (2007) Synthesis of graphene-based nanosheets via chemical reduction of exfoliated graphite oxide. Carbon 45:1558–1565
131. Zhao X, Zhang Q, Chen D (2010) Enhanced mechanical properties of graphene-based poly(vinyl alcohol) composites. Macromolecules 43:2357–2363
132. Stankovich S, Dikin DA, Dommett GHB, Kohlhaas KM, Zimney EJ, Stach EA (2006) Graphene-based composite materials. Nature 442:282–286
133. Liu N, Luo F, Wu H, Liu Y, Zhang C, Chen J (2008) One step ionicliquidassisted electrochemical synthesis of ionic-liquid-functionalized graphene sheets directly from graphene. Adv Funct Mater 18:1518–1525

134. Mu Q, Feng S (2007) Thermal conductivity of graphite=silicone rubber prepared by solution intercalation. Thermochim Acta 462:70–75
135. Hu HT, Wang JC, Wan L, Liu FM, Zheng H (2010) Preparation and properties of graphene nanosheets—polystyrene nanocomposites via insitu emulsion polymerization. Chem Phys Lett 484:247–253
136. Kuilla T, Bhadra S, Yao D, Kim NH, Bose S, Lee JH (2010) Recent advances in graphene based polymer composites. Prog Poly Sci 35(11):1350–1375
137. Peponi L, Tercjak A, Verdejo R, Lopez-Manchado MA, Mondragon I, Kenny JM (2009) Confinement of functionalized graphene sheets by triblock copolymers. J Phys Chem 113:17973–17978
138. Wang DW, Li F, Zhao J, Ren W, Chen ZG, Tan J (2009) Fabrication of graphene/polyaniline composite paper via in situ anodic electro-polymerization for high-performance flexible electrode. ACS Nano 7:1745–1752
139. Zhao L, Zhao L, Xu Y, Qiu T, Zhi L, Shi G (2009) Polyaniline electrochromic devices with transparent graphene electrodes. Electrochim Acta 55:491–497
140. Yan J, Wei T, Fan Z, Qian W, Zhang M, Shen X (2010) Preparation of graphene nanosheet/carbon nanotube/polyaniline composite as electrode material for supercapacitors. J Power Sources 195:3041–3045
141. Mazinani S, Ajji A, Dubois C (2009) Morphology, structure and properties of conductive PS=CNT nanocomposite electrospun mat. Polym 50:3329–3342
142. Bourlinos AB, Gournis D, Petridis D, Szabo T, Szeri A, Dekany I (2003) Graphite oxide: chemical reduction to graphite and surface modification with primary aliphatic amines and amino acids. Chem Mater 19:6050–6055
143. Liang J, Huang Y, Zhang L, Wang Y, Ma Y, Guo T (2009) Molecular-level dispersion of graphene into poly(vinyl alcohol) and effective reinforcement of their nanocomposites. Adv Funct Mater 19:2297–2302
144. Lee YR, Raghu AV, Jeong HM, Kim BK (2009) Properties of waterborne polyurethane/functionalized graphene sheet nanocomposites prepared by an in situ method. Macromol Chem Phys 210(15):1247–1254
145. Liang J, Xu Y, Huang Y, Zhang L, Wang Y, Ma Y (2009) Infraredtriggered actuators from graphene-based nanocomposites. J Phys Chem 113:9921–9927
146. Ansari S, Giannelis EP (2009) Functionalized graphene sheetpoly(vinylidene fluoride) conductive nanocomposites. J Polym Sci Part B Polym Phys 47:888–897
147. Xu Y, Wang Y, Jiajie L, Huang Y, Ma Y, Wan X (2009) A hybrid material of graphene and poly(3,4-ethyldioxythiophene) with high conductivity, flexibility, and transparency. Nano Res 2:343–348
148. Kim H, Macosko CW (2009) Processing–property relationships of polycarbonate/graphene nanocomposites. Polym 50:3797–3809
149. Cussler EL, Hughes SE, Ward WJ III, Aris R (1988) Barrier membrane. J Membr Sci 38:161–174
150. Halperin BI, Feng S, Sen PN (1985) Differences between lattice and continuum percolation transport exponents. Phys Rev Lett 54:2391–2394
151. Moussa M, El-Kady MF, Zhao Z, Majewski P, Ma J (2016) Recent progress and performance evaluation for polyaniline/graphene nanocomposites as supercapacitor electrodes. Nanotechnology (27):442001
152. Tai Z, Yan X, Xue Q (2012) Three-dimensional graphene/polyaniline composite hydrogel as supercapacitor electrode. J Electrochem Soc 159(10):A1702
153. Liu H, Wang Y, Gou X, Qi T, Yang J, Ding Y (2013) Three-dimensional graphene/polyaniline composite material for high-performance supercapacitor applications. Mater Sci Eng 5:293–298
154. Kulkarni SB, Patil UM, Shackery I, Sohn JS, Lee S, Park B, Jun S (2014) High-performance supercapacitor electrode based on a polyaniline nanofibers/3D graphene framework as an efficient charge transporter. J Mater Chem A 14:4989–4998

155. Zhang Y, Huang Y, Yang G, Bu F, Li K, Shakir I, Xu Y (2017) Dispersion–assembly approach to synthesize three-dimensional graphene/polymer composite aerogel as a powerful organic cathode for rechargeable Li and Na batteries. ACS Appl Mater Interf 18:15549–15556
156. Li WW, Yu HQ, He Z (2014) Towards sustainable wastewater treatment by using microbial fuel cells-centered technologies. Energy Environ Sci 3:911–924
157. Xie X, Criddle C, Cui Y (2015) Design and fabrication of bioelectrodes for microbial bioelectrochemical systems. Energy Environ Sci 12:3418–3441
158. Zou L, Qiao Y, Wu ZY, Wu XS, Xie JL, Yu SH, Li CM (2016) Tailoring unique mesopores of hierarchically porous structures for fast direct electrochemistry in microbial fuel cells. Adv Energy Mater 4:1501535
159. Yang Y, Dong R, Zhu Y, Li H, Zhang H, Fan X, Chang H (2020) High-performance direct hydrogen peroxide fuel cells (DHPFCs) with silver nanowire-graphene hybrid aerogel as highly-conductive mesoporous electrodes. Chem Eng J 381:122749
160. Du Q, An J, Li J, Zhou L, Li N, Wang X (2017) Polydopamine as a new modification material to accelerate startup and promote anode performance in microbial fuel cells. J Power Sources 343:477–482
161. Tang C, Zhang Q (2016) Can metal–nitrogen–carbon catalysts satisfy oxygen electrochemistry? J Mater Chem A 14:4998–5001
162. Yong YC, Dong XC, Chan-Park MB, Song H, Chen P (2012) Macroporous and monolithic anode based on polyaniline hybridized three-dimensional graphene for high-performance microbial fuel cells. ACS Nano 6(3):2394–2400
163. Xia H, Hong C, Li B, Zhao B, Lin Z, Zheng M, Aldoshin SM (2015) Facile synthesis of hematite quantum-dot/functionalized graphene-sheet composites as advanced anode materials for asymmetric supercapacitors. Adv Funct Mater 25(4):627–635

Chapter 11
Core–Shell Structured Nanomaterials for High-Performance Dielectric Applications

Anupam Sahoo, Sangita Kumari Swain, and Sukanta Kumar Swain

Abstract Green energy is a relatively new kind of energy that is gaining considerable interest from nations all over the globe as a solution to the problem of environmental pollution on a global scale as well as the current energy crisis (Luo et al. in Adv Energy Mater 9(5):1803204, 2019 [1], Wang et al. in J Mater Chem A 8(22):11124–11132, 2020 [2], Wang et al. in Nano Energy 78:105247, 2020 [3], Zheng et al. in Compos Sci Technol 222:109379, 2022 [4]). Owing to numerous applications in the electronic and electrical industries, such as transistors, actuators, and capacitors, dielectric materials having low dielectric loss and high dielectric constant have gained growing consideration in recent years (Zheng et al. in Compos Sci Technol 222:109379, 2022 [4], He et al. in Compos A Appl Sci Manuf 93:137–143, 2017 [5], Huang and Jiang in Adv Mater 27(3):546–554, 2015 [6], Wang et al. in IEEE Trans Dielectr Electr Insul 17(4):1036–1042, 2010 [7]). In this chapter, various types of core–shell nanoparticles are discussed for high-performance dielectric applications.

Keywords Core–shell nanoparticles · Dielectric materials · Polymer nanocomposites · High-k materials

11.1 Introduction

High dielectric constant materials (high-k) possess various implications in organic thin-film electroluminescent devices [8], organic field effect transistors (OFETs) [9–11], actuators, and [12, 13] energy storage devices [14–16], and electrical stress

A. Sahoo (✉)
Department of Chemistry, Sri Krushna Chandra Gajapati (Autonomous) College, Paralakhemundi, Odisha, India
e-mail: anupamchem0@gmail.com

S. K. Swain
Centre of Excellence, Berhampur University, Berhampur, Odisha, India

S. K. Swain
Department of ECE, Indian Institute of Information Technology, Ranchi, Jharkhand, India

S. Moharana et al. (eds.), *Emerging Nanodielectric Materials for Energy Storage*, Nanostructure Science and Technology, https://doi.org/10.1007/978-3-031-40938-7_11

control applications[17–19]. High-k materials have the ability to significantly lower the surface electric stress and maintain it at a level that is lower than the breakdown strength of air that may finally lead to the elimination of the surface discharge. A seamless high-k material has the characteristics such as high dielectric constant, high breakdown strength, minimal dielectric loss, and superior processability. These are the general features of an ideal high-k material. In general, polymers based on organic molecules have a high breakdown strength and are simple to produce, however the vast majority of organic polymers have disadvantage of low dielectric constant, e.g., around 2.0–5.0. Despite the fact that many have very high dielectric constants, the dielectric loss of these materials is often rather considerable, especially when subjected to strong electric fields [20]. In addition to having enormous dielectric constants that range from ~1000 to 10,000, dielectric ceramics like $BaTiO_3$ have lower breakdown strengths and/or substantial dielectric loss [21]. Additionally, their lack of flexibility and low processability severely restrict their use. In general, it is challenging to discover a single material that combines all the required qualities for real-world applications. Therefore, it may be anticipated that combining the benefits of ceramics with polymers would be a successful strategy for developing new high-k materials.

When compared with microparticles, the interlinked area that polymer material has with nanoparticles is much higher. Nanoparticles impart larger interconnected areas in comparison with micro-sized particles, with the polymer nanocomposites. As a result, nanoparticles are the potential materials that exhibit significantly higher polarization levels, mechanical enhancements, and breakdown strength. Additionally, they provide more opportunities to tune and optimize the characteristics of the polymer matrix. Further, the smaller size of nanomaterials makes them feasible to lower the fabrication dimensions of devices based on polymer nanocomposites. This is a feature that is of the utmost significance for the ongoing process of shrinking the size of electronic devices. In the case of embedded planar capacitors, for instance, the dielectric film must be very thin in order to provide a higher capacitance; normally, this is accomplished using films on the micrometer scale. When measured on this scale, an enhancement in the size of the high-k particles from 100 nm to 2 μm results in a considerable reduction in the capacitance density of polymer composite films [22]. These exciting factors make nanoparticles most significant for the development of high-k dielectric materials.

Although the incorporation of nanoparticles into the design of high-k polymer composites has a number of opportunities for improvement, the realization of high-performance nanocomposites presents a number of obstacles. The synthesis of homogenously dispersed nanoparticles and the tuning of interfaces between polymer and nanoparticle are typical examples of existing problems. Both of these factors play significant roles in producing appropriate electrical and other characteristics. Surface modification of nanoparticles by coupling agents [18], hydroxylation [23], surfactants, phosphoric acids, and other organic compounds [24–26] has been extensively employed to enhance the performance of polymer nanocomposites to achieve a high-k material. However, these approaches still have certain drawbacks in terms of achieving the complete potential of a high-performance polymer matrix. This is due to the fact that the surface modifiers themselves do not often exhibit huge participation in the enhancement of the dielectric characteristics of the nanocomposites.

Over the past few years, considerable efforts has been invested in the synthesis and design of core–shell nanoparticles, with the goal of realizing the full potential that nanoparticles have to offer in terms of improving the characteristics of polymer nanocomposites. Figure 11.1 provides many generic ways for manufacturing high-k polymer nanocomposites utilizing core–shell methodologies. As an example, this figure uses $BaTiO_3$ as its subject. These approaches may be broken down into the following four categories:

(i) Core–shell nanoparticles that have been fabricated by "grafting from" [27–29]
(ii) Core–shell nanoparticles that have been fabricated by "grafting to" [30–33]
(iii) The use of organic–inorganic based core–shell nanoparticles as filler materials [34, 35] and
(iv) The use of other varieties of core–shell nanoparticles as filler materials [36–39].

This chapter focuses mostly on reviewing these developing approaches. In order to successfully use high-k polymer nanocomposites, it is essential to strike an appropriate balance between the dielectric loss, dielectric constant, and breakdown strength[40–42]. For this reason, a focus is made by virtue of core–shell techniques in attaining lower dielectric loss and strong breakdown strength while simultaneously preserving the high dielectric constant of the various nanocomposite

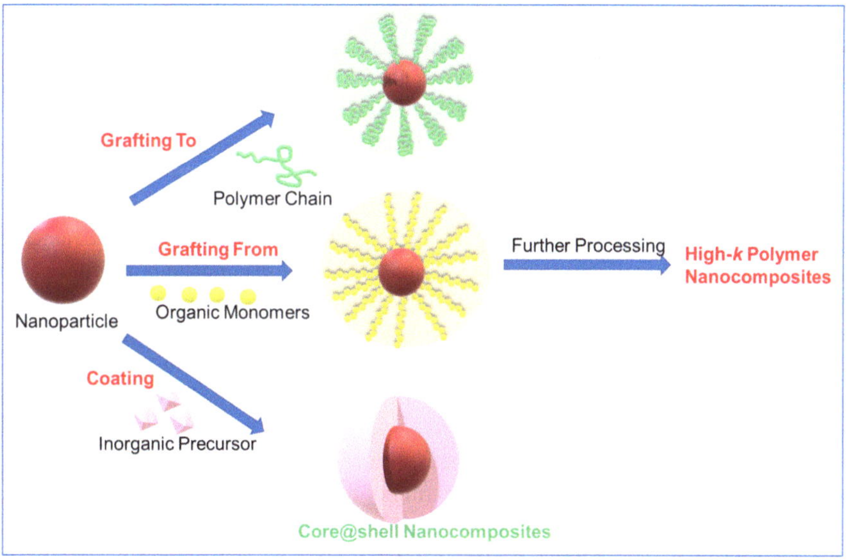

Fig. 11.1 Methods adopted for the synthesis of core–shell nanoparticles for high-k polymer nanocomposite matrix

matrices. However, it is important to know about the advantages and classes of core–shell nanoparticles. The following section describes a brief idea about core–shell nanoparticles.

11.2 Classification of Core–Shell Nanomaterials

Nanoparticles may be classified into simple, core/shell, and composite categories, depending on whether they are composed of a single substance or many components. In general, nanoparticles that are classified as simple are constructed using just a single kind of material. On the other hand, particles classified as composite or core/shell are made up of two or more types of materials. The core–shell type of nanoparticles may be broadly characterized as nanoparticles that consist of an inner core material and an exterior shell material (outer layer material). These may be made up of a broad variety of distinct combinations that are interacting closely with one another, such as components that are organic/organic, organic/inorganic, inorganic/organic, or inorganic/organic. The final application and purpose often have a significant impact on the material that is selected to function as the shell of the core–shell nanoparticle. Figure 11.2 provides a schematic representation of many distinct types of core–shell nanoparticles. Concentric spherical core/shell nanoparticles are considered to be the most common kind of nanoparticle (Fig. 11.2a), which consists of a basic spherical core particle that is entirely covered by a shell made of a different substance. Due to the fact that they each possess their own unique set of characteristics, various core/shell nanoparticles have also sparked a significant amount of research interest. When a core is not spherical, as illustrated in Fig. 11.2b, different shaped core/shell nanoparticles will often occur.

Multiple core core–shell particles may be created, as shown in Fig. 11.2c when a single layer of shell material is deposited concurrently over a large number of relatively small core particles. This can result in the formation of multiple core/shell particles. Nanoshells are shown as having alternating coatings of dielectric core material and metal shell material onto each other in Fig. 11.2d. These nanoshells are concentric in nature. These nanoshells have an A/B/A structure configuration. In this structure, nanoscale spacer layers made of dielectric material are employed to separate concentric layers of metallic material. These sorts of nanoparticles are known as nanomaterial or multi-layered metal-dielectric nanostructures. Such materials possess unique plasmonic properties.

The plasmonic features of these types of particles are the major reason for their significance. After applying a bilayer coating of the core material and then using the proper procedure to remove the first layer, it is also feasible to fabricate a movable core particle that is contained inside a hollow spherical shell material (shown in Fig. 11.2e). This can be done after a double-layer coating of the core material [43].

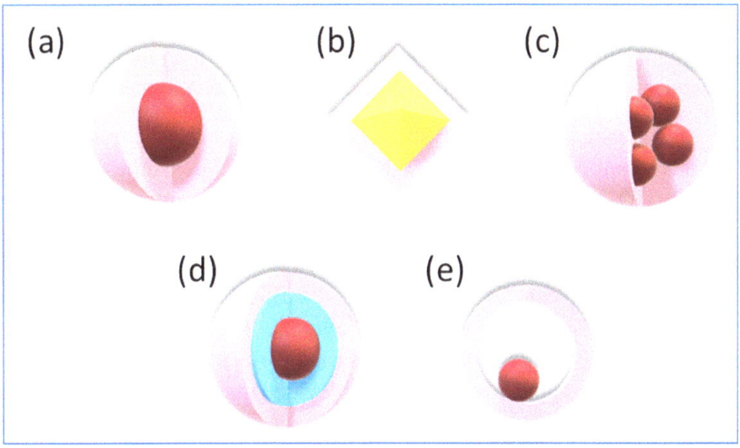

Fig. 11.2 Various types of core–shell nanoparticles; **a** Spherical, **b** Hexagonal, **c** Multicore single shell, **d** Multi-layered nanomaterial, **e** Movable core in the shell

11.3 Approaches for the Development of Core–Shell Polymer Nanocomposites with High-k

11.3.1 Core–Shell Nanoparticles Produced Using the "Grafting-From" Technique

This technique is based on the in situ polymerization of monomers on the surfaces of nanoparticles that have been functionalized with an initiator. This results in the production of nanocomposites. The incorporation of an adequate number of starting sites onto the surfaces of the nanoparticles is the essential component of this technique. The controlled or live radical polymerization approach, which includes techniques such as

a. Reversible addition-fragmentation chain transfer polymerization (RAFT)
b. Atom transfer radical polymerization (ATRP),

These are the potent "grafting from" methods. These methods offer many benefits as follows [27–29, 35].

(i) The shell layer may be formed as a matrix to form highly stable core–shell materials with high-quality and highly filled materials free of voids, pores, and defects.
(ii) The nanoparticle aggregation has been prevented by the outer layer (shell) of the core material.
(iii) The feed ratio of the monomer and the functionalized initiator nanoparticles may be modified to change the nanoparticle concentration.

(iv) Any polymer chains are firmly attached to the surfaces of the nanoparticles, creating a sturdy nanoparticle/matrix composite with an intact interface.
(v) A wide variety of monomers may be converted into polymers.

Jiang and Huang utilized polystyrene (PS) and poly (methyl methacrylate) [PMMA] nanocomposites and adopted RAFT and ATRP to synthesize high-k core–shell PS@$BaTiO_3$ and PMMA@$BaTiO_3$ nanoparticles correspondingly [28, 29]. The schematic representation has been shown in Fig. 11.3. In both these types of nanoparticles homogeneous dispersion has been observed. The dielectric constant was increased from 2.80 to 24 with the 48 vol% $BaTiO_3$in PS@$BaTiO_3$ nanocomposite while compared with pure PS. The dielectric loss was mostly as low as pure PS. The result depicted that the dielectric loss and dielectric constant remained stable over a wide scale of frequencies. This result provided the potential for this material to be used in various functional devices and exhibits a stable and insensitive operation with varying frequencies.

Electrical vehicles, consumer electronics pulse-power systems, and electric grids make extensive use of capacitors made from biaxially oriented polypropylene (BOPP) film because of the unique blend of properties that it possesses. These properties include high breakdown strength, low energy loss, and low-capacity loss even under conditions like high sealability, frequencies, flexibility, and lightweight. However, due to the lower dielectric constant of polypropylene (PP), the use of BOPP is restricted from being utilized in potential future applications that may call

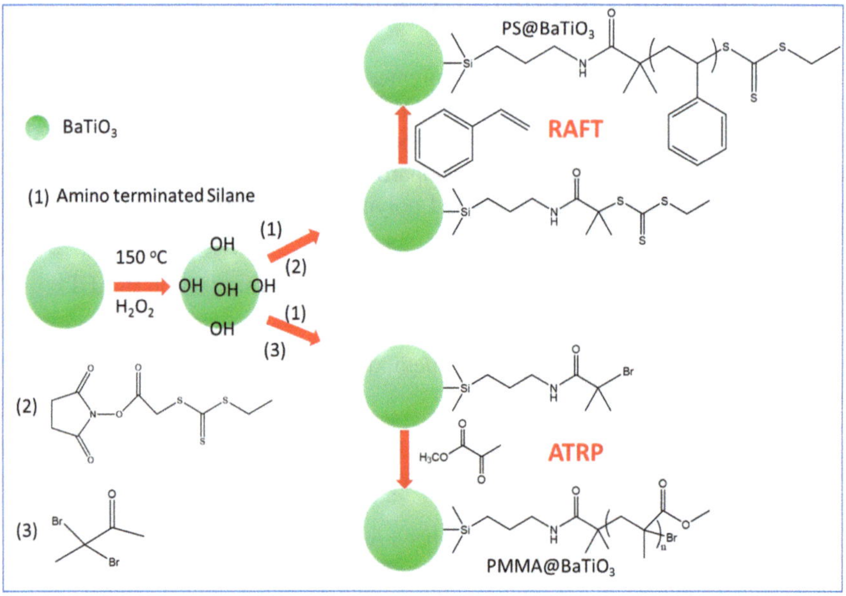

Fig. 11.3 Schematic representation of the preparation procedure for high-k PMMA and PS nanomaterials by RAFT and ATRP polymerization techniques, correspondingly

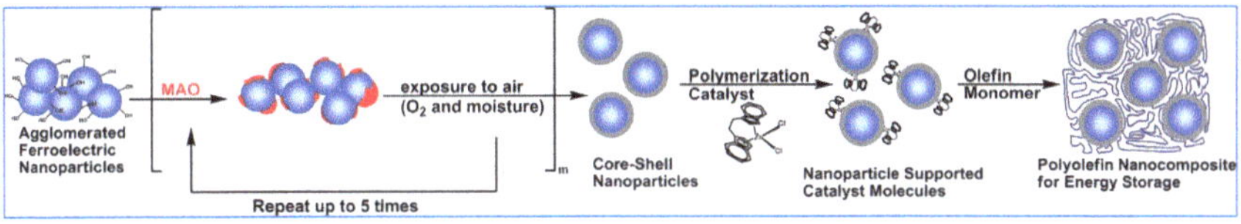

Fig. 11.4 Schematic presentation of synthetic routes of PP core–shell nanocomposite matrix [32]. Reproduced with permission from Ref. [32]. Copyright 2010, American Chemical Society

for the speedy transmission of significant amounts of energy. An efficient approach for the fabrication of PP-based high-k and low-loss nanocomposites was reported to have been developed by the Marks research group in the year 2007 [32, 44–47]. In this procedure, initially the nanoparticles (e.g., Al, TiO_2, $SrTiO_3$, $BaTiO_3$, MgO, ZrO_2, and $Ba_{0.5}Sr_{0.5}TiO_3$) were treated with a co-catalyst methylaluminoxane (MAO) which enabled the formation of Al-O bond with covalent character (Fig. 11.4).

Further, the nanoparticles that had been treated with MAO were put through a reaction with a catalyst acethylenebisindenyl zirconium chloride ($EBIZrCl_2$) for metallocene olefin polymerization. The surface functionalization of MAO on nanoparticles utilizing $EBIZrCl_2$ led to the development of species active for polymerization, being anchored on the surfaces of the nanoparticles. These polymerization-active species had the ability to initiate the in situ propylene polymerization, which ultimately resulted in the production of isotactic PP (^{iso}PP) nanocomposites. In comparison with pure PP, the ^{iso}PP nanocomposites that were synthesized, displayed significantly increased energy storage capacity as a result of their increased dielectric constant, as well as their suitably high breakdown strength.

Even though this approach only works with a small number of monomers, it offers a number of special advantages:

(i) The production of nanocomposites is quite feasible.
(ii) Large local hydrostatic pressures caused by the propagating polyolefin chains from the active catalyst centers may prevent nanoparticle agglomeration in the final nanocomposites.

Most significantly, this technology offers a practical approach to reducing the amount of electrical mismatch that exists between the matrix and the nanoparticles. Specifically, the nanoparticles are covered with a material (Al_2O_3) having a dielectric constant that is lower than that of the nanoparticles but higher than that of the matrix themselves. It is worth noting that the large dielectric constant mismatch is the primary cause of the reduction in the breakdown strength of high-k polymer composites. Keeping this in mind, the incorporation of Al_2O_3 may prove to be advantageous in preserving the high breakdown strength of nanomaterials. In specifically, Al-^{iso} PP nanocomposites show a remarkable breakdown strength[47]. According to the findings that were published by Marks, even at a high concentration (12.4 vol%) of metal nanoparticles, the Al-^{iso} PP composites exhibit a high breakdown strength of almost 80 MV m^{-1}. Even when the particle concentration is well below the percolation threshold, polymer composites filled with conductive particles typically display a very lower breakdown strength as a consequence of the electrical conduction that occurs as a result of inter-particle tunneling. This is the case even when the particle concentration is extremely low [48]. In this case, the ^{iso}PP nanocomposite matrix having a high amount of Al nanoparticles shows high breakdown strength. This may be due to the surface modification by Al_2O_3 layer, which hinders the tunneling current among particles. This is the case because these nanocomposites have a higher percentage ratio of Al nanoparticles in the composite.

11.3.2 Core–Shell Structure Fabricated Using the "Grafting-To" Method

This technique includes the production of nanomaterials by grafting the pre-synthesized polymers onto the surface of the nanoparticles. This grafting takes place as a result of an interaction between the functional groups on the surfaces of nanoparticles and the polymer end-groups. In contrast to the "grafting from" method, the "grafting to" method enables us to exercise greater command over the molecular composition as well as the molecular mass of the polymer molecules. This gives us the ability to tailor the ultimate nanocomposites to achieve the desired level of performance.

Click chemistry is a universal approach that can be utilized to join reactants. It possesses the benefits of insensitivity to solvents, high efficacy, and mild reaction parameters. Click chemistry was developed in the 1970s. As a result, click reactions are an adaptable approach for the grafting of polymer molecules to the surfaces of nanoparticles. Vaia and Tchoul et al. have used a Cu(I)-catalyzed azide-alkyne click reaction (CuAAC) to create a sequence of core–shell PS@TiO_2 nanocomposite matrix [30].

The core–shell PS nanocomposites having 27 vol% TiO_2 had a dielectric constant of 6.4 at 1000 Hz, while the dielectric loss tangent was as low as 0.625%. The nanocomposites were made with 100 kg mol^{-1} of PS. In order to obtain high carrier mobility and low leakage current in thin-film-transistors based on organic semiconductors, the composite can be utilized as the gate dielectric in these devices [31].

Due to its high complexation with azides and triazoles, the CuBr catalyst utilized in the CuAAC process cannot be readily recovered from the nanocomposites. This strong complexation has the potential to create strong frequency-dependent dielectric characteristics as well as substantial dielectric loss. In contrast to the CuAAC reaction process, the thiol-ene click chemistry is very effective and does not produce any undesirable by-products. Further, it does not need the transition metals for the catalytic process to progress. The fabrication of core–shell polymer nanocomposites is made significantly easier by the use of the thiol-ene click reaction. Utilizing thiol-ene reaction method, Huang and Jiang developed polymer@$BaTiO_3$ nanocomposite matrices. The preparation procedure is depicted in Fig. 11.5a [49].

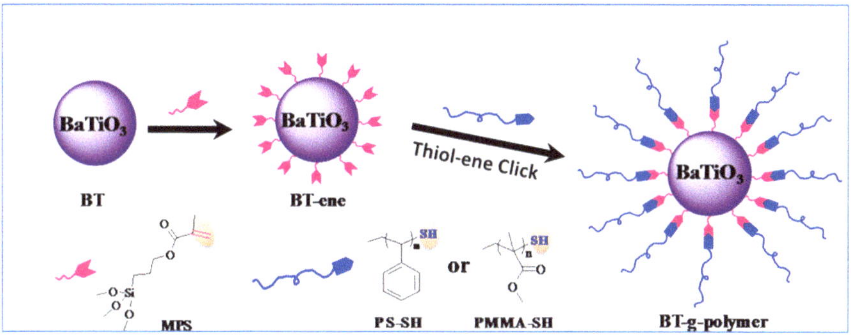

Fig. 11.5 Schematic presentation of synthetic procedure for PS@$BaTiO_3$ and PMMA@$BaTiO_3$ nanomaterials utilizing the thiol-ene click reaction [49]. Adopted with permission from Ref. [49] copyright@2014, American Chemical Society

By employing the thiol-ene click reaction macromolecules are grafted onto vinyl-functionalized $BaTiO_3$nanoparticle surface to synthesize core–shell polymer@$BaTiO_3$ nanoparticles. Using RAFT polymerization, initially, thiol-ended PMMA or PS macromolecules of various molecular mass were developed. Except for extremely low frequencies, the nanocomposites displayed a considerably improved dielectric constant, although the dielectric loss remained just as low when compared with pure polymer. This investigation also demonstrated that the dielectric properties of the polymer@$BaTiO_3$core–shellcomposite depended upon two factors. These factors include the grafting density of the core–shell structured nanoparticles and the molecular mass of the polymer chains, particularly at low frequencies. In particular, the dielectric loss was shown to be related to the molecular mass of the polymer molecules. High molecular mass polymer chains frequently have a propensity to cause poor grafting density and hence show a significant dielectric loss in nanocomposites [50].

Except for click reaction, the technique of producing core–shell structured high-k nanocomposites may make use of any alternative reaction to facilitate a linkage between nanoparticle surface and organic molecule chains. Maliakal et al. are able to affix oleic acid-terminated TiO_2 nanoparticle surfaces and phosphonate end-functionalized PS and effectively create core–shell high-k PS@TiO_2 nanocomposite dielectrics. This was done in an effort to find organic thin-film semiconductor transistors with high-k [31].

When the volume percentage of TiO_2 reached 18.2% the nanocomposite matrix exhibited a high dielectric constant (9.4), and the transistor that was produced demonstrated good mobility, which was desirable. Kim and Jung synthesized core–shell nanomaterials by surrounding $BaTiO_3$ with a PS-based copolymer. The name of the copolymer was polystyrene-block-poly (styrene-co-vinylbenzylchloride) [PS-b-PSVBC]. This resulted in a novel core–shell material in which the positively charged PSVBC was protected by the negatively charged PS shell. In such a case, the charged PSVBC shell resulted in improvements in the dielectric constant, while the PS shell operated to lower the amount of leakage current and minimize the formation of breakdown channels. Both effects were achieved via the combined efforts of the two shells. The increase in the dielectric constant may be traced back to the primary reason, which is the increased interfacial polarization that can be found between the PSVBC shell and the PS shell. In conclusion, the shielding of the PS shell on the charged PSVBC shell is responsible for the decrease in leakage currents as well as the rise in breakdown strength. The nanocomposite materials demonstrated a substantially enhanced energy storage density, and the PS-b-PSVBC composite matrix composed of 75 weight percent $BaTiO_3$ was discovered to have the greatest energy density determined theoretically. The energy density was found to be 9.7 J cm^{-3}. The nanocomposites also showed a significant reduction in the amount of space required to store the energy [33].

When it comes to the preparation of high-k nanocomposites on a large scale, the use of polymers that are readily available on the market as shells is the more appealing option. In recent work, Huang and Jiang generated polymer@$BaTiO_3$core–shell nanocomposites by adopting the "grafting to" technique with high-k. They accomplished this by employing poly (vinylidene fluoride-co-hexafluoropropylene) [PVDF-HFP] functionalized with commercially available poly (glycidyl methacrylate) [PGMA] [51]. The PGMA-functionalized PVDF-HFP generated a strong and durable layer on the surface of $BaTiO_3$, providing a homogeneous distribution of $BaTiO_3$ in PVDF-HFP matrix. Additionally, the energy density and dielectric constant of the nanocomposite matrix improved with increasing $BaTiO_3$ concentration, although the dielectric loss dropped moderately. This was seen despite the fact that the dielectric loss increased to some extent.

11.3.3 Nanoparticles Comprised of an Organic Core Surrounded by an Inorganic Shell Used as Fillers

This method refers to the synthesis of polymer matrix by inserting core–shell inorganic–organic nanomaterial into a nanocomposite, analogous to the usual solution-mixing and melt-mixing processes for preparing composites. The ability to accurately control the concentration of inorganic nanoparticles within the nanocomposites as well as the qualities that are closely connected with the nanoparticle concentration is one of the most significant benefits of this technology. The importance of this method lies in the fact that it enables an individual to gain a basic understanding of the function exhibited by the interface in determining the electrical characteristics of the nanocomposite matrix. Huang and co-worker explored the effect of matrix/shell physical interface interactions on the dielectric properties of nanocomposites composed of fluoro-polymer@$BaTiO_3$ nanoparticles and ferroelectric polymers [35]. They started with preparing core–shell fluoro-polymer@$BaTiO_3$ nanoparticles, which not only had differing thicknesses of polymer shell but also had different molecular structures. This was accomplished by inserting two different kinds of monomers based on fluoroalkyl acrylate by RAFT polymerization. These are specifically, trifluoroethyl acrylate (TFEA), and 1H, 1H, 2H, 2H-heptadecafluorodecyl acrylate (HFDA). The nanocomposites were made by first combining the fluoro-polymer with the $BaTiO_3$ nanoparticles, and then adding the PVDF-HFP. The authors discovered that the shell structure of the fluoro-polymer@$BaTiO_3$ was closely connected with the dielectric characteristics and energy storage capacity of the nanocomposite matrix. The authors hypothesized that the ordered chain structure of PHFDA led to lower inter-chain interactions and lower chain mobility between PHFDA@$BaTiO_3$ and the PVDF-HFP composite matrix, whereas the irregular arrangement of the chain structure of PTFEA led to higher inter-chain interactions and higher chain mobility between PTFEA@$BaTiO_3$ and the PVDF-HFP composite matrix. This led to a more robust suppression of the space charge. Because of this, the PTFEA@$BaTiO_3$ nanocomposites were able to attain a reduced dielectric loss, a higher breakdown strength, and a greater capacity for energy storage. The effect that the shell/matrix chemical interaction plays (also known as interfacial bonding) in determining the dielectric characteristics of high-k polymer nanocomposites was another aspect that Huang and Jiang looked at. They did this by comparing the dielectric characteristics of single-core double-shell nanocomposite matrix (NC I) and single-core single-shell nanoparticle-filled nanocomposites (NC II). In this case, the matrix of NC II and the outer shell of NC I shared an identical chemical structure [52].

Functionalizing hyperbranched aromatic polyamides (HBP) onto the surfaces of the $BaTiO_3$ nanoparticles was the initial step in the process of making single-core single-shell nanoparticles, which they referred to as HBP@$BaTiO_3$. Then, they grew PMMA on the surface of HBP@$BaTiO_3$ nanoparticles in order to construct nanocomposite I (also known as PMMA@HBP@$BaTiO_3$). This was done by using ATRP. The nanocomposite known as nanocomposite II (PMMA/HBP@$BaTiO_3$) was created by combining core–shell HBP@$BaTiO_3$ nanocomposite with PMMA [52].

Both nanocomposites displayed a nanoparticle dispersion that was homogeneous across their structures and strong adhesion between their interfaces. Dielectric analysis, on the other hand, revealed that the two types of nanocomposites displayed significantly distinct sets of properties. The PMMA@HBP@$BaTiO_3$ nanomaterials show high electrical insulation. This exhibited low electrical conductivity at low frequencies. Electrical conductivities were substantially frequency dependent, whereas dielectric properties were only mildly frequency dependent. Because of their high dielectric constant, these materials are particularly appealing for use in applications involving energy storage and gate dielectrics. On the other hand, the highly loaded PMMA/HBP@$BaTiO_3$ nanocomposites exhibited dielectric characteristics that were comparable with those of percolative composites.

To be more specific, the electrical conductivity of the nanocomposites was not affected by frequency. However, the dielectric constant and dielectric loss of the nanocomposite exhibited a significant increase at low frequencies, while simultaneously demonstrating a rapid decrease as the frequency was increased. In spite of the extremely high dielectric constant that was attained, the PMMA/HBP@$BaTiO_3$ nanocomposites are not desirable for use in any power energy storage or dielectric application. This is because the nanocomposites have substantial electrical conductivities and dielectric loss. In a different study, Huang and Jiang created mixes of fluoro-tripolymer and HBP@$BaTiO_3$. It was observed that the core–shell HBP@$BaTiO_3$ nanoparticles were not compatible with the fluoro-tripolymer matrix [53].

The researchers investigated that when the composites were subjected to relatively high concentrations of HBP@$BaTiO_3$ (i.e., concentrations that were more than 20% by volume), the composites started to demonstrate properties of dielectric materials that were comparable with those of percolative nanocomposites. According to the findings of the two observations that were just discussed, the establishment of interfacial interaction is typically the most important feature that plays a role in realizing required dielectric characteristics in composites when the matrix and the introduced inclusion have a significant difference in their electrical conductivity. In the instance of single-core double-shell nanomaterials, the inner shell was covalently bound to the outer shell. The outer shell was capable to prevent the passage of charge carriers by acting as a shield on the charge-rich inner shell. This caused poor interfacial polarization and a lower leakage current density, which in turn enhanced the dielectric properties of the nanocomposites and resulted in a low dielectric loss.

A number of the core–shell polymer@$BaTiO_3$ documented in the literature still now, have been based on commercially accessible $BaTiO_3$ nanoparticles, which is a technique that is quite viable and might potentially save time. However, these approaches are not able to control the size of the nanoparticles, and as a result, their electrical properties are size-dependent. More recently, the research team led by Lin presented an exciting technique to manufacture core–shell nanocomposites, comprising the core–shell polymer@$BaTiO_3$ [54, 55]. They started by preparing star-like gigantic molecules, as shown in Fig. 11.6. For this purpose, poly (acrylic-acid)-block-polystyrene (PAA-b-PS) was generated from a core comprised of cyclodextrin using ATRP and process of hydrolysis. Further, they added the inorganic precursor, and by selecting the appropriate solvent, they were able to force the precursor to concentrate in the hydrophilic core poly (acrylic-acid) of the PAA-b-PS. Condensation at elevated temperatures finally produced single-crystalline $BaTiO_3$ nanoparticles that were the same size as that of the PAA core. By utilizing this method, it was possible to regulate the size of the crystalline $BaTiO_3$ nanoparticles by adjusting the core size of the PAA. It was found that two different types of core–shell nanoparticles with varying $BaTiO_3$ sizes (11 and 27 nm) were effectively synthesized. These core–shell nanoparticles resulted in high dielectric constants than that of formerly published core–shell nanomaterials. It was primarily because of the higher size (approximately 50–140 nm) of commercially accessible $BaTiO_3$ nanoparticles [34, 54].

It is interesting to note that the core–shell nanoparticles are also known to be used as fillers during the synthesis of high-k polymer nanocomposites based on deblock copolymer (PS-b-PMMA). PS@$BaTiO_3$ was injected selectively into the cylindrical PS nanodomains in PS-b-PMMA, yielding in nanocomposites with vertically oriented PS nanocylinders filled with PS@$BaTiO_3$. This is in contrast to the nanocomposites that were created using conventional mixing techniques, in which

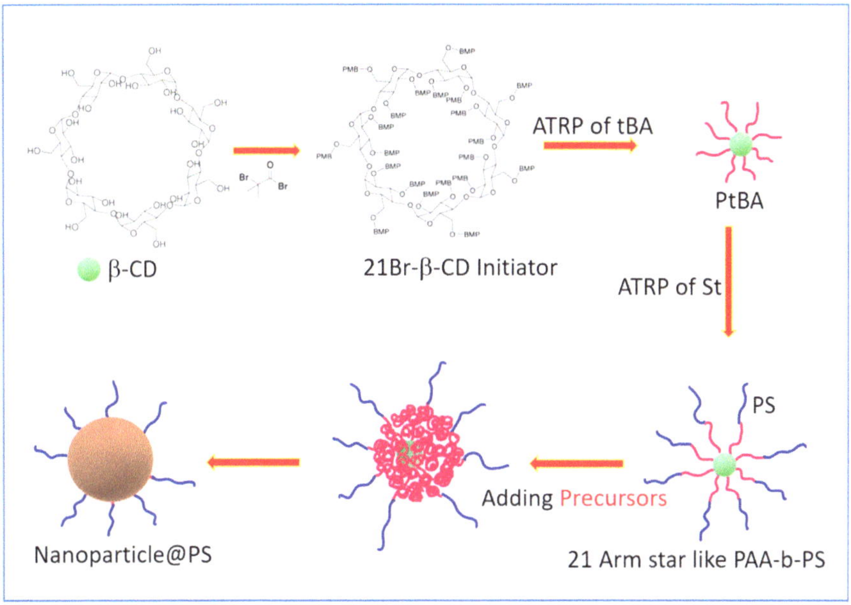

Fig. 11.6 Scheme displaying synthetic approach of PS@$BaTiO_3$ core–shell nanocomposite

the nanomaterials were arbitrarily dispersed throughout a polymer matrix. The PS-b-PMMA/PS@$BaTiO_3$ nanocomposites exhibited high dielectric constants (greater than 18) independent of frequency. Additionally, a negligible dielectric loss was observed even at a high-frequency range of 2–15 GHz. These nanocomposites also showed tremendous potential for dielectric and energy storage applications[34].

11.3.4 Core–Shell Nanoparticles of Other Types as Fillers

In addition to core–shell structured inorganic–organic nanoparticles, to create high-k polymer nanocomposites, different varieties of core–shell nanocomposites have also been utilized as filler materials. These nanoparticles include inorganic-inorganic (e.g., SiO_2 @ $BaTiO_3$ and TiO_2@$BaTiO_3$) [38, 56], organic-organic (e.g., PDVB@PANI, and poly divinyl benene@polyaniline) hybrids, metal-inorganic (e.g., TiO_2 @ Ag), and Al_2O_3 @Al [36, 39], and metal–organic (e.g., C@Ag) [37, 57, 58]. The core–shell structured TiO_2 @ $BaTiO_3$ nanoparticles as well as the associated PVDF nanocomposites were synthesized by Yao and Rahimabady [56]. They made a startling discovery when they compared the dielectric characteristics of the PVDF/TiO_2@$BaTiO_3$ nanocomposites with those of the PVDF/$BaTiO_3$ nanocomposites: the PVDF/TiO_2 @$BaTiO_3$ nanocomposites exhibited a considerable improvement. Specifically, the PVDF/TiO_2 @$BaTiO_3$ nanocomposites exhibited an increase in both the dielectric constant and the breakdown strength. Discharged energy density of the nanocomposite with 30 vol% TiO_2 @$BaTiO_3$ was 12.2 J cm^{-3} at 340 MV m^{-1}, which is almost three times as high as the discharged energy density of pure PVDF (4.1 J cm^{-3}). In polymer nanocomposites containing high-k particles, the large increase in dielectric constants is often achieved at the price of a reduction in breakdown strength. This is a conclusion that has been widely accepted based on research. Considering this interpretation, the findings of Yao and Rahimabady's study provide a significant piece of the puzzle necessary to reconcile this well-known discrepancy. The insertion of a TiO_2 layer, which resulted in a reduction in the local electric field distortion, was credited by the authors as the cause of the improvement in dielectric performance. The "buffer layer" hypothesis that had been developed by the Marks group is basically compatible with this finding. Wong and Xu are considered to be pioneers in the fabrication of core–shell structured Al_2O_3@Al nanoparticles for use in low-loss, high-k polymer nanocomposites[39].

It has been claimed that the dielectric constant may reach the highest value as 60 when the proportion of Al nanoparticles (about 100 nm in size) in the material is 50 weight percent, while the dielectric loss can reach a lower value of 0.02. The Al core which is electrically conductive causes a rise in electric field strength inside the polymer nanocomposite, which adds to a rise in the dielectric constant of nanocomposite matrix. As a consequence of the shielding Al_2O_3 layer that is located surrounding the Al core, electron transport between the Al nanoparticles is restricted, which in turn results in a minimal dielectric loss. Since Ag is an excellent electrical conductor, Ag-filled polymer composites function as conventional percolation systems. When the Ag nanoparticle loading is far below the percolation threshold in these systems, the dielectric constant of the composites grows slowly, and when it is close to the percolation threshold, it increases quickly. Since Ag is an excellent electrical conductor, Ag-filled polymer composites may function as characteristic percolation systems. After passing the percolation threshold, the composites eventually become electrically conductive, and after that, there is a sudden drop in the dielectric constant [59].

The creation of conductive routes between neighboring Ag nanoparticles is inhibited after the Ag nanoparticles have been enclosed by an insulating polymer shell. Additionally, the tunneling currents between neighboring Ag nanoparticles decrease as the shell thickness increases. Nanocomposite with an Ag core encapsulated by an insulating shell behave as insulating material even a considerable amount of Ag present after the shell width reaches a threshold value at which the tunneling current reduces to a minimal value. In this scenario, the nanocomposites will demonstrate both a high dielectric constant as well as a low dielectric loss at the same time. When the concentration of Ag nanoparticles was higher than 20 vol%, Shen and Nan discovered that the dielectric constant increased by more than two times in magnitude when the epoxy nanocomposite is filled with core–shellC@Ag nanoparticles as fillers. On the other hand, a little enhancement (i.e., from 2 to 4%) in the dielectric loss tangent was observed [37, 57].

Additionally, it is important to note that the shell thickness has an effect on the dielectric characteristics of the epoxy/C@Ag nanocomposites and that this effect may be modified. Coulomb blockade and quantum confinement effects may take place if the diameter of a nanocrystal is shrunk to a size that is sufficiently small [60]. In this scenario, a metallic particle may acquire insulating properties against electrical current. In addition, there will be a restriction placed on the movement of electrons from one particle to its adjacent particle. In this scenario, one may make use of the coulomb blockade effects and quantum confinement of ultra-small metal nanomaterials in order to reduce the amount of leakage current and to improve the breakdown strength of nanocomposites. Based on this concept, Huang and Xie discovered that in contrast with $BaTiO_3$ nanomaterials, nanocomposite matrix packed with $BaTiO_3$ decorated with ultra-small Ag (such as Ag@$BaTiO_3$) hybrid nanoparticles had increased breakdown strength and lower dielectric loss [61]. Scientists inferred that the presence of coulomb-blockade effect in Ag@$BaTiO_3$ nanocomposites based on the dramatically reduced electrical conductivities of the nanocomposite. Core–shell nanoparticles that are based on inorganic and metal cores often have a high density, which might cause the target devices to become unnecessarily heavy. The use of completely organic fillers is encouraged as a means of lowering the device's overall weight while keeping the device's functioning intact. Opris and Molberg developed high-k poly diomethyl siloxane (PDMS) nanomaterials by utilizing polyaniline functionalized poly divinyl benzene (PDVB@PANI) core–shell particles [58].

Surprisingly, when dielectric constant PDMS/PDVB@PANI composites were compared with pure PDMS, an enhancement in dielectric constant was observed, however, exhibited a breakdown strength that was equivalent to that of pure PDMS. For instance, the dielectric constant rose from 2.3 for pure PDMS to 7.6 for composites containing 31.7% PDVB@PANI nanoparticles at 100 Hz, whereas the breakdown strength of the composites only exhibited a minor drop (going from 66.1 to 64.1 $V\ m^{-1}$) throughout this same time period.

Strawberry-like core–shell $BaTiO_3$-polydopamine-Ag (BT-PDA-Ag) hybrid nanocomposites were produced by Xingyi Huang et al., along with the related poly (vinylidenefluoride-co-hexafluroro propylene) P(VDF-HFP)/BT-PDA-Ag nanocomposites (Fig. 11.7). The PDA shell not only made it easier to disperse the BT nanoparticles homogeneously throughout the P (VDF-HFP) matrix, but it also improved the interfacial interaction between the polymer matrix and the nanoparticle. In order to properly introduce the Coulomb-blockade effect into the nanocomposites, the nano-Ag was painted in a homogeneous manner onto the PDA shell. In comparison with P (VDF-HFP)/BT-PDA and P (VDF-HFP)/BT, the polymer matrix with BT-PDA-Ag displays exceptional electric properties. These excellent electric properties include reduced remnant polarization, suppressed dielectric loss, enhanced energy efficiency, and energy density as well as breakdown strength. Therefore, the nanocomposites that were produced as a consequence have a great deal of promise for use in dielectric and energy storage applications. This discovery also paves the way for the development of novel methods for the preparation of nanocomposites that have a high energy density and a low dielectric loss [62].

$BaTiO_3$ nanowires with a TiO_2 shell layer encapsulation were produced by Xingyi Huang and co-workers in order to construct polymer nanocomposites with a high capacity for energy storage. According to the findings of the research, the shell

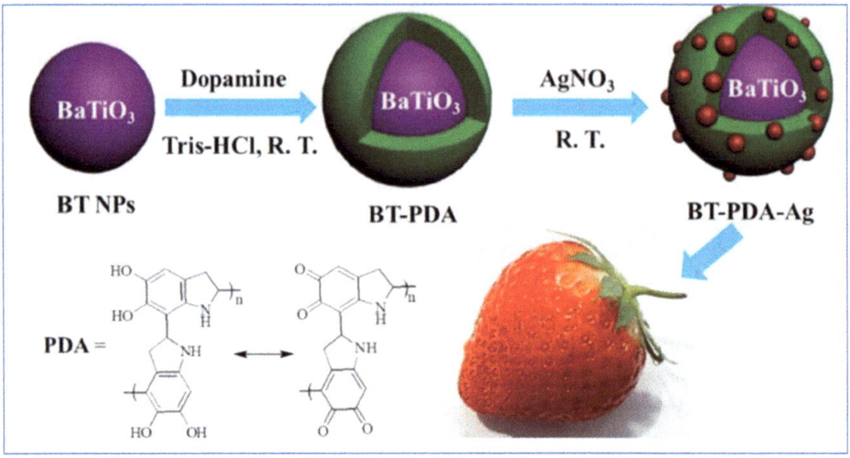

Fig. 11.7 Schematic representation of strawberry like BT-PDA-Ag nanoparticles [62]. Adopted with permission from Ref. [62] copyright@ 2015 WILEY–VCH Verlag GmbH & Co. KGaA, Weinheim

width of the buffer layer on the nanofillers plays an essential role in the process of customizing the performance of the nanocomposites that are created in this way. By using nanowires with a core–shell structure rather than bare one, it was possible to drastically reduce both the dielectric loss and the leakage current. Additionally, the introduction of TiO_2 shell layers helped to improve the reduction in the breakdown strength of these nanocomposites that included raw nanowires. The decrease of dielectric loss and the amplification of electric fields in polymer nanocomposites was another area where the finite element analysis validated the superiority of TiO_2 shell layer-encased nanowires. In light of this fact, the nanocomposites that had TiO_2 shell layers that enclosed $BaTiO_3$ NWs demonstrated a significantly enhanced capacity for the storage of energy in contrast to those that contained naked nanowires. The extensive analysis of the core–shell nanowires' electrical properties revealed that the performance of the nanocomposites does not rise in a linear fashion with increasing shell thickness. Instead, the performance of the nanocomposites was shown to be something that could be optimized by varying the shell thickness. This opens the door for the construction of high-energy-density polymer nanocomposites that are appropriate for use in applications involving the next generation of energy storage capacitors (Fig. 11.8) [63].

Core–shell BT@Al_2O_3 nanoparticles were developed by Yao Wanga and Yuan Deng et al., and then integrated into the PVDF matrix to generate flexible composite films. This was accomplished via a heterogeneous nucleation approach. There were discernible impacts of the uniform Al_2O_3 shell on the nanocomposite's resultant dielectric properties, and these effects were detected. The nanocomposite with bare BT NPs had a dielectric loss of about 0.02 and a conductivity of about 2.3×10^{-8} Sm^{-1}, whereas the composite with 20 vol% BT@ Al_2O_3 NPs had conductivity that was about 2.3×10^{-8} Sm^{-1}. This represents a decrease of about 48% and

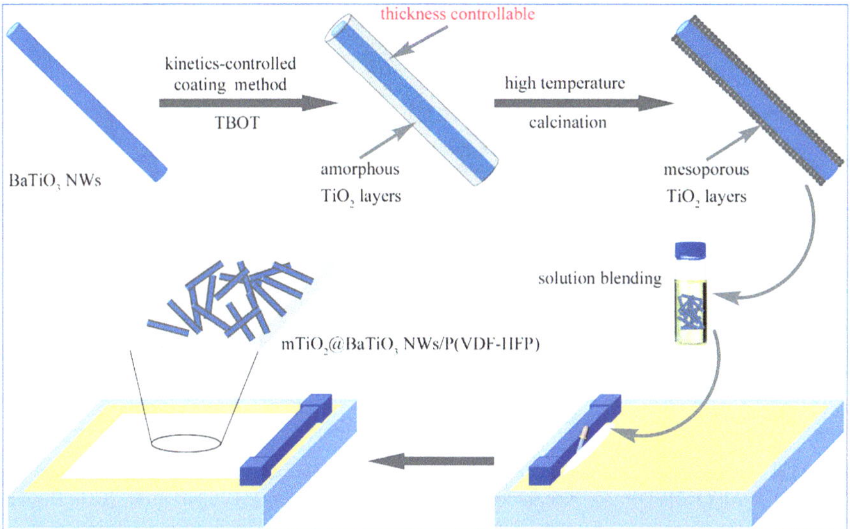

Fig. 11.8 Schematic diagram of the preparation of $mTiO_2$@$BaTiO_3$ NWs followed by synthesis of nanocomposite film [63]. Adopted with permission from Ref. [63] copyright@2017, Royal Society of Chemistry

78.8%, respectively, from the nanocomposite with bare BT NPs. The use of an Al_2O_3 shell layer has been shown to effectively reduce interfacial polarization as well as electric conduction and electric field concentration in composites. This has led to an increase in dielectric breakdown strength as well as a reduction in energy loss without causing degradation in the dielectric constant or discharged energy storage density. They revealed an efficient approach for obtaining dielectric materials with fairly low dielectric loss and decreased energy storage loss. These are extremely significant factors for assessing the performances of high-power systems, and the method demonstrates how it may be done [5] (Fig. 11.9).

Sandwiched polymer ceramic-based dielectrics were developed by Chung Ming Leung and co-workers. These dielectrics show the possibility for use in energy storage devices owing to their large power density, excellent flexibility, and cheaper cost. The poor energy storage density, however, continues to be the barrier in this process. These nanocomposite films are designed by keeping the poly (vinylidene fluoride) (PVDF) as the middle layer integrated with $BaTiO_3$ (BT) coated Ag nanomaterials (Ag@BT NPs). The bottom/top layers are composed of PVDF-embedded 2D layered MXene nanosheets. This results in the formation of sandwiched polymer-ceramic nanocomposite films. It is interesting to note that the composite best performed with a reversible energy storage density (U_eW_{re}) as high as 22.3 J/cm^3 and an efficiency of 77% at a low electric field of 270 MV/m that are reached for the film with a middle layer that contains 5 weight percent DA@Ag@BT/PVDF. The high-energy storage ability may be attributed to the strong interfacial coupling, ionic contact, and higher breakdown strength owing to the Coulomb-blockade effect in the hierarchical

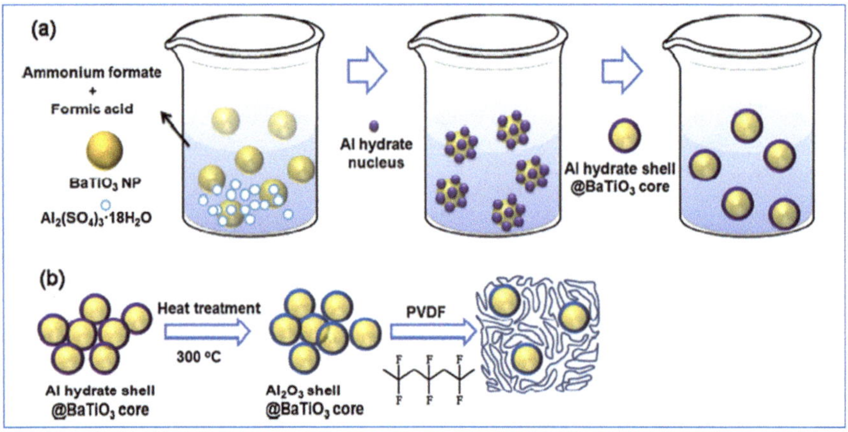

Fig. 11.9 Schematic presentation of **a** Synthesis of $BaTiO_3@Al_2O_3$ nanoparticles and **b** $BaTiO_3@Al_2O_3$/PVDF [5]. Adopted with permission from Ref. [5] copyright@ 2016, Elsevier Ltd.

structures that include a variety of fillers. In addition, the results of the phase-field simulation demonstrated that enough quantity of DA@Ag@BT NPs can effectively avoid the creation of conductive networks, suppress dielectric loss, and increase breakdown strength. This hierarchical technique offers an appropriate route for the fabrication of high-performance polymer nanocomposites (Fig. 11.10) [64].

Poly (p-phenylenebenzobisoxazole) (PBO) is a kind of polymer that maintains its structural integrity at high temperatures. However, its employment in high-performance dielectrics is limited due to its poor processability as well as its low inherent dielectric constant. Xiaoyun Liu and Qixin Zhuang produced PBO using the precursor technique, and the preparation of polymer films is carried out using solution casting, which is a more time- and labor-efficient approach than the

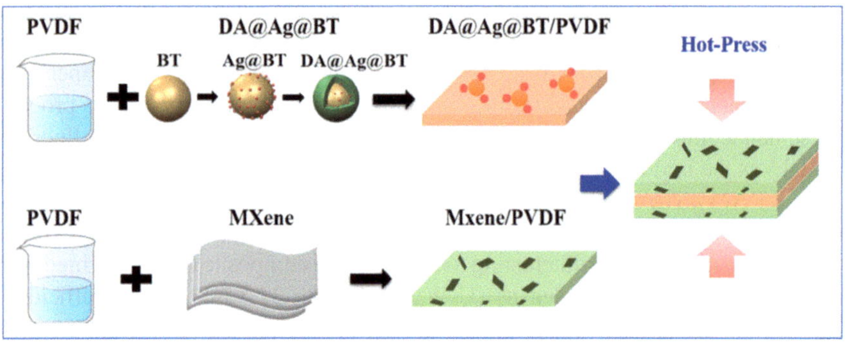

Fig. 11.10 Schematic representation of the synthesis of DA@Ag@BT, MXene, and sandwiched structure of MP-xDA@Ag@BT/PVDF [64]. Adopted with permission from Ref. [64] copyright@ 2022, Elsevier Ltd. and Techna Group S.r.l

conventional way of processing. In addition to this, a hierarchically organized one-dimensional Al_2O_3@$NaNbO_3$ is fabricated and then injected into the PBO matrix. The nanocomposite has significantly improved dielectric characteristics and breakdown strength as a result of both its high aspect ratio and its unique structure, which is Al_2O_3@$NaNbO_3$. The nanocomposite that is filled with 3 volume percent Al_2O_3@$NaNbO_3$exhibits the maximum breakdown strength of 286 kV mm^{-1}. In addition, maintains a high dielectric constant of 6.68 as well as a low dielectric loss of 0.035 at 1 kHz. Further, the energy density enhanced to 1.64 J cm^{-3} at 250 kV mm^{-1}. This is about 200% higher than the pure PBO and the nanocomposite has a low dielectric loss. In addition, the dielectric characteristics of 3 vol% Al_2O_3@$NaNbO_3$-PBO demonstrate remarkable thermal stability from 25 to 250 °C. This great thermal stability may be due to the heat resistance of both PBO and Al_2O_3@$NaNbO_3$, which are both present in the material. This study demonstrates a potentially useful method for the fabrication of polymeric dielectrics that may be used in harsh conditions [65].

In the constantly emerging areas, the synthesis of multifunctional epoxy resins using environmentally friendly processes is significant (AI, 5G, IoT, electrical and electronics appliances). In order to accomplish this goal, a multifunctional nanocomposite was manufactured by Yiting Xu et al. by utilizing electrostatic interactions and hydrogen bonding to functionalize a coating of bio-based polyelectrolyte (containing chitosan and phytate) on the surface of boron nitride in an environmentally friendly manner. This coating was done in order to create a multifunctional nanoparticle with multiple applications. Most significantly, the building process makes no use of any organic solvents and does not rely on any outside sources of heat. Further, this nanoparticle with several functions is used in the process of functionally modifying epoxy resin. The LOI value of the functionalized epoxy resin is increased to 32.8% at a reduced adding quantity of 7 weight% because of the synergistic action of polyelectrolyte and boron nitride at the nanoscale. In addition, when subjected to a vertical combustion test, the modified epoxy resin demonstrated a quick extinguishing phenomenon that lasted for just 0.5 s after the fire source was turned off. When compared with clean EP, the peak heat release rate (pHRR) and total smoke production (TSP) both show reductions of 54.3% and 33.0%, respectively. Additionally, the multifunctional nanoparticles confer excellent thermal conductivity, mechanical characteristics, and dielectric properties on the modified epoxy. Overall, the results of this study provide a sustainable approach to the production of epoxy resins with several functionalities (Fig. 11.11) [66].

Due to the high dielectric constant that may be attained around the percolation threshold, conductive filler-filled polymer composites have shown a great deal of promise. However, the substantial dielectric loss that is connected with these composites prevents their widespread use in reality. In recent years, considerable effort has been put toward encapsulating conductive fillers within an insulating shell, with the intention of limiting the dielectric loss. This conundrum begs the issue of whether or not core–shell structured fillers are ultimately useful for the dielectric performance of composites, which is an area that has received less investigation. For this purpose, Zhi-Min Dang evaluated the properties of polymer nanocomposites that include a variety

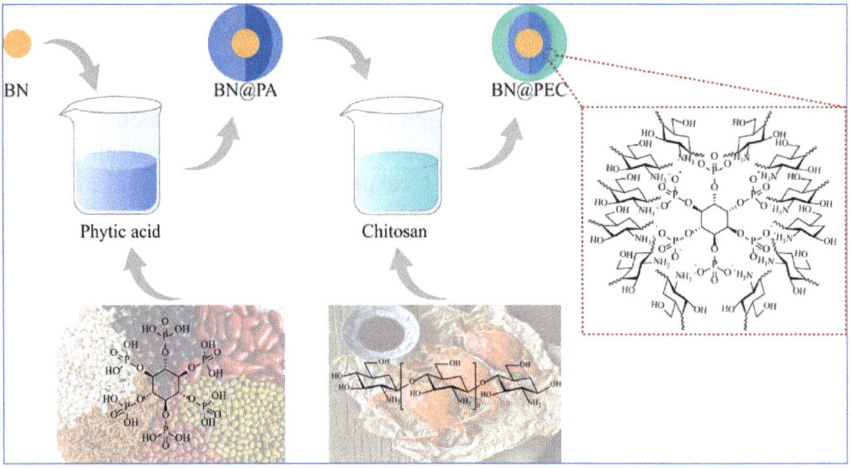

Fig. 11.11 Schematic representation of a green synthetic procedure for BN@PEC [66]. Adopted with permission from Ref. [66] copyright@ 2022, Elsevier Ltd.

of Al@Al_2O_3 nanofillers with varying shell thicknesses. It demonstrates that a rapid intra-particle polarization and a sluggish inter-particle polarization both contribute to the high dielectric constant of percolative composites. The formation of an insulating shell makes it possible to exercise separate control over the two polarizations, which, in ordinary percolative composites, would otherwise be linked (meaning that they would increase or decrease simultaneously). The core–shell structured nanocomposites are able to obtain a high dielectric constant, while simultaneously having a low dielectric loss, which is a significant improvement over the unmodified nanofiller composites. This is accomplished by promoting polarization within the particle and suppressing polarization between the particles. In addition to that, both the thermal conductivity and the high field resistivity have been increased, which brings about an operating temperature that is both consistent and low. This study presents a new paradigm for the design of percolative polymer composites, which have the potential to increase thermomechanical characteristics in addition to having a high dielectric constant and a low dielectric loss [67].

Inorganic nanoparticles with a high dielectric constant (ε_r) have been used to strengthen dielectric polymer nanocomposites, which have been the subject of much research for energy storage functions in modern electronics and electrical devices. Although the integration of inorganic nanoparticles with a high ε_r-value can enhance the ε_r-value of the composite matrix to a certain level, it will also certainly decrease the inclusive breakdown strength (E_b) of the complex matrix. This eventually hampers the effective enhancement of the energy storage efficacy of the composites. Xin Zhang et al. developed a method to fabricate high-ε_r $BaTiO_3$ (BTO) nanoparticles with polyimide polymer (PI) shells (PI@BTO) through an in situ polymerization technique using poly etherimide (PEI)-based nanocomposites. Because of the modified PI shell, the interaction between the organic or inorganic interface has been improved, which has led to a uniform distribution of nanoparticles throughout the PEI composite matrix. Specifically, the normal electrostatic interaction between the PEI matrix and polymer chains present in the PI shell increases the E_b of the PEI/PI@BTO nanocomposite in comparison with pure PEI. This, in turn, results in a high charge–discharge efficiency (η) of more than 80% and a high-energy storage density (U_e) of 6.2 J/cm^3 in the PEI nanocomposites, which is an improvement of 150% than that of pure PEI. The work proposes a facile and efficient approach for surface functionalization to increase the compatibility between the interfaces and E_b of polymer nanocomposite matrix. Furthermore, the remarkably improved performances in energy storage of the synthesized nanocomposite matrix provide great potential for the fabrication of high-performance capacitors to meet the growing demands for more compact, energy-efficient, and cost-effective electrical device electronics[68].

Filling an organic dielectric with inorganic nanoparticles that have a high dielectric constant is an efficient method that may be used to improve the energy storage capability of the dielectric. On the other hand, the dielectric difference between polymer and ceramic produces early breakdown, which in turn restricts the storage density of ceramic/polymer nanocomposites when they are used for dielectric capacitors. Liang Cao and Huixing Lin et al. used magnesium oxide (MgO) as a buffer barrier because of its high medium dielectric constant and insulation. This was done to diminish the difference in dielectric characteristics that were present between the poly (vinylidene fluoride-hexafluoropropylene) (P (VDFHFP)) substrate and $BaTiO_3$ (BT) nanoparticles. A straightforward chemical precipitation procedure was used to coat the spherical BT with the foreign oxide, resulting in the formation of a $BaTiO_3$@MgO (BT@MgO) core–shell nanostructure. This structure has been meticulously analyzed by EDS and TEM. As a result of the BT-MgO heterogeneous interfacial region's provision of carrier channels and promotion of charge movement, both the dielectric constant and the potential shift have been greatly improved. The BT@MgO/P (VDF-HFP) nanocomposite with a filling ratio of 1 volume percent gave the greatest energy density U_d. The U_d value extends up to 5.6 J/cm^3, which is 55.6% and 40.0% larger than that of the BT-filled counterpart and host matrix with the same filler quantity. The BT@MgO core–shell nanostructure is an example of an alternate method that may effectively increase the energy storage functioning of polymer/ceramic composite dielectrics. Due to MgO's high insulation and medium dielectric constant, it has been decided to use it as a buffer barrier between the BT nanoparticles and the P (VDF-HFP) substrate in this research. This helped to reduce the dielectric mismatch that existed between the two. The core–shell $BaTiO_3$@MgO nanoparticles have a heterogeneous interfacial area that is composed of BTMgO. This region functions to create pathways for carriers and enhance charge mobility. Therefore, the inclusion of an MgO shell considerably improved the energy storage qualities. The BT@MgO-reinforced composite dielectric with a 1 volume % filling ratio achieved the highest possible U_d value for energy density. At 4235 kV/cm this value reached 5.6 J/cm^3, which is larger by 40.0% than that of the host matrix and 55.6% more than that of BT-loaded dielectric with similar filler composition. In the meanwhile, it maintains a high-energy usage efficacy, with a value of 51.3%

for the coefficient. The core–shell-structured BT@MgO nanofiller is an example of an alternate approach that may effectively enhance the energy storage function of polymer/ceramic dielectric composites [69].

11.4 Conclusion

In conclusion, this chapter discusses the various methods of preparation of high-k core–shell nanomaterials for dielectric applications. Core–shell techniques have made significant strides in enhancing nanocomposites with high-k for dielectric applications and energy storage, however, these techniques suffer from certain problems. These problems include producing high-k nanocomposites that concurrently display a low dielectric loss, high breakdown strength, and other desired features at large production scales while preserving fine control over the nanostructures of the core–shell nanoparticles. Utilizing one-core multiple-shell systems, including nanoparticles with desired nanoeffects in the shells, and developing advanced synthesis processes for core–shell nanomaterials may give technological means to transcend these limitations.

References

1. Luo S, Yu J, Yu S, Sun R, Cao L, Liao WH, Wong CP (2019) Significantly enhanced electrostatic energy storage performance of flexible polymer composites by introducing highly insulating-ferroelectric microhybrids as fillers. Adv Energy Mater 9(5):1803204
2. Wang PJ, Zhou D, Guo HH, Liu WF, Su JZ, Fu MS, Singh C, Trukhanov S, Trukhanov A (2020) Ultrahigh enhancement rate of the energy density of flexible polymer nanocomposites using core–shell $BaTiO_3$@ MgO structures as the filler. J Mater Chem A 8(22):11124–11132
3. Wang P-J, Zhou D, Li J, Pang L-X, Liu W-F, Su J-Z, Singh C, Trukhanov S, Trukhanov A (2020) Significantly enhanced electrostatic energy storage performance of P (VDF-HFP)/$BaTiO_3$-Bi (Li0. 5Nb0. 5) O3 nanocomposites. Nano Energy 78:105247
4. Zheng W, Ren L, Zhao X, Li H, Xie Z, Li Y, Wang C, Yu L, Yang L, Liao R (2022) Tuning interfacial relaxations in P(VDF-HFP) with Al_2O_3@ZrO_2 core-shell nanofillers for enhanced dielectric and energy storage performance. Compos Sci Technol 222:109379
5. He D, Wang Y, Chen X, Deng Y (2017) Core–shell structured $BaTiO_3$@Al_2O_3 nanoparticles in polymer composites for dielectric loss suppression and breakdown strength enhancement. Compos A Appl Sci Manuf 93:137–143
6. Huang X, Jiang P (2015) Core–shell structured high-k polymer nanocomposites for energy storage and dielectric applications. Adv Mater 27(3):546–554
7. Wang Y, Zhou X, Chen Q, Chu B, Zhang Q (2010) Recent development of high energy density polymers for dielectric capacitors. IEEE Trans Dielectr Electr Insul 17(4):1036–1042
8. Chen Y, Xia Y, Sun H, Smith GM, Yang D, Ma D, Carroll DL (2014) Solution-processed highly efficient alternating current-driven field-induced polymer electroluminescent devices employing high-k relaxor ferroelectric polymer dielectric. Adv Funct Mater 24(11):1501–1508
9. Ortiz RP, Facchetti A, Marks TJ (2010) High-k organic, inorganic, and hybrid dielectrics for low-voltage organic field-effect transistors. Chem Rev 110(1):205–239
10. Li J, Sun Z, Yan F (2012) Solution processable low-voltage organic thin film transistors with high-k relaxor ferroelectric polymer as gate insulator. Adv Mater 24(1):88–93

11. Kim Y-J, Kim J, Kim YS, Lee J-K (2013) Erratum to "TiO_2-poly (4-vinylphenol) nanocomposite dielectrics for organic thin film transistors. Org Electron 14:3406–3414; 2(15):640 (2014)
12. Zhang QM, Bharti V, Zhao X (1998) Giant electrostriction and relaxor ferroelectric behavior in electron-irradiated poly (vinylidene fluoride-trifluoroethylene) copolymer. Science 280(5372):2101–2104
13. Zhang Q, Li H, Poh M, Xia F, Cheng Z-Y, Xu H, Huang CJN (2002) An all-organic composite actuator material with a high dielectric constant. Nature 419(6904):284–287
14. Dang ZM, Yuan JK, Yao SH, Liao RJ (2013) Flexible nanodielectric materials with high permittivity for power energy storage. Adv Materi 25(44):6334–6365
15. Chu B, Zhou X, Ren K, Neese B, Lin M, Wang Q, Bauer F, Zhang QJS (2006) A dielectric polymer with high electric energy density and fast discharge speed. Science 313(5785):334–336
16. Alam MA, Azarian MH, Pecht MG (2012) Effects of moisture absorption on the electrical parameters of embedded capacitors with epoxy-$BaTiO_3$ nanocomposite dielectric. J Mater Sci Mater Electron 23:1504–1510
17. Dang Z-M, Yuan J-K, Zha J-W, Zhou T, Li S-T, Hu G-H (2012) Fundamentals, processes and applications of high-permittivity polymer–matrix composites. Progr Mater Sci 57(4):660–723
18. Huang X, Xie L, Yang K, Wu C, Jiang P, Li S, Wu S, Tatsumi K, Tanaka T (2014) Role of interface in highly filled epoxy/$BaTiO_3$ nanocomposites. Part I-correlation between nanoparticle surface chemistry and nanocomposite dielectric property. IEEE Trans Dielectr Electr Insul 21(2):467–479
19. Sharifi E, Jayaram SH, Cherney EA (2010) Temperature and electric field dependence of stress grading on form-wound motor coils. IEEE Trans Dielectr Electr Insul 17(1):264–270
20. Hwang SK, Bae I, Cho SM, Kim RH, Jung HJ, Park C (2013) High performance multi-level non-volatile polymer memory with solution-blended ferroelectric polymer/high-k insulators for low voltage operation. Adv Funct Mater 23(44):5484–5493
21. Arbatti M, Shan X, Cheng ZY (2007) Ceramic–polymer composites with high dielectric constant. Adv Mater 19(10):1369–1372
22. Das RN, Lauffer JM, Markovich VR (2008) Fabrication, integration and reliability of nanocomposite based embedded capacitors in microelectronics packaging. J Mater Chem 18(5):537–544
23. Li J, Seok SI, Chu B, Dogan F, Zhang Q, Wang Q (2009) Nanocomposites of ferroelectric polymers with TiO2 nanoparticles exhibiting significantly enhanced electrical energy density. Adv Mater 21(2):217–221
24. Kim P, Jones SC, Hotchkiss PJ, Haddock JN, Kippelen B, Marder SR, Perry JW (2007) Phosphonic acid-modified barium titanate polymer nanocomposites with high permittivity and dielectric strength. Adv Mater 19(7):1001–1005
25. Kim P, Doss NM, Tillotson JP, Hotchkiss PJ, Pan MJ, Marder SR, Li J, Calame JP, Perry JW (2009) High energy density nanocomposites based on surface-modified $BaTiO_3$ and a ferroelectric polymer. ACS Nano 3(9):2581–2592
26. Hu P, Shen Y, Guan Y, Zhang X, Lin Y, Zhang Q, Nan CW (2014) Topological-structure modulated polymer nanocomposites exhibiting highly enhanced dielectric strength and energy density. Adv Funct Mater 24(21):3172–3178
27. Paniagua SA, Kim Y, Henry K, Kumar R, Perry JW, Marder SR (2014) Surface-initiated polymerization from barium titanate nanoparticles for hybrid dielectric capacitors. ACS Appl Mater Interfaces 6(5):3477–3482
28. Xie L, Huang X, Wu C, Jiang P (2011) Core-shell structured poly (methyl methacrylate)/ $BaTiO_3$ nanocomposites prepared by in situ atom transfer radical polymerization: a route to high dielectric constant materials with the inherent low loss of the base polymer. J Mater Chem 21(16):5897–5906
29. Yang K, Huang X, Xie L, Wu C, Jiang P, Tanaka T (2012) Core–shell structured polystyrene/ BaTiO3 hybrid nanodielectrics prepared by in situ RAFT polymerization: a route to high dielectric constant and low loss materials with weak frequency dependence. Macromolecular Rapid Commun 33(22):1921–1926

30. Tchoul MN, Fillery SP, Koerner H, Drummy LF, Oyerokun FT, Mirau PA, Durstock MF, Vaia RA (2010) Assemblies of titanium dioxide-polystyrene hybrid nanoparticles for dielectric applications. Chem Mater 22(5):1749–1759
31. Maliakal A, Katz H, Cotts PM, Subramoney S, Mirau P (2005) Inorganic oxide core, polymer shell nanocomposite as a high K gate dielectric for flexible electronics applications. J Am Chem Soc 127(42):14655–14662
32. Li Z, Fredin LA, Tewari P, DiBenedetto SA, Lanagan MT, Ratner MA, Marks TJ (2010) In situ catalytic encapsulation of core-shell nanoparticles having variable shell thickness: dielectric and energy storage properties of high-permittivity metal oxide nanocomposites. Chem Mater 22(18):5154–5164
33. Jung HM, Kang JH, Yang SY, Won JC, Kim YS (2010) Barium titanate nanoparticles with diblock copolymer shielding layers for high-energy density nanocomposites. Chem Mater 22(2):450–456
34. Pang X, He Y, Jiang B, Iocozzia J, Zhao L, Guo H, Liu J, Akinc M, Bowler N, Tan XJN (2013) Block copolymer/ferroelectric nanoparticle nanocomposites. Nanoscale 5(18):8695–8702
35. Yang K, Huang X, Huang Y, Xie L, Jiang P (2013) Fluoro-polymer@ $BaTiO_3$ hybrid nanoparticles prepared via RAFT polymerization: toward ferroelectric polymer nanocomposites with high dielectric constant and low dielectric loss for energy storage application. Chem Mater 25(11):2327–2338
36. Dang ZM, You SS, Zha JW, Song HT, Li ST (2010) Effect of shell-layer thickness on dielectric properties in Ag@ TiO_2 core@ shell nanoparticles filled ferroelectric poly (vinylidene fluoride) composites. Phys Status Solidi (A) 207(3):739–742
37. Shen Y, Lin YH, Nan CW (2007) Interfacial effect on dielectric properties of polymer nanocomposites filled with core/shell-structured particles. Adv Funct Mater 17(14):2405–2410
38. Yu K, Niu Y, Bai Y, Zhou Y, Wang H (2013) Poly (vinylidene fluoride) polymer based nanocomposites with significantly reduced energy loss by filling with core-shell structured $BaTiO_3/SiO_2$ nanoparticles. Appl Phys Lett 102(10):102903
39. Xu J, Wong CP (2005) Low-loss percolative dielectric composite. Appl Phys Lett 87(8):082907
40. Wang YU, Tan DQ (2011) Computational study of filler microstructure and effective property relations in dielectric composites. J Appl Phys 109(10):104102
41. Balasubramanian B, Kraemer KL, Reding NA, Skomski R, Ducharme S, Sellmyer DJ (2010) Synthesis of monodisperse TiO_2−paraffin core−shell nanoparticles for improved dielectric properties. ACS Nano 4(4):1893–1900
42. Ducharme S (2009) An inside-out approach to storing electrostatic energy. ACS Nano 3(9):2447–2450
43. Ghosh Chaudhuri R, Paria S (2012) Core/shell nanoparticles: classes, properties, synthesis mechanisms, characterization, and applications. Chem Rev 112(4):2373–2433
44. Guo N, DiBenedetto SA, Tewari P, Lanagan MT, Ratner MA, Marks TJ (2010) Nanoparticle, size, shape, and interfacial effects on leakage current density, permittivity, and breakdown strength of metal oxide− polyolefin nanocomposites: experiment and theory. Chem Mater 22(4):1567–1578
45. Guo N, DiBenedetto SA, Kwon DK, Wang L, Russell MT, Lanagan MT, Facchetti A, Marks TJ (2007) Supported metallocene catalysis for in situ synthesis of high energy density metal oxide nanocomposites. J Am Chem Soc 129(4):766–767
46. Fredin LA, Li Z, Ratner MA, Lanagan MT, Marks TJ (2012) Enhanced energy storage and suppressed dielectric loss in oxide core–shell–polyolefin nanocomposites by moderating internal surface area and increasing shell thickness. Adv Mater 24(44):5946–5953
47. Fredin LA, Li Z, Lanagan MT, Ratner MA, Marks TJ (2013) Substantial recoverable energy storage in percolative metallic aluminum-polypropylene nanocomposites. Adv Funct Mater 23(28):3560–3569
48. Stoyanov H, Mc Carthy D, Kollosche M, Kofod G (2009) Dielectric properties and electric breakdown strength of a subpercolative composite of carbon black in thermoplastic copolymer. Appl Phys Lett 94(23):232905

49. Yang K, Huang X, Zhu M, Xie L, Tanaka T, Jiang P (2014) Combining RAFT polymerization and thiol–ene click reaction for core–shell structured polymer@ $BaTiO_3$ nanodielectrics with high dielectric constant, low dielectric loss, and high energy storage capability. ACS Appl Mater Interfaces 6(3):1812–1822
50. Pirzada B, Sabir S (2018) Polymer-based nanocomposites for significantly enhanced dielectric properties and energy storage capability. In: Polymer-based nanocomposites for energy and environmental applications. Elsevier, pp 131–183
51. Xie L, Huang X, Yang K, Li S, Jiang P (2014) "Grafting to" route to PVDF-HFP-GMA/$BaTiO_3$ nanocomposites with high dielectric constant and high thermal conductivity for energy storage and thermal management applications. J Mater Chem A 2(15):5244–5251
52. Xie L, Huang X, Huang Y, Yang K, Jiang P (2013) Core@ double-shell structured BaTiO3–polymer nanocomposites with high dielectric constant and low dielectric loss for energy storage application. J Phys Chem C 117(44):22525–22537
53. Xie L, Huang X, Huang Y, Yang K, Jiang P (2013) Core-shell structured hyperbranched aromatic polyamide/BaTiO3 hybrid filler for poly (vinylidene fluoride-trifluoroethylene-chlorofluoroethylene) nanocomposites with the dielectric constant comparable to that of percolative composites. ACS Appl Mater Interfaces 5(5):1747–1756
54. Pang X, Zhao L, Han W, Xin X, Lin Z (2013) A general and robust strategy for the synthesis of nearly monodisperse colloidal nanocrystals. Nature Nanotechnol 8(6):426–431
55. Guo HZ, Mudryk Y, Ahmad MI, Pang XC, Zhao L, Akinc M, Pecharsky VK, Bowler N, Lin ZQ, Tan X (2012) Structure evolution and dielectric behavior of polystyrene-capped barium titanate nanoparticles. J Mater Chem 22(45):23944–23951
56. Rahimabady M, Mirshekarloo MS, Yao K, Lu L (2013) Dielectric behaviors and high energy storage density of nanocomposites with core–shell $BaTiO_3$@ TiO_2 in poly (vinylidene fluoride-hexafluoropropylene). Phys Chem Chem Phys 15(38):16242–16248
57. Shen Y, Lin Y, Li M, Nan CW (2007) High dielectric performance of polymer composite films induced by a percolating interparticle barrier layer. Adv Mater 19(10):1418–1422
58. Molberg M, Crespy D, Rupper P, Nüesch F, Månson JA, Löwe C, Opris DM (2010) High breakdown field dielectric elastomer actuators using encapsulated polyaniline as high dielectric constant filler. Adv Funct Mater 20(19):3280–3291
59. Zhu L, Wang QJM (2012) Novel ferroelectric polymers for high energy density and low loss dielectrics. Macromolecules 45(7):2937–2954
60. Balberg I (2011) Electrical transport mechanisms in three dimensional ensembles of silicon quantum dots. J Appl Phys 110(6):13
61. Xie L, Huang X, Li BW, Zhi C, Tanaka T, Jiang P (2013) Core–satellite Ag@ $BaTiO_3$ nanoassemblies for fabrication of polymer nanocomposites with high discharged energy density, high breakdown strength and low dielectric loss. Phys Chem Chem Phys 15(40):17560–17569
62. Yang K, Huang X, He J, Jiang P (2015) Strawberry-like core–shell Ag@ polydopamine@ $BaTiO_3$ hybrid nanoparticles for high-k polymer nanocomposites with high energy density and low dielectric loss. Adv Mater Interfaces 2(17):1500361
63. Wang G, Huang Y, Wang Y, Jiang P, Huang X (2017) Substantial enhancement of energy storage capability in polymer nanocomposites by encapsulation of $BaTiO_3$ NWs with variable shell thickness. Phys Chem Chem Phys 19(31):21058–21068
64. Zhu J, Wang D, Liu Z, Leung CM, Chen J, Zeng M, Lu X, Gao X, Liu J-M (2022) Superior energy storage of sandwiched PVDF films by separate introduction of core-shell Ag@BT nanoparticles and 2D MXene nanosheets. Ceram Int 48(13):19274–19282
65. Li J, Jiang J, Cheng Q, Cui Z, Liu X, Zuo P, Zhuang Q (2022) Construction of a flexible 1D core–shell Al_2O_3@$NaNbO_3$ nanowire/poly(p-phenylene benzobisoxazole) nanocomposite with stable and enhanced dielectric properties in an ultra-wide temperature range. J Mater Chem C 10(2):716–725
66. Xia L, Wang X, Ren T, Luo L, Li D, Dai J, Xu Y, Yuan C, Zeng B, Dai L (2022) Green construction of multi-functional fire resistant epoxy resins based on boron nitride with core-shell structure. Polym Degrad Stab 203:110059

67. Cheng S, Zhou Y, Li Y, Yuan C, Yang M, Fu J, Hu J, He J, Li Q (2021) Polymer dielectrics sandwiched by medium-dielectric-constant nanoscale deposition layers for high-temperature capacitive energy storage. Energy Storage Mater 42:445–453
68. Zeng J, Yan J, Li B-W, Zhang X (2022) Improved breakdown strength and energy storage performances of PEI-based nanocomposite with core-shell structured PI@$BaTiO_3$ nanofillers. Ceram Int 48(14):20526–20533
69. Chen J, Huang F, Zhang C, Meng F, Cao L, Lin H (2022) Enhanced energy storage density in poly(vinylidene fluoride-hexafluoropropylene) nanocomposites by filling with core-shell structured $BaTiO_3$@MgO nanoparticals. J Energy Storage 53:105163

Chapter 12
Transition Metal Oxide-Based Nanomaterials for Advanced Energy Storage

Priyambada Mallick, Srikanta Moharana, L. Biswal, and Santosh Ku Satpathy

Abstract With the improvement of the global economy, one of the greatest challenges in the world is to provide energy requirements, which are exponentially rising and is projected to triple at the end of the century. Therefore, the improvement of renewable energy sources and electrochemical energy storage systems is essential. Fuel cells, rechargeable supercapacitors, and lithium-ion batteries are three intriguing contenders in the field of energy and power densities with long life spans, lightweight, and eco-friendly. The most promising materials for such devices with high performance or high conversion efficiency are transition metal oxide (TMO)-based nanomaterials because of their distinctive properties like environmental friendliness, innovative size effects, improved kinetics, greatly enhanced conductivity, activity, excellent redox properties, electrochemical stability, high specific power, and sustainability. TMO-based nanomaterials as electrode materials typically exhibit exceptional capacitance and very high energy densities. Due to distinctive configurations of d electron, the synergistic effect of multi-metal atoms, low cost, and outstanding electrochemical stability nature of TMO-based nanomaterials, the energy storage mechanism of electrochemical supercapacitors, batteries, and fuel cells can be enhanced significantly. However, the current issues and potential future of TMO-based nanomaterials in energy conversion and storage applications depend on our ability to comprehend the innovative synthesis, application performance, and catalytic mechanism of these nanomaterials. In this chapter, TMO-based nanomaterials such as NiO, SnO_2 RuO_2, MnO_2, V_2O_5, Co_3O_4, Mn_3O_4, TiO_2–V_2O_5, and $NiMn_2O_4$, etc., and their morphological and electrochemical properties will be studied extensively. Detailed descriptions of energy storage mechanisms and key concepts will be provided in order to fully grasp the promising impact of metal oxides in energy storage devices.

P. Mallick · S. Moharana · S. K. Satpathy (✉)
School of Applied Sciences, Centurion University of Technology and Management, Bhubaneswar, Odisha, India
e-mail: santosh.satpathy@cutm.ac.in

L. Biswal
Department of Physics, School of Applied Sciences, KIIT Deemed to be University, Bhubaneswar, Odisha 751024, India

S. Moharana et al. (eds.), *Emerging Nanodielectric Materials for Energy Storage*, Nanostructure Science and Technology, https://doi.org/10.1007/978-3-031-40938-7_12

This chapter aims to highlight the synthesis, characteristics, applications as well as the recent development, accomplishments, flaws, challenges, potential solutions, and future aspects of TMO-based nanomaterials in advanced energy storage systems such as rechargeable supercapacitors, lithium-ion batteries (LIBs), PV cells, and so on. In particular, new advancements in the synthesis of TMO-based nanomaterials and their uses in energy storage and conversion will be discussed.

Keywords Transition metal oxide · Super capacitor · Batteries · Fuel cells · Advanced energy storage system

12.1 Introduction

To meet the rapid development of the world's energy requirements, we have to give priority on green or renewable energy sources including solar, wind, tidal, etc., rather than fossil fuels in order to conserve natural energy, reduce energy consumption, and reduce environmental pollution [1]. Because the utilization of fossil fuels rises the average temperature of the earth by emitting carbon dioxide (CO_2), methane (CH_4), nitrogen oxide (N_2O), etc., which are very dangerous to living and nonliving beings of the Earth [2]. The world's energy requirement increases the motivation for research on advanced energy storage systems. Therefore, the development of renewable energy sources and electrochemical energy storage systems with lightweight, low cost, and eco-friendly is essential. Among all the non-conventional energy devices, fuel cells, rechargeable supercapacitors, and lithium-ion batteries are three intriguing contenders in the field of energy and power densities with long life spans. Having high specific power, medium energy density, excellent cycling life, long durability, electrochemical stability, and rapid charging or discharging rate, these electrochemical energy devices are used in portable electronic devices and hybrid vehicles, where electrochemical reactions convert chemical energy to electrical energy. The most promising materials for such devices with high performance or high conversion efficiency are transition metal oxide (TMO)-based nanomaterials [3, 4] because of their distinctive properties like environmental friendliness, innovative size effects, large specific surface area, improved kinetics, greatly enhanced electrical conductivity, activity, electrochemical stability, high specific power, and sustainability. The discharging time, specific energy, and specific power of some advanced energy storage system are given in Table 12.1.

This chapter mainly focuses on the recent development of transition metal oxides (TMO)-based nanomaterials and their excellent properties as well as the effect of these transition metal oxide-based nanomaterials on the advancement of society. This chapter also focuses on the recent researchs on the synthesis and electrochemical performance of transition metal oxide-based nanomaterials. Finally, the limitations, prospects, and suggestions for improving the performance of TMO-based nanomaterials in energy storage systems are discussed.

Table 12.1 Specific energy, specific power, and discharging time of various energy storage systems [1]

Energy storage devices	Specific energy (Wh/kg)	Specific power (kW/kg)	Discharging time
Fuel cells	100–1000	0.01–0.1	10 h
Conventional batteries	10–100	0.05–1	1 h
Supercapacitors	1–10	0.5–10	1 s
Conventional capacitors	0.01–0.5	10–100	0.03 s

12.2 Transition Metal Oxide Nanomaterials

Nanomaterial having size 1–100 nm possesses comparatively better mechanical, electronic, magnetic, thermal, catalytic, and optical properties [5] with high surface area, surface energy than bulk materials with macroscopic size. However, the most difficult problem is nanoparticle aggregation, which effectively increases the strain for electrolyte ion diffusion within the nanoparticles at the electrode surface. With nanotechnology, any substance or material can be reduced in weight with increased stability and functionality. Although proton diffusion and electron hopping within nanoparticles are intrinsic properties of nanoparticles, the resistance caused by intra-particle electron hopping can be reduced by loading the nanoparticles on a stable metal oxide. It has also been observed practically that loading a stable metal oxide affects the increase of a proton's diffusion barrier within active materials, which results from the loss of effective sites [2]. From the literature survey, it was found that the development of advanced high-performance energy storage systems depends highly on the electrode materials, synthesis techniques, nature of the used electrolyte, and range of voltage potential [6]. Transition metal oxide nanomaterials (TMOs) as functional materials which are composed of negative oxygen ions and positive transition metals such as ruthenium oxide (RuO_2), titanium oxide (TiO_2), manganese oxide (MnO_2), tin oxide (SnO_2), nickel oxide (NiO), zinc oxide (ZnO), cobalt oxide (Co_3O_4), and vanadium oxide (V_2O_5) are new-generation stable and robust materials. These materials attracted vast attention as electrode materials from researchers, scientists, and industrialists due to their variable oxidation states, various structures, good chemical stability, high electrical conductivity, small resistance, wide band gap, excellent energy density, and high specific capacitance (100–3000 F/g) as given in Table 12.2. Besides these oxides, other metal oxides including $BiFeO_3$ [7], $PbZrO_3$, $BaTiO_3$, $PbTiO_3$, etc., rare earth/transition metals such as Gd, Sm [8], Ni, La, Nd, Mg, Co, and Cu doped oxides [9, 10] and transition metal ferrites including $ZnFe_2O_4$ [11], $MgFe_2O_4$, $CoFe_2O_4$ [12], $CuFe_2O_4$ [13], etc., are most investigated materials for advanced energy storage system in the field of non-volatile memory, electronic devices, and power backup systems. However, the practical application of TMO-based devices is still a long way off, owing to the lack of controllable preparation techniques and a thorough understanding of the growth mechanisms involved in the synthesis of TMO nanomaterials. As a result, numerous scientific challenges remain

Table 12.2 Comparison of the theoretical specific capacitances of different TMOs

Transition metal oxides	Specific capacitance (F/g)
RuO_2	1400
MnO_2	1370
V_2O_5	375
NiO	2584
Co_3O_4	1780
TiO_2–V_2O_5	526
SnO_2	420

to overcome in order to fully exploit the exceptional performance of TMO nanomaterials [14]. Unfortunately, the easy agglomeration of TMOs at high mass loading, poor cyclability, and conductivity properties is the factors that impede achieving a high specific capacity in TMOs. Thus, incorporating conductive materials, such as EDLC materials, with TMOs can significantly improve ionic and electrical conductivity due to the synergistic effect of various components [1].

12.2.1 *Ruthenium Oxide (RuO_2)*

Ruthenium oxide with oxidation state +4 is the most used nanomaterial in the field of advanced energy storage systems due to its high specific capacitance (1400–2200 F/g), high ionic conductivity, rapidly reversible redox reactions, high reversible oxidation states, excellent electrical conductivity, high chemical and thermal stability, high rate capability, and long cyclic life. RuO_2 is widely used as positive electrode material for supercapacitors and Li-ion batteries. Although the preparation of RuO_2 is costly, this oxide has the potential to improve the specific energy of advanced energy systems, especially for supercapacitors. The reversible storage of protons in interaction with the electrolyte causes charge transfer on the RuO_2 surface, which is influenced by the crystalline phase of the oxide. RuO_2 can form two preferential crystalline structures, the tetragonal with space groups of P42/mnm and cubic systems with space groups of $Pa\overline{3}$. Doping of rare earth and transition metals or other nanostructures with RuO_2 can suppress the resistance of charge transfer which results in the improvement of the efficiency of RuO_2.

Park et al. [15] obtained Ru and RuO_2 thin films by thermal metallorganic CVD using ruthenocene precursor in the presence of oxygen and concluded that with an increasing flow rate of oxygen, the phase transition from Ru to RuO_2 occurred with the increasing deposition rate. At a fixed substrate temperature and flow rate of oxygen, a decrease in precursor partial pressure led to RuO_2 formation.

Dao et al. [16] observed that ruthenium oxide-based nanocomposite (RGO) has 136.7 F/g specific capacitance at a scan rate of 20 mVs^{-1} which is suitable for energy storage applications. Nisha et al. [17] synthesized RuO_2 successfully by using a bottom-up approach and found a crystallite size of 9.5 nm from the XRD analysis. Figure 12.1a and b represent the curves between current—voltage and potential—time which were obtained from the analysis of cyclic voltammetry (CV). It is observed from the study of electrochemical properties that the value of specific capacitance decreases from 209 to 115 F/g with increasing scanning rate as shown in Fig. 12.1c. According to the galvanostatic charge–discharge (GCD) study, there is a gradual decrease in the corresponding specific capacitance with current densities ranging from 167 to 45 F/g, as shown in Fig. 12.1d. Due to charge transfer and ion diffusion over the working electrode (RuO_2 coated over the carbon sheet), RuO_2 has long-term cyclic stability (98% retained after 1000 cycles of charge–discharge cycles), as shown in Fig. 12.1e. This demonstrated that the prepared RuO_2 material has long-term electrochemical stability as well as a high degree of charge–discharge reversibility, proving excellent stability and performance.

12.2.2 Manganese Oxide (MnO_2)

As a transition metal element, manganese exists in a variety of stable oxides (MnO, Mn_3O_4, Mn_2O_3, and MnO_2). Manganese dioxide (MnO_2) is one of the most widely and competitively used faradaic materials which can be used as electrodes in the battery as well as super capacitance due to its high specific capacity, high electronic and ionic conductivity, cost-effective, absorption of toxic ions, excellent operating potential, small toxicity, natural abundance, easily available, and good environmental compatibility. However, due to its poor conductivity and slow ion transport rate, its application in supercapacitors is severely limited. The structural parameters are important in determining and optimizing the electrochemical behaviors of MnO_2 for advanced energy storage systems. For the synthesis of MnO_2, several techniques are adopted including sol-gel, co-precipitation, electrochemical deposition, hydrothermal, etc. The manganese oxide can store charge in two ways: (i) the addition of electrolytic cations (C^+, Li^+, Na^+, and K^+) into reactions mainstream of the electrode (ii) adsorption of electrolyte cations on the MnO_2 electrode surface. But because of its low ionic and electrical conductivity, the practical application and rate capability of MnO_2-based nanomaterials were affected [1]. Though the theoretical capacitance of MnO_2 is 1370 F/g, experimentally it exhibits specific capacitance within the range of 50–200 F/g. But by doping other metal elements such as Co, Fe, Sn, Ag, and Au or oxides like RuO_2, NiO, Co_3O_4, etc., in MnO_2, the specific capacitance and electrical conductivity increase which in turn enhances the charge and energy storage capacity. The investigation of MnO_2 anode is relatively complex due to the existence of various forms such as α, β, γ, ε, and δ type [18] with specific capacitance ranging between 100 and 300 F/g.

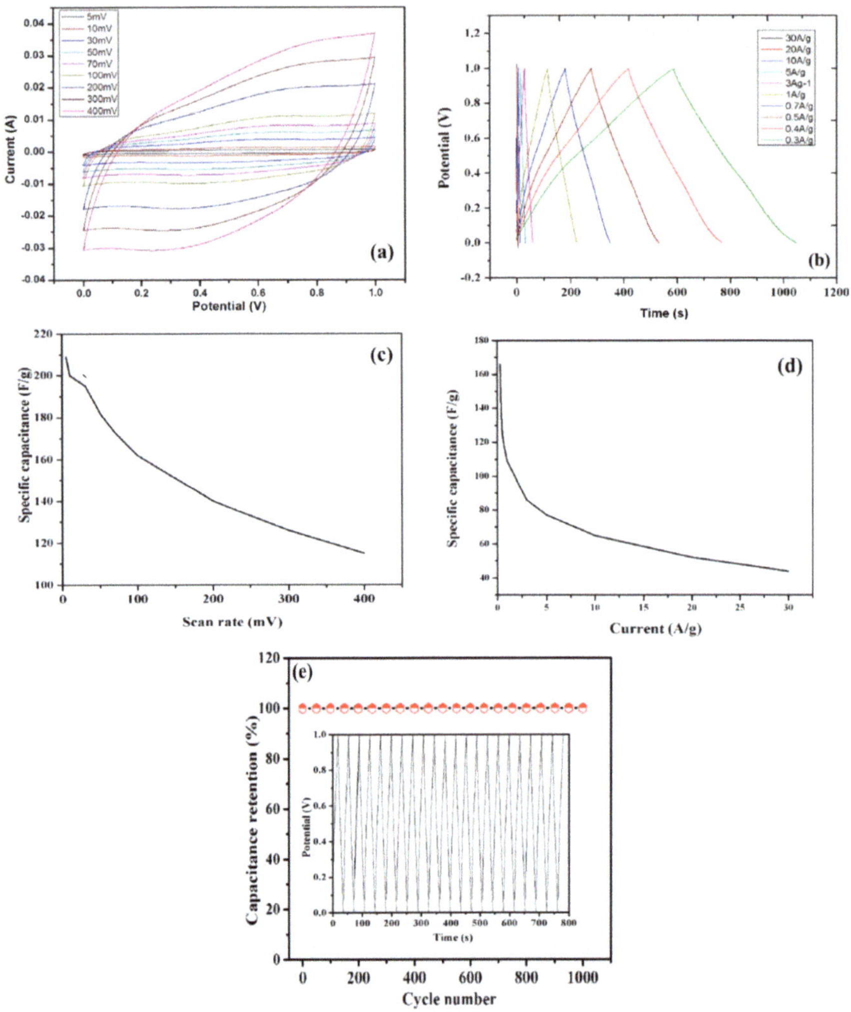

Fig. 12.1 (**a**) Current versus potential graph from CV study, (**b**) charging discharging study, (**c**) specific capacitance versus scan rate graph, (**d**) specific capacitance versus current graph, (**e**) cyclic stability graph. Reprinted the figure with Copyright (2020) permission from Elsevier [17]

Meng and his team et al. [19] prepared α-MnO_2-based nanocomposite using the solid-state technique. From the cyclic Voltammograms, as shown in Fig. 12.2, a good cycling performance with a reversible capacity of 610 mAh g^{-1} over more than 200 cycles at a 99% coulombic efficiency is confirmed which makes it suitable as a LIB anode.

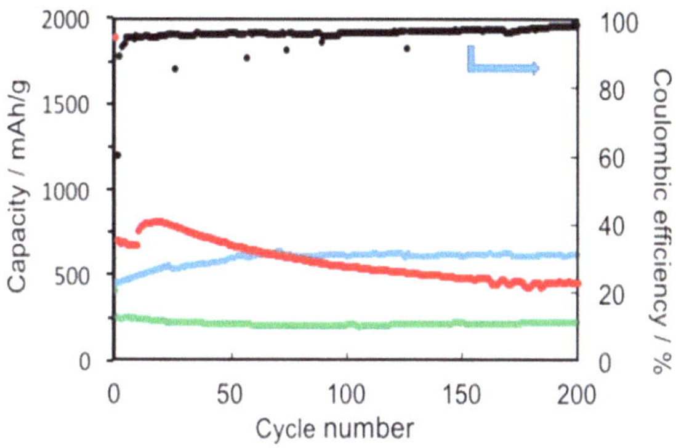

Fig. 12.2 Variation of capacity and columbic efficiency with cycle number for α-MnO_2 (green), Ni-α-MnO_2 (red), Ni-α-MnO_2/RGO (blue). Reprinted with permission from [19]. Copyright (2021) American Chemical Society

12.2.3 Vanadium Pentoxide (V_2O_5)

Among the promising cathode materials and the redox-active TMOs for next-generation advanced energy systems, vanadium oxide/pentoxide is considered to be the most promising material and has gained much attention because of its high energy density, various oxidation states (V^{2+} to V^{5+}), unique layered structure, better cycle life, low cost, easy synthesis process, abundant sources, non-toxic in nature, adequate safety, and high specific capacity [20]. The theoretical capacity of V_2O_5 is approximately 294 mAh/g, which is significantly higher than the theoretical capacities of commonly used cathode materials. But its practical application is limited due to its poor cycling stability, and low electrical conductivity which can be overcome by controlling synthesis approaches, and morphology of the nanostructures. Vanadium oxide is not only used for battery applications but also used as an electrolyte in supercapacitors.

Sakunthala et al. [21] reported that by doping niobium (Nb) and tantalum (Ta) ions with V_2O_5, the electrochemical performance of V_2O_5 was enhanced with 96% chemical stability. Xie et al. [22] observed from CV graphs as shown in Fig. 12.3 that V_2O_5 attains a high value of specific capacitance, i.e., 350 F/g with 75% retains capacitance showing great potential for high-performance applications in thin-film energy storage devices.

Boruah et al. [23] and his team proposed that vanadium pentoxide can be used as photo cathode material in the advanced energy storage system that can be charged directly up to 2.82 V along with 2.6% conversion efficiency and good cycling stability by light without any external solar cell or power sources which are higher than other systems.

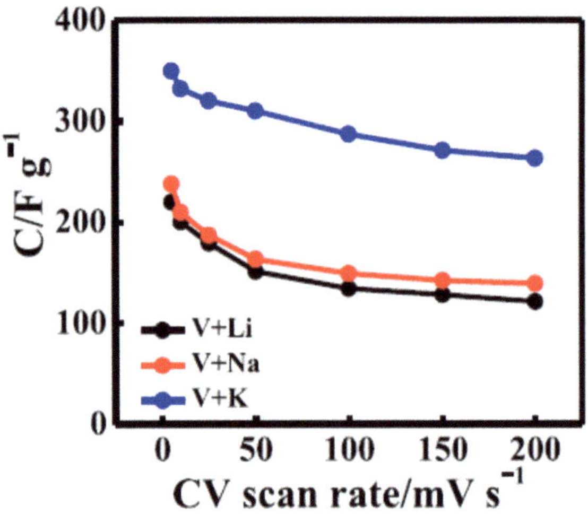

Fig. 12.3 Specific capacitance versus scan rate graph of V_2O_5 oxides. Reprinted from [22]. Copyright (2018) with permission from Elsevier

12.2.4 Cobalt Oxide (Co_3O_4)

Cobalt oxide (Co_3O_4) is a cubic structure with AB_2O_4 spinel configuration with a theoretical super capacitance of about 3560 F/g. Because of exceptionally good reversible redox properties, stable catalytic behavior, theoretically high specific capacitance, small rate capability, excellent corrosion, and chemical stability, good electronic conductivity, low cost, high surface area to volume ratio, short life cycle, and simple synthesis, cobalt oxide (Co_3O_4) has been proven especially appealing as an ideal electrode material in the recent years, potentially replacing the use of high-cost and ecologically unfriendly RuO_2. The electrochemical properties of Co_3O_4 are highly affected by the structural morphology of the material and the electronic states of the metal (Co^{3+} or Co^{2+}), and thus, various strategies have been dedicated to the development of various forms of Co_3O_4 nanocomposites to obtain the improvement in the charge storage performance. The main disadvantage of Co_3O_4 is its poor cyclability and structural rigidity due to swelling and shrinking during cycling, which can be overcome by doping other elements or metal oxides such as cadmium (Cd), nickel (Ni), and nitrogen (N), etc. [24].

The porous Co_3O_4 was fabricated by a hydrothermal process [25] and considered as a potential electro-active material for a pseudo-super capacitor. The study of cyclic Voltammograms showed that by decreasing calcination temperature from 500 to 300 °C, the specific capacitance increases from 101.3 to 226.3 F/g at 10 mV/s with 76% of retention capacity after 5000 cycles. Xie et al. [26] prepared Co_3O_4 by hydrothermal technique and confirmed from the electrochemical studies that the prepared materials can be useful in the fabrication of energy storage systems because of their excellent specific capacitance values of up to 263 F/g at a charge–discharge current density of 1 A/g with 10.6% decay in the available capacity and electrochemical stability with ~98.8% coulombic efficiency. The specific capacitance of cobalt oxide can be enhanced up to 761.25 F/g [27] by synthesizing the metal oxides using the sol-gel technique which can be applied for energy storage applications.

12.2.5 *Titanium Oxide (TiO_2)*

TiO_2, an n-type oxide is one of the most popular investigated anode materials for an advanced storage system with a lower capacity at higher potential, high resistive, high dielectric constant, good stability, high charging/discharging potential, low toxicity, less expensive, easily available, eco-friendly, and other electrochemical behaviors. Having these excellent properties, these nanomaterials can be applied in various areas including solar cells, gas sensors, and most importantly lithium-ion batteries. However, due to its low electronic and ionic conductivities, there arise some serious problems that can be overcome by developing nanostructure engineering. Titanium dioxide-based nanomaterials can be in several forms including amorphous, rutile, and anatase which are based on the fabrication techniques and preparing temperature.

Lee et al. [28] prepared TiO_2 nanostructures for lithium-ion storage with a storage capacity of 228 mAh/g over 100 cycles and reported that TiO_2 can be used as a suitable electrode material for energy storage mechanism due to long-term cycling and excellent energy storage capacity. Ray et al. [29] prepared TiO_2–V_2O_5 by wet chemical method successfully and reported that by increasing the ratio of vanadium in the samples, the specific capacitance of the prepared sample is increased up to 310 F/g at 2 mV/s scan rate as shown in Fig. 12.4 which can be used as an electrode material for a supercapacitor as well as a lithium-ion battery.

The poor electrical conductivity nature of TiO_2 limits its practical device applications. The combination of TiO_2 with low valence atoms or oxides including chromium (Cr), aluminum (Al), cobalt (Co) [30], and vanadium oxide (V_2O_5) results in an overall improvement of electrochemical activity and cycle stability because of the decrease in the intra-particle electron-hopping resistance. Due to the three-dimensional, mesoporous, interlinked tube-like ordered architecture of TiO_2–V_2O_5 nanocomposite, it has a large surface area by the formation of Ti–O–V bonds. Thus, TiO_2–V_2O_5 nanocomposite may be a viable candidate for supercapacitor electrodes in the near future [2]. Fomekong et al. [30] reported that by doping Co in titanium oxide, the n-type TiO_2 exhibits p-type conductive properties and alters the electronic properties as well cycle stability.

12.2.6 *Nickel Oxide-Based Nanomaterials (NiO)*

Like other TMO-based nanomaterials, nickel oxide (NiO), a p-type semiconductor having a cubic crystal phase is a suitable electrode material for fuel cells, supercapacitors, lithium-ion batteries, etc., due to its non-toxicity, low preparation cost,

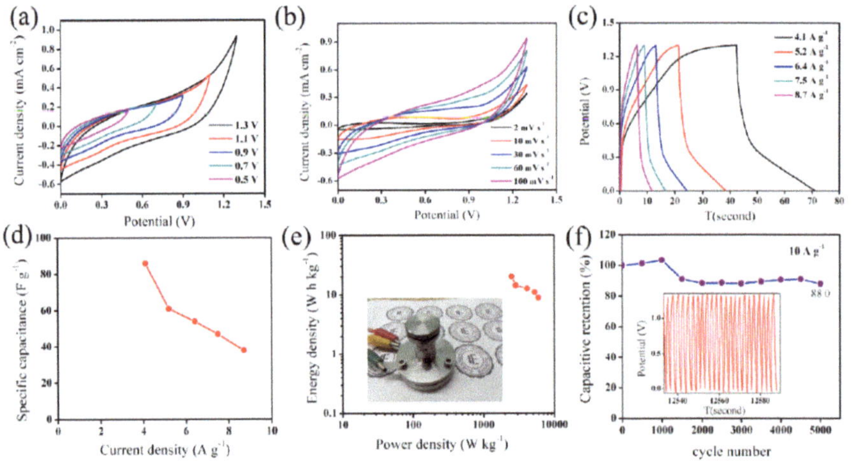

Fig. 12.4 (**a**) Electrochemical performance, (**b**) current density versus potential curves at different scan rates, (**c**) charging/discharging plots at different current densities, (**d**) specific capacitance versus current density, (**e**) energy density versus power density plot, (**f**) cycling stability at the different current density of TiO_2–V_2O_5. Reprinted with permission from [29]. Copyright (2018) American Chemical Society

short diffusion process, high surface area, wide band gap, a wide range of work function, excellent capacitance, thermal, chemical, mechanical, and catalytic properties [31]. NiO-based nanomaterial can be easily prepared by various techniques including hydrothermal, chemical vapor deposition, electrodeposition processes, etc. [1]. But its practical applications restrict due to the poor intrinsic conductivity.

The coexistence of Ni and Mn in the lattice of this binary oxide has been shown to improve the electrochemical charge storage capability of the electrodes due to increased electronic or ionic conductivity [2]. One of the most attractive ternaries of single-phase metal oxide is $NiMn_2O_4$ having excellent electrochemical and electrical conductivity properties at low cost and is non-toxic in nature. This electrochemical nature helps to obtain high power and energy density which makes these metal oxides suitable to use as electrodes for the advanced energy storage system. Also, these materials offer a higher surface-to-volume ratio and shorter electron-ions transport channels which makes them more appropriate than traditional bulk materials as electrode materials for advanced energy storage systems including supercapacitors, batteries, etc.

NiO nanoparticles were synthesized using the green route by Lamba et al. [32] for use as electrode materials in supercapacitors and reported from the study of cyclic Voltagramms (CV) that NiO has good capacitive properties with a specific capacitance of 85.31 F/g. The NiO nanoparticle was fabricated by hydrothermal technique and S.D Dash et al. [33] investigate from the CV study as shown in Fig. 12.5 that NiO can have specific capacitance up to 132 F/g with 75% retention, i.e., excellent

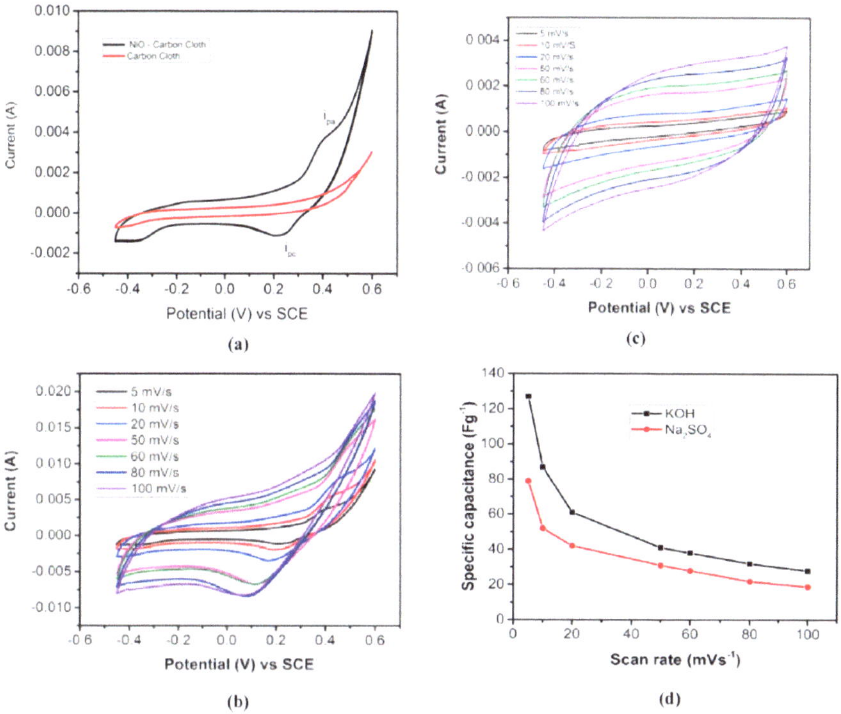

Fig. 12.5 CV curves and specific capacitance versus scan rate curves of NiO nanoparticle. Reprinted the figure with Copyright (2020) permission from Elsevier et al. [33]

stability by taking KOH electrolyte making it as a potential candidate for energy storage applications.

12.2.7 *Tin Oxide (SnO_2)*

Because of the low cost and high power density, SnO_2-based nanomaterials have drawn a lot of attention as promising electrode materials for supercapacitors. SnO_2 has potential applications in a wide range of fields including sensors [9] and solar cells [10]. Because of its superior electrochemical performance, SnO_2 appears to be a promising electrode material for energy storage devices such as lithium-ion batteries and supercapacitors [11]. Though SnO_2 has a better response to the specific capacitance of supercapacitors, its applicability is still limited due to poor electrolyte ion transport within the SnO_2 matrix and poor electrical conductivity.

Velmurugan et al. [34] prepared SnO_2-based nanomaterials using wet chemical methods and concluded from the study of electrochemical behavior that by adding graphene, the tin oxide can exhibit high specific capacitance up to 818.6 F/g with

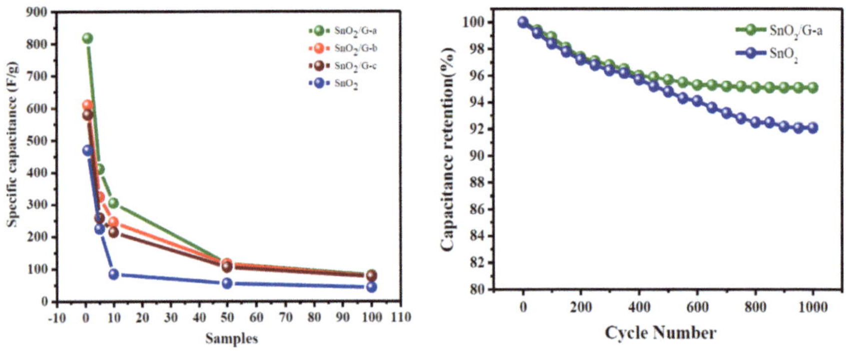

Fig. 12.6 Specific capacitance versus scan rate and capacitance retention versus cycle number curves. Reprinted the figure with Copyright (2016) permission from Elsevier et al. [34]

95.1% retention capacity as shown in Fig. 12.6 which may be applicable for high-performance advanced energy storage systems.

12.3 Fabrication of TMO-Based Nanomaterials

Indeed, transition metal oxide nanostructures have become the focus of recent researches due to their potential use in high-performance energy storage devices. Furthermore, the morphology, particle, and crystallite size of these nanostructures appear to influence their physical and electrochemical properties. The method of synthesis should be eco-friendly, easy handling, and low preparation cost. As a result, fabricating metal oxide nanomaterials with controlled morphology and size that also have good electrical conductivity is a challenge for their use in energy storage devices like batteries, solar cells, and electrochemical supercapacitors with high efficiency [35]. There are various methods for preparing TMO-based nanomaterials, namely physical processes including low-pressure gas evaporation, sputtering, plasma-induced, high energy ball milling process, etc., and chemical techniques including sol-gel, electrospinning (ES), co-precipitation method, spraying, laser ablation, chemical vapor deposition (CVD), and hydrothermal synthesis, etc., which can have a significant impact on both dimensions and the nature of nanomaterials. The main purpose of synthesis is to produce single-phase nanoparticles and to reduce the preparation cost. However, some techniques are quite complex, time-consuming, and expensive, which can limit the large-scale production and reproducibility of the as-prepared nanomaterials. Thus, it is significant to develop a simple and time-saving route to fabricate nanomaterials. In this section, widely adopted chemical methods including sol-gel, hydrothermal, co-precipitation, and chemical vapor deposition techniques are summarized below because these are straightforward, lead to tunable properties (diameter, morphology, and composition), and can be scaled-up for industrial production.

12.3.1 Sol-Gel Synthesis

Although the sol-gel process is one of the outdated bottom-up chemical synthesis approaches in which components of atomic dimensions are assembled to form nanomaterials, the importance of this process in material fabrication has been growing rapidly since the development of this technique in 1940 [8]. Sol-gel processes have a remarkable advantage for preparing superfine metal oxides (MO) as well as nanopowders of non-oxide materials having the same size with high purity at a low-temperature range and low cost. It is also able to produce two or more nanoparticles simultaneously [36]. Here, sol stands for dispersed solid colloidal ions or particles in a solvent of solution which still contains water or solvent, and gel is a 3D porous interconnected network structure in the liquid phase which is produced when the solvent from the sol begins to evaporate [9].

Several steps are taken in this process to convert sol into a gel. Dispersed solid nanoparticles (sols with diameters ranging from 1 to 100 nm) are mixed in a homogeneous liquid medium and agglomerated to form a continuous three-dimensional network (gel) in the liquid phase with pore diameters in the sub-micrometer domain. The final product is produced by following several irreversible chemical reactions. The reaction rate of this technique can be raised by using a proper catalyst which can be removed after the reaction [36]. The benefit of this technique is that particle size and homogeneity can be controlled. It is a method of wet chemical synthesis used to produce fibers, thin films, powders of ceramic, dielectric, and glass materials. In this method, metal alkoxides and chlorides are typically utilized as precursors. During the synthesis technique, pH must be managed because this process is sensitive to it and it will help to prevent precipitation and create a homogeneous gel. The synthesis process is controlled by many factors including pH, temperature, concentration, type of solvent, etc. Siswanto et al. [37] reported that in sol-gel synthesis, pH is an important factor to control the size of nanoparticles. He noticed that with the rise in pH from 7 to 12, the crystallite size of zinc oxide increases from 10.94 to 74.04 nm.

The different stages of the sol-gel process from solution to nanosphere and nanorod are shown in Fig. 12.7.

R. S. Dubey and his team successfully synthesized titanium oxide nanoparticles by sol-gel as well as solvothermal technique and confirmed pure crystal phase having crystallite size 13 nm and 11 nm corresponding to sol-gel and solvothermal techniques from the study of X-ray diffraction (XRD) peaks without any impurities as shown in Fig. 12.8.

Priyadarshini and her team et al. [27] successfully prepared cobalt oxide nanoparticles by using this technique revealing the cubic phase (Fd3m) of crystallite size 13.76 nm from the XRD technique and confirmed that the sol-gel technique is a better choice for the preparation of transition metal oxides in the energy storage system.

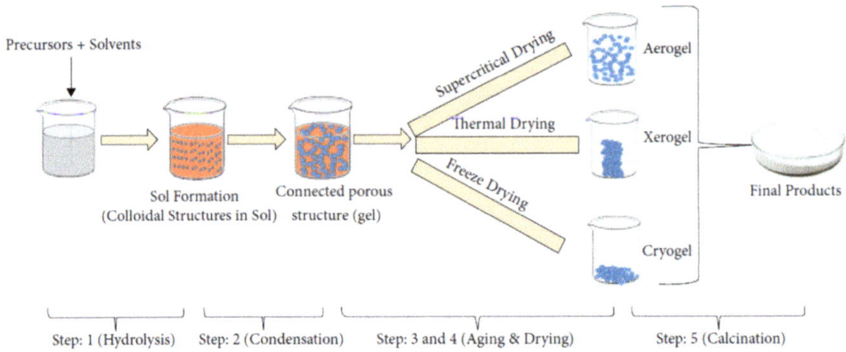

Fig. 12.7 Different steps of sol-gel method. Copyright © 2021 Dmitry Bokov et al. [36]

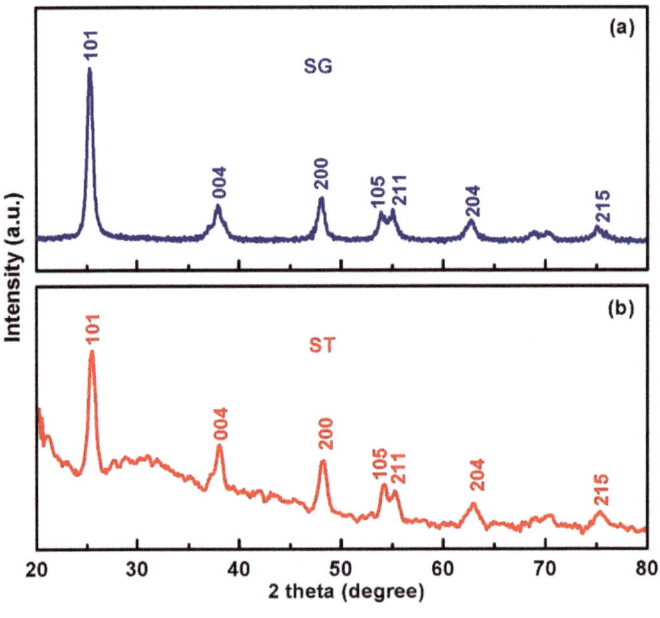

Fig. 12.8 X-ray diffraction patterns of TiO_2 nanoparticle using **a** sol-gel technique, **b** solvothermal technique. Reprinted the figure with Copyright (2019) permission from Elsevier et al. [38]

TiO_2–V_2O_5 nanomaterial was prepared using the sol-gel technique by Ngaotrakanwiwat et al. [39] and the excellent energy storage ability with the highest initial charging rate is confirmed.

12.3.2 Chemical Vapor Deposition (CVD)

Chemical vapor deposition (CVD) is a most promising technique for the direct deposition of different precursors such as thin films of metals on specific structures and substrates because of the various advantages such that short processing time, flexibility, deposition of high-purity nanomaterials, narrow size distribution, etc. This technique is mostly used in the semiconductor industry for the synthesis of nanometric layers of inorganic materials on the surface of 3D substrates and for depositing thin films of various materials. It is an important technology for creating a thin, high-purity coating. At lower temperatures, different types of catalysts are used for the deposition of metallic coatings. As the solid films are deposited on surfaces, it is called chemical vapor synthesis or chemical vapor condensation. There are three successive steps involved in this process: (i) introduction of the volatile or vaporized precursors (may be solid, liquid, or gas under ambient conditions) by carrier gas to the reactor chamber; (ii) adsorb of the substrate to one or more volatile at a high temperature which favors homogeneous nucleation and the formation of intermediate products; and (iii) decomposition of these products on the heated substrate followed by heterogeneous nucleation and growth of the solid layer/grains, and the formation of volatile by-products under certain conditions (temperature, pressure, substrate, etc.) and their removal from the chamber by the carrier gas.

The important factors such as type and quality of precursors, property of solvent, reaction temperature and pressure in the chamber, chemical properties of substrate and gas carrier, time and rate of deposition influenced the CVD techniques and the quality of the products. The most important factor which affects the crystal structure, size, and morphology of the particle is the temperature at the time of preparation. Laser chemical vapor deposition (LCVD) is a CVD variation that uses a laser beam to heat the surface of a substrate to a suitable temperature for localized synthesis. The ability to instantly modify and control the growth rate is what distinguishes LCVD from conventional CVD. In 2014, the synthesis of cobalt oxide using LCVD techniques has been reported, and various factors affecting the density and crystallization of cobalt oxide fractal shapes consisting of the cubic CoO and Co_3O_4 phases, as well as the hexagonal Co phase, have been investigated [40].

Cobalt oxide with cubic crystal phase was synthesized using a novel precursor (Cobalt β-diketonate) through chemical vapor deposition technique by Armelao [41] and concluded that the used precursor is a versatile and suitable for the preparation of cobalt oxide. It was also observed that by controlling pressure and temperature of reactor and substrate, the other hybrids of cobalt including CoO, Co_3O_4, CoO + Co_3O_4 etc can be obtained. Maruyama et al. [42] prepared amorphous nickel oxide by using CVD techniques and from the study of Coulometry, electrochromic efficiency was found to be 44 cm^2/C. Lee et al. [43] with his team prepared vanadium pentoxide (V_2O_5) successfully by using the CVD technique and reported that the prepared sample exhibits amorphous nature at lower temperatures while at the higher temperature, the materials exhibit a single crystalline phase without any impurities.

The particle and crystallite size was calculated from the XRD peaks and observed that both values increase with increasing preparation temperature.

12.3.3 Co-precipitation Technique

Among the various synthetic approaches, the co-precipitation method is regarded as one of the best and potentially advantageous in comparison to other methods for producing pure phase formation of transition metal oxide-based nanomaterials without any impurities at low temperatures. Synthesis of metal oxide nanoparticles using this technique is increasing day by day due to excellent properties such as simple processing, less agglomeration, controllable particle size with high purity, and small path for diffusion. [35]. The supersaturation state of certain species soluble in the solvent is induced using this method. The supersaturating conditions required for the precipitation of the species are typically caused by a chemical reaction. A precipitation agent is a solution containing soluble metal cations and another solution called a co-precipitation solution. Co-precipitation agents can be either inorganic or organic in nature. Temperature, mixing rate, pH, and concentration are all adjusted to achieve the desired products and their physical properties, such as product size, morphology, particle size distribution, and so on. Following precipitation, the precipitate can be removed via centrifugation, filtration, or decantation and then dissolved in acids or an organic solvent, such as isobutyl methyl ketone, to be measured.

Nickel oxide nanoparticles were prepared by using this technique by Khandagale and her team [44] and amorphous nature was identified from the XRD data. Using this method, Buraso and his team et al. [45] synthesized titanium oxide taking titanium isopropoxide as the initial material. He noticed that the particle and crystallite size increased, while the energy band gap decreased with increasing calcination temperature.

12.3.4 Hydrothermal Synthesis

The hydrothermal process gained much more attention for the synthesis of nanomaterials as in this technique grain size, the shape of the particle, crystalline state, and surface morphology of materials can be changed by controlling various factors such as the property of solvent, the concentration of reactants, temperature, aging and pH time, and additives [4]. Hydrothermal synthesis (also known as solvothermal) is an easy and fast process with low cost, controlled shape, and size to prepare ultrafine nanomaterials under a hydrothermal environment (i.e., neither the solution is non-aqueous nor aqueous) at low temperature in the range of 100–374 °C without the use of milling and calcination process. The basic principle of this technique is a reaction in an aqueous solution or suspension of the precursors at high temperatures and pressure [11]. This high pressure decreases the boiling point and increases the solubility of the precursors providing better growth of crystals.

This process possesses excellent consistency and particle uniformity with appropriate size which possesses higher mechanical or electrical properties. In a hydrothermal method, nanoparticles are created through hydrolysis processes that take place at high temperatures and vapor pressure. In this process, materials are obtained by the crystallization of phases under high temperatures and pressure. The method can be used at room temperature and below critical pressure. This technique is dependent on the solubility of minerals in hot, high-pressure water. Factors affecting the hydrothermal synthesis process are the initial pH of the solution, duration, temperature, and pressure in the autoclave. Typically, an autoclave is used to perform the hydrothermal technique. In this process, a metal precursor solution is mixed in a hydrolyzing solution and kept at a high temperature for 10–24 h in an autoclave, then cooling it down to ambient temperature for crystallization. The resulting crystals are collected using filtration, washing, and drying. Large particle size nanocrystals can be synthesized using the hydrothermal method which is useful to increase the dielectric constant of the material and hence, increases the dielectric properties of the synthesized materials. Nanomaterials synthesized using this method tend to withstand high frequencies and high temperatures, and therefore, can be used in semiconductor devices. The schematic diagram for the preparation of nickel oxide nanoparticles by using the hydrothermal technique is shown in Fig. 12.9.

After preparing titanium dioxides using four different techniques including sol-gel, hydrolysis, co-precipitation, and hydrothermal, Liu and his team et al. [46] concluded that the hydrothermal process is the best synthesis technique than others as it provides the least grain size with excellent photo-catalytic behavior which can be applicable for advanced energy storage systems. Feng et al. [6] successfully prepared

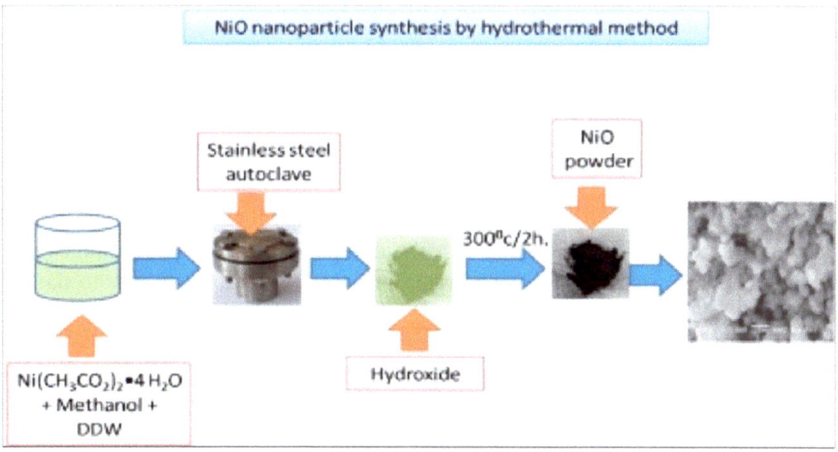

Fig. 12.9 Hydrothermal techniques for the preparation of NiO nanoparticles. Reprinted the figure with Copyright (2020) permission from Elsevier et al. [33]

two α-MnO_2 using the hydrothermal method and the same crystallographic phases are confirmed by XRD patterns which are critical for evaluating the relationship between electrochemical performances and morphologies of MnO_2 crystals as anodes for lithium-ion batteries. Vanadium pentaoxide (V_2O_5) was also prepared using this technique by Li et al. [47], and the orthorhombic phase with space group Pmmn was confirmed from the XRD data which makes it suitable for supercapacitor applications.

12.4 TMO-Based Nanomaterials in the Advanced Energy Storage System

The expansion of the global economy leads to increased consumption of fossil fuels, which has two major consequences. The first is the acceleration in the depletion of fossil fuel resources, and the second is environmental problems caused by increased greenhouse emissions and pollution of water and air. All of this contributes to the need for alternative, clean, and sustainable energy sources, as well as new energy storage technologies. To meet the future requirements of advanced energy storage systems, TMO-based nanomaterials can provide high energy, high power density, low cost, eco-friendliness, and fast availability. The growing demand for affordable, environmentally friendly energy for the world's growing population has sparked intense research into efficient energy storage systems. The majority of research has concentrated on lithium-ion batteries and supercapacitors. Lithium-ion batteries with high energy density, long durability, and a consistent rate of charge/discharge are widely used in daily life as portable batteries for phones, computers, and watches, as well as backup power devices. Supercapacitor research as advanced energy systems is gaining traction due to their high power density (10 kW kg^{-1}), fast charge/discharge rate, and long cycle life (>10^4 cycles) [24].

12.4.1 Lithium Ion Batteries (LIBs)

Rechargeable non-aqueous lithium-ion batteries (LIBs) have advanced rapidly in the past twenty years which can store electrical energy in the form of mechanical energy with much-improved energy density, no memory effect, wide voltage range, low maintenance, good performance rate, and are widely used in day to day life such as a portable electronic, electrical vehicle, medical devices, and backup power devices. However, high costs and safety concerns make current LIBs less appealing for large-scale stationary EES, where cost, safety, and cycle life, in addition to energy density, become important considerations [48]. Synthesizing transition metal oxide-based nanomaterials with unique structures can enhance safety, storage capacity, and other storage properties and also able to reduce the cost of lithium-ion batteries [49].

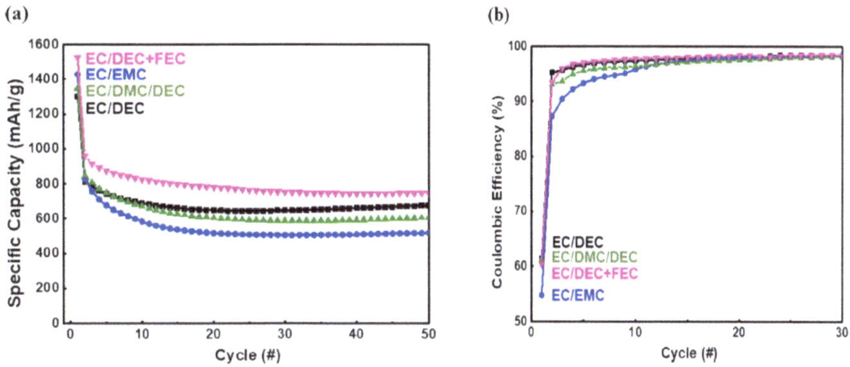

Fig. 12.10 Electrochemical performance of MnO_2 using four different electrolytes where EC stands for ethylene carbonate, DEC is diethyl carbonate, DMC refers to dimethyl carbonate and EMC stands for ethyl methyl carbonate. Reprinted (adapted) with permission from [54]. Copyright (2019)

Thus, TMO-based nanomaterials can be the most promising negative electrodes for next-generation LIBs.

Ei-Deen et al. [50] found that titanium dioxide of 20 nm particle size prepared by sol-gel method is a promising electrode material having a specific capacity of 174 mAh/g and improved electrochemical performance for lithium-ion battery. Later Q. He et al. [51] proposed that by doping nitrogen in TiO_2, the nanomaterials show a specific capacity of 350 mAh/g which makes them suitable electrode materials for lithium-ion battery. In 2019, Rana and his team et al. [52] transition metal oxides especially hybrid material of MnO_2 are very much suitable to achieve high Li storage capacity having a specific capacity value above 1100 and 500 mAh/g at a discharge current density of 25 mA/g and 5 A/g, respectively, with coulombic efficiency of 97.5%. Xu et al. [53] reported that ruthenium oxide can also be served as electrodes for lithium-ion batteries because of their large cyclability, low sensitivity, and high volumetric charge capacity. Tsai et al. [54] fabricated MnO_2 as anodes in lithium-ion battery by taking four different electrolytes and observed that its discharge capacity rises from 813 to 962 mAh/g. The electrochemical performance of MnO_2 using four different electrolytes is shown in Fig. 12.10.

12.4.2 Supercapacitors (SCs)

From the last decade of the twentieth century, supercapacitors also known as electrochemical capacitors or ultra-capacitors are gaining much more attention in the energy sector for the development of energy storage systems because of their high power density, large storage capacity, high conductivity and tunable specific capacitance, fast-charged/ discharged rate, long cycle life, lightweight, easy operational

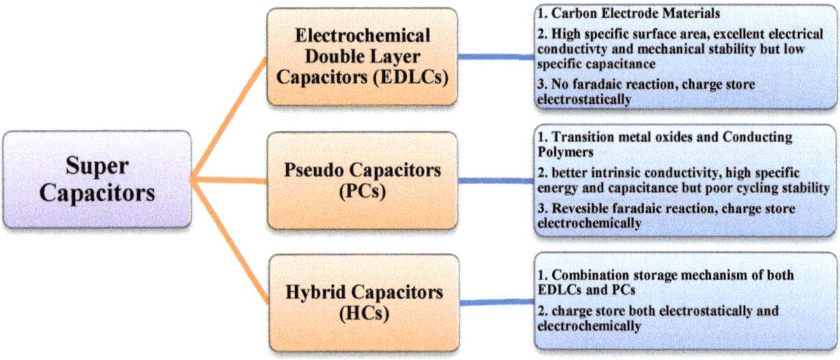

Fig. 12.11 Classification of supercapacitors with its advantages and disadvantages

technique. With these excellent properties, supercapacitors can be used in the field of electronics, especially in computers and memory storage devices, power backup devices, transport devices, public securities, medical and military services [55], etc. By the production of electrolyte ions on the surface of electro-active materials through reversible redox reactions, the supercapacitors acquire a distinct ability to store energy. According to the charge and energy storage mechanism, supercapacitors are classified [2] into (i) electrochemical double-layer capacitors (EDLCs), (ii) pseudo-supercapacitors (PCs), and (iii) hybrid supercapacitors (HCs) as shown in Fig. 12.11.

As the energy density of a supercapacitor depends on capacitance and operating voltage, so by increasing the potential window and using an electrode material with high capacitance, the energy storage capacity of the supercapacitor can be enhanced [56]. The choice of electrode materials is what makes it possible for supercapacitors to have great electrochemical performance. The electrode materials used in super capacitors should have mesoporous behavior with a high specific surface area. From the literature review, it is confirmed that TMO-based nanomaterials are more suitable electrodes for supercapacitors than others due to their high specific power and capacitance [2], multiple valency states, dielectric properties, electrical conductivity [1], stability, polarizability, long life cycles, easy scaling, eco-friendly, and sustainability [57]. Several studies on the physicochemical and electrochemical properties of TMO-based nanomaterials, such as RuO_2, MnO_2, V_2O_5, NiO, and Co_3O_4 are made for the development of high-efficiency SCs which depends on their capacitive behavior, oxidation states, charging and discharging rate, chemical composition, the crystalline phase, particle size, ion exchange capacity, and inherent thermal and chemical stability, etc. By combining expensive metal oxides with other low-cost metal oxides, the cost of electrode materials can be reduced and the electrochemical performance of the material can be enhanced.

Dung et al. [18] prepared Co and Fe-doped manganese oxide using the sol-gel technique for supercapacitor and reported that Co-doped MnO_2 exhibits a specific capacitance of 186.2 F/g with 84% stability while Fe-doped MnO_2 shows high specific capacitance of 298.4% with 83% stability. Mevada et al. [58] synthesized SnO_2–RuO_2-based nanocomposite to reduce the cost of the materials. From the electrochemical study, it was observed that the composite attains maximum specific energy of 0.91 Wh/kg with 87.5% of capacitance retention which makes it a promising electrode for supercapacitor applications. Kuo et al. [59] synthesized a supercapacitor electrode by combining tin oxide and ruthenium oxide and the specific capacitance of the composite is found to be 670 F/g with a specific energy of 0.5 Wh/kg.

12.5 Future Prospective

The main disadvantage of TMOs is their low conductivity, which prevents them from achieving a high theoretical specific capacitance value. The main component of advanced energy storage systems including supercapacitors and battery is the electrodes. Thus, electrodes with sufficient electrical conductivity, adequate mechanical properties, and cost-effectiveness should be researched and encouraged. TMO-based nanomaterials may be mixed with other metal oxide, carbon-based nanomaterials, polymers, etc., to enhance the properties such as surface area, surface activities, charge transfer processes, and to reduce electron–hole recombination which will provide greater efficiency in practical applications, especially in advanced storage system. Recently, TMOs derivatives have been shown to deliver high electrical conductivity, they should be modified into various types that can serve several multifunctionalities. Separators are normally electrical insulators, ionically conducting, and withstand high mechanical stresses caused by expansion, compression, etc. It is the interface between the electrodes but many separators nowadays hardly serve the multipurpose function needed, hence must have excellent interfacial adhesion to provide sufficient mechanical stability in conjunction with multifunctional matrices for high-performance supercapacitors.

The cost of supercapacitors is still very high because the technology required for their production is still sophisticated; hence, it is difficult for it to replace batteries effectively irrespective of its numerous advantages nowadays. We strongly recommend the following future prospects below:

(a) Better simple and low-cost fabrication techniques to control the surface morphology and the issue of self-discharging.
(b) Scientists and researchers should focus on improving cycling stability, lifespan, and fast charging time of advanced energy storage systems.
(c) TMO-based hybrid materials need to be examined effectively because of their high specific surface area, which improves storage systems.

12.6 Conclusion

To meet the requirement of sustainable and renewable energy storage for a variety of applications, such as electronic devices, hybrid electric vehicles as well in large industrial machinery, advanced energy storage systems such as batteries, PV cells, and supercapacitors have been acknowledged as ideal storage devices in fulfilling future demands. The discovery of supercapacitors and lithium-ion battery as some substitute systems to meet the requirement of storing renewable energy appeared enticing because it possesses numerous properties including high power density, cycle stability, fast rate of charge or discharge, and other electrochemical performance which suit them for many applications such as (i) emergency power back up systems, (ii) electronic devices, (iii) battery buffers, (iv) electric vehicles, and (v) medical devices.

The performance of advanced energy storage systems is largely affected by the selection of electrode materials. TMOs are the most favored electrode materials due to their enlarged specific surface area, long durability, better chemical stability, higher energy density, specific capacity, and comparatively high conductive than polymers which can greatly enhance ion/electron transport than other materials and polymers. In this chapter, we have summarized recent developments of TMOs-based electrode materials for lithium-ion batteries and supercapacitors including fabrication, and electrochemical performances. RuO_2 is the ideal TMOs electrode material, but high cost and toxicity are its fatal shortcomings. Co_3O_4, MnO_2, TiO_2, vanadium pentoxide, and other metal oxides are supposed to substitute RuO_2. However, existing problems like low conductivity hinder their practical application in electrode materials.

However, the design and optimum conditions for preparing nanomaterials with superior features remain a great challenge, and therefore, some factors including the morphology, pore size, and surface area of the nanomaterials should be taken into consideration. In this chapter, recent progress in TMO-based nanomaterials has been pointed out and reviewed. Although TMO-based nanomaterials can deliver a good electrochemical performance, these individual electrode materials still suffer from some limitations including the poor conductivity of TMO, which contributes to a low specific capacitance and causes the structural destruction of the TMO materials during the doping-developing process, thus leading to an inferior stability performance. Although the ideal synthesis of TMO-based nanomaterials is believed to overcome the disadvantages of each component, further improvements in the electrochemical performances of the materials, that is, specific capacitance, specific energy, and long-term stability are needed for scalable and flexible energy storage devices. It is highly recommended to fabricate TMO-based nanomaterials by combining two electrode materials with different potential windows. This hybrid electrode design could broaden the potential window, thus exhibiting high specific energy. In short, tremendous efforts in advancing TMO-based nanomaterials should be made to enlarge the scopes of energy storage systems into flexible, stretchable, and compressible devices.

References

1. Abdah MAAM, Azman NHN, Kulandaivalu S, Sulaiman Y (2020) Review of the use of transition-metal-oxide and conducting polymer-based fibres for high-performance supercapacitors. Mater Des 186:108199
2. Ray A, Roy A, Saha S, Das S (2019) Transition metal oxide-based nano-materials for energy storage application. Sci Technol Adv Appl Supercapacitors, 1–17
3. Mallick P, Satpathy SK, Behera B (2022) Investigation on the structural, electrical and thermistor parameters of La-doped $BiFeO_3$–$PbZrO_3$ for energy storage devices. Bull Mater Sci 45(4):1–11
4. Mallick P, Champati L, Biswal SK, Satpathy SK (2022) Experimental investigation on structural, dielectric, and transport properties of copper doped $SrTiO_3$. Integr Ferroelectr 230(1):156–162
5. BahariMollaMahaleh Y, Sadrnezhaad SK, Hosseini D (2008) NiO nanoparticles synthesis by chemical precipitation and effect of applied surfactant on distribution of particle size. J Nanomater
6. Majumdar D, Maiyalagan T, Jiang Z (2019) Recent progress in ruthenium oxide-based composites for supercapacitor applications. ChemElectroChem 6(17):4343–4372
7. Sahu AK, Satpathy SK, Rout SK, Behera B (2020) Dielectric and frequency dependent transport properties of gadolinium doped bismuth ferrite. Trans Electr Electron Mater 21(2):217–226
8. Mallick P, Satpathy SK, Behera B (2022) Study of structural, dielectric, electrical, and magnetic properties of samarium-doped double perovskite material for thermistor applications. Braz J Phys 52(6):1–15
9. Sahu AK, Mallick P, Satpathy SK, Behera B (2021) Effect on structural, electrical and temperature sensing behavior of neodymium doped bismuth ferrite. Adv Mater Lett 12(7):1–7
10. Mallick P, Sahu AK, Biswal SK, Satpathy SK, Behera B (2022) Investigation on structural, dielectric, thermistor parameters and negative temperature coefficient behaviour of Nd, Gd, and la-doped bismuth ferrite. Trans Electr Electron Mater, 1–13
11. Mohanty D, Mallick P, Biswal SK, Behera B, Mohapatra RK, Behera A, Satpathy SK (2020) Investigation of structural, dielectric and electrical properties of $ZnFe_2O_4$ composite. Mater Today Proc 33:4971–4975
12. Mohanty D, Naik AU, Nayak PK, Behera B, Satpathy SK (2021) Structural and electrical properties of magnesium-doped $CoFe_2O_4$. Powder Metall Met Ceram 59(9):507–514
13. Mohanty D, Satpathy SK, Behera B, Mohapatra RK (2020) Dielectric and frequency dependent transport properties in magnesium doped $CuFe_2O_4$ composite. Mater Today Proc 33:5226–5231
14. Diao F, Wang Y (2018) Transition metal oxide nanostructures: premeditated fabrication and applications in electronic and photonic devices. J Mater Sci 53(6):4334–4359
15. Park SE, Kim HM, Kim KB, Min SH (2000) Metallorganic chemical vapor deposition of Ru and RuO_2 using ruthenocene precursor and oxygen gas. J Electrochem Soc 147(1):203
16. Dao VD, Hoa ND, Vu NH, Quang DV, Van Hieu N, Dung TTN, Viet NX, Hung CM, Choi HS (2019) A facile synthesis of ruthenium/reduced graphene oxide nanocomposite for effective electrochemical applications. Sol Energy 191:420–426
17. Nisha B, Vidyalakshmi Y, Razack SA (2020) Enhanced formation of ruthenium oxide nanoparticles through green synthesis for highly efficient supercapacitor applications. Adv Powder Technol 31(3):1001–1006
18. Dang TD, Le TTH, Hoang TBT, Mai TT (2015) Synthesis of nanostructured manganese oxides based materials and application for supercapacitor. Adv Nat Sci Nanosci Nanotechnol 6(2):025011
19. Meng Y, Liu Y, He J, Sun X, Palmieri A, Gu Y, Zheng X, Dang Y, Huang X, Mustain W, Suib SL (2021) Large scale synthesis of manganese oxide/reduced graphene oxide composites as anode materials for long cycle lithium ion batteries. ACS Appl Energy Mater 4(6):5424–5433
20. Yao J, Li Y, Massé RC, Uchaker E, Cao G (2018) Revitalized interest in vanadium pentoxide as cathode material for lithium-ion batteries and beyond. Energy Storage Mater 11:205–259

21. Sakunthala A, Reddy MV, Selvasekarapandian S, Chowdari BVR, Selvin PC (2011) Energy storage studies of bare and doped vanadium pentoxide, $(V_{1.95}M_{0.05})O_5$, M = Nb, Ta, for lithium ion batteries. Energy Environ Sci 4(5):1712–1725
22. Xie JD, Li HY, Wu TY, Chang JK, Gandomi YA (2018) Electrochemical energy storage of nanocrystalline vanadium oxide thin films prepared from various plating solutions for supercapacitors. ElectrochimicaActa 273:257–263
23. Boruah BD, Wen B, De Volder M (2021) Light rechargeable lithium-ion batteries using V_2O_5 cathodes. Nano Lett 21(8):3527–3532
24. Zallouz S, Réty B, Vidal L, Le Meins JM, MateiGhimbeu C (2021) Co_3O_4 nanoparticles embedded in mesoporous carbon for supercapacitor applications. ACS Appl Nano Mater 4(5):5022–5037
25. Jang GS, Ameen S, Akhtar MS, Kim E, Shin HS (2017) Electrochemical investigations of hydrothermally synthesized porous cobalt oxide (Co_3O_4) nanorods: supercapacitor application. ChemistrySelect 2(28):8941–8949
26. Xie L, Li K, Sun G, Hu Z, Lv C, Wang J, Zhang C (2013) Preparation and electrochemical performance of the layered cobalt oxide (Co_3O_4) as supercapacitor electrode material. J Solid State Electrochem 17(1):55–61
27. Priyadharsini CI, Marimuthu G, Pazhanivel T, Anbarasan PM, Aroulmoji V, Siva V, Mohana L (2020) Sol-gel synthesis of Co_3O_4 nanoparticles as an electrode material for supercapacitor applications. J Sol-Gel Sci Technol 96(2):416–422
28. Lee DH, Lee BH, Sinha AK, Park JH, Kim MS, Park J, Shin H, Lee KS, Sung YE, Hyeon T (2018) Engineering titanium dioxide nanostructures for enhanced lithium-ion storage. J Am Chem Soc 140(48):16676–16684
29. Ray A, Roy A, Sadhukhan P, Chowdhury SR, Maji P, Bhattachrya SK, Das S (2018) Electrochemical properties of TiO_2-V_2O_5 nanocomposites as a high performance supercapacitors electrode material. Appl Surf Sci 443:581–591
30. LontioFomekong R, Saruhan B (2019) Synthesis of Co^{3+} doped TiO_2 by co-precipitation route and its gas sensing properties. Frontiers in Materials 6:252
31. Napari M, Huq TN, Hoye RL, MacManus-Driscoll JL (2021) Nickel oxide thin films grown by chemical deposition techniques: potential and challenges in next-generation rigid and flexible device applications. InfoMat 3(5):536–576
32. Lamba P, Singh P, Singh P, Kumar A, Singh P, Kumar Y, Gupta M (2021) Bioinspired synthesis of nickel oxide nanoparticles as electrode material for supercapacitor applications. Ionics 27(12):5263–5276
33. Dhas SD, Maldar PS, Patil MD, Nagare AB, Waikar MR, Sonkawade RG, Moholkar AV (2020) Synthesis of NiO nanoparticles for supercapacitor application as an efficient electrode material. Vacuum 181:109646
34. Velmurugan V, Srinivasarao U, Ramachandran R, Saranya M, Grace AN (2016) Synthesis of tin oxide/graphene (SnO_2/G) nanocomposite and its electrochemical properties for supercapacitor applications. Mater Res Bull 84:145–151
35. Yadav S, Rani N, Saini K (2022) A review on transition metal oxides based nanocomposites, their synthesis techniques, different morphologies and potential applications. IOP Conf Ser Mater Sci Eng 1225(1):012004
36. Bokov D, TurkiJalil A, Chupradit S, Suksatan W, Javed Ansari M, Shewael IH, Valiev GH, Kianfar E (2021) Nanomaterial by sol-gel method: synthesis and application. Adv Mater Sci Eng
37. Rochman NT, Akwalia PR (2017) Fabrication and characterization of Zinc Oxide (ZnO) nanoparticle by sol-gel method. J Phys Conf Ser 853(1):012041
38. Dubey RS, Krishnamurthy KV, Singh S (2019) Experimental studies of TiO_2 nanoparticles synthesized by sol-gel and solvothermal routes for DSSCs application. Results Phys 14:102390
39. Ngaotrakanwiwat P, Meeyoo V (2012) TiO_2–V_2O_5 nanocomposites as alternative energy storage substances for photocatalysts. J Nanosci Nanotechnol 12(1):828–833
40. Haniam P, Kunsombat C, Chiangga S, Songsasen A (2014) Synthesis of cobalt oxides thin films fractal structures by laser chemical vapor deposition. Sci World J

41. Barreca D, Massignan C, Daolio S, Fabrizio M, Piccirillo C, Armelao L, Tondello E (2001) Composition and microstructure of cobalt oxide thin films obtained from a novel cobalt (II) precursor by chemical vapor deposition. Chem Mater 13(2):588–593
42. Maruyama T, Arai S (1993) The electrochromic properties of nickel oxide thin films prepared by chemical vapor deposition. Sol Energy Mater Sol Cells 30(3):257–262
43. Le HA, Chin S, Park E, Bae G, Jurng J (2012) Chemical vapor synthesis and physico-chemical properties of V_2O_5 nanoparticles. Chem Vap Deposition 18(1–3):6–9
44. Pooja K, Dipali S (2017) Synthesis and characterization of nickel oxide nanoparticles by using co-precipitation method. Int J Adv Res 5(5):1333–1338
45. Buraso W, Lachom V, Siriya P, Laokul P (2018) Synthesis of TiO_2 nanoparticles via a simple precipitation method and photocatalytic performance. Mater Res Express 5(11):115003
46. Liu Z, Wang R, Kan F, Jiang F (2014). Synthesis and characterization of TiO_2 nanoparticles. Asian J Chem 26(3)
47. Li M, Ai T, Kou L, Song J, Bao W, Wang Y, Wei X, Li W, Deng Z, Zou X, Wang H (2022) Synthesis and electrochemical performance of V_2O_5 nanosheets for supercapacitor. AIP Adv 12(5):055203
48. Teng X (2019) Transition metal oxides nanomaterials for aqueous electrochemical energy storage (No. DOE-UNH-SC0010286). XiaoweiTeng/University of New Hampshire
49. Wu HB, Chen JS, Hng HH, Lou XWD (2012) Nanostructured metal oxide-based materials as advanced anodes for lithium-ion batteries. Nanoscale 4(8):2526–2542
50. El-Deen SS, Hashem AM, Abdel Ghany AE, Indris S, Ehrenberg H, Mauger A, Julien CM (2018) Anatase TiO_2 nanoparticles for lithium-ion batteries. Ionics 24(10):2925–2934
51. He Q, Sun Z, Shi X, Wu W, Cheng J, Zhuo R, Zhang Z, Wang J (2021) Electrochemical performance enhancement of nitrogen-doped TiO_2 for lithium-ion batteries investigated by a film electrode model. Energy Fuels 35(3):2717–2726
52. Rana M, Avvaru VS, Boaretto N, Víctor A, Marcilla R, Etacheri V, Vilatela JJ (2019) High rate hybrid MnO_2@ CNT fabric anodes for Li-ion batteries: properties and a lithium storage mechanism study by in situ synchrotron X-ray scattering. J Mater Chem A 7(46):26596–26606
53. Xu L, Thompson CV (2020) Mechanisms of the cyclic (de) lithiation of RuO_2. J Mater Chem A 8(41):21872–21881
54. Tsai YC, Kuo CT, Liu SF, Lee YT, Yew TR (2021) Effect of different electrolytes on MnO_2 anodes in lithium-ion batteries. J Phys Chem C 125(2):1221–1233
55. Obodo RM, Nwanya AC, Hassina T, Kebede MA, Ahmad I, Maaza M, Ezema FI (2019) Transition metal oxide-based nanomaterials for high energy and power density supercapacitor. In: Electrochemical devices for energy storage applications. CRC Press, pp 131–150
56. Liang R, Du Y, Xiao P, Cheng J, Yuan S, Chen Y, Yuan J, Chen J (2021) Transition metal oxide electrode materials for supercapacitors: a review of recent developments. Nanomaterials 11(5):1248
57. Quispe-Garrido V, Cerron-Calle GA, Bazan-Aguilar A, Ruiz-Montoya JG, López EO, Baena-Moncada AM (2021) Advances in the design and application of transition metal oxide-based supercapacitors. Open Chem 19(1):709–725
58. Mevada C, Mukhopadhyay M (2020) High mass loading tin oxide-ruthenium oxide-based nanocomposite electrode for supercapacitor application. J Energy Storage 31:101587
59. Kuo SL, Wu NL (2003) Composite supercapacitor containing tin oxide and electroplated ruthenium oxide. Electrochem Solid-State Lett 6(5):A85

Chapter 13
Fluoropolymer-Based Nanodielectrics for Energy Storage Application

Anindita Mukherjee, Anupam Ghosh, and Barnali Dasgupta Ghosh

Abstract Growing demand for advanced energy storage systems (ESSs) leads to the synthesis of novel materials with the ability to store high-power density and electrical energy. Film capacitors based on fluoropolymers like poly(vinylidene fluoride) (PVDF) and its copolymers (PVDF-HFP, and PVDF-TrFe) have enormous applications for energy storage portable devices due to high dielectric constant and breakdown strength with low discharge efficiency. The electroactive structure of fluoropolymers initiates a micro-capacitance and develops high polarizability. Thus, this chapter investigates the properties, applications, and opportunities of fluoropolymers with high-energy storage performance. Also, it elucidates the structural dependence due to the inherent polarity (C–H and C–F bond) of polymers to the dielectric constant (charge storage mechanism), ferroelectric polarization, and low leakage current for a high capacitive response. This chapter will mostly stress PVDF, due to its low cost, easy availability, stability, and flexibility for enormous electrical applications. The introduction of high k dielectric materials into polymers increases their dielectric constant, energy storage density, and thermal stability to appreciable value. Therefore, it is essential to investigate the capacitive properties of PVDF-based composites loaded with ceramics and metal oxide. Overall, the chapter provides the readers with the elaborated significance of fluoropolymers and composites as a capacitor for future high-field applications.

Keywords Fluoropolymers · Polarizability · Energy storage performance · Breakdown strength · Applications

---

A. Mukherjee · B. Dasgupta Ghosh (✉)
Department of Chemistry, Birla Institute of Technology Mesra, Ranchi 835215, India
e-mail: barnalidg@bitmesra.ac.in; barnali.iitkgp@gmail.com

A. Ghosh
Department of Management, Birla Institute of Technology Mesra, Ranchi 835215, India

S. Moharana et al. (eds.), *Emerging Nanodielectric Materials for Energy Storage*, Nanostructure Science and Technology, https://doi.org/10.1007/978-3-031-40938-7_13

13.1 Introduction

The alternate solution of nonrenewable energy sources is one of the very promising research areas in terms of both academia and industries. The solution for this is the development of energy storage devices like chemical batteries, dielectric capacitors, and supercapacitors. Compared to supercapacitors and chemical batteries dielectric capacitors are mostly preferred [1]. It possesses an ever-increasing demand for energy storage devices, their usage leads to the sustainable growth of modern society. It is one of the crucial parts of our daily lives since the manufacture of such dielectric capacitors is highly cost-effective. A dielectric capacitor has a role in the energy-storing ability of a capacitor. Also, it exhibits ultrafast charging ability and this leads to its tremendous usage in pulsed power generation, medical defibrillators, high-frequency inverters, and hybrid electric vehicles. Dielectric films, due to its flexibility and facile fabrication impart an important integrated part within electrical energy storage devices (EESD) [2]. The discharge energy density is represented through Eq. (13.1).

$$U_e = \int_{D_{\max}}^{0} E \mathrm{d}D \tag{13.1}$$

The efficiency is computed as in Eq. (13.2):

$$\eta = \frac{U_{\text{discharged}}}{U_{\text{discharged}} + U_{\text{loss}}} \tag{13.2}$$

However, for linear dielectric materials, the energy density is given through Eq. (13.3).

$$U_e = \frac{1}{2}\varepsilon_r \varepsilon_0 E_b^2 \tag{13.3}$$

As per the above equation, the discharge energy density depends upon the breakdown strength, dielectric permittivity, and dielectric loss (tan δ). Dielectric permittivity is the intrinsic property of a dielectric material. However, low breakdown strength could develop mechanical, thermal, and electrical issues. Also, the discharged energy density is square to the power of breakdown strength, proving the importance of breakdown strength for high energy storage performance. Overall, the material should have high permittivity along with high breakdown strength [3]. Thus, fluoropolymers are effective for high capacitive performance. Its electroactive structure produces high permittivity and its monophase structure, creates high breakdown strength (Fig. 13.1) [4]. According to Eq. (13.2), the efficiency increases with the increase in the ($U_{\text{discharged}} - U_{\text{loss}}$) value, due to the respective increase of discharge density as compared to energy loss. The high increment of discharge energy density with respect to energy loss density would decrease the denominator and would thereby increase the efficiency. Single-phase dielectric capacitor reduces the dielectric loss and energy loss density as well. Thus, the usage of the polymer as a dielectric capacitor increases the overall efficiency of the capacitor. Even though the polymers exhibit a high breakdown strength, still possess very low permittivity. A high dielectric constant and permittivity are required for the high polarization represented in the P-E loop. Similar to the charging and discharging phenomena of the capacitor, the P-E loop also represents the capacitance behavior of a dielectric capacitor. High polarization increases the energy storage density of the capacitor (represented through an orange part in Fig. 13.1). Therefore, the application of polymer with electroactive features is the solution for the manufacture of energy storage devices with capacitors of high energy storage density. The electroactivity is mostly present in polymers with ionic bonds in their chain. For example, fluoropolymer like PVDF, P(VDF-HFP), and PVDF-TrFE has a high dielectric permittivity as compared to other amorphous and semicrystalline polymers. Due to their high polarizability, fluoropolymers possess efficient electrical properties like high piezoelectric, dielectric, and ferroelectric properties. Still, the permittivity needs to be enhanced within such polymer to meet the requirement of high-energy storage portable devices. The polarization or the permittivity could be increased through poling. Poling like corona poling is an effective method for the synthesis of polymer with a highly dipole-oriented structure. It involves the usage of both electric fields and heating for orientation. A high electric field ionizes the gas within the poling chamber. These gases induce the polymer dipole to orient in a certain axial direction. The heat treatment mobilizes the dipole. After the heating, cooling is done so that it permanently freezes the dipole in its specific axial direction. Another method is the addition of insulative fillers like ZnO, $BaTiO_3$, and PZT. The morphology of the grain of such insulative filler has an enormous effect on energy storage density as the asymmetric structure (nanorod, nanowires, nanoplates, and nanograins) produces higher polarizability. The mentioned shape possesses a high aspect ratio. High aspect ratio filler exhibits high surface area interaction with the polymer. Thus, the incorporation or loading of even a small amount of such filler increases the dielectric constant. Therefore, the combination of the high-aspect filler with fluoropolymers produces composites with high energy storage density. However, the inclusion of filler diminishes its overall efficiency as a capacitor and increases the dielectric loss. The breakdown strength and dielectric loss could be modified for low dissipation through surface functionalization of the filler where its outside surface is surrounded by organic parts to decrease the agglomeration and increase the interaction of the filler with the polymer matrix.

In summary, this chapter focuses on the energy storage properties of fluoropolymer. This chapter gives an idea about the structure of monomers, different phases, and the effect of fluorine on the polarizability as well as the dielectric, piezoelectric, ferroelectric, and energy storage properties. Also, the chapter elaborates on the dependence of the energy storage property of a PVDF film in terms of its dielectric constant and breakdown strength. The manufacture or fabrication process has a huge role in the synthesis of the film with high thermal stability, energy storage density,

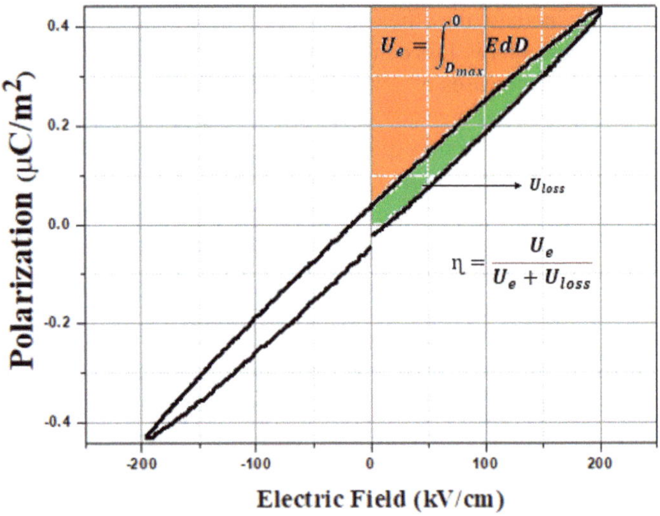

Fig. 13.1 Representation of discharged energy density (orange part), energy loss density (green part), energy efficiency, and discharge energy density for the linear dielectric

and low dielectric loss. Here three such methods are discussed electrospinning, solution casting, and hot-press method. Finally, the energy storage properties of some earlier research works are also represented. These include PVDF, P(VDF-HFP), and PVDF-TrFE composites with asymmetric fillers exhibiting the properties discussed above. However, the properties will mostly focus on PVDF composites, due to their high dielectric constant and easy availability.

13.2 Fluoropolymers

The replacement of hydrogen with fluorine creates a tremendous change in the properties of fluoropolymer. Fluorine helps to improve thermal and thermo oxidative stability, mechanical property, and chemical resistance in comparison to their non-fluorinated analogues [5]. Therefore, fluoropolymers can be used in a wide range of applications due to their superior properties. Improvement in thermal resistance properties helps to widen its application from low-temperature to high-temperature regions. Fluoropolymers show higher flexibility, it does not easily break under pressure. So can be easily mouldable and different complicated shape products can be obtained with the application of cables and covers. Due to the chemical inertness of fluoropolymers such polymers could be used in a harsh and aggressive environment. Similarly, it has its application in the medical field due to its biocompatibility feature. Among the polymer, fluoropolymers exhibit the highest energy storage density due to their high dielectric constant with high breakdown strength. The energy storage performance is dependent on the structural feature of such polymers. Thus, this section discusses the structural properties of PVDF and PVDF-based copolymers.

PVDF is a semicrystalline polymer with the repeat unit of $–CF_2–CH_2–$. The structure consists of negatively charged fluorine and positively charged hydrogen, thereby the structure is inherently polar [6]. Figure 13.2a describes the crystalline form of PVDF. According to the relative orientation of atoms, the nine crystallines are divided into four categories of α, β, and γ Here β has an all-trans chain with the same parallel orientation of CH_2 and CF_2, while α form has antiparallel TGTG[1] confirmation with nonpolar structure. The crystal transformation of the α phase to β phase could occur from mechanical stretching and annealing. These external processes rotate the C–C bond in PVDF and thereby affecting the relative orientation. On a molecular level, the processes destroy the spherulite structure of PVDF and convert it into the stretched part with highly aligned oriented-lamelle as shown in Fig. 13.2b. Such a high polar structure creates strong micro-capacitance and increases the dielectric constant and energy storage density.

Poly (vinylidene fluoride-co-hexafluoropropylene) (PVDF-HFP) is a copolymer of PVDF with extra four fluorine atoms. This makes it hydrophobic and less crystalline [7]. However, the presence of fluorine atoms and hydrogen atoms develops a strong dipole moment (Fig. 13.3). Introduction of TrFE into VDF crystallizes to form PVDF-TrFE polymer. The introduction of TrFE greater than 11%, creates enough defect by the bulky fluorine group preventing the formation of Trans-Gauche confirmation. Overall TrFE addition transforms directly the α phase to the β phase without any mechanical stretching and annealing [8].

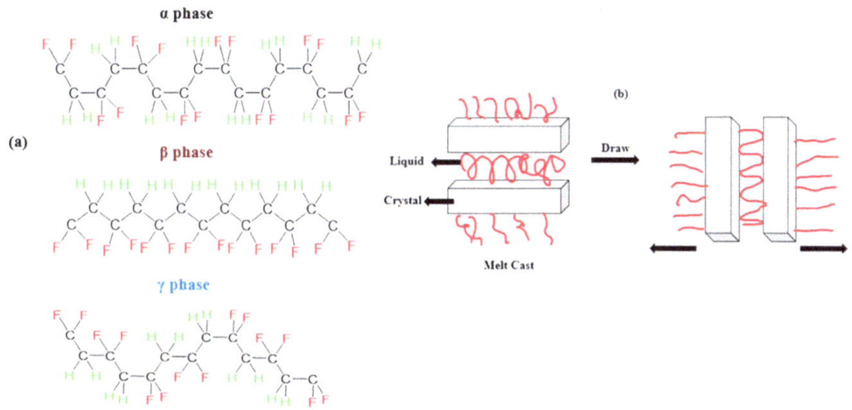

Fig. 13.2 **a** α, β, γ forms of PVDF and **b** formation of oriented lamelle through a drawing of a spherulite

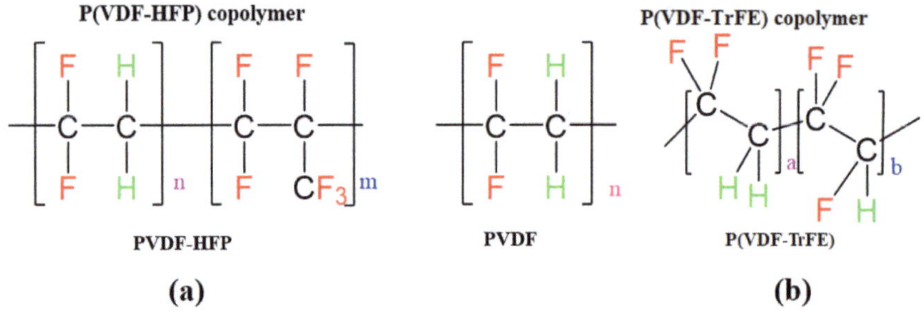

Fig. 13.3 Structure of **a** PVDF-HFP and **b** PVDF-TrFE

13.2.1 Synthesis of Fluoropolymer

The synthesis of PVDF involves free radical polymerization of 1,1-difluoroethylene. The free radical polymerization initiates with a highly active initiator. Such initiators are usually free radical since it is susceptible to a chain reaction. The radicals react with the required monomer here as 1,1-difluoroethylene and form a new radical. This new radical is similar to the initiator, is highly reactive, and thereby undergoes a chain-growth polymerization. The required polymerization leads to the synthesis of polyvinylidene fluoride. The respective copolymers of PVDF (PVDF-HFP and PVDF-TrFE) are produced through radical copolymerization reaction [9]. Similar to free radical polymerization, here a radical monomer is initially formed for either part of the copolymer. The free radical monomer being reactive enough, attacks the monomer of another part forming a new radical. This process continues as the chain growth copolymerization forms the final copolymers of PVDF-HFP and PVDF-TrFE. The initial radical of radical copolymerization is usually formed in the presence of an initiator, UV, or other radiation. The mode of attack and the basic reactivity and its kinetics depend upon the groups or the functional groups present in the monomer. For example, the highly electronegative fluorine group present in 1,1-difluoroethylene makes the monomer susceptible to attack from an electron-donating monomer or radical.

13.2.2 Electrical Properties of Fluoropolymer

As discussed above structural properties of PVDF and its copolymers influences the polarity of the overall polymer. The polarity of the polymer increases the dielectric constant, ferroelectric, and energy storage properties. Therefore, this section focuses on the piezoelectric, ferroelectric, dielectric, and energy storage properties of PVDF and its copolymers.

13.2.2.1 Piezoelectric Properties

Piezoelectricity is the generation of electricity through pressure (direct piezoelectric effect) and the development of strain in the material on the application of an electric field (indirect piezoelectric effect). Bilal Zaarour et al. synthesized THF-treated electrospun PVDF for developing a piezoelectric nanogenerator (PNG). He synthesized three different films according to the amount of treatment. Here, THF(100%)-treated electrospun showed the highest voltage about 2.98 V [10]. Simadri Badatya et al. fabricated graphene quantum dots reinforced PVDF-HFP composite and investigated the piezoelectric output from the compressive force, wrist movement, and breathing condition. Here graphene quantum dots reinforcement within polymer matrix showed an excellent piezoelectric response of about 6 V and a current of 25 nA (Fig. 13.4) [11]. Pi et al., fabricated thin film PVDF-TrFE through a spin coating technique and investigated the piezoelectric response on variation in strain% applied within the film. The output voltage was about 7 V and the short circuit current was about 58 nA in the given piezoelectric nanogenerator (PVDF-TrFE) (Fig. 13.5) [12].

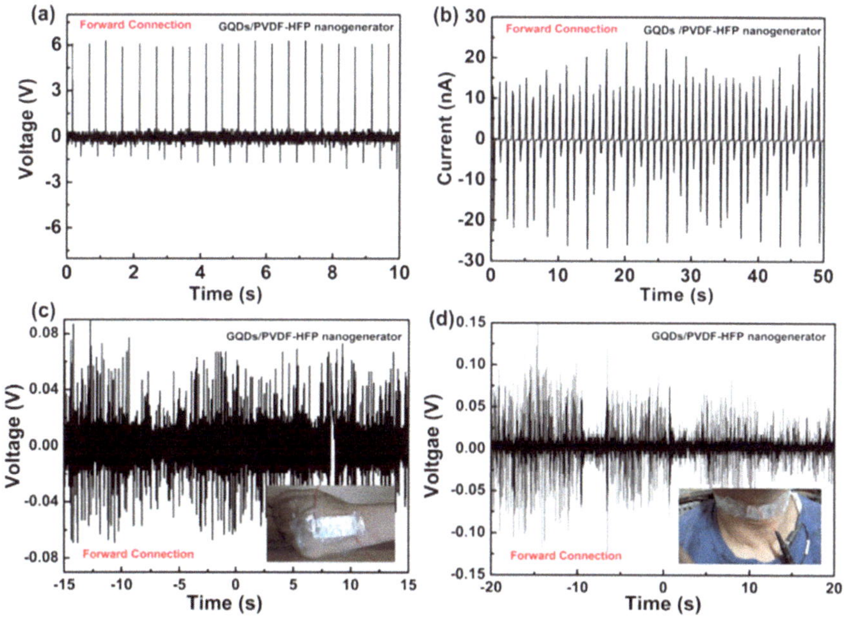

Fig. 13.4 Piezoelectric **a** output voltage, **b** output current on compressive force, output voltage, **c** on wrist motion, **d** neck movement (Badatya et al. [11]. Reproduced with permission)

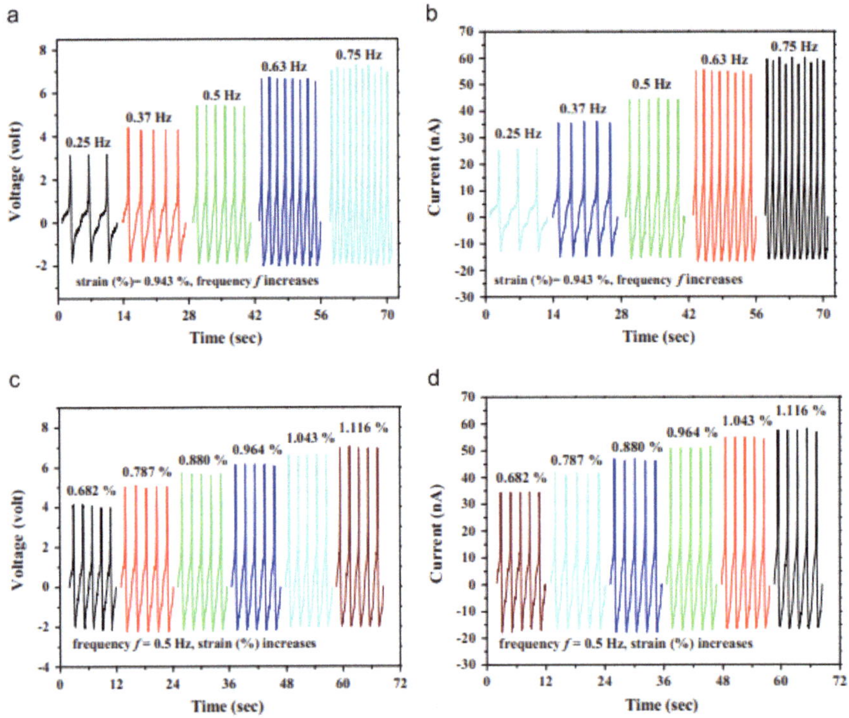

Fig. 13.5 Piezoelectric **a** output voltage, **b** output current on constant strain, but with variation in frequency, **c** output voltage, **d** output current on constant frequency, but with strain variation (Pi et al. [12]. Reproduced with permission)

13.2.2.2 Ferroelectric Properties

Ferroelectricity is the presence of polarization in the absence of an electric field. These are represented in terms of remanent polarization (P_r), maximum polarization (P_m), and coercive electric field (E_c). Maya Sharma et al. synthesized PVDF and investigated its ferroelectric properties through mechanical rolling, poling, annealing, and a combination of these processes. Here, the influence of mechanical rolling and the combination of rolling and annealing showed the highest ferroelectric properties [13]. Smaranika Dash et al., synthesized surface hydroxylated BZT/BCT embedded in PVDF-HFP with variation in weight% from 0 to 40 wt%. The films were of 75 μm thickness, and further the respective ferroelectric properties investigation proved 15 wt% loading as the optimum concentration due to excellent ferroelectric properties. It also proved that 15 wt% showed the highest interfacial interaction between polymer and filler (Fig. 13.6) [14]. Nan Meng et al., blended PVDF/PVDF-TrFE and investigated its ferroelectric properties. Low crystallinity, processing cost, and low Curie point were tackled by such polymer melt blending. This approach tackled the immiscible nature of PVDF and PVDF-TrFE and produced a strong

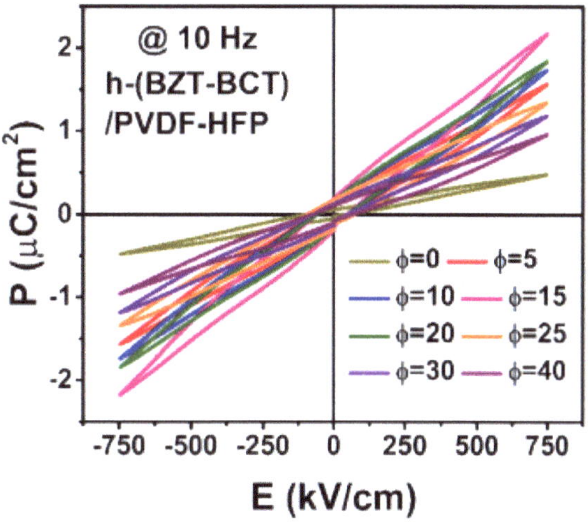

Fig. 13.6 P-E loop of h-BZT-BCT (0–40) wt%/PVDF-HFP with variation in weight% of h-BZT-BCT (Dash et al. [14]. Reproduced with permission)

interaction in the melt. It enhanced the ferroelectric properties by increasing the remanent polarization from 0.040 C m^{-2} to 0.077 C m^{-2} [15].

13.2.2.3 Dielectric Properties

Dielectric properties comprise of two components of dielectric constant and dielectric loss. The combination of these dielectric components determine the energy storage properties of a dielectric polymer composite [16, 17]. Vineet Tiwari et al, synthesized PVDF polymer films and investigated the crystallinity, β phase content and dielectric properties through annealing and quenching temperature conditions. PVDF showed the highest β phase content corresponding to an annealed temperature of 80 °C and quenching temperature of 20 °C. Also, the dielectric constant of the synthesized PVDF represented the highest around 60 [18]. Feng et al. fabricated PVDF-HFP with embedded BT nanowires and BT nanoparticles through in-situ polymerization. The high aspect ratio of BT nanowires developed an excellent interfacial polarization between filler and polymer as compared to the contribution of BT nanoparticles. Therefore, BT nanowires enhanced the dielectric properties tremendously as compared to BT nanoparticles [19]. Arshad et al., fabricated PVDF-TrFE composite films incorporated with (1–7) wt% MgO through a spin-coating technique. Here, PVDF-TrFE/MgO (7 wt%) showed the highest β content with no observable defects. However, the dielectric properties were highest for PVDF-TrFE/MgO (3 wt%) (Fig. 13.7) [20].

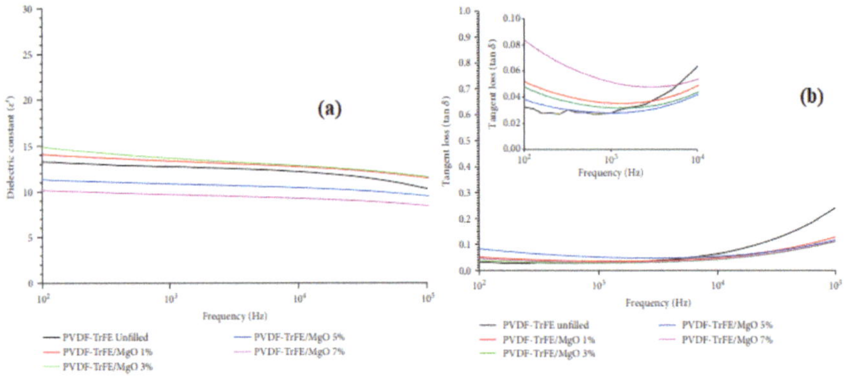

Fig. 13.7 **a** Dielectric constant versus frequency of PVDF-TrFE/MgO (0–7 wt%) and **b** dielectric loss versus frequency of PVDF-TrFE/MgO (0–7 wt%) (Arshad et al. [20]. Reproduced with permission)

13.2.2.4 Energy Storage Properties

Wenjing Li et al, fabricated α, β, and γ PVDF films and investigated the energy storage properties of these films. The β PVDF showed the lowest breakdown strength with an energy density of 1.5 J/cm^3 for an electric field of 150 MV/cm. The α phased PVDF represented enhanced breakdown strength but also showed a high energy loss density below 350 MV/cm. However, γ PVDF showed breakdown strength beyond 350 MV/cm with a high energy storage density of 14 J/cm^3, which makes it desirable for high-field capacitor application (Fig. 13.8) [21]. Kishor Kumar et al., fabricated PVDF-HFP composites with high energy storage density and increased breakdown strength through loading of Lanthanum Zirconium Oxide (LZO) using a spin coating technique. A computational model represented the optimum quantity of LZO for high energy density with a low dielectric loss so that there is no further wastage of filler. Here, PVDF-HFP/LZO (15 wt%) showed a better energy storage density of about 16.5 J/cm^3 at 530 MV/m [22]. Hong Zhu et al., synthesized ArPTU/PVDF-TrFE-CTFE through solution blending followed by the Hot-Press method. The required blending improved the energy density and decreased the dielectric loss of the composite films. Here, ArPTU/PVDF-TrFE-CTFE (15/85) showed a maximum energy density of about 19.2 J/cm^3 at 700 MV/m, along with an energy storage efficiency of 85% [23].

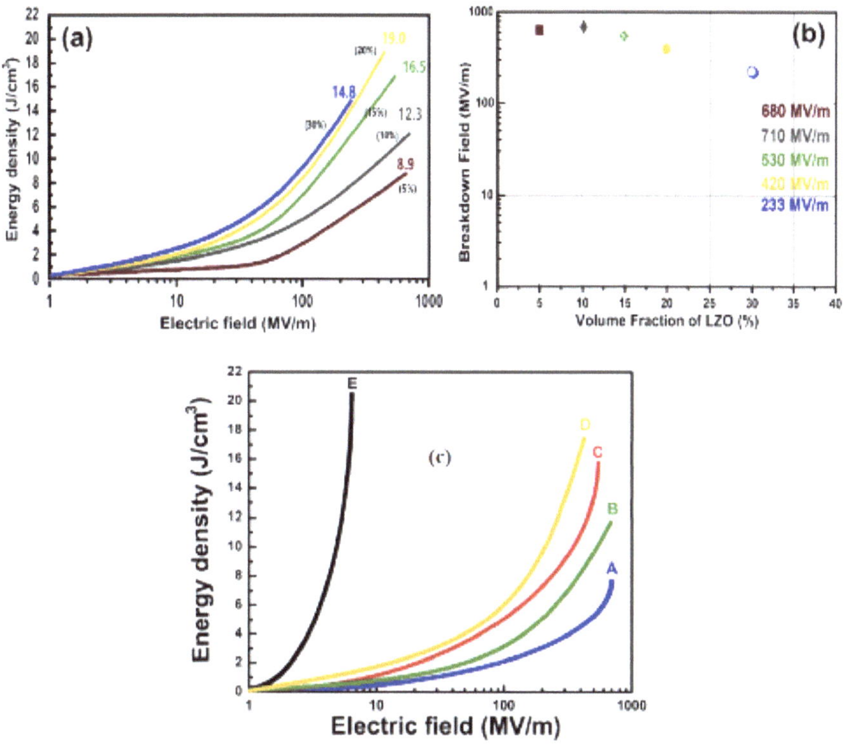

Fig. 13.8 Computational results of **a** energy storage density, **b** breakdown strength of PVDF-HFP/LZO and **c** experimental results of energy storage density of PVDF-HFP/LZO (Kishor and Kalathi [22]. Reproduced with permission)

13.3 Dependence of Film Thickness and Filler Size on Dielectric Properties

The breakdown strength of the sample depends upon the film thickness. The film with lower thickness exhibits high breakdown strength. However, the dielectric constant possesses different properties than the breakdown strength [24]. For example, when the thickness of PVDF-TrFE decreases lower than 100 nm, the crystallization process gets hindered (Fig. 13.9a) [25]. The electric field becomes more uneven with the increment in film thickness over the edges. An uneven electric increases defect and heat dissipation which further leads to a decrease in breakdown strength. The breakdown strength dependence on thickness was carried out by synthesizing polymer films from the range of 8–76 μm and investigating the respective dielectric strength. It represented an increase in dielectric strength with a decrease in film thickness [26]. The space charge dynamic model under DC electric field determines the dependence on film thickness. The simulation results provided the concept of a rate-dependent decrease of breakdown strength with the increment in thickness of polymer film

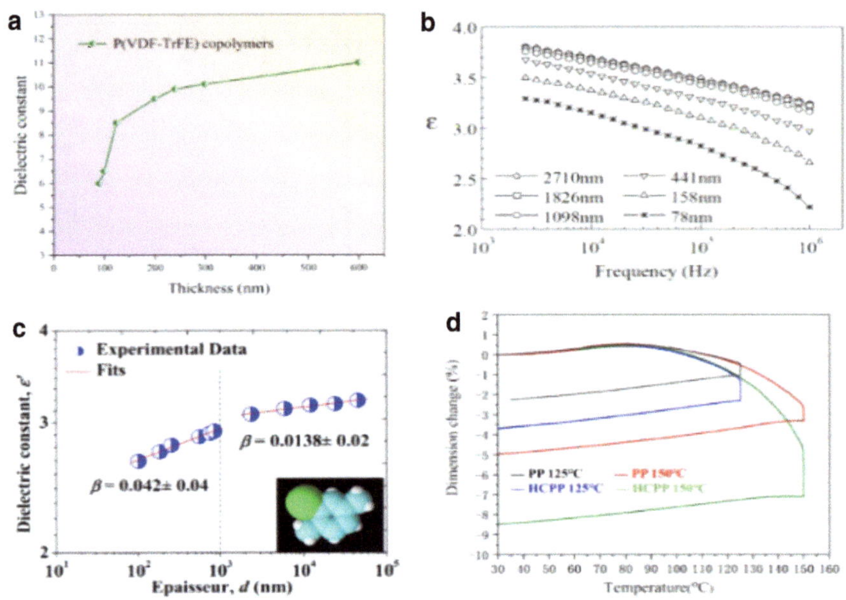

Fig. 13.9 **a** Dependence of dielectric constant of PVDF-TrFE with thickness of film, **b** dependence of dielectric constant with a thickness of polymer film, **c** dielectric dependence of polymer film to thickness through dielectric spectroscopy, **d** thermal stability in accordance with the dimensional change in polymer film (Wu et al. [24]. Reproduced with permission)

[27]. Also, when the thickness decreases to (10–100 μm), the breakdown strength becomes either constant or decreases in certain cases [28, 29].

K. Dransfeld et al., synthesized thin PVDF-TrFE films for a thickness of about 0.2–1.8 μm and investigated the dielectric constant of each film. The experimental results represented a decrease in the dielectric constant with the decrease in the given polymer film thickness [30]. It was observed that when the polymer film thickness decreases below 1000 nm, the dielectric constant decreases drastically. Above the 1000 nm thickness of the film, the dielectric constant remains almost independent of the thickness of the film (Fig. 13.9b) [31]. The experimental results by Kahouli et al. showed that the dielectric constant increases as a power law with the thickness of the film. Thickness below 1000 nm, increases the voids and discontinuities which decreases the dielectric constant (Fig. 13.9c) [32]. Stretching operation in transverse and machine directions produces thermally unstable polymer films. Thus, a variation in dielectric constant and breakdown strength is observed with an increase in stretching operation (Fig. 13.9d) [24].

The size of the filler also has a significant contribution to energy storage properties. Mendes et.al, fabricated PVDF/$BaTiO_3$ composites and compared the dielectric properties results with nanosized filler to submicron size filler. The results signified enhanced dielectric properties of composites with nanosized filler due to increment in space-charge interfacial interaction between polymer matrix and filler [33].

13.4 Fabrication Method for High-Energy Storage Performance

As discussed above for a dielectric film with high-energy storage, the thickness of the films should be controlled for a high dielectric constant with high breakdown strength. Accordingly, the fabrication method must be selected carefully for the manufacturing of polymer film with efficient dielectric properties. Thus, this section discusses the important fabrication techniques for energy storage properties.

13.4.1 Solution Casting

Solution casting is based on the principal of Stokes' law. It involves the addition and intermixing of polymer to a homogeneous solution of filler in a specific solvent. The filler forms a van der Waals interaction with the polymer chain. The homogeneous solution so formed is casted in a clean petri-dish, followed by the evaporation of the solvent. The solvent swells the polymer chain and undergoes polymer chain breakage. Also, it decreases the viscosity and maintains the ordering of filler or micro-domains in the polymer composite solution The desorption of solvent is accompanied by increasing van der Waals' forces between the polymer chain and filler and decreases the entropy of the solution. Solution evaporation intercalates a new polymer chain along with filler forming a polymer composite [34].

L. Sunet al. synthesized PVDF and PVDF/CNF (1, 3, 5, 7, and 15 wt%) films through a solution casting method with a thickness of 70 μm. Here, 1.5 g of PVDF was added in 10 ml of DMF and magnetic stirred at 70 °C. CNF was ultrasonicated to remove the air bubbles and improve the dispersion of filler into the polymer. The intermixed solution of CNF and PVDF was solution casted at 70 °C to completely evaporate the solvent [35]. Md. Ataur Rahmanet al. fabricated PVDF and PVDF-graphene nanocomposites through the solution casting method. The dried GO was ultra-sonicated for 2 h at about 60 °C. Separately 2.5 g of PVDF was magnetically stirred in 50 ml DMF for 2 h at 30 °C. A varying concentration of GO (0–0.5 wt%) solution was added to the PVDF solution. The overall composite solution was stirred for perfect dispersion of GO in PVDF at 500 rpm for 1 h at about 60 °C. The homogeneous solution was casted in a clean petri dish using a casting knife at about 70 °C to evaporate the solvent [36]. Ye Yang et al. synthesized polydopamine-modified $BaTiO_3$/PVDF composite films through the solution casting method. Here, mass fractions of PDA@BTO (5, 9, 13, 17, 21 wt%) w.r.t PVDF were dispersed in DMF and ultrasonicated for 30 min. Later on, a stoichiometric amount of PVDF was added and stirred at 60 °C for about 4 h for the homogenous mixture of PVDF composite solution. The solution was casted into clean petri-dish at 90 °C for 1 h (Fig. 13.10) [37]. Srikanta Moharana et al. fabricated Ag–GO–PVDF-HFP composite films through the solution casting method. Ag–GO was ultrasonicated in DMF for 2.5 h to avoid any agglomeration. Now, PVDF-HFP was stirred and dissolved in

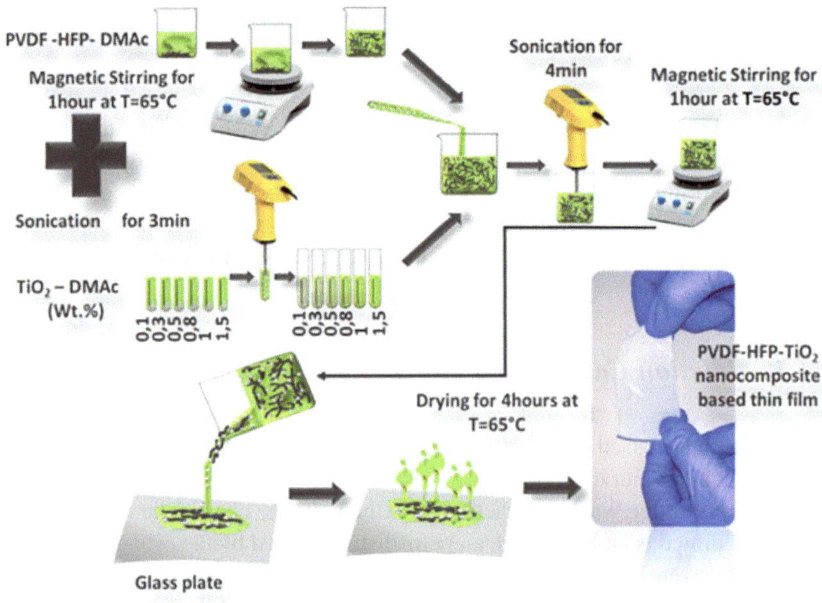

Fig. 13.10 Procedure for the synthesis of PVDF-HFP/TiO_2 nanocomposite films (Chakhchaoui et al. [39]. Reproduced with permission)

DMF. The Ag–GO solution was dispersed into PVDF-HFP solution for a homogenous composite solution. The composite solution was casted into a petri-dish at 80 °C in an oven [38]. Nabil Chakhchaoui synthesized PVDF-HFP and PVDF-HFP/ TiO_2 composite films through the solution casting method. Here TiO_2 was ultrasonicated in DMAc for 3 min. Separately PVDF-HFP was dissolved in DMAc at 65 °C for 1 h. The mixture of both solutions was dissolved for another 1 h at 65 °C. The composite films were casted in glass plates at a temperature of 65 °C [39]. L. Wu et al. designed PVDF-TrFE/graphene composite films with high piezoelectricity through the solution casting method. The solution casting procedure increased the piezoelectric voltage to 12.43 V [40]. Rajesh Sahoo et al. synthesized PVDF-TrFE/$Z_{(1-x)}Fe_{(x)}O$ composite films through the solution casting method. It involved the dissolution of 10 g of PVDF-TrFe at 60 °C for 2 h. Later on, $Z_{(1-x)}Fe_{(x)}O$ was added to the polymer solution maintaining the 1 wt% of PVDF-TrFE. The composite solution was casted in petridish and dried at 80 °C in an oven [41].

13.4.2 Hot Press Method

The procedure of hot-press incorporates melting of the polymer near its melting point followed by cooling for the final polymer film with enhanced energy storage

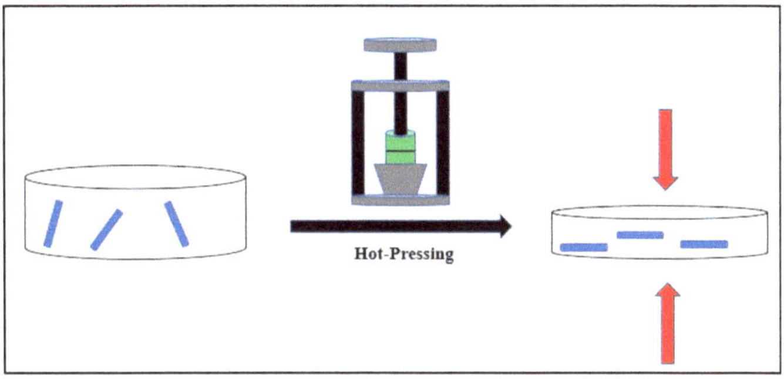

Fig. 13.11 Hot-pressing process

density [42]. Here, the temperature and pressure play a crucial role in enhanced electrical application in polymer film through the Hot-Press method (Fig. 13.11). Remelting of polymer film near its melting point produces a polymer film with an increased internal cross-linking and adhesion. It reduces porosity in the polymer film which in turn improves the internal crystal structure. Improved crystal structure decreases the dielectric loss of the polymer film and increases the breakdown strength. However, the important step to consider in this procedure is that the temperature shouldn't be very high. High temperature causes shrinkage of polymer fibres and affects the crystal structure of polymer film. High temperature creates internal defects and thereby forms a film of poor quality with high dielectric loss. High dielectric loss possessing polymer film is unable to bear high breakdown strength. Therefore, the energy storage density performance is reduced. Thus, for the synthesis of PVDF and PVDF composites with high-energy storage properties the Hot-Press technique should be performed at optimized pressure and temperature [43].

Weiyan Li et al. fabricated MXene/PVDF composite films. Firstly, the PVDF solution was stirred at 60 °C in DMF. The polymer solution was later spin-coated at 300 rpm followed by drying at 80 °C in an oven for 8 h. The spin coating method is followed by spray painting MXene solution through a nozzle size of 0.2 mm and pressure of 50 psi. The dried MXene/PVDF and PVDF composite films were detached from the glass substrates. Finally, the MXene/PVDF and PVDF composite films were overlapped into each other as per the stiochimetricx MXene/yPVDF (where, x = layer numbers of MXene and y = layer numbers of PVDF) through the Hot-Press method at 180 °C for 15 min [44]. Yunjin Fu et al. prepared PVDF/mica composite films with volume ratios of 98:2, 95:5, 92:8, and 90:10. The composite solution of mica nanosheets and PVDF was stirred for 3 h followed by ultrasonication. Later on, the solution was vacuumed for removing the bubbles formed in the solution. The solution was casted and later hot pressed by following preheating at 180 °C for 20 min. This leads to the formation of dense and homogeneous PVDF/mica composite films. The pressure was maintained at about 20 MPa [45]. N. Angulakshmi et al. synthesized PVdF-HFP/chitin/$LiClO_4$ composite films through the Hot-Press method. Earlier the films were vacuum dried at 50 °C for 5 days. Later on, the stoichiometric amount of films were taken and hot pressed for 20 min under a pressure of 20 MPa at 110 °C. The synthetic procedure produces homogeneous and strong membrane films with a thickness of 30–50 μm [46]. Boor Singh Lalia et al. fabricated PVDF-HFP through the hot-pressed method. PVDF-HFP was taken separately at about 10, 12, and 15 wt% in a solution of a binary mixture of acetone and dimethylacetamide within a ratio of 7:3. The respective solution was put into the glass syringe for obtaining electrospun PVDF-HFP fibres. The films were later hot press through household iron. The temperature of the surface of iron was greater than the melting temperature of PVDF-HFP. The films were covered with A4 size paper so that the surface of iron does not affect directly the membrane of the polymer and does not damage the fibres of the polymer. The iron was pressed on both the bottom and top surface of the film and moved for about 1–2 s across [47]. Qinghua Yang et al. fabricated $PbTiO_3$/P(VDF-TrFE) composite film through the Hot-Press method. The initial composite film of P(VDF-TrFE) and $PbTiO_3$ was synthesized through solution casting. The nanowires of $PbTiO_3$ were dispersed in DMF and ultrasonicated. P(VDF-TrFE) was added to the solution and stirred for 3 h at 70 °C. The homogeneous solution was casted in a glass substrate and kept in an oven at 80 °C for 12 h. The dried films were cut for about 0.02 gm and mounted in a cylindrical mold. It was kept for about 20 min at 120 °C. This was followed by the application of 3 MPa pressure for 15 min [48]. K.H. Lam et al. fabricated 65PMN–35PT/P(VDF-TrFE) 0–3 bulk composite films through the Hot-Pressing method. The polymer powder was dissolved in MEK solvent and later on, the crystal powder was added to it. The solution was ultrasonicated for homogeneous dispersion of powder into the matrix. Finally, the mixture was casted at 80 °C in the oven for overnight. For the preparation of bulk composite, the dried films were cut into small pieces. These pieces were molded at 230 °C and later hot pressed with a pressure of 0.1 MPa at 120 °C for a duration of 2 h [49].

13.4.3 Electrospinning Method

The procedure of the electrospinning process involves the formation of nanofibers through the passage of polymer solution in the form of a jet in the presence of the applied electric field. Under the influence of an electric field, the polymer solution is held at the end of the capillary tube through its surface tension. The applied electric field generates charge around the polymer solution held through surface tension. The charge induced in the solution undergoes mutual repulsion with the applied electric field. This creates opposing forces between each other. With the increase in the intensity of the electric field, the mutual repulsion increases, and thereby the shape of the hemispherical drop elongates and changes into a conical shape called Taylor Cone. But, at a certain critical value of the electric field, the mutual repulsion between the forces is overcome and a jet of the solution is ejected. However, now the electric field controls the motion of the polymer solution jet. In the presence of air,

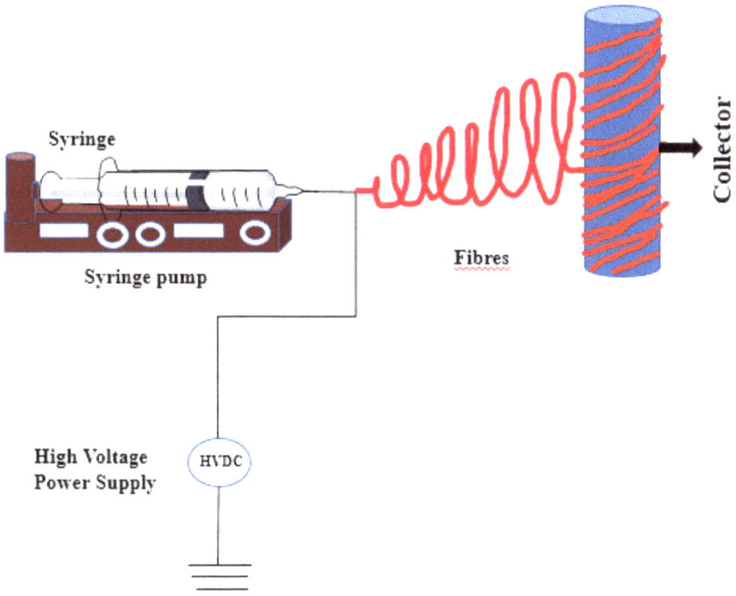

Fig. 13.12 Electrospinning procedure

the solvent present in the polymer solution evaporates and this leads to the formation of charged polymer fibres lying on the metal screen or the collector (Fig. 13.12) [50].

Avinash Baji fabricated nano-structured $BaTiO_3$/PVDF composite films through a two-step electrospinning process. In the first step, the $BaTiO_3$ was synthesized through the sol-gel process, followed by its synthesis of $BaTiO_3$ fibres through the electrospinning technique. The so obtained fibres of filler are dispersed into PVDF solution and electrospun to obtain $BaTiO_3$/PVDF composite films. The solvent used here is DMF. The electrospinning was carried out under an electric field of 15 kV with a spacing of 15 cm between the collector and needle along with a feed rate of 0.12 mm/min [51]. Chang Mou Wu synthesized PVDF/CNT nanofibers through an electrospinning technique. The PVDF pellets were dissolved in DMF at 100 °C for 2 h. Separately CNT was ultrasonicated in DMF. A stoichiometric amount of CNT was added to the solution. The solution was stirred at 100 °C and later on, acetone was added to it to undergo easy evaporation. The applied voltage was about 20 kV. The working distance is about 200 mm here and the flow rate is about 0.5 mL/h [52].

13.5 Energy Storage Application

PVDF and its copolymers have an electroactive structure, which produces a high energy storage density. The flexibility and reliability of such polymers lead to their integration as dielectric capacitors. However, the dielectric permittivity is enhanced with the introduction of filler within its matrix. Thus, it is important to investigate the energy storage performance of such PVDF and its copolymers composites for its future energy storage application in portable devices. AbhishekSasmal et al. synthesized $SmFeO_3$ nanoparticles through the Sol-Gel method. The gel was dried and calcinated to form SFO ($SmFeO_3$) powder. After this 1 wt%, 5 wt%, 10 wt% of SFO were added in different batches of PVDF solution in DMF. The solution was later on casted and hot-pressed for 20 min. The addition of SFO nanoparticles didn't increase the electrical performance as expected. This was due to the formation of a local electric field in addition to SFO according to the approximation model. But the loading of 10 wt% SFO within the matrix increased the energy storage efficiency as compared to neat PVDF [53].

Xiaoying Xiong et al. demonstrated the influence of the BST@SiO_2 nanotubes prepared through energy storage performance on the energy storage properties of PVDF. The modification of BST through tubular structure and control of the SiO_2 coating thickness around BTO showed tremendous improvement in dielectric permittivity and also reduced the dielectric loss. Here, low filler content in PVDF enhanced and improved the breakdown strength and reduced the dielectric loss compared to the neat PVDF, which is due to the generation of homogenous electric field distribution by high aspect ratio filler BST@SiO_2 and the high interface between polymer and filler. The composites were prepared through a simple solution-casting method. In comparison with the pure PVDF (U_d = 11.25 J/cm^3 at 465 MV/m, η = 59.87%), loading of PVDF with 2 vol% BST@SiO_2, enhanced the energy storage density (U_d) to about ~18.08 J/cm^3 (at breakdown strength of 525 MV/m) and appreciable efficiency (η) of 70.06%. Therefore, the novel approach of introducing BST@SiO nanotubes filler with a high aspect ratio in PVDF satisfied the demand for efficient energy storage composite capacitors (Fig. 13.13) [54].

Chi Qingguo et al. fabricated sandwich-structured PVDF composites loaded with BZT-BCT nanofibers having one-dimensional morphology deposited with $CoFe_2O_4$ nanoparticles. BZT-BCT was synthesized through electrospinning, while CFO@BZT-BCT was synthesized through hydrothermal synthesis. Later on, the inorganic filler was surface modified through PDA. The surface modification improved the interfacial polarization. The final sandwich structure was fabricated through the Hot-Press method. The Improved energy storage density of the synthesized composite was originated due to the low redistributed electric field between the middle layer present in the sandwich fabricated structure, according to the CSF theory. Here, P/5 vol% BZT-BCT NFs-P/P showed the highest energy storage density of about ~11.3 J/cm^3 and dielectric constant of about 20.1 at 100 Hz. It showed an appreciable efficiency of 55.5%. The multi-layer fabricated sandwich structure

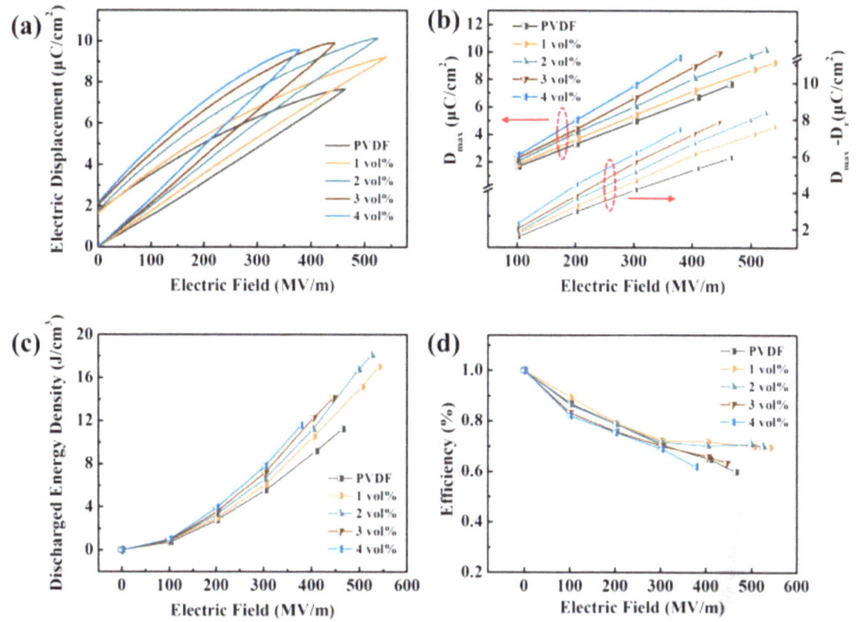

Fig. 13.13 **a** Unipolar hysteresis loop, **b** D_{max} and D_{max}–D_r versus electric breakdown strength, **c** discharged energy density, and **d** energy storage efficiency of BST@SiO_2/PVDF composites (Xiong et al. [54]. Reproduced with permission)

hindered the growth of leakage energy density and improved the electrical breakdown strength of the composite (Fig. 13.14) [55].

Abhishek Sasmal et al. reported the synthesis of PVDF-ZnO@$ZnSnO_3$ for its energy storage properties. $ZnSnO_3$ and ZnO@$ZnSnO_3$ were synthesized through hydrothermal technique. Both of these fillers were incorporated in the PVDF polymer matrix with the variation in wt% of 5%, 10%, and 15%. $ZnSnO_3$ incorporated polymer composite films were termed 5ZS, 10ZS, and 15ZS. Similarly, the inclusion of ZnO@$ZnSnO_3$ incorporated polymer composite films was termed 5ZNZS, 10ZNZS, and 15ZZNZS. The composite films were fabricated through the drop-casting method, addition of the fillers increased the properties of the composite films like polarity, energy harvesting capabilities, dielectric, and energy storage performance. In comparison to the PVDF-ZS composite films, the incorporation of 10 wt% ZnO@$ZnSnO_3$ enhanced the total electroactive properties up to ~87%, energy storage density to 134.98 μJcm^{-3} at 1 Hz. This reduction was approximately 6.6% of PVDF (Fig. 13.15) [56].

Wang et al. showed a strategy to enhance the dielectric permittivity of PVDF-HFP by loading silver nanoparticles organically modified with montmorillonite (OMMT) nanoplatelets. The modification involved first exfoliating pristine OMMT powder in DMF solution followed by continuous stirring for 24 h, then it was exposed to centrifugation at 2500 rpm for at least 30 min, the latter was filtered and dried at 80 °C

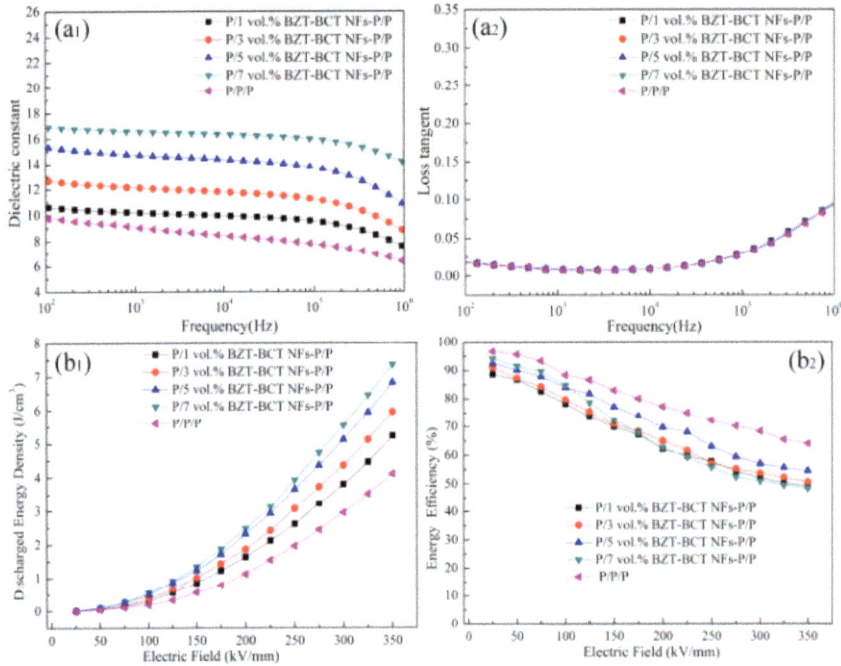

Fig. 13.14 $\mathbf{a_1}$ Dielectric constant, $\mathbf{a_2}$ dielectric loss, $\mathbf{b_1}$ discharged energy density and $\mathbf{b_2}$ storage efficiency of PVDF/CFO@BZT-BCT NFs-PVDF/PVDF composite (Chi et al. [55]. Reproduced with permission)

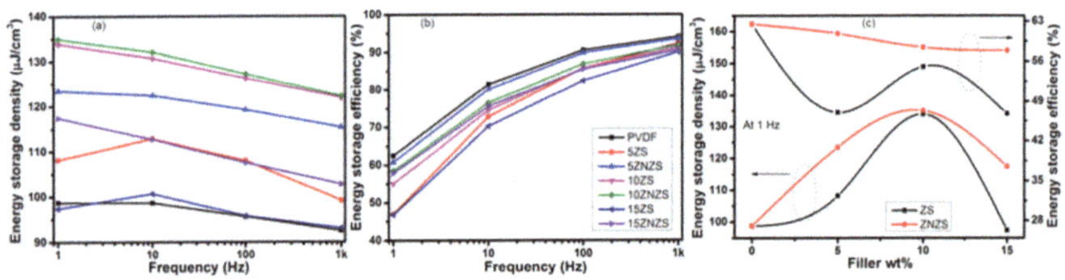

Fig. 13.15 Frequency-dependent: **a** energy storage density, **b** efficiency, and **c** filler-dependent energy storage density and energy storage efficiency (Sasmal et al. [56]. Reproduced with permission)

to obtain the Ag–OMMT nanoplates which are then incorporated into PVDF-HFP in order to increase the breakdown strength and dielectric constant. The silver nanoparticles were set down on the surfaces and sandwiched between the galleries of OMMT nanoplates by an in-situ reduction in exfoliated OMMT dispersions. The subsequent OMMT nanoplate was incorporated into the PVDF-HFP which result in the preparation of PVDF-HFP/Ag–OMMT composites. The introduction of nanoplates initiated the formation of nano-capacitors. Nano-capacitors are responsible for the increased

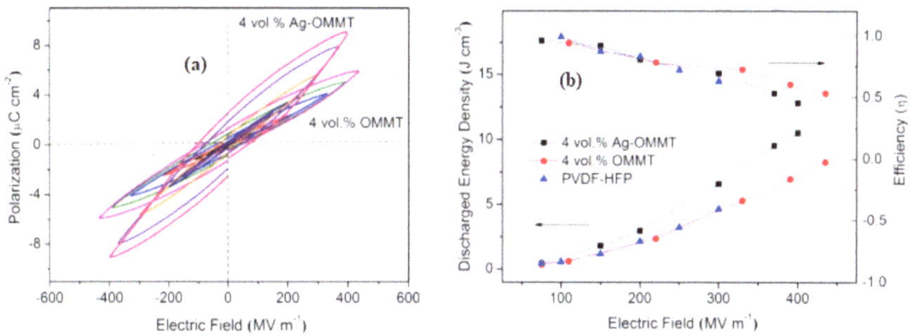

Fig. 13.16 **a** Hysteresis loop and **b** energy storage performance of Ag–OMMT/PVDF-HFP composites (Wang et al. [57]. Reproduced with permission)

dielectric constant of the whole system. Also, the morphology of nanoplates is very crucial for reduced dielectric loss and increased polarization. The high aspect ratio of nanoplates enhances the interfacial interaction between polymer and inorganic filler. If the filler is organically modified, then the interaction is more enhanced leading to high energy storage properties. This phenomenon is explained in this work in terms of the "electric barrier effect of OMMT modification." The enhancement of both breakdown strength and dielectric constant has been accomplished upon combining the Ag-OMMT nanoplates with PVDF-HFP where 4 vol% of Ag–OMMT and found to reach 10.51 J cm^{-3} at 400 MV m^{-1} (Fig. 13.16) [57].

Shewli Pratihar et al. fabricated a nanocomposite with enhanced dielectric permittivity, ferroelectric and piezoelectric properties. The enhancement was due to the addition of a third phase in the MWCNT in ZnO/PVDF composites. ZnO was synthesized through the drop-casting method. While the PVDF composite was synthesized through the drop-casting method. Here, 15 wt% ZnO/0.1 wt% MWCNT/PVDF composite showed the highest energy storage density nearly about 110 μJ/cm^3. The energy storage density is computed through the ferroelectric P-E loop. Usually, the fillers are randomly oriented. But, the application of an electric field orients the filler and enhances the polarization properties in the P-E loop. The enhancement of polarization increases the area for energy storage density. It was observed that the incorporation of ZnO nanorod increased polarization. However, the polarization was more pronounced through the addition of MWCNT [58]. Lei Gao et al. introduced and developed a novel scheme that improved interfacial interaction between filler and polymer. The improved interfacial interaction improved the energy storage density of the system. Most of the organic coupling agents undergo van-der Waals interaction with the inorganic filler. The most effective method is the hydroxylation of inorganic filler by pretreatment with hydrogen peroxide followed by the addition of coupling agents like DN-101. This procedure is followed for the surface modification of BT in this work. The surface-modified BT was incorporated in the PVDF solution and the composite solution was cast to obtain the final flexible composite film (D-h-BT/PVDF) with high energy density. As a result, the final composite exhibited enhanced dielectric polarization with improved breakdown strength. The dielectric polarization of D-h-BT/PVDF was about 6.41 μC cm^{-2}. The surface modification reduced the agglomeration within the filler particles. Thus, the breakdown strength was improved to 240 MV m^{-1}. The synergistic effect between the coupling agent and hydroxylated BT improved the energy storage density to 9.01 J cm^{-3} and discharged energy density to 4.31 J cm^{-3}[59]. Yu Fenga et al. developed a PVDF-HFP composite containing a double layer of polymer and incorporated it with BTO nanowires. According to the formula of calculation of efficiency of energy storage performance of a dielectric capacitor, the role of energy loss density supersedes the recoverable energy storage density. Therefore, it is crucial for the reduction of energy loss density for high energy storage efficiency. Thus, here the PVDF-HFP double layer was fabricated. The dielectric constant of the double layer is lower than the dielectric constant of the single layer. But the breakdown strength of such fabricated structure is much higher than that of a single layer of PVDF-HFP. The BTO nanowires were prepared through the hydrothermal method. The BTO nanowires were incorporated into the polymer according to 2.5, 5, and 10 vol% stoichiometry. The initial solution was prepared through physical blending. This was followed by the Hot-Pressing technique. The breakdown strength was calculated theoretically through the finite element method. It was observed that the experimental result was similar to FEA results. The experimental results showed high breakdown strength of the double layer as compared to the single layer. Thus, the double layer possesses enhanced energy storage properties [60].

Yang Yu et al. made the usage of MnO_2 filler to enhance the energy storage properties of PVDF composites. The inorganic MnO_2 filler was added with variations of 5, 10, 15, 20 wt% within PVDF. The mixture of filler and polymer were stirred and casted in a petri-dish and dried. The composite films were hot-pressed for 1 h under a pressure of 10 MPa. The dielectric constant of composite with 15 wt% loading of MnO_2 was found to be 27 at 100 Hz. However, the composite with 10 wt% loading presented the maximum breakdown strength. As a result, the energy storage density of MnO_2 10%/PVDF was high about 1.67 J/cm^3. Also, the energy storage efficiency was about high as 81.9%. Thus, the loading of MnO_2 improved the dielectric constant, polarization, breakdown strength and with low dielectric loss [61] (Table 13.1).

13.6 Conclusion

In summary, this chapter represents the energy storage properties of fluoropolymers. It elucidated the principal and equations representing the energy storage properties of dielectric polymer composites. It demonstrated the high permittivity of fluoropolymers due to their intrinsic high dipole structure. High intrinsic dipole structures developed within the fluoropolymers make them suitable for enormous electrical applications like piezoelectric, dielectric, ferroelectric, and energy storage properties. Such energy storage properties or dielectric properties depend upon the thickness of the film. It elaborated on the fact that the two important parts of electric storage i.e., breakdown strength and dielectric permittivity varies with the thickness of the

Table 13.1 Energy storage density, efficiency and dielectric permittivity of some PVDF, PVDF-based copolymer and its composites

S. No.	Sample name	Energy storage density (U_d)	Efficiency (η) (%)	Dielectric constant/ permittivity (ε')	References
1	(10 wt%) $SmFeO_3$–PVDF	–	43.62	–	[53]
2	2 vol% BST@SiO_2/PVDF	18.08 J/cm^3	70.06	–	[54]
3	P/5 vol% BZT-BCT NFs-PVDF/PVDF	11.3 J/cm^3	55.5	20.1	[55]
4	10ZNZS/PVDF	134.98 μJ cm^{-3}	58.41	–	[56]
5	4 vol% of Ag–OMMT/ PVDF-HFP	10.51 J cm^{-3}	–	–	[57]
6	15 wt% ZnO/ 0.1 wt%MWCNT/ PVDF	~110 μJ/cm^3	–	–	[58]
7	D-h-BT/PVDF	4.31 J cm^{-3}	–	–	[59]
8	10 wt% MnO_2/ PVDF	1.67 J/cm^3	81.9	27	[61]

synthesized film. The properties of a synthesized polymer film or composite mainly depend upon the fabrication process. Fabrication undergoes variations in thickness as well as defects and voids in the material. Thus, some of the major fabrication techniques like electrospinning, solution casting and Hot-Press method are discussed here. Finally, the energy storage properties are elucidated for the fluoropolymers. The investigation of the energy storage-based performance of PVDF and its based composites is important due to its enhancement in daily usage. It has become an integral part of various electronic devices like energy storage piezoelectric devices. The important part of such a device is to have a high energy storage density with low energy dissipation. Therefore, fluoropolymers and their composites play a crucial role in the fabrication of portable electronic devices with high capacitive response for energy storage applications.

13.7 Challenges and Author's Viewpoint

Dielectric-based capacitor film is highly useful in energy storage applications compared to chemical batteries and electrochemical capacitors. This is due to the high power density and low toxicity of the capacitor. However, still, the most important challenge faced during the fabrication of fluoropolymers by scientists is the low energy density of such dielectric-based capacitors. A high energy storage density is required to replace the electrochemical-based capacitor. Another issue is the dissipation factor of such fabricated film. For the longevity of the working of the dielectric film, it should exhibit low dielectric loss and high breakdown strength. So, even after the long hours of usage of the capacitor, the heat dissipation is low and the capacitor could be used safely.

Scientists still are working for improved energy storage density materials. For this, there is a requirement for constructing a dielectric film with enhanced polarization and dielectric constant. High polarization increases the integral area for energy storage density. Such fabrication requires the incorporation of high permittivity filler in a fluoropolymer matrix. It would induce the transformation of thermodynamically stable and non-polar α phase to highly polar β phase. Also, it depends on the morphology of the filler being introduced, i.e., its respective alignment as 1D or spherical. A highly aligned 1D filler possesses a uniaxial orientation of its macrodomain dipole and a high surface interaction area. This facilitates the increased heterogeneity in the composite and micro-capacitors at the interface of the film. Later on, it needs to be kept in mind to fabricate a dielectric film with reduced leakage current and high breakdown strength. Such factors are also responsible for a high integral area of energy storage density by reducing the hysteresis loss and increasing the breakdown voltage endurance. The synthesis procedure usually involves the surface functionalization of the inorganic filler by an organic core-shell, to improve the interfacial adhesion and compatibility between organic fluoropolymer and inorganic filler. By governing the above points, a highly compatible dielectric material can be fabricated, which could be easily used for high-energy storage portable devices.

References

1. Chen C, Xing J, Cui Y et al (2020) Designing of ferroelectric/linear dielectric bilayer films: an effective way to improve the energy storage performances of polymer-based capacitors. J Phys Chem C 124
2. Xie B, Zhu Y, Marwat MA et al (2018) Tailoring the energy storage performance of polymer nanocomposites with aspect ratio optimized 1D nanofillers. J Mater Chem A Mater 6
3. Zhang X, Li BW, Dong L et al (2018) Superior energy storage performances of polymer nanocomposites via modification of filler/polymer interfaces. Adv Mater Interfaces 5
4. Shalu S, Roy S, Mukherjee A et al (2022) Effect of Mn-doping on the morphological and electrical properties of $(Ba_{0.7}Sr_{0.3})$ $(Mn_xTi_{1-x})O_3$ materials for energy storage application. Ceram Int 48:25816–25825
5. Dasgupta B, Banerjee S (2010) A study of gas transport properties of semifluorinated poly (ether imide) membranes containing cardodiphenylfluorene moieties. J MembSci 362:58–67
6. Vinogradov A, Holloway F (1999) Electro-mechanical properties of the piezoelectric polymer PVDF. Ferroelectrics 226
7. Polat K (2020) Energy harvesting from a thin polymeric film based on PVDF-HFP and PMMA blend. Appl Phys A Mater Sci Process 126
8. Oliveira F, Leterrier Y, Sereda O et al (2014) Process influences on the structure, piezoelectric, and gas-barrier properties of PVDF-TrFE copolymer. J PolymSci B Polym Phys 52
9. Feng CY, Khulbe KC, Matsuura T, Ismail AF (2013) Recent progresses in polymeric hollow fiber membrane preparation, characterization and applications. Sep Purif Technol 111

10. Zaarour B (2021) Enhanced piezoelectricity of PVDF nanofibers via a plasticizer treatment for energy harvesting. Mater Res Express 8
11. Badatya S, Kumar A, Sharma C et al (2021) Transparent flexible graphene quantum dot-(PVDF-HFP) piezoelectric nanogenerator. Mater Lett 290
12. Pi Z, Zhang J, Wen C et al (2014) Flexible piezoelectric nanogenerator made of poly(vinylidenefluoride-co-trifluoroethylene) (PVDF-TrFE) thin film. Nano Energy 7
13. Sharma M, Madras G, Bose S (2014) Process induced electroactive β-polymorph in PVDF: effect on dielectric and ferroelectric properties. In: Physical chemistry chemical physics 16
14. Dash S, Thakur VN, Kumar A et al (2021) Enhancing functional properties of PVDF-HFP/BZT-BCT polymer-ceramic composites by surface hydroxylation of ceramic fillers. Ceram Int 47
15. Meng N, Zhu X, Mao R et al (2017) Nanoscale interfacial electroactivity in PVDF/PVDF-TrFE blended films with enhanced dielectric and ferroelectric properties. J Mater Chem C Mater 5
16. Shalu S, Kar P, Krupka J, Dasgupta Ghosh B (2018) Synthesis, characterization, thermal, dynamic mechanical, and dielectric studies of $Ba_{0.7}Sr_{0.3}TiO_3$/polystyrene composites. Polym Compos 39
17. Shalu S, Dasgupta Ghosh B (2019) Synthesis, characterisation and dielectric properties of low-loss Zr-doped barium strontium titanate materials. Adv Appl Ceram 118
18. Tiwari V, Srivastava G (2014) Effect of thermal processing conditions on the structure and dielectric properties of PVDF films. J Polym Res 21
19. Feng Y, Li WL, Hou YF et al (2015) Enhanced dielectric properties of PVDF-HFP/$BaTiO_3$-nanowire composites induced by interfacial polarization and wire-shape. J Mater Chem C Mater 3
20. Arshad AN, Wahid MHM, Rusop M et al (2019) Dielectric and structural properties of poly(vinylidene fluoride) (PVDF) and poly(vinylidene fluoride-trifluoroethylene) (PVDF-TrFE) filled with magnesium oxide nanofillers. J Nanomater
21. Li W, Meng Q, Zheng Y, et al (2010) Electric energy storage properties of poly(vinylidene fluoride). Appl Phys Lett 96
22. Kishor KK, Kalathi JT (2020) Investigation on the dielectric performance of PVDF-HFP/LZO composites. J Alloys Compd 843
23. Zhu H, Liu Z, Wang F (2017) Improved dielectric properties and energy storage density of poly(vinylidene fluoride-co-trifluoroethylene-co-chlorotrifluoroethylene) composite films with aromatic polythiourea. J Mater Sci 52
24. Wu X, Chen X, Zhang QM, Tan DQ (2022) Advanced dielectric polymers for energy storage. Energy Storage Mater 44
25. Zhang QM, Xu H, Fang F et al (2001) Critical thickness of crystallization and discontinuous change in ferroelectric behavior with thickness in ferroelectric polymer thin films. J Appl Phys 89.
26. Cygan S, Laghari JR (1987) Dependence of the electric strength on thickness area and volume of polypropylene. IEEE Trans Electr Insul EI-22
27. Chen G, Zhao J, Li S, Zhong L (2012) Origin of thickness dependent dc electrical breakdown in dielectrics. Appl Phys Lett 100
28. Neusel C, Schneider GA (2014) Size-dependence of the dielectric breakdown strength from nano- to millimeter scale. J Mech Phys Solids 63
29. Tan DQ (2020) The search for enhanced dielectric strength of polymer-based dielectrics: a focused review on polymer nanocomposites. J Appl Polym Sci 137
30. Glatz-Reichenbach J, Li-Jie LJ, Schilling D et al (1990) Dielectric and piezoelectric properties of very thin films of VDF-TrFE copolymers. Ferroelectrics 109.
31. Liang T, Makita Y, Kimura S (2001) Effect of film thickness on the electrical properties of polyimide thin films. Polymer (Guildf) 42
32. Kahouli A (2012) Effect of film thickness on structural, morphology, dielectric and electrical properties of parylene C films. J Appl Phys 112
33. Mendes SF, Costa CM, Caparros C et al (2012) Effect of filler size and concentration on the structure and properties of poly(vinylidene fluoride)/$BaTiO_3$ nanocomposites. J Mater Sci 47

34. Das R, Pattanayak AJ, Swain SK (2018) Polymer nanocomposites for sensor devices. In: Polymer-based nanocomposites for energy and environmental applications. Woodhead Publishing series in composites science and engineering, Chapter 6 :185-203
35. Sun LL, Li B, Zhao Y et al (2010) Structure-induced high dielectric constant and low loss of CNF/PVDF composites with heterogeneous CNF distribution. Nanotechnology 21
36. Rahman MA, Chung GS (2013) Synthesis of PVDF-graphene nanocomposites and their properties. J Alloys Compd 581:724–730
37. Yang Y, Pan H, Xie G et al (2020) Flexible piezoelectric pressure sensor based on polydopamine-modified BaTiO3/PVDF composite film for human motion monitoring. Sens Actuators A Phys 301:111789
38. Moharana S, Mahaling RN (2017) Silver (Ag)–Graphene oxide (GO)—Poly (vinylidene fluoride-co-hexafluoropropylene) (PVDF-HFP) nanostructured composites with high dielectric constant and low dielectric loss. Chem Phys Lett 680:31–36
39. Chakhchaoui N, Farhan R, Eddiai A et al (2019) Improvement of the electroactive b-phase nucleation and piezoelectric properties of PVDF-HFP thin films influenced by TiO_2 nanoparticles. Mater Today Proc 39
40. Wu L, Jing M, Liu Y et al (2019) Power generation by PVDF-TrFE/graphene nanocomposite films. Compos B Eng 164:703–709
41. Sahoo R, Mishra S, Unnikrishnan L et al (2020) Enhanced dielectric and piezoelectric properties of Fe-doped ZnO/PVDF-TrFE composite films. Mater Sci Semicond Process 117:105173
42. Shalu S, Dasgupta K, Roy S et al (2021) Thermal, mechanical, and dielectric properties of low loss $PbZr_{0.3}Ti_{0.7}O_3$/polystyrene composites prepared by hot-press method. Polym Compos 42
43. Lu L, Ding W, Liu J, Yang B (2020) Flexible PVDF based piezoelectric nanogenerators. Nano Energy 78
44. Li W, Song Z, Zhong J et al (2019) Multilayer-structured transparent MXene/PVDF film with excellent dielectric and energy storage performance. J Mater Chem C Mater 7
45. Fu Y, Wang Y, Wang S et al (2019) Enhanced breakdown strength and energy storage of PVDF-based dielectric composites by incorporating exfoliated mica nanosheets. Polym Compos 40.
46. Angulakshmi N, Thomas S, Nahm KS et al (2011) Electrochemical and mechanical properties of nanochitin-incorporated PVDF-HFP-based polymer electrolytes for lithium batteries. Ionics (Kiel) 17
47. Lalia BS, Guillen-Burrieza E, Arafat HA, Hashaikeh R (2013) Fabrication and characterization of polyvinylidenefluoride-co-hexafluoropropylene (PVDF-HFP) electrospun membranes for direct contact membrane distillation. J Memb Sci 428
48. Yang Q, Shi Z, Ma D et al (2018) Flexible $PbTiO_3$-nanowires/ P(VDF-TrFE) composite films and their dielectric, ferroelectric and pyroelectric properties. Ceram Int 44:14850–14856
49. Lam KH, Chan HLW, Luo HS, et al (2003) Dielectric properties of 65PMN-35PT/P(VDF-TrFE) 0–3 composites. In: Microelectronic engineering
50. Doshi J, Reneker DH (1995) Electrospinning process and applications of electrospun fibers. J Electrostat 35
51. Baji A, Mai YW, Li Q, Liu Y (2011) Nanoscale investigation of ferroelectric properties in electrospun barium titanate/polyvinylidene fluoride composite fibers using piezoresponse force microscopy. Compos Sci Technol 71
52. Wu CM, Chou MH (2016) Polymorphism, piezoelectricity and sound absorption of electrospun PVDF membranes with and without carbon nanotubes. Compos Sci Technol 127
53. Sasmal A, Sen S, Devi PS (2020) Synthesis and characterization of $SmFeO_3$ and its effect on the electrical and energy storage properties of PVDF. Mater Res Bull 130:110941
54. Xiong X, Zhang Q, Zhang Z et al (2021) Superior energy storage performance of PVDF-based composites induced by a novel nanotube structural BST@SiO_2 filler. Compos Part A Appl Sci Manuf 145
55. Chi Q, Ma T, Zhang Y et al (2018) Excellent energy storage of sandwich-structured PVDF-based composite at low electric field by introduction of the hybrid $CoFe_2O_4$@BZT-BCT nanofibers. ACS Sustain Chem Eng 6

56. Sasmal A, Medda SK, Devi PS, Sen S (2020) Nano-ZnO decorated $ZnSnO_3$ as efficient fillers in PVDF matrixes: toward simultaneous enhancement of energy storage density and efficiency and improved energy harvesting activity. Nanoscale 12
57. Wang H, Xie H, Wang S et al (2018) Enhanced dielectric property and energy storage density of PVDF-HFP based dielectric composites by incorporation of silver nanoparticles-decorated exfoliated montmorillonite nanoplatelets. Compos Part A Appl Sci Manuf 108
58. Pratihar S, Patra A, Sasmal A et al (2021) Enhanced dielectric, ferroelectric, energy storage and mechanical energy harvesting performance of ZnO-PVDF composites induced by MWCNTs as an additive third phase. Soft Matter 17
59. Gao L, He J, Hu J, Li Y (2014) Large enhancement in polarization response and energy storage properties of poly(vinylidene fluoride) by improving the interface effect in nanocomposites. J Phys Chem C 118
60. Feng Y, Li JL, Li WL et al (2019) Effect of $BaTiO_3$ nanowire distribution on the dielectric and energy storage performance of double-layer PVDF-based composites. Compos Part A Appl Sci Manuf 125:105524
61. Yu Y, Wang X, Wang X et al (2020) Enhanced energy storage efficiency in PVDF based composite films using MnO_2 nano-fillers. J Mater Sci Mater Electron 31

Chapter 14
Application of Organic–Inorganic Nanodielectrics for Energy Storage

Nupur Saxena, P. Sakthivel, D. Sridharan, and Pragati Kumar

Abstract Organic–inorganic nanodielectric materials are frequently employed for energy storage due to their superior electrical, thermal, and mechanical capabilities. While organic dielectrics, which have a grace failure mechanism and a self-healing feature, contributed to the high capacitance of electrolytic capacitors and the miniaturization of multilayer ceramic capacitor (MLCC) construction, the high dielectric constant (K) of inorganic dielectrics allowed for reliable operation and higher capacitance stability. Polypropylene (PP), a nonpolar polymer dielectric, has steadfastly maintained its position as the industry benchmark despite significant efforts being made to investigate these methods. In addition to having the lowest dielectric loss and the highest dielectric strength, this dielectric also offers manufacturing maturity in its widely used film form. There may be an improvement in the dielectric constant (ε_r) and dielectric breakdown strength (E_b) of dielectric materials, which would be advantageous for a capacitor. It is seen that the size of the fillers is causing the improved characteristics. The creation of an improved capacitor made of metalized electrodes sandwiched between dielectric films or sheets is one of the most significant uses. For the implementation of more compact energy storage and conversion devices, advancements in the energy density and operating temperature of capacitors have become crucial.

N. Saxena
Department of Physics, Indian Institute of Technology (IIT), Jammu, Jammu 181221, India
e-mail: n1saxena@gmail.com

P. Sakthivel
Centre for Materials Science, Department of Physics, Science and Humanities, Faculty of Engineering, Karpagam Academy of Higher Education, Coimbatore, Tamil Nadu 641021, India
e-mail: sakthivel.p@kahedu.edu.in

D. Sridharan
Vioma Motors Pvt Ltd., Mumbai, Maharashtra 400028, India

P. Kumar (✉)
Nano Materials and Device Lab, Department of Nanoscience and Materials, Central University of Jammu, Samba, Jammu, J & K 181143, India
e-mail: pkumar.phy@gmail.com

S. Moharana et al. (eds.), *Emerging Nanodielectric Materials for Energy Storage*,
Nanostructure Science and Technology, https://doi.org/10.1007/978-3-031-40938-7_14

Keywords Organic–inorganic nanodielectrics · Energy storage · Supercapacitor · Lithium battery · High-voltage direct current cables

14.1 Introduction

Energy conversion and storage are critical for our future sustainable energy model, and they are fields where electrochemistry is conventionally foremost and hybrid materials are emerging as strong candidates. Batteries and supercapacitors (SCs) are electrochemical energy storage devices. Regular capacitors employ a purely electrophysical mechanism to store nominal energy while providing it quickly, that is, at high current and high power. Due to their large surface area and short charge separation distance, double-layer SCs were an inventive way to harness charge separation between electrode and electrolyte interphase and store more energy than conventional capacitors.

Many novel nanocomposites (NC) materials have recently succeeded by incorporating macromolecular species into 2D-layered transition metal oxides. Over the last few years, conducting polymer-based NC has established a lot of consideration for their ability to profitably integrate many appropriate properties of both conducting polymer and transition metal oxide [1, 2]. As a result, the organic–inorganic hybrids are combined at the nanoscale level; numerous novel properties emerge, such as improved electrochemical storage ability, due to synergistic effects [3].

Over the last decade, researchers have paid close attention to organic–inorganic NC, composed of more than two components of various natures. Subsequently, their use in convenient energy storage systems and devices such as batteries and SCs has gained critical importance in everyday life in recent years. SCs are broadly used in electrical power backup systems, portable electronic devices, and electric vehicles due to their higher power density and longer cycle life when compared with Lithium-ion batteries (LIB) [4]. SCs are mainly classified into two types based on their charge-storage mechanism: electrochemical double layer supercapacitor (EDLSC), which has capacitance due to charge separation at the electrode/electrolyte interface, and pseudo-SC, which have capacitance on account of surface Faradaic redox reactions [5]. Carbon-based electrodes are used in EDLC, whereas polymers and transition metal oxides are used in pseudocapacitors. The recent focus of supercapacitor research is on advancing hybrid nanocomposite materials with enhanced energy density. In addition, asymmetric electrode formation increases supercapacitor energy density by expanding the device's overall potential window [6].

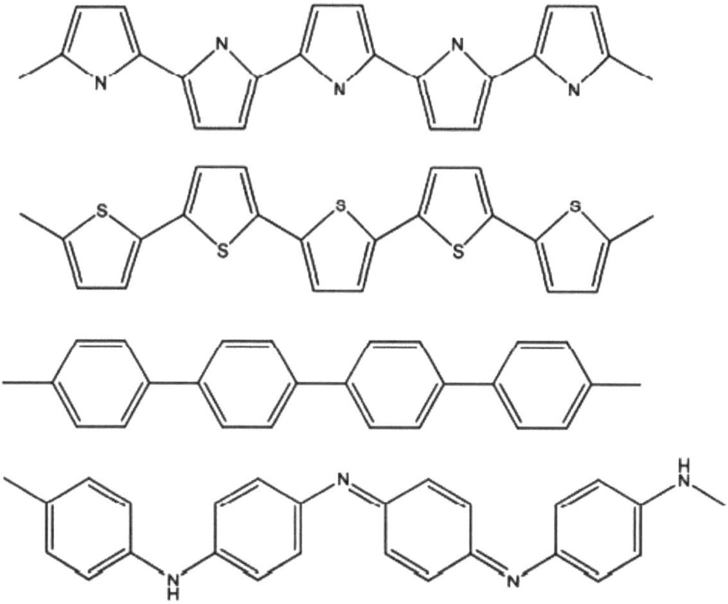

Fig. 14.1 Structures of conducting polymers (from top to bottom): polypyrrole, polythiophene, polyparaphenylene, polyaniline

14.1.1 Conducting Polymers

Conducting-organic polymers have been extensively researched as promising novel materials, with reports on their potential use for rechargeable batteries [7] and electrochemical capacitors [8]. Conducting polymers such as polyaniline (PANI), polythiophenes, polypyrroles, polyphenylenevinylenes, and others are particularly appealing for electronic applications (Fig. 14.1). Furthermore, conducting polymers can be hybridized with transition metal nanoparticles as well as nano-carbons.

14.1.2 Transition Metal Oxide

A comparison of the use of transition metal oxides as sole and their nanocomposites with other materials as anodes in rechargeable lithium batteries and supercapacitors shows that when these materials form nanocomposites, many properties such as improved charge storage capacity emerge due to combined properties of constituent materials [9]. Vanadium pentoxide (V_2O_5), MoO_3, and WO_3 are common oxidizing hosts in conducting polymer-based nanocomposite for supercapacitors.

14.1.3 Reduced Graphene Oxide (rGO) and CNT

On the other hand, reduced graphene oxide (rGO) is an ideal supercapacitor electrode material due to excellent electrical conductivity, large surface area, excellent rate capacity, adaptable electronic characteristics, and ease of synthesis methods. The 3D MoS_2 with rGO and CNT complex composite was used as an electrode material for a stretchable supercapacitor. With the introduction of CNTs as beneficial components, the group of probable hybrid combinations for energy storage is closed with hybrids prepared of polymers and transition metal oxides, and hybrids made of conducting polymers and CNTs [10].

14.1.4 Synthesis of Organic–Inorganic (O–I) Nanocomposites

Hybrid O–I materials have employed a vital role in the growth of cutting-edge functional materials, such as innovative materials in the energy field in recent energy technology. Polymers have many advantages, including ease of production and handling, chemical and structural range, not high weight, and flexibility. As a result, hybrid inorganic/organic nanocomposites composed of conducting polymers comprising metallic nanoparticles, CNTs, and graphene have substantial potential for use in supercapacitors and non-volatile memory devices.

Nanostructured materials have different characteristics depending on their size and shape. Semiconductor particles in magnetic materials are all founded to be size-dependent properties. Simple mixing, and nanofiller synthesis in a polymer, can all be used to create nanocomposites. Nanoparticles coated with a polymer are significantly more robustly opposed to accumulation than bare particles because of a significant reduction in surface energy. The existence of polymers on the surface of a nanoparticle can modify. As a result, the better nanocomposite conductivity. Van der Waals interactions, electrical double-layer interactions, and steric interactions [39] are the three primary forces between particles in nanocomposites.

Metals are well known for being the most excellent conductive materials on the planet as their high free electron density. Metallic nanoparticles (MNP) are used in various applications, including surface-enhanced catalysis, microelectronics, LEDs, PV cells, and biological applications. Furthermore, the synthesis method plays a critical role to finalize the various properties of nanoparticles. Many physical and chemical nanoparticle synthesis techniques have been developed to date. Pure nanoparticles clump together easily. They are protected from aggregation by surfactants and polymers. Composites of metal nanoparticles and polymers are essential in Nanoelectronics devices because charge transfer interaction is critical to their functionality. Polymers with N or S elements in their repeating unit are interesting contenders for fabricating organic/nanoparticle nanocomposites.

Nanocomposite's application potential as an electrode material for electrochemical supercapacitors has been examined, with the better enrichment of double layer capacitance of poly(3,4-ethylene dioxythiophene) (PEDOT–MoO_3) nanocomposites (300 mFg^{-1}) compared to pristine MoO_3 (40 mFg^{-1}). The enhanced electrochemical performance is ascribed to the intercalation of electronically conducting PEDOT between MoO_3 layers with increased bidimensionality and surface area.

Hybrid materials for polyoxometalates and supercapacitors are combined with conducting polymers. Polyoxometalates are usually prepared by W or Mo oxides and are the most similar to an oxide quantum dot. They are electroactive oxides that have been reduced to their smallest size. Incorporating this metal oxide into conducting polymers resulted in the progress of novel electroactive hybrids for supercapacitors [11]. The hybrid $PMo_{12}O_{40}$/PANI as an active electrode material in a supercapacitor is demonstrated significant advantages. The results exhibit the capacitance evolves during cycling up to 120 F/g and the increased capacitance over the first 300 cycles observed. This electrode is caused by progressive penetration with the electrolyte, which indicates the performance of supercapacitor material with optimization of the microstructure, resulting in a maximal electrode–electrolyte interface. Examples of this type were hybrids based on V_2O_5 xerogels and various intercalated conducting polymers, such as polyaniline (PANI) [12], polythiophene (PTH), and polypyrrole (PPy), and were offered as cathode material for rechargeable LIBs [13]. MoO_3in batteries was also investigated as interactive matrices to form hybrids with polyphenylenevinylene (PPV) [14] and PANI.

14.1.4.1 PEDOT–V_2O_5 Nanocomposite

The PEDOT–V_2O_5 nanocomposite is synthesized and characterized by various analytical tools. The XRD studies on nanocomposite reveal the understated structural variations upon intercalation [54]. The interlayer spacing is directly related to the most substantial peak observed at the low angle corresponding to the (0 0 1) plane of the layered V_2O_5 structure. A strong diffuse scattering noticeably forms the V_2O_5 diffraction arrays, and the d-spacing rises from 4.3 to 14.1 Å in the composites. This implies that the expansion continues, creating a polymer monolayer and the V_2O_5 films. It is noticed that the hybrids' ($h\ k\ l$) pattern corresponds to those of pristine V_2O_5. HRTEM reveals that the V_2O_5 host comprises several conducting polymer nanoribbons. The low scattering power results in high contrast for white lines, each about 1.4 nm long and located among double dark fringes of vanadate layers. It is observed the highly crystalline V_2O_5 in this hybrid material is separated by alternating conducting polymer nanoribbons.

After redox polymer intercalation, XPS is a surface-specific technique for studying the polymer–V_2O_5 interaction and vanadium (V^{5+}/V^{4+}) oxidation state. XPS results identify the V 2p and O 1s core levels of polymer intercalated nanocomposite. The spectra are observed at 517.2 eV, and a low Binding Energy is visible, along with V^{5+}. The low BE at 515.7 eV, ascribed to V^{4+}, is visible after the deconvolution of the 2p core levels [15]. The nature of charge transport is determined by the relative carriers' mobility, as revealed by the conductivity of PEDOT–V_2O_5 is 102 times greater than V_2O_5.

The PEDOT–V_2O_5 nanocomposite has a higher capacity during the charging than during the discharging. It causes the existence of V^{4+}, which can be easily oxidized electrochemically, as previously observed in PPy–V_2O_5 and PTH–V_2O_5 nanocomposites. It is worth noting that the PEDOT–V_2O_5 nanocomposite has the highest reversibility (330 mAh/g). The nanocomposites have a higher capacity when compared with pure V_2O_5, which is due to increased conductivity, resulting in improved dimensionality.

14.1.4.2 PEDOT–MoO_3 Nanocomposite

The PEDOT–MoO_3 nanocomposite shows promise as a material for non-aqueous supercapacitors. Polymer intercalated PEDOT–MoO_3 nanocomposite X-ray diffraction pattern. However, a rise in the interlayer distance of the pure oxide, MoO_3, for PEDOT–MoO_3 composite from $d = 6.93$ (0 2 0) to 13.46 A (0 0 1) shows significant assimilation of polymeric material between the layers. As a result, considerable bonding interaction between organic and inorganic components is estimated, most likely as hydrogen bonding. The XRD confirmed the composite forms a homogeneous matrix with a lamellar morphology, and the combination of PEDOT into the MoO_3 results in surface modification. More importantly, the S 2p core level from PEDOT peaks at 164 eV, typical for thiophene sulfur. However, after intercalation, there is a widening on the lower BE side with the evolution of a second peak. The ethylenedioxy group is responsible for the C 1 s peak around 286 eV, and the thiophene carbon atoms are responsible for the lower BE peak at 284 eV. PEDOT–MoO_3 composite, 3d5/2 core level at 232.5 eV is distinctive for MoO_3 [60], and XPS suggests oxidation of PEDOT after intercalation into MoO_3 as well as charge transfer from sulfur to MoO_3. Accordingly, the XPS studies agree well with XRD and other characterization techniques. While pristine MoO_3 is an insulator with an ambient temperature conductivity of 3.78×10^{-5} S/cm, the ambient temperature conductivity of the PEDOT–MoO_3 nanocomposite is 1.82×10^{-1} S/cm, which is four orders of magnitude higher than pristine MoO_3 [16].

The CVs of pristine MoO_3 and PEDOT–MoO_3 nanocomposite were recorded with electrolyte 1 M $LiClO_4$, demonstrating a significant variation in CV response caused by polymer inclusion. It is worth noting that the EDLC rises from pristine $MoO_3 \sim 40$ mFg^{-1} to after polymer intercalation $\sim$300 Fg^{-1}. The crystalline MoO_3 undergoes a well-known phase transformation in the first cycle scan, that is, when the potential variations from -1.5 to 0.25 V versus the Ag/AgCl reference electrode, and the stabilization indicates that the structural change is stable. Furthermore, the Vander Waals interactions between the interlamellar layers agree for faster Li^+ ion insertion in the ribbons than in MoO_3. In contrast, there is no evidence of irreversible structural modification in the PEDOT–MoO_3 hybrids, but the broad cathodic peak resembles that of 2D inorganic layered compounds. PEDOT–MoO_3 nanocomposite is used as nonaqueous type supercapacitors and the hybrid approach employed at this point offers a capable method of fabricating low-cost supercapacitor electrodes.

14.1.4.3 Poly-(5-Cyanoindole) P5ICN–WO_3 Nanocomposite

P5ICN was polymerized with WO_3 after being composited. When compared to pure WO_3, the surface of the P5ICN/ WO_3 material exhibited more wrinkles, effectively increasing the surface and proving the successful preparation of the composite material. Electrochromic-supercapacitor device (ESD) is a type of multifunctional supercapacitor that can be used to visually evaluate the device's level of energy storage through color change. With the increasing importance of energy and environmental issues, supercapacitors are regarded as promising energy storage devices due to their high-power density and long cycle life [12]. Electronic devices in recent times, researchers have paid increasing consideration to intelligent systems and electronic devices. Multifunctional supercapacitors, such as flexible supercapacitors, self-healing supercapacitors, and electrochromic supercapacitors, have been reported to meet these demands [17]. The new multifunctional electrode material has developed the recent research focus on energy storage devices. The electrochromic supercapacitor material can vary in color through the device's charging and discharging processes. Both functions can be combined into a single device to create an electrochromic-supercapacitor device (ESD) [18].

ESD electrode materials comprise metal oxides and conducting polymers [19]. WO_3 has been employed due to its high theoretical capacitance, better cyclic stability, and high color contrast [20]. Many researchers have focused their efforts in recent years on improving the electrochromic properties of WO_3 to know its application in ESD [21], as WO_3 constituents can change color between colorless and blue in the oxidation and reduction process. However, it is hoped that the electrode will have further-color deviations in applying intelligent supercapacitors to enable optical observation of altered energy storage devices. ESD made from conducting polymer has recently developed a hot topic due to their rich electrochromic-color variations and high-energy storage capacity [22]. Polyindole, a good conducting polymer material, exhibits good electrochemical activity and better thermal stability [23].

FT-IR spectra were used to characterize the prepared P5ICN/WO_3 hybrid materials. The stretching vibration of W–O–W was observed in the 700–1000 cm^{-1} range. The small amount of water may be responsible for 3446 and 1621 cm^{-1} [24]. XPS was used to characterize the prepared P5ICN/WO_3 composites. The interaction of tungsten and nitrogen atoms may be responsible for the peak at 398.1 eV. The peaks at 35.7 eV and 37.8 eV corresponded to the W^{6+} state caused by W $4f_{7/2}$ and W $4f_{5/2}$, respectively. The W4f XPS spectrum of P5ICN/WO_3 was observed after composited of P5ICN and WO_3. After composition, the peaks of W $4f_{7/2}$ and W $4f_{5/2}$ shifted to 35.1 eV and 37.2 eV, respectively. The interaction between P5ICN and WO_3 could explain the peak shift phenomenon [25].

P5ICN/WO_3 can switch between blue-green, yellow, and green. The synthesized ESD, made of P5ICN/WO_3, has a high energy storage capacity and visible-color variations. The ESD has a specific capacitance of 38.6 $mFcm^{-2}$ and a coloration efficiency of 548 cm^2 C^{-1}. Furthermore, the intelligent ESD's stored energy level can be visually monitored via rapid, reversible color changes. As a result, the ESD can light up an LED in practical applications. In addition, the ESD demonstrates good electrochromic stability and is to be developed for other applications of electronic devices.

14.1.4.4 Polymer–Graphene Nanocomposites

Various carbon-based nanoparticles have been developed in recent years. Because of their chemical inertness, hardness, and better electrical and thermal properties, such as fullerenes, carbon nanoparticles, onion-like particles, nanotubes, graphenes, and nanofoams have sparked widespread scientific interest and applications [26]. The absorption spectrum of graphene nanosheets is broad, with a maximum of 320 nm, similar to that of graphene. It is caused by π–π^* transitions in aromatic sp^2 domains. Graphene absorbs 2.3% of incident light in a single layer. Measurements revealed a significant difference between a graphene sheet's in plane and transverse conductivities. Mechanical graphite cleavage, chemical vacuum deposition, epitaxial growth, and microwave and laser exfoliations are well-known processes [27].

The simple techniques to create graphene–polymer composites include mixing, blending, and in-situ polymerization. It is simple and requires no special equipment to prepare polymer composites by solution mixing method. The polymer–graphene interface is critical in the preparation of advanced graphene-based nanocomposites. Hydrophobic graphene is incompatible with organic polymers and does not form homogeneous composites. The poor solubility of graphene in organic solvents, particularly nonpolar ones, severely limits its processability, acting as a bottleneck for graphene incorporation into most polymer matrices. It should also be noted that recent research indicates that graphene's hydrophobicity is caused by airborne hydrocarbon contamination. A clean graphite surface is mildly hydrophilic with a water contact angle of 64° [17]. The π–π interactions are what caused the graphene sheets to be restacked. When the electron densities of both π systems are very similar, dispersion interactions predominate. When one system is electron-rich and another one is electron-deficient, the resulting complexes are forced by initiation interactions, as when the negative charge is transferred from benzene to hexafluorobenzene [28]. N-methyl pyrrolidone and ortho dichlorobenzene exhibit the best graphene interaction and solubility along with good solubility in the mixture of perfluorinated organic and water–ethanol. Besides, many aromatic molecules such as porphyrin, tetracyanoquinodimethane, tryptophan, 1,5-naphthalene diamine, 9,10-dimethylanthracene, and ionic liquids significantly improve graphene solubility. Non-covalent functionalization is also done with inorganic materials. According to recent research, graphene can promote nanoscale clusters of metals with manageable size distributions, feasibly opening up a novel method for the synthesis of nanoparticles.

Noble metals such as gold, platinum, palladium, and silver are used in the new graphene-based nanocomposites; on the other hand, there is increasing attention on the use of metals such as cobalt, zinc, etc. A stable graphene–copper nanocomposite can quickly form 100 nm thin films that are stable, transparent, and highly conducting. The superconductivity of tin nanoparticle-activated graphene is observed [29]. Graphene nanomaterial composites are in high demand for solar cells, Li-ion batteries, and supercapacitors due to their excellent electronics, optics, etc. properties. TiO_2, ZnO, SnO_2, MnO_2, Co_3O_4, Fe_3O_4, Fe_2O_3, NiO, Cu_2O, RuO_2, CdS, and CdSe are just a few of the semiconductor nanomaterials that have synthesized with graphene-based templates. The quenching of the fluorescence intensity from CdSe/ZnS nanocrystals by a factor of 70 indicates this energy transfer. An "insulating layer" can prevent fluorescence quenching by graphene oxide.

The formation of the composite is electrostatic forces between the two constituents. The presence of organic aliphatic molecules improves the solubility of composite in organic solvents such as toluene. Because of this insulating layer, the composite exhibited a distinct fluorescence spectrum following quantum dot photoexcitation. The thickness of the organic molecule layer typically adjoining the quantum dots is critical in achieving effective energy transfer efficiencies into graphene [20]. Covalent functionalizations create stable graphene inks and polymer composites. However, while graphene dispersion in the polymer is significantly enhanced, conductivity frequently reduces when covalent functionalization is used. The polymer-graphene nanocomposite conductivity is higher than pure polymer but remains considerably lower than that of pristine graphene. PANI is a popular host polymer due to its high capacitive property and inexpensive. The nafion-graphene nanocomposite produced good outcomes with the resistance 1030 Ω/sq, however, the transmittance was reduced to 50.5% [30]. It is clear that the presence of additional contacts between carbon nanomaterial and polymer via metal nanoparticles significantly improves conductivity. The schematic diagram of various CV responses of capacitors is shown in Fig. 14.2.

14.1.5 Electrochemical Measurements

14.1.5.1 Cyclic Voltammetry (CV)

Based on a surface charge transfer process, the CV curves of the nanocomposite electrode materials indicate pseudocapacitive behavior and that the oxidation reaction is greater than the reduction reaction. The nonrectangular form of the CV curve shows the material's pseudocapacitive behavior. In addition, the rise in current on both anodic and cathodic sides with increasing scan rate demonstrates the faradic redox reactions on the electrode. Figure 14.3 depicts schematic illustration of organic, inorganic nanocomposites for supercapacitors.

Using Eq. 14.1 [31], the specific capacitance values of the nanocomposite samples were calculated from the CV curves

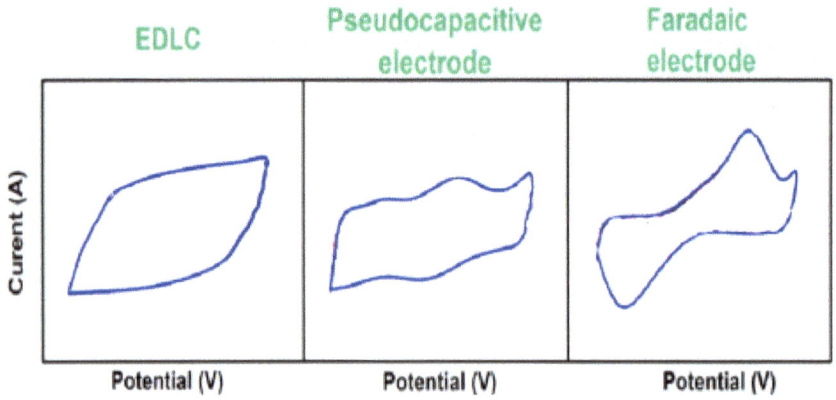

Fig. 14.2 Schematic diagram of cyclic voltammetry response of EDLC, pseudocapacitive electrode, and Faradaic electrode

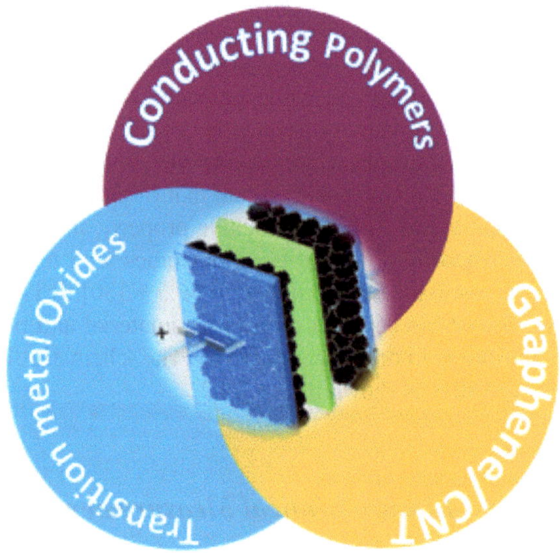

Fig. 14.3 Schematic illustration of organic–inorganic nanocomposites for supercapacitors

$$\text{Specific Capacitance } (C) = 2[ms(\Delta v)]^{-1} \int I\,(V)\,\mathrm{d}v \qquad (14.1)$$

where m is the mass of the active material (g), s is the scan rate (Vs^{-1}), Δv is the potential window (V), and I dv represents the area.

14.1.5.2 Electrochemical Impedance Spectroscopy (EIS)

EIS measurements were performed in an electrolyte over a frequency range of 0.01 Hz to 100 kHz to determine the electrochemical properties of the supercapacitor electrodes. According to the Nyquist plots (NP), the electrode has a semicircle arc and straight line in the high and low-frequency regions respectively [32]. At lower frequencies, the vertical shape indicates pseudocapacitive behavior and indicates ion diffusion in the electrode structure. The NP straight lines have a steeper response, representing faster ion movement and behavior shown to ideal capacitors. The semicircle arc of the composite electrode in the high-frequency region is negligible, indicating that the former has improved conductivity due to the incorporation of rGO and metal oxides [33].

14.1.5.3 Galvanostatic Charge–Discharge Curves

Over a potential range, the galvanostatic charge–discharge performances of nanocomposite electrodes are measured. The tests are performed in an aqueous electrolyte with a current density of 6 Ag^{-1}. The pseudocapacitive behavior of sample caused by redox reactions is confirmed by nonlinear discharge curves. However, the nanocomposite has a long discharge time at constant current density, indicating that the capacitance response has improved. Using Eq. 14.2 the specific capacitance of the electrode materials was measured.

$$\text{Specific capacitance } (C) = I/m.\mathrm{d}t/\mathrm{d}v \tag{14.2}$$

where I is the current dv/dt is the potential scan rate, and m is the mass of active material in the electrode.

14.2 Organic–Inorganic Nanocomposites for Supercapacitors

In contrast to batteries (volumetric charge storage), supercapacitors store charge on the surface of electrodes, and thus, they can offer high-power densities due to their capability to release energy more simply from the sub-surface layer rather than the bulk. Furthermore, supercapacitors have remarkable cycling ability and a short charging time because charging-discharging takes place at the surface, which does not cause severe structural changes in the electroactive materials. Because of these unique characteristics, supercapacitors are considered one of the most promising energy storage devices. Supercapacitors are classified into two types based on their operation mechanism: EDLCs and pseudocapacitors. The energy in EDLCs is stored electrostatically at the interface of the electrode and electrolyte in the double layer, whereas charging storage in pseudocapacitors proceeds in the redox reactions on the electrode. Carbon-based materials like carbon allotropes, carbon compounds along with their nanocomposites, metal oxides/hydroxides, as well as conducting polymers are the three significant electrode materials used in supercapacitors. Every material possesses its own distinct set of merits and demerits. Conducting polymers have high capacitance, good conductivity, low costs, and ease of fabrication but poor mechanical stability and cycle life. Graphene and conducting polymers can be amalgamated to form a hybrid supercapacitor, which combines EDLC and pseudocapacitors. The development of capacitors, which include a Faradaic electrode and a capacitor-type electrode to associate with the benefits, is an encouraging way to permit improved cell voltage and energy density [34]. Nanocomposites prepared by graphene are considered to have a bright prospect in supercapacitors.

The components and their interfacial features determine nanocomposite electrode characteristics. The performance of nanorod-PANI-graphene composites was considerably developed. The maximum specific capacitance value of the graphene oxide/PANI composite was 1136 Fg^{-1}. Recently, conductive graphene/SnO_2 nanocomposites with higher and more stable specific capacitance than graphene were created. It was made from polyaniline/SnO_2 nanocomposites, which contained SnO_2 nanoparticles embedded in a polyaniline matrix. This composite material has a threefold higher energy storage density than sole SnO_2 and a specific capacitance decay rate of only 4.5% after 500 cycles [35].

14.3 Nanocomposite XLPE Insulated HV Direct Current Cables

14.3.1 Exploration of XLPE

By cross-linking polyethylene, high-density cross-linked polyethylene (XHDPE) or cross-linked low-density polyethylene (XLDPE) is created. These materials have strong mechanical qualities, a longer useful life, and an increased working limit at high temperatures. It might result from the development of 3D structures. The polymer's tensile strength, heat stability, and 198 R are all enhanced by this structure. Jose Varghese et al. enhances chemical resistance and hardness. There are three ways to trigger PE cross-linking: Radiation, adding peroxide or silane, or both. The use of silane, which Dow Corning patented in the late 1960s, did not alter the base polymer's crystalline properties as reported earlier and also enhanced the flexibility via cross-linking (Si–O–Si) rather than making the process more challenging [36]. This is particularly intriguing since it ensures that each Si atom possesses three active sites, enabling it to attach to up to six network bridges and six polymer chains. However, when mechanical, chemical, and thermal resistances are needed, manufacturing condition optimization is helpful for better performance. Cross-linked polyethylene is currently utilized in several industries, including the cable and wire industry, hot water pipe manufacturing, and steam-proof food packaging. In addition to this, the majority of XLPE investigations rely on cross-linking LDPE with peroxide or silane. Cross-linked polyethylene (XLPE) replaces oil paper insulated systems and high voltage AC cables because of its high working temperature, performance, and well-controlled extrusion technique. The base matrix is polyethylene (PE), which can be cross-linked using peroxide. Cross-linking is thought to be important, however low-density polyethylene (LDPE) melts more readily at temperatures around 100 °C, which reduces its mechanical stability [37]. Contrarily, linear high-density polyethylene (HDPE) exhibits a high melting point and may operate at a higher temperature. Today, HV cable insulation uses XLPE on a large scale. Today, a radical reaction with an organic peroxide was one of the most popular cross-linking processes. The developments in the field of electrical insulation are illustrated in Fig. 14.4.

However, the residual cross-linking byproducts reveal some electrical property loss, leading to space charge accumulation. Degassing is required for XLPE insulating cables to lessen these pollutants. Because of its superior thermo-mechanical and dielectric properties, XLPE has been employed extensively as power line insulation. High voltage direct current (HVDC) transmission systems, XLPE has the

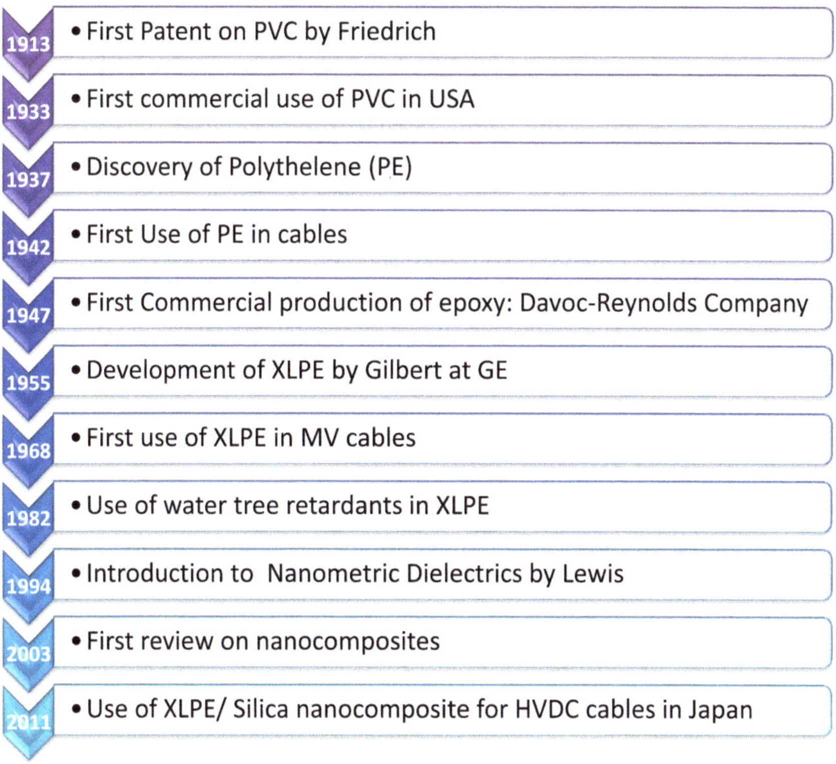

Fig. 14.4 Developments in the field of electrical insulation

inherent disadvantage of HVDC cable insulation because the cross-linking by-products can significantly reduce the electrical lifetime of cable insulation and generate large space charge accumulation. Since XLPE is a thermoset plastic, it cannot be recycled or used again after the cables have been retired, which results in significant reuse costs and environmental issues. As a result, recyclable high-performance thermoplastic cable insulation is widely coveted.

14.3.2 Significance of XLPE

Because of the inclusion of nanoparticles, the electrical characteristics of XLPE nanocomposites have improved greatly. According to literature reviews, nanoparticles are a strong option for use in DC and AC cables, insulations, and semiconductor screens. SiO_2 and various kinds of nano clay are examples of nanoparticles that are utilized in both AC and DC cable applications. In comparison to virgin polymers, the addition of silica nanoparticles in the polymer significantly increased the AC and DC breakdown voltage by 15%. As a result, medium and low-voltage cables are made of SiO_2-based XLPE nanocomposites. Alumina/silicate/nanoclay based nanocomposites are effective for low-voltage insulation applications, while SiC-based XLPE nanocomposites are ideal for DC cable insulation. The XLPE matrix's thermal and chemical stability and mechanical strength were both increased by the addition of SiO_2, Al_2O_3, and TiO_2 nanoparticles. According to reports, carbon black-based nanocomposites are employed for semiconductor screens whereas MgO-based nanocomposites are used for DC cable applications [38]. To confirm the space charge trapping qualities and the long-term performance of these new materials before they are made available for use in power cables, appropriate levels for the design field, as well as cost-effectiveness, must be defined [39]. This is true although several XLPE-based nanocomposites outperform the unfilled PE polymer in terms of electrical and thermal properties.

14.3.3 Types of HVDC Cables

The types of HVDC cables are shown in Fig. 14.5. It is suggested to use extruded polymeric cables as a power grid since they are economical and environmentally friendly. Networks will be able to boost their operational voltage levels so they can resist the ensuing rise in electric stress through the usage of HVAC and HVDC cables. It is preferred to use an XLPE matrix with fewer contaminants. The General Electric Global Research Center developed a brand-new insulating material to support direct current (DC) applications. It incorporates nanoclay particles to support the N-EPR insulating performance of ethylene propylene rubber (EPR). Space charge propagation is present in N-EPR when compared to XLPE insulation. The structural design of N-EPR insulation allows for the restriction of electrical enhancement. In addition

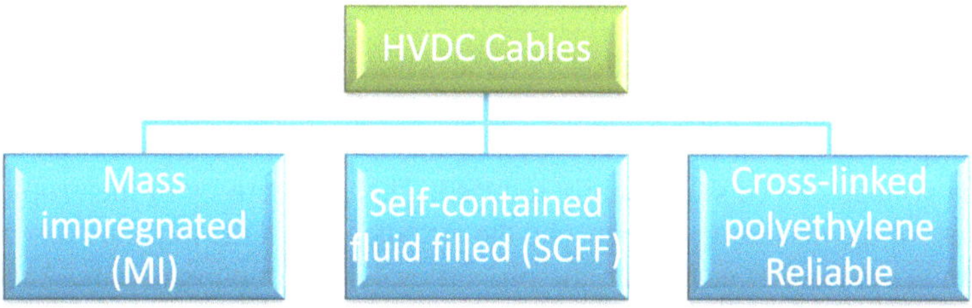

Fig. 14.5 Types of HVDC cables

to these benefits, N-EPR offers greater power density and skips the time-consuming and expensive degassing required for XLPE insulation. Due to its excellent dielectric qualities, processing viability, acceptable chemical properties, and eco-friendliness, cross-linked polyethylene (XLPE) is frequently used for cable insulation. These characteristics aid in the distribution and transmission of electrical energy [40]. It is a good contender for the cable industry due to its strong resilience to thermal shocks and tuneable thickness. For the application of XLPE in subterranean electric cables, these qualities were used. In addition to its benefits, cross-linked polyethylene (XLPE) is better able to withstand heavy filler loadings than uncross-linked polyethylene, which becomes brittle when fillers are added. Insulation material's dielectric, electrical, and thermal characteristics very slightly improve when NPs are added. The particles are bound and trapped inside the polymer matrix as a result of cross-links forming [41]. As a result, XLPE would give reinforcement at quantities of filler that are undesirable and cause the polymer to crack. The qualities of polyethylene can be improved through cross-linking, and these improved properties can be used in cable installation, heat-resistive food packaging (till 200 °C), thermal insulation foams, and chemical-resistant sealing.

14.3.4 Role of XLPE

The polyethylene with nanoparticles incorporation can be treated using the etching process. The dispersion of nanoparticles, spherulite structure, and the prevention of the overlap between interaction zones can all be improved by properly modifying the surface using an effective coupling agent. The usage of commercial (AC grade) XLPE and its nanocomposite for DC applications have also been tested. According to reports, adding nanofiller (such as nanosilica) increased the material's breakdown strength in the DC step test [42]. At a concentration of 1 wt%, SiC nanoparticles are said to prevent the buildup of space charges. Further increases in concentration inhibited the effect of NPs on the generation of space charges [43]. There will be a lot of space between NPs when there is a low concentration of them. Previous studies on MgO XLPE composites indicated the presence of deep traps surrounding the MgO nanoparticles, which drew charges to the MgO NPs. An NP can be made electrically neutral by surrounding it with either electrons or holes [44]. Another element that influences the aging of the insulation material is moisture mixed with electrical stress. The partial discharge of the substance that results from this action can be followed by disintegration. When water and contaminants on the insulating surface combine, the flash voltage can be significantly reduced. The conductivity of the droplets, the roughness of the insulation surface, and the arrangement of the moisture droplets are the main variables that determine the properties of insulation materials. According to [45], the flashover voltage decreases as the moisture content's conductivity rises. MgO/XLPE nanocomposites (0.5 wt%) were reported to have better flashover performances than PE. This explains why the flash overvoltage is improved by the NP addition. This is a result of NPs restricting the migration of surface charges. The type of NPs affects how many are present in the polymer matrix. As was mentioned above at the optimal concentration, an insulating material's properties can decline. The dispersion of NPs can be improved by a surface-modifying insulating material to increase the bond between nanofillers and the polymer matrix. As a result, breakdown strength (AC/DC), electrical and water tree resistance, permittivity, and dielectric loss will all significantly improve. The Partial discharge—A schematic representation is shown in Fig. 14.6

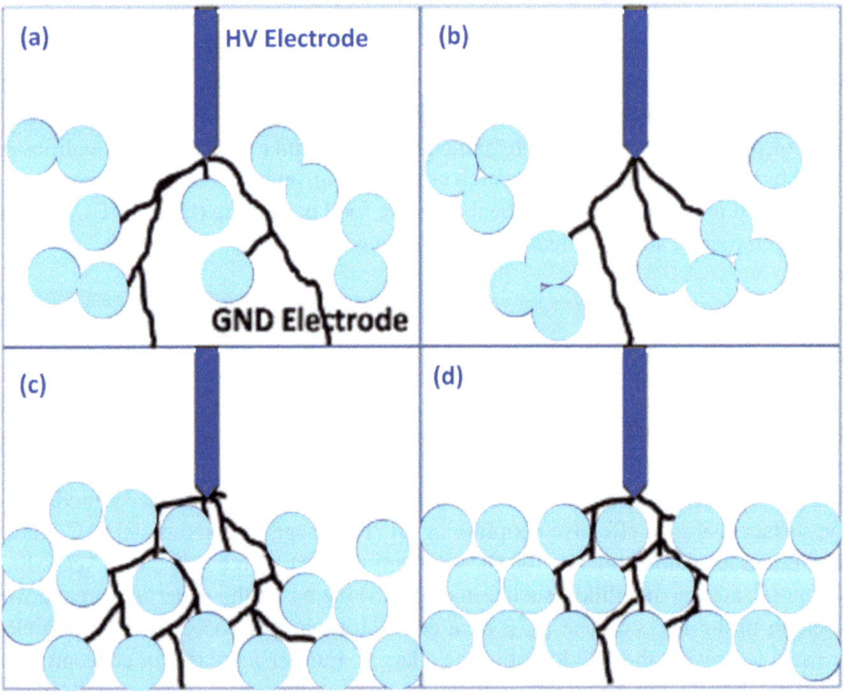

Fig. 14.6 Partial discharge—A schematic representation: **a** unmodified, **b** agglomerated, **c** surface modified, **d** ideally dispersed XPLE nanocomposites. Adopted with permission from Ref. [40] © 2021, Springer Nature

Compared to pure XLPE, XLPE nanocomposite exhibits increased crystallinity. Gibbs Thomson equation provides a relationship between the melting point (T_{m}) and lamellar thickness (L).

$$L = \frac{2\sigma_e T_{\mathrm{m}}^0}{(T_{\mathrm{m}}^0 - T_{\mathrm{m}})\Delta H_0} \tag{14.3}$$

where σ_e = free surface energy, T_{m}^0 = equilibrium melting temperature (414.6 K), T_{m} = melting temperature of the tested material. Equation demonstrates that the thickness of the lamellar layer grows with melting temperature. The crystallinity subsequently rises as a result. It is an accepted idea that the increasing lamellar thickness causes the higher value of crystallinity to indicate the more densely connected chains (i.e., intact spherulite structures) [46].

Therefore, to break the chains, a strong electrical field is needed. After thermal aging in pure XLPE, the lamellar thickness is greatly reduced (because of the lower crystallinity). As a result of the reduced chemical bonds, this reduces the DC breakdown strength. However, the crystallinity is somewhat reduced in the XLPE nanocomposite. In comparison to pure XLPE, the deep trap density is much higher. As a result, the charges would continue to be trapped in the deep traps and would need a very strong electric field to release them. It is hypothesized that the XLPE nanocomposite's driving parameter is the high density of the deep traps [47].

14.4 Nanocomposite Solid Insulated Switchgear

Partial discharges frequently occur at triple points in the gas–solid insulating systems used in high-voltage equipment like switchgear when the electric field approaches the initiation threshold field.

Regulations prohibiting the use of SF_6 have made it more urgent to investigate alternate dielectrics, such as CO_2 and vacuum [48], which are employed in gas circuit breaker technology. Researchers have looked into the characteristics and impacts of alternative gases on surface flashover. The final length that creeping discharges take in SF_6 is shorter as compared to other gases, showing a dependence between gas and discharges, according to Sadaoui [49], who observed the creeping discharges under DC in SF_6, N_2, and CO_2. By comparing the flashover voltages of SF_6/N_2 mixtures with those of air, N_2, and N_2/O_2 mixtures, Rokunohe [50] examined these mixtures. It was demonstrated that pressure as well as gas dielectric strength had a significant impact on surface flashover voltages. The DC flashover voltage of polymers like PTFE, PMMA, and nylon in N_2 and SF_6 was measured in earlier work. In different matrices including epoxy, silicone rubber, and polyethylene, nanoparticles' unique nanometric structure, and particular effects produce novel dielectric characteristics. Our previous study resulted in the production of nanocomposites with improved dielectric breakdown strength, lower permittivity, and greater surface flashover voltages in a vacuum. Epoxy resin is essential for the electrical insulation of machinery like gas-insulated switchgear (GIS) and motors [51]. However, the buildup of space charges or the concentration of electric fields in electrical apparatus leads epoxy resin to age more quickly. By adding the right amount of semiconducting fillers, like silicon carbide (SiC), to the polymer matrix, epoxy resin is created with an on-linear conductivity property. This kind of composite, also known as nonlinear resistive field grading composite or stress grading composite, is extremely useful for preventing the buildup of space charges and homogenizing the distribution of the electric field in various pieces of machinery or components, such as motor ends, GIS spacers, and bushings.

14.5 Organic–Inorganic Hybrids in Lithium Batteries

Scarceness of non-renewable fuels, serious environmental problems, and increasing energy demand are enforced to develop efficient, cost-effective, and environment-caring green energy conversion/storage devices. Lithium batteries, which are typical electrochemical energy storage devices that can convert and store electricity in a chemical form, may be one of the most striking and promising solutions. However, poor stability, sluggish reaction kinetics, limited active sites, high price, etc. are the numerous key factors that result in their unsatisfactory performance and thus hindered their real-world execution [52–55]. Therefore, to improve their performance researchers are focusing on the development of appropriate materials for efficient electrodes, separators, electrolytes, etc. Among various functional materials, organic–inorganic hybrids have attracted much attention as they offered specific properties due to the synergetic effect along with the characteristics of both constituents. The organic part expands the range of available matrices, provides low density, and facilitates tunable physical or chemical properties, whereas, the inorganic part enables good stability and electro-catalytic active species [56]. The suitable choice and proper hybridization of organic and inorganic precursors deliver desirable properties. The organic–inorganic hybrids are persisting in consistent devotion for the last few decades due to their specific properties, like; high structural and compositional diversity that enable requirements as per application, large surface area which offers a high ratio of accessible active sites, and tunable porous structure that can accelerate the diffusion of electrolyte/reactants and act as a sieve/host, etc. Besides, they deliver diminish damages during insertion/extraction caused by large volume variation of reactants. Such advantages improve the comprehensive performance of Li batteries including the storage capacity, temporal stability, and so on. To date, several organic–inorganic nanocomposites have been studied and utilized as electrodes/separators/electrolytes materials in lithium batteries [57–59].

By the nature of the interface and their importance, organic–inorganic hybrids are classified into two families say Class I and Class II [60]. The class I hybrids are those in which the interaction forces between organic and inorganic parts are week like van der Waals forces, electrostatic attraction, hydrogen bonds, etc. Contrary, the primary bonds like covalent or ionic bonds alone or jointly with secondary bonds connect the organic and inorganic parts in the class II hybrids. However, the properties of class II hybrids are preliminarily governed by strong bonds. In general, Class I hybrids possess the advantages of relatively easy and feasibility for removal of the organic part on demand without damaging the whole structure along with the drawbacks of stability and durability that limit their usage. In contrast, many superior properties are offered by class II hybrids due to the strong bindings between the organic and inorganic parts. Thus, the well-defined and controlled processing of class II hybrids facilitates the desired structure and properties, control over the hydrophilic and hydrophobic balance, etc. Therefore, extensive devotion is paid to the synthesis, study, and application of Class II hybrids during the past few decades.

Both the conventional and flexible lithium batteries need four main components: an electrolyte, a separator, an anode, and a cathode as illustrated in Fig. 14.7. Each component should be low-cost, non-toxic, safe, and lightweight. In addition, flexible electrodes must be bendable and should have low self-discharge, high capacity, stable. This section presents an overview of broadly used hybrid materials for the development of the aforementioned components.

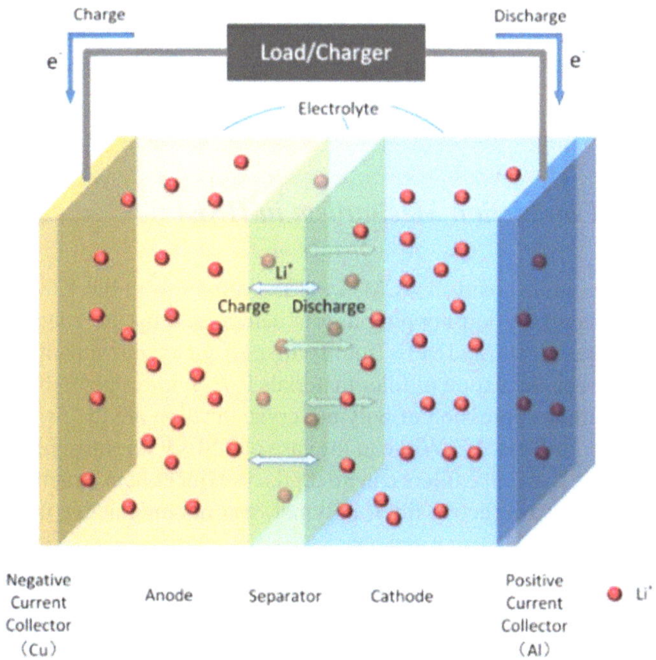

Fig. 14.7 Schematic diagram of electrodes for Li battery. Adopted from Ref. [61] © 2018, (IEEE Access an open-access journal of IEEE)

14.5.1 Electrolytes

Despite the high ionic conductivity of liquid and liquid-based electrolytes, their utility as electrolytes in lithium batteries is limited because of their low mechanical stability, short range of operational temperature, safety issues (i.e. explosion, leakage, electrode corrosion), and capacity fade over time [62, 63]. On the other hand, solid electrolytes (SEs) may avoid dendrite growth, leakage, short-circuiting, and electronic conduction during the crossover of Li^+ between anode and cathode [64] and offer higher thermal stability, energy and power density, and cycle life [65], therefore are suitable replacement of liquid electrodes. Typically, SSEs are categorized into two; inorganic solid electrolytes (ISEs) and solid polymer electrolytes (SPEs). The practical applications of polymeric SPEs are limited due to low ionic conductivity at ambient temperature, poor mechanical strength, and a narrow operational window. Whereas the brittleness, grain boundary resistance, chemical, and electrochemical stability with the electrodes i.e. poor contact with electrodes, high cost, and lack of processability limits the utilities of inorganic ISEs [66]. In contrast, organic–inorganic hybrids offer advantages for each component with the removal of their flaws due to specific synergetic effect and thereby demonstrates multifold as illustrated in Fig. 14.8 [67]. In practice, the ionic conductivity and structure stability of polymeric SSEs is enhanced through the filling of a polymer matrix with inorganic or organic–inorganic composite materials. To date, several inorganic fillers have been utilized for the fabrication of organic–inorganic-based hybrid solid electrolytes (HSEs) for lithium batteries [68–74].

14.5.2 Ion Conduction Mechanism in HSEs

The ion conduction mechanism (ICM) in HSEs differs from the ICM for both the ISEs (usually occurs by ion hopping and is determined by the concentration and distribution of the defects) and SPEs (jointly arises from ion movement and jumping) and also depends on the nature of filler materials [67]. For example, the Lewis acid–base theory and the suppression of polymer crystallization are the two mechanisms proposed for inert filler HSEs. While in the case of HSEs with active fillers ICMs are intricated because the active fillers can affect the polymer chain structure and create additional Li^+ pathways (active filler, polymer, and polymer/active filler interface) simultaneously [67].

The Lewis acid-base theory for HSE systems considered Li-ion as Lewis acid, the solvating groups of polymers and ClO_4^- anions as Lewis bases, and the fillers as Lewis acid or base depending on their surface chemistry. The acidic surface may attract the polymer solvating groups or ClO_4^-, release associated Li^+ cations, and the basic surface attracts Li-ions and polymer functional groups, improving the dissociation of Li^+–ClO_4^- ion pairs [75]. The enhanced conductivity due to increment in

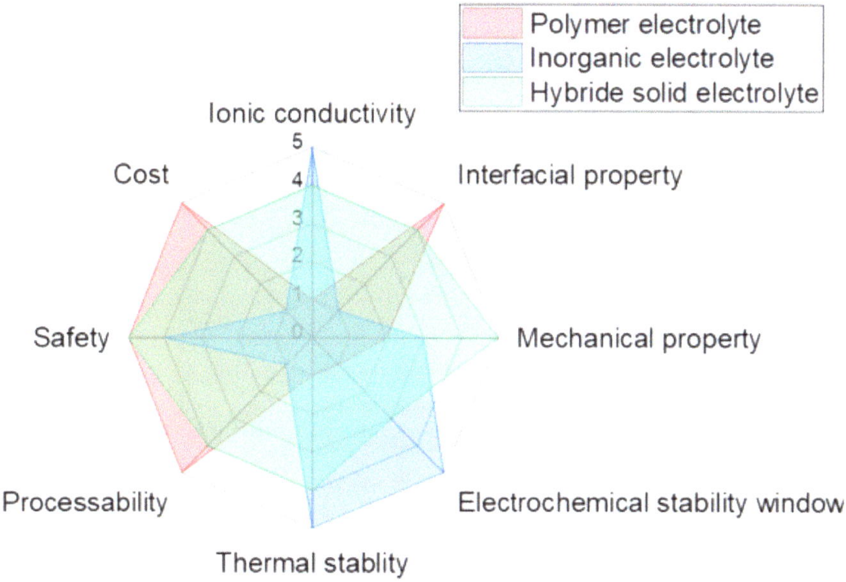

Fig. 14.8 Ranking of properties of solid electrolytes (Best = 5, Worst = 1). Adopted from Ref. [67] (Frontiers in Energy Research an open-access journal of Frontiers)

segmental dynamics induced by crystallinity suppression of polymer in some polymers (like PEO) based electrolyte systems is noticed [76, 77]. Besides, these two ICMs in HSEs with inert fillers, the possibility of the governance of other mechanisms is also under consideration based on some observations. As an illustration, the ionic conductivity for a PEG-based electrolyte is found to be increased via the addition of neutral, acidic, or basic inert filler Al_2O_3 [78]. Similarly, the addition of fumed silica particles into PEG-based electrolytes with lithium salts increases the conductivity of the composite electrolyte, irrespectively whether PEG is polar or non-polar [79]. Contrary, Hanson et al. [80] reported that the lithium-ion diffusivities in PEO/TiO_2 reduce with enhanced particle loading. The contradictions discussed above may show the contribution of other causes or ICMs operating simultaneously, which are still not clear. On the other hand, active fillers like $Li_7La_3Zr_2O_{12}$ (LLZO) [81, 82], $Li_{3x}La_{2/3-x}TiO_3$ (LLTO) [83, 84], and $Li_{1+x}Al_xTi_{2-x}(PO_4)_3$(LATP) [85–87] create new Li^+ pathways that govern ICMs together with Li-ion concentration and ion mobility.

14.5.3 Separator

In lithium batteries, the separator is a material that prevents any possible electrical short circuits within the device and avoids the problem of thermal runaway, burning,

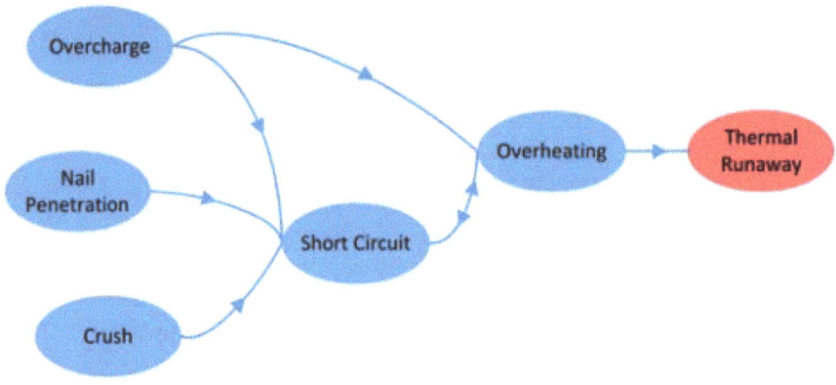

Fig. 14.9 Relationship of different abuse conditions. Adopted from Ref. [61] © 2018, (IEEE Access an open access journal of IEEE)

and even explosions [88]. The relationship of different abuse conditions that results in thermal runway is illustrated here in Fig. 14.9.

The high ionic conductivity, large electrolyte wettability, and superior electrolyte uptake are the key features of the perfect separator [89]. Instead of poor wettability towards polar liquid electrolytes, polyolefin-based materials are still the commonly used separators in LIBs [89]. Therefore several separators including organic–inorganic nanocomposites were engineered and employed for better performance of LIBs [72, 89–92]. Currently, the most practical approaches used for the fabrication of hybrid separators are the coating of inorganic submicron nanomaterials onto the surface of PE or PP membranes [93, 94] and direct surface modification methods like grafting methods or the layer-by-layer (LbL) self-assembly process [89, 95–97]. For instant, Xu et al. [89] developed a poly(-acrylic acid) (PAA)/ZrO_2 modified polyethylene (PE) separator using the LbL self-assembly method. They reported that the $LiCoO_2$/Li half-cell assembled with $(PAA/ZrO_2)_1$ PAA-modified PE and $(PAA/ZrO_2)_3$-modified PE separators show improved discharge capacities as compared to pristine PE upon diverse discharge current densities from 0.2 to 7 C. Further, the Nyquist plots indicate much better stability of the $(PAA/ZrO_2)_3$-modified PE than those of pristine PE and $(PAA/ZrO_2)_1$ PAA-modified PE. Whereas PP separators coated with Al_2O_3 or TiO_2 using atomic layer deposition (ALD) showed reduced thermal shrinkage and improved wettability without any increment in overall separator thickness [98, 99]. To address the issue of mechanical brittleness, the introduction of GO to integrate with the organic–inorganic hybrids fabricate robust and flexible membranes was carried out. For instance, Biaet al. fabricated a MOF@GO separator with a size window around 9 Å by using $Cu_3(BTC)_2$(HKUST-1) and GO that selectively sieves Li^+ and block polysulfides significantly [55]. Such a separator possessed high capacity (799 mA h g after 500 cycles at C/2) and had a low capacity decay rate. An effective separators was also prepared by replacing $Cu_3(BTC)_2$(HKUST-1) by $Zn_3(BTC)_2$ (Zn-HKUST-1) [100]. Further, Li et al. [101] developed four different MOF-based separators through integrating CNTs and demonstrated that rather than the internal pore size of the MOFs, the chemical stability and packing morphology of MOF particles are more imperative in determining the performance of the separator. Indeed few efforts have been made to design novel and efficient membranes, but still, it needs large efforts.

14.5.4 Anode

As we know the anode i.e. the −ve electrode discharges lithium ions into the electrolyte, is transported to the cathode (+ve electrode), and then absorbed as shown in Fig. 14.1. This is the energy discharging process in lithium-ion batteries (LIBs). Thus anode plays a very critical role in rechargeable batteries and based on its properties and morphology, the overall performance of the whole battery changes remarkably. In commercial (LIB) cell, earlier used Li metals as an anode materials was replaced by graphite since 1991due to safety concern. Unfortunately, the low energy density, cost, stability, and safety with graphite anode is also hindered from further applications in transportation and large-scale stationary energy storage [102]. As a replacement of graphite, Si could be a promising anode material, owing to its high capacity (~3500 mAh g^{-1} at room temperature, which is10 times better than that of commercial graphite anodes), low discharging potential, low cost due to abundance on the earth, and nontoxicity due to environmental friendly. However, the commercialization of Si anode manly stuck due to its huge volume changes and technical issues allied during lithiation and delithiation [103]. Likewise common problems with the other inorganic materials are their low capacity utilization, lower rate capability, lesser cycling stability, or even complete electrochemical inactivity in LIBs. Though, they shows great interest as anode materials due to various advantageous features like good crystallinity, excellent conductivity and redox properties. Additionally, they offer easy pathway for the transportation of Li^+ ions and electrons that allows better accommodation of ions. Contrary, electrodes based on organic materials that can be broadly categorized in to conducting polymers (polyacetylene, polyaniline, polypyrrole, polythiophene), organosulfurs, organic salts, nitroxideradicals, nonconjugated or conjugated polymers and carbonyl compounds. The carbonyl compounds generally possess classic redoxactivity due to their reversible redox stability, adjustable structure, high theoretical capacity, and multi-electro reaction, whereas they suffer with the dissolution issue in organic electrolytes that result in the fading capacity during the cycling and the inherent poor electrical conductivity that slow down their reaction kinetics, high-voltage hysteresis and high irreversible capacity. In contrast, conducting polymers typically exhibit good electrical conductivity and high coulombic efficiency and hence they can be cycled hundreds or thousands of times but having common issues of low capacity in terms of lithium insertion and extraction process i.e. the inclination in their characteristic discharge curve. Thus, the problems with inorganic and organic anode materials can be minimize by the use of hybrid materials of two. There are significant works on hybrid of organic and inorganic compounds that includes conducting polymers and carbonyl compounds/ inorganic (metal alloys, metal oxides, ceramics etc.) composites and carbon materials (carbon nanotube, carbon fibers, graphene, carbon clothe, carbon fabric, carbon textile, expanded graphite paper, and carbon foam etc.)/inorganic composites, metal organic frame works (MOFs)/inorganic etc. [104–110]. For instance, layered transition metal sulfide (MoS_2) has S–Mo–S layer that exhibit a structure similar to graphite along with van der Waals forces and thus can facilitate Li^+ insertion/extraction. But possess poor conductivity between two adjacent S–Mo–S sheets that can be enhanced by making its composite with PANI and thus designed anode materials exhibiting high capacity and good cyclability for Li-ion batteries [111]. Similarly, the issue of the poor conductivity and large volume variation during the battery's charge–discharge process with promising anode materials like silicon, tin, and antimony can be overcome by the designing the anode of these materials with MOFs [103, 112, 113]. Likewise, the most common approaches, to omit the large volume variation issue with transition metal dichalcogenides and phosphides during their utility as anode in LIBs, is fabricating their porous structures in the presence of conductive carbon materials. For example, the anode of one-dimensional FeS_2@C porous nanowires facilitated transfer of electrons and ions along with mitigation of volumetric shrinkage/expansion and hence delivered much improved specific capacities of 889 mA h/g and 521 mA h/g at 0.1 A/g and 10 A/g, respectively [114].

14.5.5 Cathode

The inorganic and organic materials used for the development of cathodes in lithium batteries possess the certain limitation in a similar manner as they used for the development of other components for the lithium batteries. As an illustration, $LiCoO_2$ was the first commercially used cathode material in LIBs. However, the use of $LiCoO_2$ is limited due to number of factors that includes environmental pollution, overcharging, unsafe, high cost and the poor storage capacity [1]. Subsequently, many alternative inorganic materials like $LiNiO_2$, $LiMn_2O_4$, $LiNi_{(1-x-y)}Mn_xCo_yO_2$ (NMC) and $LiNi_{(1-x-y)}Co_xAl_yO_2$ (NCA) were studied. However, each of the limited due to certain challenges like challenge in manufacturing, limited temperature and rate capability or problem of deformation during the cyclic process, fast degradation of capacity or safety and long cyclic life etc. on the other hand lithium iron phosphate ($LiFePO_4$) cathodes exhibit long cycle life, higher thermal stability, and low cost along with environmental friendly and safer. Therefore, broadly used as an inorganic cathode material. However, the low electron conduction and inadequate Li-ion diffusion limits its performance. Researchers dilutes the aforementioned issues by reducing LFP's particle size, multi-element doping and hybridizing it with organic material. Till date, numerous organic hybrid materials including carbon and its allotropes, and polymers etc. have been used to design the cathode with improving performance of lithium batteries. Figure 14.10 illustrates the various carbon allotropes that have been hybrid with $LiFePO_4$ (LFP). For instant Jeon et al.

synthesized LFP/r-GO hybrid without using an organic solvent and demonstrated that at high current rates, hybrid cathode can store larger amount of charges and possess high cycling stability than conventional LFP cathodes with capacity of 37 mA h g^{-1} at a very high current rate (2040 mA g^{-1}). Whereas at such a high current rate, the conventional LFP-based electrodes fabricated with poly(vinylidene fluoride) and carbon black in an organic solvent, exhibited negligible capacity (<1 mA h g^{-1}). They observed that the capacity retention levels of the hybrid and conventional LFP/PVdF/carbon electrode after 1000 cycles at 3400 mA g^{-1} from 1.5 to 4.5 V (vs. Li/Li^{+}) were 88% and 38% respectively [115]. While, Chen et al. designed cathode using nanocrystalline $LiFePO_4$/graphene-carbon nanotubes (LFP-G-CNT) composite and noticed that designed network provides highly conductive paths for electron transfer during the intercalation/deintercalation procedure, facilitates electron migration throughout the secondary particles, accelerates the penetration of the liquid electrolyte into the LFP-G-CNT composite in all directions and enhanced the diffusion of Li ions. Further, a high initial discharge capacity of 168.4 mAh g^{-1} at 0.1 C and 103.7 mAh g^{-1} at 40 C with an excellent cycling stability was observed for the composite with a low content of G and CNT [116]. Whereas, Lian and coworkers used composite of polyaniline (PANI)/active carbon (a-C)/LFP as cathode and observed remarkable improvement in the capacity and rate performance of PANI/a-C/LFP cathode in comparison to a-C/LFP cathode [117].

Alternatively, Li-rich cathode materials (LRM) like $x\mathrm{Li_2MnO_3}\cdot(1-x)\mathrm{LiTMO_2}$ (TM = Ni, Mn, Co, etc.), have been regarded as one of the most promising candidates

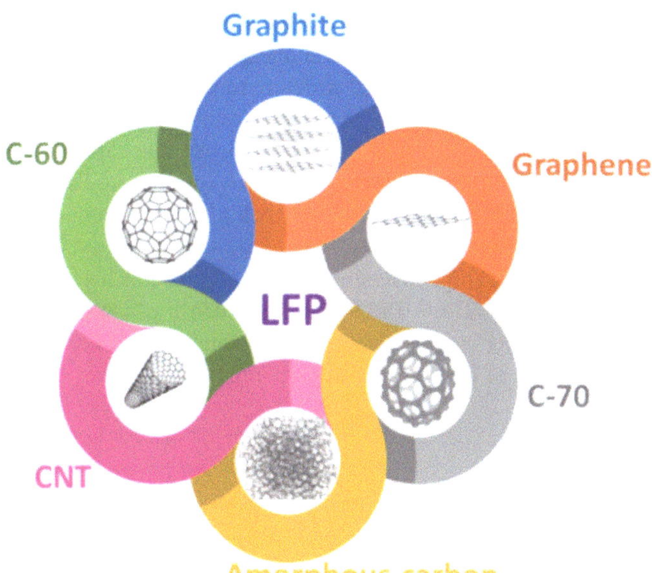

Fig. 14.10 Schematic of possible carbon allotropes for LFP. Adopted from Ref. [118] (Batteries an open access journal of MDPI)

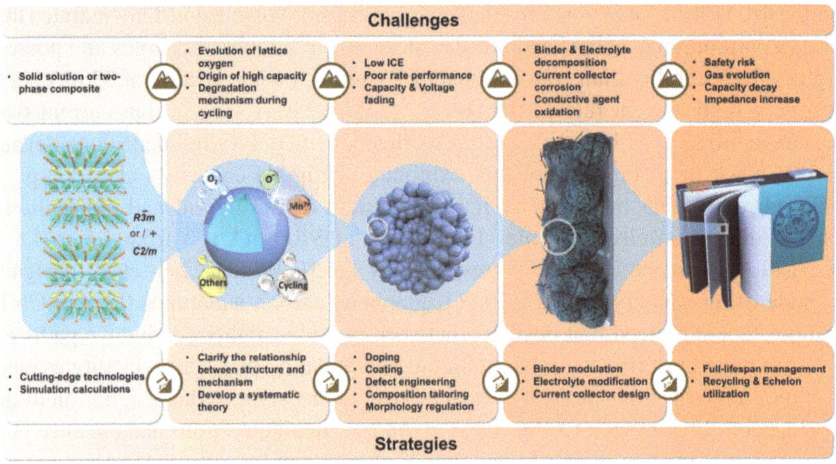

Fig. 14.11 Schematic diagram summarizing the challenges and strategies of LRM cathode materials in different stages from crystal structure to practical application. Adapted with the permission ref. [121] © 2021 Wiley-VCH GmbH

for next-generation cathode materials for rechargeable LIBs due to their prominent specific capacity particularly when $x = 0.5$. Besides, LRM cathodes are relatively low cost, environmental friendly, and high thermal stability with the reduced use of expensive Co. However, inherent features like low initial Coulombic efficiency (ICE), poor rate capacity, and serious voltage fading of LRM restricts their use in realtime applications. Figure 14.11 summarizes the challenges and strategies of LRM cathode materials at various stages from crystal structure to practical application. The challenges could be addressed with the use of composites of LRM with organic materials. For example; Wu and co-workers illustrated the better rate capability and cycling properties of PEDOT:PSS/LRM ($Li_{1.2}Ni_{0.2}Mn_{0.6}O_2$) composite than LRM with an excellent initial discharge capacity of 286.5 mA h g^{-1} at a current density of 0.1 C and discharge capacity remained 146.9 mA h g^{-1} at 1 C after 100 cycles [119]. While, Zheng et al. demonstrated a high reversible capacity (289.5 mA h g^{-1} at 0.1 C), excellent rate capability (263.6, 218.9, 182.9, and 108.6 mA h g^{-1} at 0.5, 1, 5, and 10 C, respectively), and superior cycling stability (with a high capacity retention of 88.5% at 5 C after 500 cycles) of fluorine-doped carbon/LRM ($Li_{1.2}Mn_{0.54}Ni_{0.13}Co_{0.13}O_2$) composite [120].

14.6 Conclusions

The increasing energy demands around the globe require multiple sources of energy to fulfill them. In this row, the requirement for energy storage is a great challenge to overcome. To deal with this challenge, organic–inorganic nanocomposites play a decisive role. This chapter outlines the developments in the field of organic–inorganic nanodielectrics-based applications in energy storage. Several nanocomposites viz. PEDOT–V_2O_5, PEDOT–MoO_3, P5ICN–WO_3 and polymer–graphene are discussed for their synthesis and properties. Further, the measurement methods and expressions are detailed. The applications of these nanodielectrics in supercapacitors, XLPE insulated high voltage direct current cables, insulated switchgear and lithium batteries are presented in detail. It may be concluded that semiconductor like SiC filled polymer are the outstanding nanocomposite for reducing space charge and enhancing the uniform distribution of electric field in various machinery components. Carbon-based materials are a good choice for the cathode in lithium batteries. A comprehensive review of various nanodielectrics used for electrolyte, separator, anode and cathode in lithium batteries is presented to provide an analytical approach.

References

1. Conway BE (1991) J Electrochem Soc 138:1539
2. Jeong YU, Manthiram A (2001) J Electrochem Soc 148:A189–A193
3. Murugan AV, Kwon CW, Campet G, Kale BB, Mandale AB, Sainker SR, Gopinath CS, Vijayamohanan K (2004) J Phys Chem B 108:10736–10742
4. Bulakhe RN, Nguyen VH, Shim JJ (2017) New J Chem 41:1473–1482
5. Bai Y, Rakhi RB, Chen W, Alshareef HN (2013) J Power Sources 233:313–319
6. Chen W, Rakhi RB, Hu L, Xie X, Cui Y, Alshareef HN (2011) Nano Lett 11:5165–5172
7. Gomez-Romero P (2001) Adv Mater 13:163–174
8. Ryu KS, Kim KM, Park NG, Park YJ, Chang SH (2002) J Power Sources 103:305–309
9. Murugan AV, Viswanath AK, Campet G, Gopinath CS, Vijayamohanan K (2005) Appl Phys Lett 87
10. Baibarac M, Lira-Cantu M, Oro-Sole J, Casan-Pastor N, Gomez-Romero P (2006) Small 2:1075–1082
11. Cuentas-Gallegos AK, Lira-Cantú M, Casañ-Pastor N, Gómez-Romero P (2005) Adv Func Mater 15:1125–1133
12. Powell AV, Kosidowski L, McDowall A (2001) J Mater Chem 11:1086–1091
13. Lira-Cantu M, GoHmez-Romero P (1999) J Solid State Chem 147:601
14. Boyano I, Bengoechea M, de Meatza I, Miguel O, Cantero I, Ochoteco E, Grande H, Lira-Cantú M, Gomez-Romero P (2007) J Power Sources 174:1206–1211
15. Nazar LF, Zhang Z, Zinkweg D (1992) J Am Chem Soc 114:6239–6240
16. Wagner CD (1979) Handbook of x-ray photoelectron spectroscopy: a reference book of standard data for use in x-ray photoelectron spectroscopy. Perkin-Elmer
17. Mane AT, Navale ST, Pawar RC, Lee CS, Patil VB (2015) Synth Met 199:187–195
18. Yan J, Li S, Lan B, Wu Y, Lee PS (2019) Adv Funct Mater 30
19. Wang F, Wu X, Yuan X, Liu Z, Zhang Y, Fu L, Zhu Y, Zhou Q, Wu Y, Huang W (2017) Chem Soc Rev 46:6816–6854
20. Li Z, Wang Y, Kozbial A, Shenoy G, Zhou F, McGinley R, Ireland P, Morganstein B, Kunkel A, Surwade SP, Li L, Liu H (2013) Nat Mater 12:925–931
21. Zhu M, Huang Y, Huang Y, Meng W, Gong Q, Li G, Zhi C (2015) J Mater Chem A 3:21321–21327
22. Wang R, Yao M, Niu Z (2019) InfoMat 2:113–125
23. Huo X, Shen W, Li R, Zhang M, Guo M (2020) Scripta Mater 174:1–5
24. Wang WQ, Wang XL, Xia XH, Yao ZJ, Zhong Y, Tu JP (2018) Nanoscale 10:8162–8169
25. Dai Y, Li W, Zhao R, Huang Q, Xu N, Yuan F, Zhang C (2019) Electrochim Acta 318:322–332

26. Nie G, Zhou L, Yang H (2011) J Mater Chem 21
27. Yang G, Takei T, Yanagida S, Kumada N (2019) Appl Surf Sci 498
28. Novoselov KS, Fal'ko VI, Colombo L, Gellert PR, Schwab MG, Kim K (2012) Nature 490:192–200
29. Kim H, Abdala AA, Macosko CW (2010) Macromolecules 43:6515–6530
30. Jenny Malig AWIS, Wagner P, Wallace GG, Officer DL, Guldi DM (2012) Chem Commun 48:8745–8747
31. Ajayi OA, Anderson NC, Cotlet M, Petrone N, Gu T, Wolcott A, Gesuele F, Hone J, Owen JS, Wong CW (2014) Appl Phys Lett 104
32. Potts JR, Dreyer DR, Bielawski CW, Ruoff RS (2011) Polymer 52:5–25
33. Ni G-X, Zheng Y, Bae S, Tan CY, Kahya O, Wu J, Hong BH, Yao K, Ozyilmaz B (2012) ACS Nano 6:3935–3942
34. Jamal M, Razeeb KM, Shao H, Islam J, Akhter I, Furukawa H, Khosla A (2019) Sci Rep 9:4659
35. Ramachandran R, Mani V, Chen S-M, Saraswathi R, Lou B-S (2013) Int J Electrochem Sci 8:11680–11694
36. Jeanene Willcox P, Howie DW, Schmidt-Rohr K, Hoagland DA, Gido SP, Pudjijanto S, Kleiner LW, Venkatraman S (2000) Journal of Polymer Science: Part B: Polymer Physics, 37 (2000) 3438–3454.
37. Andritsch T, Vaughan A, Stevens GC (2017) IEEE Electr Insul Mag 33
38. Thomas J, Joseph B, Jose JP, Maria HJ, Main P, Rahman AA, Francis B, Ahmad Z, Thomas S (2019) Ind Eng Chem Res 58:20863–20879
39. Montanari GC (2011) IEEE Trans Dielectr Electr Insul 18:339–364
40. Jose Varghese R, Vidya L, Joseph TM, Gudimalla A, Harini Bhuvaneswari G, Thomas S (2021) Potential applications of XLPE nanocomposites in the field of cable insulation. In: Thomas J, Thomas S, Ahmad Z (ed) Crosslinkable polyethylene based blends and nanocomposites. Springer Nature Singapore Pte Ltd., Singapore, pp 197–213
41. Yang J, Han CR, Duan JF, Xu F, Sun RC (2013) Nanoscale 5:10858–10863
42. Rahim NH, Lau KY, Muhamad NA, Mohamad N, Rahman WAWA, Vaughan AS (2019) IEEE Trans Dielectr Electr Insul 26:284–291
43. Basnet P, Chatterjee S (2020) Nano-structures & Nano-objects 22
44. Sima W, Shi J, Yang Q, Huang S, Cao X (2015) IEEE Trans Dielectr Electr Insul 22:380–390
45. Ahmadi-Joneidi I, Majzoobi A, Shayegani-Akmal AA, Mohseni H (2013) IEEE Trans Dielectrics Electr Insul 20:212–220
46. Ashish Sharad P, Kumar KS (2017) Nanocomposites 3:30–41
47. Paramane A, Chen X, Dai C, Guan H, Yu L, Tanaka Y (2020) Polym Compos 41:1936–1949
48. Kharal, Kamran, Ullah, Saleem, Alvi, Processes, 7 (2019).
49. Shihu Y, Shengtao L, Shihang W, Yin H, Nazir MT, Phung BT (2018) IEEE Trans Dielectr Electr Insul 25:1567–1576
50. Rokunohe T, Yagihashi Y, Endo F, Oomori T (2006) Electrical Engineering in Japan 155:9–17
51. Wang Y, Yan J, Yang Z, Liu T, Zhao Y, Li J (2019) Energies 12
52. Yang Y, Zheng G, Cui Y (2013) Chem Soc Rev 42:3018–3032
53. Ma TY, Dai S, Qiao SZ (2016) Mater Today 19:265–273
54. Jiao Y, Zheng Y, Jaroniec M, Qiao SZ (2015) Chem Soc Rev 44:2060–2086
55. Bai S, Liu X, Zhu K, Wu S, Zhou H (2016) Nat Energy 1:16094
56. Nicole L, Rozes L, Sanchez C (2010) Adv Mater 22:3208–3214
57. Mehtab T, Yasin G, Arif M, Shakeel M, Korai RM, Nadeem M, Muhammad N, Lu X (2019) J Energy Storage 21:632–646
58. Wang L, Han Y, Feng X, Zhou J, Qi P, Wang B (2016) Coord Chem Rev 307:361–381
59. Xu G, Nie P, Dou H, Ding B, Li L, Zhang X (2017) Mater Today 20:191–209
60. Judeinstein P, Sanchez C (1996) J Mater Chem 6:511–525
61. Zhang J, Zhang L, Sun F, Wang Z (2018) IEEE Access 6:23848–23863
62. Agrawal RC, Pandey GP (2008) J Phys D Appl Phys 41:223001
63. Xiao Y, Wang Y, Bo S-H, Kim JC, Miara LJ, Ceder G (2019) Nat Rev Mater 5:105–126

64. Ren D, Feng X, Liu L, Hsu H, Lu L, Wang L, He X, Ouyang M (2021) Energy Storage Mater 34:563–573
65. Pomerantseva E, Bonaccorso F, Feng X, Cui Y, Gogotsi Y (2019) Science 366:969
66. Gai J, Ma F, Zhang Z, Sun D, Jin Y, Guo Y, Kim W (2019) ACS Sustain Chem Eng 7:15896–15903
67. Han L, Lehmann ML, Zhu J, Liu T, Zhou Z, Tang X, Heish C-T, Sokolov AP, Cao P, Chen XC, Saito T (2020) Front Energy Res 8:1–19
68. Adebahr J, Byrne N, Forsyth M, MacFarlane DR, Jacobsson P (2003) Electrochim Acta 48:2099–2103
69. Appetecchi GB, Passerini S (2000) Electrochim Acta 45:2139–2145
70. Kumar B (2004) J Power Sources 135:215–231
71. Zhao Y, Huang Z, Chen S, Chen B, Yang J, Zhang Q, Ding F, Chen Y, Xu X (2016) Solid State Ionics 295:65–71
72. Meyer M, Vechambre C, Viau L, Mehdi A, Fontaine O, Mourad E, Monge S, Chenal J-M, Chazeau L, Vioux A (2014) J Mater Chem A 2:12162–12165
73. Vélez JF, Procaccini RA, Aparicio M, Mosa J (2013) Electrochim Acta 110:200–207
74. Yuan C, Li J, Han P, Lai Y, Zhang Z, Liu J (2013) J Power Sources 240:653–658
75. Croce F, Persi L, Scrosati B, Serraino-Fiory F, Plichta E, Hendrickson MA (2001) Electrochim Acta 46:2457–2461
76. Tan R, Gao R, Zhao Y, Zhang M, Xu J, Yang J, Pan F (2016) ACS Appl Mater Interfaces 8:31273–31280
77. Zhu P, Yan C, Dirican M, Zhu J, Zang J, Selvan RK, Chung C-C, Jia H, Li Y, Kiyak Y, Wu N, Zhang X (2018) J Mater Chem A 6:4279–4285
78. Marcinek M, Bac A, Lipka P, Zalewska A, Zukowska G, Borkowska R, Wieczorek W (2000) J Phys Chem B 104:11088–11093
79. Fan J, Fedkiw PS (1997) J Electrochem Soc 144:399–408
80. Hanson B, Pryamitsyn V, Ganesan V (2013) ACS Macro Lett 2:1001–1005
81. Murugan R, Thangadurai V, Weppner W (2007) Angew Chem Int Ed Engl 46:7778–7781
82. Tao X, Liu Y, Liu W, Zhou G, Zhao J, Lin D, Zu C, Sheng O, Zhang W, Lee HW, Cui Y (2017) Nano Lett 17:2967–2972
83. Le HTT, Kalubarme RS, Ngo DT, Jadhav HS, Park C-J (2015) J Mater Chem A 3:22421–22431
84. Liu W, Liu N, Sun J, Hsu PC, Li Y, Lee HW, Cui Y (2015) Nano Lett 15:2740–2745
85. Chen XC, Sacci RL, Osti NC, Tyagi M, Wang Y, Palmer MJ, Dudney NJ (2019) Mol Syst Des Eng 4:379–385
86. Goodenough JB, Hong HYP, Kafalas JA (1976) Mater Res Bull 11:203–220
87. Peng J, Xiao Y, Clarkson DA, Greenbaum SG, Zawodzinski TA, Chen XC (2020) ACS Appl Polym Mater 2:1180–1189
88. Azadmanjiri J, Wang J, Berndt CC, Yu A (2018) J Mater Chem A 6:3824–3849
89. Xu W, Wang Z, Shi L, Ma Y, Yuan S, Sun L, Zhao Y, Zhang M, Zhu J (2015) ACS Appl Mater Interfaces 7:20678–20686
90. Xiao W, Gong Y, Wang H, Liu J, Yan C (2016) New J Chem 40:8778–8785
91. Moon J, Jeong JY, Kim JI, Kim S, Park JH (2019) J Power Sources 416:89–94
92. Zhao H, Deng N, Kang W, Cheng B (2020) Chem Eng J 390
93. Jeong H-S, Lee S-Y (2011) J Power Sources 196:6716–6722
94. Liu H, Xu J, Guo B, He X (2014) Ceram Int 40:14105–14110
95. Jiang X, Zhu X, Ai X, Yang H, Cao Y (2017) ACS Appl Mater Interfaces 9:25970–25975
96. Kim M, Park JH (2012) J Power Sources 212:22–27
97. Zhu X, Jiang X, Ai X, Yang H, Cao Y (2016) J Membr Sci 504:97–103
98. Liu Y, Ai K, Lu L (2014) Chem Rev 114:5057–5115
99. Chen H, Lin Q, Xu Q, Yang Y, Shao Z, Wang Y (2014) J Membr Sci 458:217–224
100. Bai S, Zhu K, Wu S, Wang Y, Yi J, Ishida M, Zhou H (2016) J Mater Chem A 4:16812–16817
101. Li M, Wan Y, Huang J-K, Assen AH, Hsiung C-E, Jiang H, Han Y, Eddaoudi M, Lai Z, Ming J, Li L-J (2017) ACS Energy Lett 2:2362–2367
102. Thackeray MM, Wolverton C, Isaacs ED (2012) Energy Environ Sci 5:7854

103. Zhu C, Han K, Geng D, Ye H, Meng X (2017) Electrochim Acta 251:710–728
104. Lu J, Chen Z, Pan F, Cui Y, Amine K (2018) Electrochem Energy Rev 1:35–53
105. Mai HD, Rafiq K, Yoo H (2017) Chemistry 23:5631–5651
106. Meng X (2017) J Mater Chem A 5:18326–18378
107. Nzereogu PU, Omah AD, Ezema FI, Iwuoha EI, Nwanya AC (2022) Appl Surf Sci Adv 9:100233
108. Sengodu P, Deshmukh AD (2015) RSC Adv 5:42109–42130
109. Wang H, Yao C-J, Nie H-J, Wang K-Z, Zhong Y-W, Chen P, Mei S, Zhang Q (1922) J Mater Chem A 8(2020):11906–11911
110. Cheng N, Ren L, Xu Z, Du Y, Dou SX (2020) Mater Today Phys 15
111. Yang L, Wang S, Mao J, Deng J, Gao Q, Tang Y, Schmidt OG (2013) Adv Mater 25:1180–1184
112. Guo Y, Zeng X, Zhang Y, Dai Z, Fan F, Huang Y, Zhang W, Zhang H, Lu J, Huo F, Yan Q, Appl ACS (2017) Mater Interfaces 9:17172–17177
113. Yu L, Liu J, Xu X, Zhang L, Hu R, Liu J, Yang L, Zhu M (2017) ACS Appl Mater Interfaces 9:2516–2525
114. Zhang F, Wang C, Huang G, Yin D, Wang L (2016) J Power Sources 328:56–64
115. Jung J-W, Cho S-H, Nam JS, Kim I-D (2020) Energy Storage Mater 24:512–528
116. Chen YT, Zhang HY, Chen YM, Qin G, Lei XL, Liu LY (2018) Mater Sci Forum 913:818–830
117. Lian J, Wang X, Zhang W, Huang Y, Xia T, Lian Y (2016) J Nanosci Nanotechnol 16:6494–6497
118. Ramasubramanian B, Sundarrajan S, Chellappan V, Reddy MV, Ramakrishna S, Zaghib K (2022) Batteries 8
119. Wu F, Liu J, Li L, Zhang X, Luo R, Ye Y, Chen R (2016) ACS Appl Mater Interfaces 8:23095–23104
120. Zheng F, Deng Q, Zhong W, Ou X, Pan Q, Liu Y, Xiong X, Yang C, Chen Y, Liu M (2018) ACS Sustain Chem Eng 6:16399–16411
121. He W, Guo W, Wu H, Lin L, Liu Q, Han X, Xie Q, Liu P, Zheng H, Wang L, Yu X, Peng DL (2021) Adv Mater 33:e2005937

Chapter 15
Metal Oxide Nanofiller-Introduced Polymer-Based Nanocomposite Dielectrics for Advanced Energy Storage Devices

Pravati Panda, Subhendu Chakroborty, Anchit Modi, and Srikanta Moharana

Abstract The modern scientific community is very interested in using polymeric nanodielectrics in capacitors and other forms of improved energy storage devices. Hybridization of inorganic metal or metal oxide NPs as nanofillers with a high dielectric constant and the polymer as a matrix with a high breakdown strength is fundamental in fabricating polymer nanodielectric films. Findings from this study show, for example, that a higher NP filler concentration results in a higher dielectric permittivity and a lower dielectric loss in the nanodielectric film, making it an attractive candidate for use in the fabrication of high-voltage storage capacitors. This chapter provides a concise overview of the methods for preparing highly flexible polymer nano dielectrics with enhanced thermostability for the fabrication of electronic and microelectronic devices, including using various metal and metal oxide NPs as cores and a variety of polymer matrices as shells.

Keywords Metal oxide · Nanofiller · Polymer · Nanocomposites · Energy storage

15.1 Introduction

Nowadays, there is a continuous depletion of energy sources like fossil fuels worldwide; hence there is a huge demand for the generation and storage of energy using renewable energy sources [1, 2]. Currently, most of the research development sectors

P. Panda
Department of Basic Sciences, RIE, Bhubaneswar, Bhubaneswar, India

S. Chakroborty (✉) · A. Modi
Department of Basic Sciences, IITM, IES University, Bhopal, Madhya Pradesh 462044, India
e-mail: drsubhendu.chakroborty@iesuniversity.ac.in

S. Moharana
School of Applied Sciences, Centurion University of Technology and Management, Bhubaneswar, Odisha, India

S. Moharana et al. (eds.), *Emerging Nanodielectric Materials for Energy Storage*, Nanostructure Science and Technology, https://doi.org/10.1007/978-3-031-40938-7_15

focus on generating plenty of energy, particularly electrical energy using tidal, wind, and solar energy resources, etc. To store the generated electricity, there is a need for highly reliable and efficient energy storage devices such as capacitors [3–5]. In this regard, dielectric capacitors in energy storage devices are a better alternative than electrochemical capacitors and batteries because of their long lifetime, high power density with high-voltage tolerance capability, and higher cycling stability [6]. All these interesting characteristics associated with dielectric capacitors make them a promising material of choice for renewable energy storage and distribution systems [7]. Capacitors possessing high dielectric constants are much needed for cable insulation, electrocaloric cooling, etc. Marketed dielectrics are generally polymer or ceramic-based [8]. Though ceramic dielectrics have their advantages, it has some limitations, such as high density, low breakdown strength with poor flexibility, low-temperature tolerance capability, etc. [9, 10]. Thus, there is a need for another kind of material that will possess opposite characteristics to the traditionally used materials.

In this connection, the scientific community suggested polymeric dielectrics as an alternative energy storage device. The associated properties, like high-voltage endurance with dielectric strength and low resistance with less dielectric loss, made dielectric polymers an incomparable material of choice for a wide range of electrical insulation and energy storage applications [8–11]. As for wire and cable insulation, the polymer coating protects the internal shielding and conductors from mechanical, external moisture, ultraviolet (UV), and ozone damage [4, 5]. For tape and capacitor insulation, polymer films/sheets work as fundamental building blocks that provide electrical current handling capability and electrical charge build-up barriers [12]. Polymeric core-shell devices containing nanofillers, generally known as nanodielectrics are highly advantageous and quite appealing for energy storage purposes and application in capacitors [13]. After many hits and trials, it was signified that the performance of metal or metal oxide-based NPs as nanofiller is more enthusiastic. Due to the small particle size and the high surface area of NPs [14–19], the bonding between polymer matrix and NPs is quite superior. Polymers with a strong polarizing tendency and exceptional effectiveness, such as poly (vinylidene fluoride) (PVDF), PVDF-based copolymers and terpolymers, and linear polymers like polyetherimide (PEI) and poly(methyl methacrylate) (PMMA), were chosen as a matrix for this application [5–7].

Moreover, it was verified that, for metallic/polymer NP dielectrics, field electron emission from metallic NPs under a high field tends to decrease the dielectric breakdown strength and, thus, increase the electronic conduction [1–3]. The charming features associated with nanodielectrics include high energy density, high capacitance density, high voltage, high current handling capability, high thermal conductivity, high temperature, lightweight, and environmental reliability. The addition of nanofillers impressively increased the flexibility, dielectric constant, breakdown strength, and thermostability with low dielectric loss of the polymer matrix [20].

This chapter briefly summarizes the preparation and application of different metal or metal oxide-doped polymer-based nanodielectrics in different energy storage devices.

15.2 Application

In order to study its potential against HV insulating systems like HVDC cables, ZnO nanoparticles were selectively distributed on polyethylene (PE) mix and thermoelastic polymer (TPE) comprising polystyrene-b-poly(ethylene-co-butylene)-b-polystyrene grafted maleic anhydride (SEBS-MA). The generated blended nanocomposites of PE/SEBS-MA/ZnO were analyzed and compared with a homopolymeric PE–ZnO nanocomposite for their mechanical and dielectric characteristics. Agglomeration of metal oxide NPs made them less compatible with the polyolefin matrix, leading to a compatibility problem. ZnO NPs were observed to be solely restricted inside the polymeric lattice and to be situated at the interface between SEBS-MA and PE, but in the case of SEBS-MA/ZnO and PE/SEBS-MA/ZnO, improved compatibility between SEBS-MA and the NPs occurred, resulting in good dispersion of the NPs. Additionally, PE/SEBS-MA/ZnO nanocomposites showed improved dielectric characteristics over PE/SEBS-MA nanocomposites because of the increased dispersion of ZnO NPs (unfilled blend). When compared with free PE/SEBS-MA, plane PE, and PE/ZnO nanocomposite, the blended nanomaterial showed reduced dielectric loss at higher power with temperatures 80 and 45 °C percent improved resistance to surface erosion value. Upon high loading of NPs, the blended nanomaterials exhibited decreased breakdown strength than unfilled material. Moreover, the blended nanocomposites displayed high mechanical flexibility with low reduction value due to the quality dispersion of the NPs (Figs. 15.1 and 15.2) [21].

Go/Au (prepared by doping of Au on GO) and GO–Cu (prepared by doping of Cu over GO) integrated over poly vinylidene fluoride (PVDF)-based new and flexible nanocomposites were generated following the solution casting process. The presence of GO/Cu and GO/Au nanofillers increased the quantity of electroactive crystalline phases of PVDF. Low dielectric loss with high dielectric constant and a high content of electroactive phases made it a suitable candidate for fabricating over flexible high-performance nanodielectric materials like electronic devices and sensors. An increase in the concentration of the nanofillers leads to a rise in the electroactive phase due to the presence of electrostatic interactions between the nanofiller and CH_2–CF_2 dipoles of PVDF, as confirmed by FTIR spectra. Along with this, it did not display any absorption band related to the non-polar α phase in the FTIR spectrum. For PVDF filled with 1% GO/Au nanofiller, the maximum value of the electroactive phase content was found to be 95% which was around 2.5 times greater than neat PVDF, as demonstrated through FTIR. Compared with GO/Cu, GO/Au-based nanocomposites possessed an increased number of electroactive phases, which XRD determined. From capacitance, inductance, and LCR (resistance) measurement, nanomaterials exhibited increasing nanofiller content with high dielectric constant with low dielectric loss, which made it an interesting material of choice to fabricate over simple and flexible high-performance nanodielectric materials (Figs. 15.3, 15.4 and 15.5) [22].

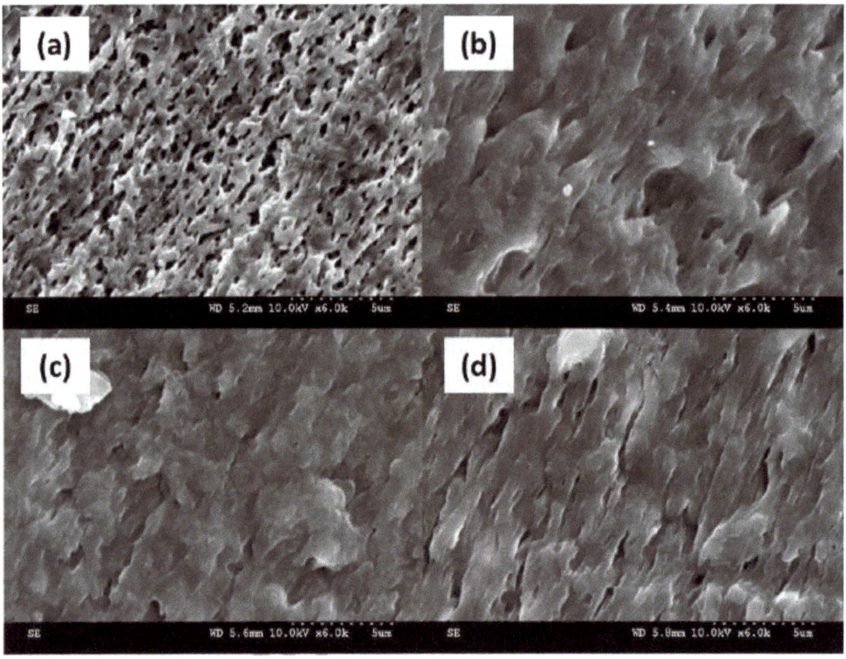

Fig. 15.1 SEM of unmodified PE–SEBS-MA (**a**), PE–SEBS-MA/1 (**b**), PE–SEBS-MA/5 (**c**), and PE–SEBS-MA/10 (**d**) composites. Reprint and Copyright permission obtained from Elsevier publication [21]

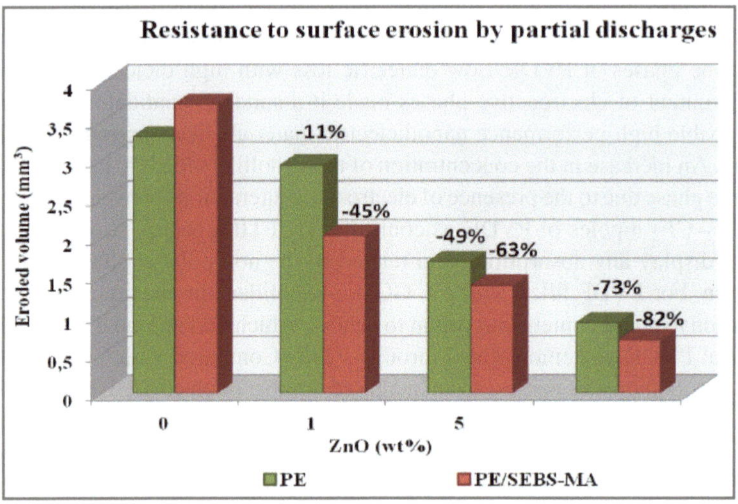

Fig. 15.2 Resistance to surface erosion through PE–ZnO as a function of PE–SEBS-MA–ZnO nanocomposites. Reprint and Copyright permission obtained from Elsevier publication [21]

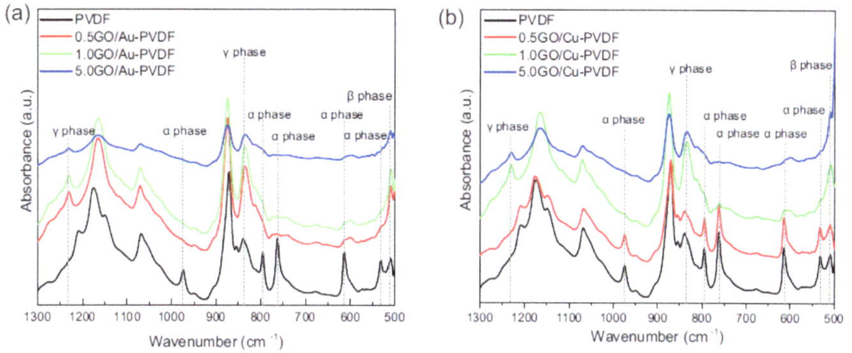

Fig. 15.3 FTIR spectra of neat PVDF and PVDF-based nanocomposites with different contents of GO–Au (**a**) and GO–Cu nanofiller (**b**). Reprint and Copyright permission obtained from Elsevier publication [22]

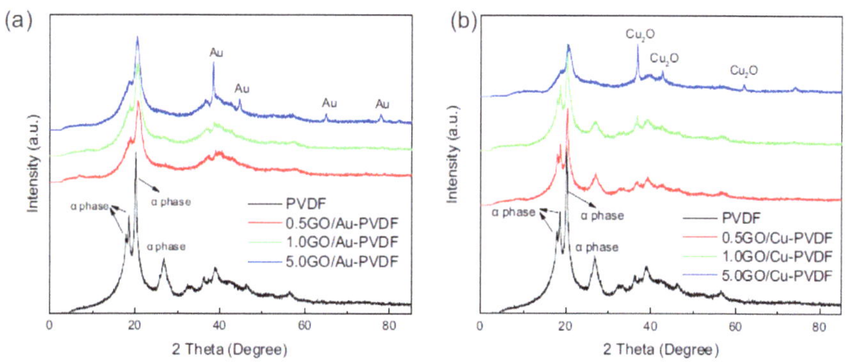

Fig. 15.4 XRD spectra of pristine PVDF and PVDF-based nanocomposites with different content of **a** GO–Au and **b** GO–Cu nanofiller. Reprint and Copyright permission obtained from Elsevier publication [22]

Dielectric polymeric nanocomposite matrices were prepared by the embedment of molybdenum disulfide (MoS_2) and polydopamine (PDA)–encapsulated hydroxylated barium titanate (BTH) over [poly (arylene ether nitrile)] (PEN) matrix. The nanofillers, i.e., MoS_2/PDA@BTH-1 were prepared through polydopamine deposition technology involving H-bonding interactions. The introduction of PDA contributed to the formation of nanocapacitor networks and multiple interfaces, enhancing interfacial polarization. Experimentally, it was found that the polymeric nanocomposite matrix exhibited a dielectric constant value of 17.3 at 1 kHz upon 15 wt% MoS_2/PDA–BTH-1 loading. The matrix also maintained low dielectric loss at the meant time. The plausible mechanism of the synergistic effect of MoS_2/PDA–BTH-1 upon the enhancement of the dielectric properties of the composite matrix

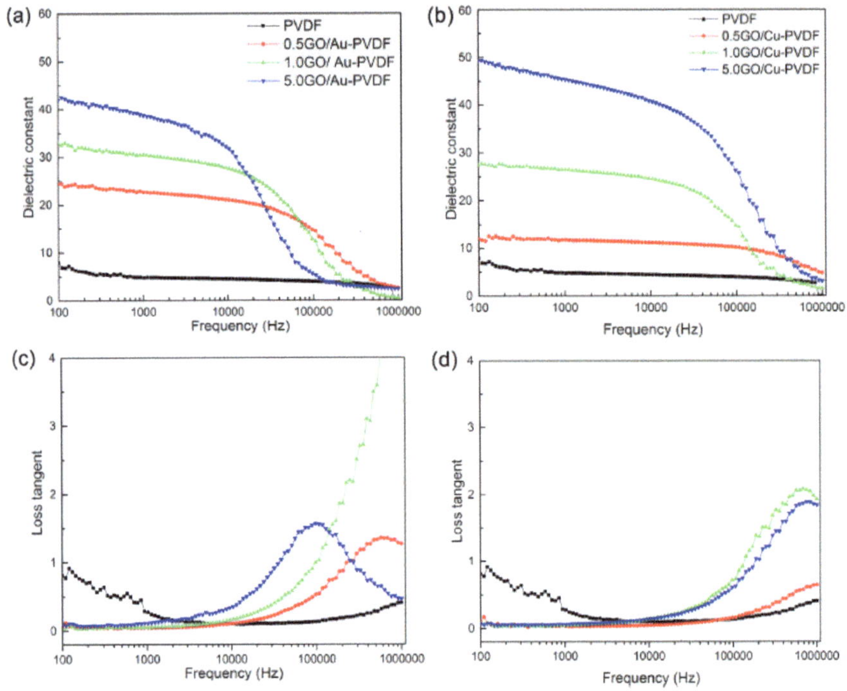

Fig. 15.5 Frequency dependence of dielectric constant and dielectric loss of pristine PVDF nanocomposites with different contents of GO–Au (**a**, **c**) and GO–Cu (**b**, **d**) nanofillers. Reprint and Copyright permission obtained from Elsevier publication [22]

was proposed, which was assumed that the dielectric constant value becomes stabilized upon reaching 158 °C as suggested by temperature-permittivity data. Thus, the developed film could be used in capacitors up to a temperature 150 °C, which was quite higher than that of most polymeric capacitors. The high-temperature tolerance nature and high dielectric constant (k) value of the developed material made it a demandable material for application in embedded devices as well as electrostatic capacitors (Figs. 15.6 and 15.7) [23].

Polymeric nanocomposite films containing carboxymethyl cellulose (CMC)-polyvinyl alcohol (PVA) (30/70 wt%) blend and Au NPs (0.36 wt%) as host polymer matrix were synthesized through the casting of aqueous solutions. The required Au NPs for synthesizing the polymeric matrix were successively generated from the leaf extract of *Morus nigra* (*M. nigra*). It was then irradiated under a nanosecond laser at a different time interval. The presence of functional groups in the CMC/PVA polymer interacted through H-bonding interactions, and simultaneously, the intensity of the spectral lines significantly increased upon irradiation and the addition of Au nanoparticles, as implied by FTIR. An increase in the absorbance intensity leading to a blue shift and appearance of SPK peak in the UV-Visible spectra confirmed the well dispersion of Au MPs upon the polymeric matrix. The TEM study confirmed that

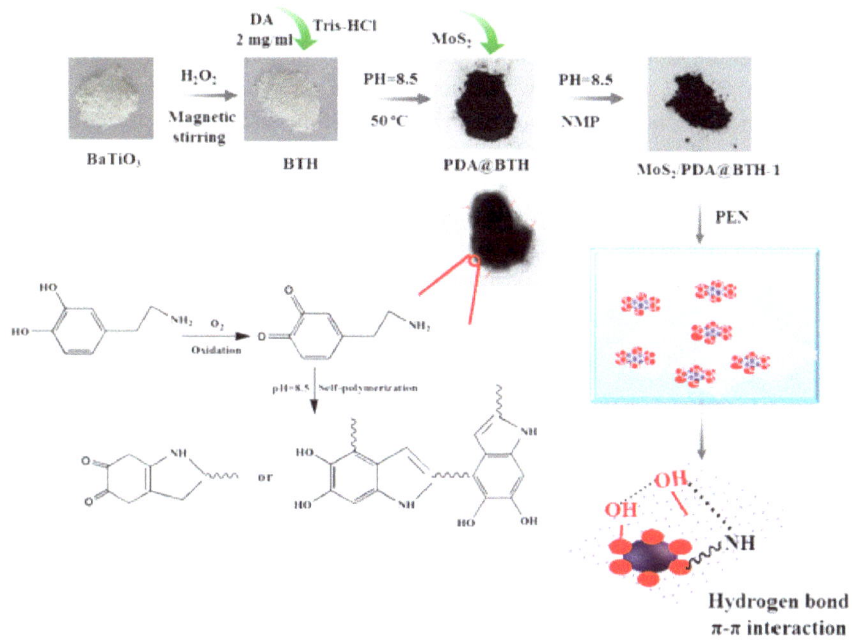

Fig. 15.6 Synthesis of MOS_2–PDA–BTH-1 and MOS_2–PDA–BTH-A–PEN nanocomposite film. Reprint and Copyright permission obtained from Elsevier publication [23]

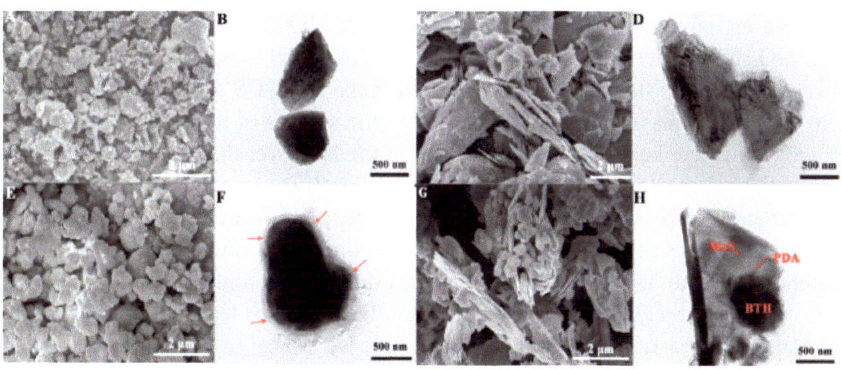

Fig. 15.7 SEM and TEM micrographs of BTH (**a**, **b**), MOS_2 (**c**, **d**), PDA@BTH (**e**, **f**), and MOS_2–PDA–BTH-1 (**g**, **h**) particles. Reprint and Copyright permission obtained from Elsevier publication [23]

laser irradiation reduced the particle size and spherical shape of Au NPs. It exhibited high dielectric constant value and dielectric permittivity, making it a valuable material for further application in optoelectronic and electronic devices [24].

Polymeric nanocomposite films were developed by the incorporation of different weight percentages of ZnO NPs upon poly (vinyl pyrrolidone) (PVP) and poly (ethylene oxide) (PEO) blend matrix. The developed blended matrix was studied for its application in preparing next-generation microelectronic and optoelectronic devices. The structure and morphology of the nano polymeric film were studied using SEM, FTIR, XRD, etc. It was found that the incorporation of different amounts of ZnO NPs inside the polymer matrix significantly influenced the polymer-polymer, porous spherulite morphology, semicrystalline structure, and polymer-nanoparticle interactions of the materials. UV-Vis spectra were used to determine the optical properties such as optical band gap, refractive index, Urbach tail energy, and ZnO surface plasmon resonance energy. Increasing the concentration of ZnO in the polymeric film decreased the optical band gap. DRS spectroscopy was used to study its implication in the flexible nanodielectrics from 20 Hz to 1 MHz. The permittivity of the complex was found to increase linearly up to 3 wt% ZnO loading with respect to an increase in temperature, then a slight decrease in the value was noticed at 5%. However, with increased temperature, the electrical conductivity and the dielectric relaxation time obeyed the Arrhenius manner. Experimentally, it was observed that the dielectric constant of the film increased upon loading of ZnO nanoparticles [25].

Dispersion of Fe_3O_4 NPs was carried out in one step over polyethylene oxide (PEO)/chitosan (CS) based nanocomposite using laser ablation at different time intervals. The as-produced nanocomposite film containing PEO/CS/Fe_3O_4 was analyzed properly using XRD, SEM, TEM, FTIR, and UV-Visible spectroscopy. Also, the authors measured the nanofilm's electrical conductivity and dielectric characteristics. The crystallinity of the nanocomposite film gets enhanced by the fabrication and well interaction of the PEO–CS matrix with Fe_3O_4 NPs. The average particle size of the Fe_3O_4 NPs generated under laser ablation was found to be 45 nm. FTIR spectra also confirmed the stable interaction present in the PEO–CS–Fe_3O_4 nanocomposite film. The appearance of the bright spots over the surface of the sample in the SEM image confirmed the successive dispersion of the NPs over the nanocomposite film. The optical characteristic of the generated material was confirmed by the appearance of an absorption peak at 218 nm, which shifted toward a higher wavelength region with respect to the rise in laser ablation time, also simultaneous decrease in the energy gap to 4.43 eV from 5.35 eV for direct transition and to 2.49 eV from 4.91 eV for indirect transition. It was also studied that the electrical conductivity of the sample increased with a simultaneous increase in Fe_3O_4 NPs. PEO–CS–Fe_3O_4 NPs achieved a high conductivity value of $\sim 8.47 \times 10^{-6}$ S/cm at a laser ablation time of 20 min. The developed nanomaterial exhibited high electrical conductance, which can be applied in different electronics (Figs. 15.8 and 15.9) [26].

Dispersion of carbon fabricated Ni NPs was carried out over epoxy-based photoresist material (SU-8), doped over metal polymer composites (MPC), and used in-depth high-voltage dielectric characterization. The integration of MPC involved three stages: deagglomeration, surface functionalization, and dispersion in a polymeric matrix following the sonochemical process. In the end, the MPC films were deposited over silicon wafers through spin-coating and finally used to fabricate metal–insulator–semiconductor (MIS) capacitors. The dielectric permittivity was observed to

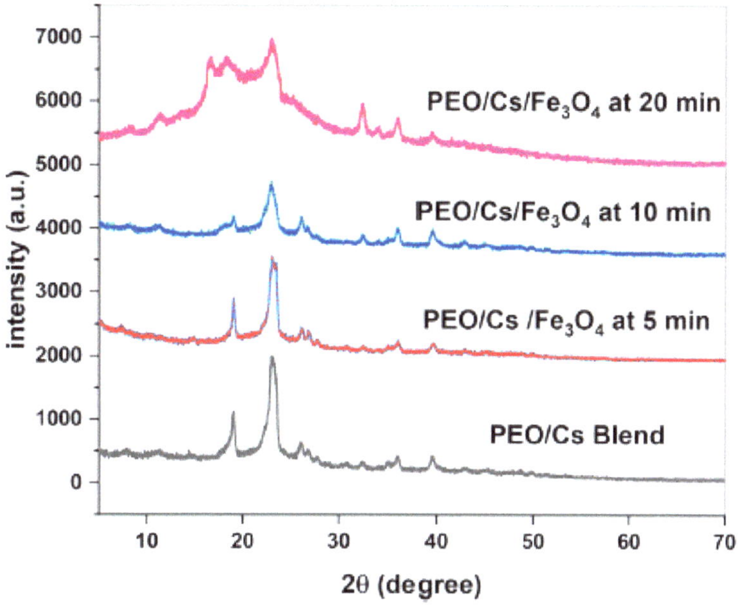

Fig. 15.8 XRD spectra of PEO–CS blend and PEO–CS–Fe_3O_4 blend at various laser ablation times. Reprint and Copyright permission obtained from Elsevier publication [26]

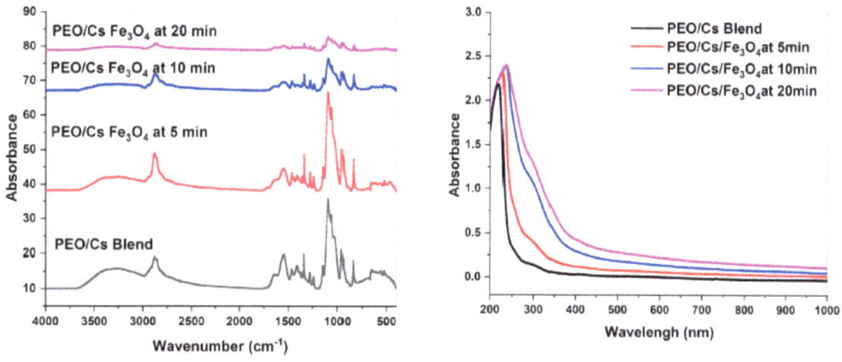

Fig. 15.9 FTIR and UV-Vis spectra of PEO–CS and PEO–CS–Fe_3O_4 blend with various ablation times. Reprint and Copyright permission obtained from Elsevier publication [26]

be enhanced to approximately 116% with 0.32 loss at 5 kHz. The measured high-voltage dielectric properties indicated the $2e^-$ field mechanism without electrical aging or dielectric breakdown less than ±0.38 MV/cm. The developed wafer-scale film fabrication process of MPCs was quite reliable and can be applied to manufacture pleasant miniaturization of high-voltage capacitors [27].

The increasing variety of uses for hybrid nanocomposites made from organic and inorganic elements has increased their popularity in recent years. The in-situ bulk polymerization method was utilized to synthesize [poly(methyl methacrylate)] (PMMA) and its binary and ternary nanocomposites using varying amounts of rGO and Fe_2O_3 NPs. PMMA; based ternary nanocomposites comprising 2:2 wt% of RGO:Fe_2O_3 NPs showed the greatest dielectric constant value up to 308 and also demonstrated a dielectric loss of 0.12 at 25 Hz, according to the results of the dielectric characteristics study conducted in the frequency range of 25 Hz to 1 MHz. The dielectric characteristics of ternary PMMA nanocomposites were predicted to be superior to those of binary nanocomposites containing only one interface, possibly as a result of the accumulation of a greater number of charges at the latter's interface. The synergistic reduction in thermal resistance of together Fe_2O_3 NPs and rGO (2:2 wt%) compared with 2 wt% of PMMA–RGO and PMMA–Fe_2O_3-based binary nanocomposites at 1.04 W/m K and 0.98 W/m K, respectively, resulted in an increased thermal conductivity value of 2.04 W/m K in the ternary nanomaterials. Therefore, the ternary nanocomposite showed promise as a heat-control mechanism component. The nanomaterial spectra data indicated a robust interaction of the nanofillers with the PMMA matrix [28].

Cast synthesis was used to create bendable nanocomposite films using carboxymethyl cellulose (CMC), polyethylene oxide (PEO), and ZnO/GO NPs as nanofiller. Nanocomposite films were analyzed for their shape, crystallinity, and structure using XRD and FTIR spectra, which showed that ZnO/GO NPs interacted effectively with the PEO/CMC mixture. The optical properties, including Urbach energy and energy band gap of the nanocomposite and pure PEO/CMC blend, were measured by UV-Vis spectroscopy. It was studied that increasing the weight percentage of ZnO–GO NPs decreased the energy band gap with an increase in the Urbach energy. AC-impedance spectroscopy was used to compute the ionic conductivity of the polymer nanocomposites. For PEO–CMC composite film, the calculated ionic conductivity was found to be $\sim 10^{-9}$ S/cm, which increased to $\sim 10^{-7}$ S/cm by adding 8% ZnO–GO nanofiller at room temperature. Increasing the frequency increased the AC electrical conductivity; however, it reduced the values of ε' and ε'' of PEO–CMC and ZnO–GO nanocomposites. Surprisingly, increasing the concentration of ZnO–GO NPs increased the ε' and ε'' values of the nanocomposites. Hence, PEO/CMC nanocomposite films prepared by using 8 wt% ZnO–GO NPs with enhanced dielectric, optical, and electrical properties can further be employed for different industrial purposes like flexible energy storage devices, i.e., optoelectronic and electric devices [29].

(CMC/PAM)/$Li_4Ti_5O_{12}$-based polymeric nanocomposite films were prepared by the fabrication of $Li_4Ti_5O_{12}$ NPs with particle size <55 nm over a carboxymethyl cellulose/polyacrylamide (CMC–PAM) blend matrix involving a solution casting process. The optical, dielectric, structural, mechanical, and thermal properties of the nanocomposites were determined through various spectral studies. The existence of coordination and hydrogen bonds in the nanocomposite samples with increased CMC/PAM structural amorphous areas was revealed through FTIR and XRD. The absorption maximum of the nanocomposite samples was significantly enhanced by the reduction of the energy band gap which was determined by UV-Vis spectra. The inclusion of $Li_4Ti_5O_{12}$ NPs, fine dispersion, and well interaction between the NPs and polymer enhanced the thermal stability of the CMC/PAM at a high temperature which was proved by the TGA study. Moreover, the addition of nanofillers improved the mechanical properties of the nanocomposite. Tuning of the dielectric parameters like modulus spectra and dielectric permittivity of the nanocomposite films can be achieved by tuning of $Li_4Ti_5O_{12}$ NP content in the CMC/PAM blend which illustrates its novel applicability in the biodegradable polymer nanodielectrics. The advantageous features like the development of polymer electrolyte, optical band gap tuner, and controllable dielectric permittivity etc., associated with the nanocomposite made it a suitable material for application in high-density energy storing/advanced microelectronic devices [30].

Polymer-perovskite-based hybrid nanocomposites are highly demanding because their integration into electronic and optoelectronic devices enhances the physical and chemical properties of the target material. $SrTiO_3$ NPs fabricated PVA–CMC-based nanocomposites were synthesized following the solution casting method. Increasing the content of $SrTiO_3$ NPs decreased the degree of crystallinity of the nanocomposites, which was demonstrated through XRD, whereas it enhanced the optical characteristics of the material as studied by UV-Vis spectra. An FTIR spectrum was used to study the influence of the $SrTiO_3$ nanoparticles upon the functional groups of the PVA–CMC hybrid system. SEM images confirmed a rise in the surface roughness with respect to the increase in the NP content. The authors also examined the dielectric dispersion and relaxation characteristics of the prepared PNC films, which signified that the polymeric nanocomposites possessed a high value of dielectric constant with a low dielectric loss than pure PVA–CMC blend. Thus, the observed outcomes depicted its further potential attribution in advanced optoelectronic devices [31].

Nanocomposites containing poly (methyl methacrylate) (PMMA) matrix and SnO_2, ZnO, and TiO_2 as nanofillers were designed and synthesized by a sonicated suspension of NPs in polymeric solutions following the casting process. DRS, XRD, and UV-Vis techniques characterized the prepared nanomaterial well. The successive formation of the nanocomposite with appropriate composition was confirmed by XRD and SEM studies. The dielectric permittivity of the PNC was also demonstrated. The band gap and the UV-Vis absorbance of the material were found to be tunable as characterized by UV-Vis. For a particular nanofiller concentration, maximum absorbance was displayed by PMMA/TiO_2 films, and it was decreased for PMMA/SnO_2 films. For equal PMMA/TiO_2 composite composition, increasing the thickness of the material reduced the energy band gap with an enhancement of UV-Vis absorbance, as observed by the thickness study. For nanocomposites containing SnO_2 and ZnO as nanofiller, the high dielectric permittivity value was found to be directly proportional to the concentration of the respective nanofillers; however, for TiO_2 NP-containing films, a decrease in the permittivity value was noticed when compared with the host matrix film at 30 °C. Moreover, enhancing the experimental electric field frequency to 1 MHz from 20 Hz gradually decreased the permittivity of the nanocomposite. Hence, the associated versatile properties make the nanocomposite film a promising candidate for further application in the development of different advanced organoelectronic, nanodielectric insulators, and optoelectronic devices and to architect the coming generation of flexible electronic devices [32].

Polymeric nanocomposites (PNCs) are considered the key moiety present in the flexible type of advanced microelectronic devices based on polymer engineering and technology. A three-phase hybrid nanocomposite consisting of SiO_2 NPs dispersed over PNC films of poly (vinylidene fluoride)/poly (ethylene oxide) (PVDF–PEO) blend was designed, synthesized, and characterized through various spectral methods. Experimentally, increasing the concentration of SiO_2 NPs decreased the relative fraction of PVDF β-phase crystals and degree of crystallinity and simultaneously modified the spherulitic morphology of the PVDF/PEO blend matrix. Moreover, the loading of 5 wt% SiO_2 nanofiller enhanced the dielectric permittivity and improved the dielectric constant at a frequency range from 20 Hz to 1 MHz. The tunable dielectric and structural properties made them a suitable candidate for successive application in appealing polymeric nanodielectrics, including organoelectronic and energy storage devices [33].

Flexible dielectric nanocomposites with excellent energy storage properties make it an alternative choice able material for its potential application in electrostatic capacitors. With an intention to eliminate the inconveniences stimulated by the interfacial polarization of two different components in the nanocomposite, authors prepared PVDF–HFP/HfO_2–BT nanocomposites by the fabrication of HfO_2–BT NPs over poly (vinylidene fluoride-co-hexafluoropropylene) film following solution casting process. Due to the high resistivity and moderate dielectric permittivity nature of the HfO_2, it acts as a buffering barrier. Its inclusion in the nanocomposite reduces dielectric mismatch between the P (VDF-HFP) matrix and BT NPs. The addition of nanofiller substantially enhanced the energy density (U_e) and breakdown strength (E_b) of the nanocomposite. The maximum U_e value of 10.7 J/cm^3 was exhibited by P (VDF-HFP)–HfO_2–BT nanocomposites at 450 MV/m, around 36% and 59% greater than that of nanocomposite filled by only BT nanoparticles and also that of pristine P (VDF-HFP). Additionally, it possessed a charge–discharge efficacy of 72%, which was quite superior to the reported values. It is concluded that adding nanofillers containing a passivation layer might be an effective route to enhance the energy storage capacity of the flexible nanodielectrics [34].

Polymeric nanocomposites were prepared by the embedment of MWCNTs/Au NPs over PEO (polyethylene oxide) samples following the casting process. The interaction between the polymer and the nanoparticles was confirmed by FTIR spectra. Reduction in the crystallinity phase of PEO due to the incorporation of MWCNTS-Au NPs as well as the semicrystalline nature of the nanocomposite was confirmed through XRD. MWCNTs having a diameter of 10–25 nm and the spherical shape of the NPs with 2–25 nm particle size was predicted by TEM analysis. The absorption maximum shifted toward the red region, which indicated the decrease in the energy band gap with good reactivity between the polymer matrix and the nanofillers, as calculated by Tauc's relation. An increase in the thermal stability of the nanocomposite was predicted by TGA analysis. Dielectric spectroscopy was used to measure the dielectric and electric spectra of the samples, which suggested its favorable application in the production of electroactive materials as well as for the preparation of electrical insulating polymeric nanodielectrics [35].

PEO/PMMA/MMT-based hybrid polymeric nanocomposite (HPNC) films were designed and synthesized by the dispersion of different composition ratios of organic (PEO and PMMA) and inorganic (MMT, montmorillonite) polymer blend matrices via solution-cast process followed by the melt-press method. The characteristics, properties, and structure of the polymer blend were analyzed by different spectral studies, i.e., SEM, XRD, DSC, dielectric spectroscopy, UV-Vis, etc. The authors also explored the influence of MMT nanofiller concentration and PEO/PMMA blend compositional ratio upon the degree of crystallinity, melting temperature of PEO crystallites, homogeneity, and variation in intercalated MMT structures, etc. It was demonstrated that the quantity of the constituents along with polymer–nanofiller, and polymer–polymer interphases in the HPNC and PB materials had a significant influence upon the UV-Vis absorbance with energy band gap, electrical conductivity, dielectric dispersion, and relaxation processes, and other optical parameters of the nanocomposites. Loading of MMT with PMMA blending enhanced the thermal stability of the PEO. It exhibited high dielectric permittivity in the lower frequency range, i.e., from 10^5 to 2×10^1 Hz, mostly due to the interfacial polarization that made them applicable in the flexible and tunable dielectrics [36].

5 wt% of different metal oxides (Al_2O_3, SnO_2, TiO_2, ZnO) NPs were used as nanofillers and were fabricated over PVDF/PEO (75/25 wt/wt%) blend considered as host to generate polymeric nanocomposite (PNC) films following casting process. The as-prepared nanocomposite films were successively characterized by XRD, FTIR, SEM, UV-Vis, DRS, etc. The influence of nanofillers upon the dielectric, optical, and structural properties of the polymeric blend was discussed. The presence of metal oxide nanoparticles significantly altered the spherulitic morphology of the polymer, created a huge number of micro- to nano-sized pores, and lowered the degree of crystallinity of the host matrix and also the β-phase content of the PVDF and crystalline phase of PEO. At ambient temperature, among the used metal oxides, Al_2O_3 NPs significantly enhanced the dielectric permittivity, whereas other nano inclusions altered the MWS relaxation and dielectric polarization processes of the PNC films in the frequency range from 20 Hz to 1 MHz. Additionally, Al_2O_3 NPs exhibited high electrical conduction, and a considerably decreased value was noticed for ZnO nanoparticles containing PNC films [37].

PEO–PMMA–SiO_2 and PEO–PMMA–SnO_2-based nanocomposite films were prepared by blending SiO_2 and SnO_2 nanofillers over PEO/PMMA polymeric matrix under ambient temperature. Experimentally, it was confirmed that, with respect to the increase in the concentration of the nanofillers and the applied electrical field frequency, the dielectric permittivity of the films gets decreased to 2.8 from 3.4. They also explored the effect of various NP sizes and dielectric constants on the dielectric permittivity of the film. The results suggested a relatively low nanodielectric loss with a dipolar relaxation at high frequency; however, the radiofrequency electrical conductivity of the HPNCs was increased linearly to 10^{-4} S/cm from $\sim 10^{-8}$ S/cm, and it was found to have not any connectivity with the concentration of the nanofiller. Hence, the film's high dielectric and electrical properties made it a material of choice for the development of radiofrequency operative flexible-type electrical and electronic components and devices [38].

Hybrid nanocomposite films were prepared by the dispersion of the 5 wt% of different metal oxides NPs like Al_2O_3, ZnO, TiO_2, or SnO_2 over the PVDF/PEO blend matrix and were characterized by various spectral studies, including DSC and RF-IA. The degrees of the crystallinity of PVDF and PEO were measured over HPNC through DSC, which were found to be dependent upon the physical characteristics of metal oxide NPs. With respect to the increase in the radio wave electric field frequency to 10^9 Hz from 10^6 Hz, a nonlinear decrease in the dielectric permittivity of the medium was observed. Among the NPs, Al_2O_3 exhibited a high dielectric permittivity value. The alternating current of the medium displayed linear progress with the frequency variation and was found not to have been influenced by the concentration of the NPs. Hence, the high dielectric constant with permittivity value, high density, lower particle size, and high flexibility made it a promising material for application in different nanodielectric insulators and technologically advanced radio frequency devices [39]. Highly functional core-shell NPs in the polymer (PVDF) were prepared using a cost-effective process and applied as nanodielectrics in different energy storage devices. For this Al NPs were chosen as the core NP, and a thin layer of Al_2O_3 worked as a capping shell and provided electrical insulation to the prepared polymeric film. It also prevented the agglomeration of Al NPs. The resulting polymeric film exhibited high dielectric permittivity, low dielectric loss, and high energy storage capacity [40].

Nanocomposite films of PVP/PEO/MoO_3 were prepared by casting MoO_3 nanoplates over PVP/PEO. XRD revealed that the addition of MoO_3 NPs upon the PVP/PEO blend increased the amorphous domain of the PNC films. The favorable interaction between PEO and PVP via H-bonding and the miscibility was confirmed by DFT/FTIR study. Also, it indicated a co-coordinative interaction existing between the C–O–C group of PEO/C=O group of PVP with MoO_3 NPs. Increasing the amount of MoO_3 NPs decreased the optical band gap of the PNC films and enhanced the electrical conductivity, as adding NPs increased the number of charge carriers. They also explored the dielectric dispersion and relaxation process of the PNC films, which were explained mechanistically using a barrier hopping model. The relaxation process followed a non-Debye one [41].

BCZT-P (VDF-HFP)-based polymeric nanocomposites were synthesized using different volume concentrations of BCZT nanofiller and studied their dielectric energy storage efficacy. Sol-gel process was followed for the synthesis of BCZT nanopowder through the citrate precursor process. XRD and TEM analysis analyzed the structural and morphological properties. To get superior ceramic interface coupling, the surface of the nanopowder was functionalized by a different aromatic ligand, naphthyl phosphate (NPh), which was validated through XPS and TGA analysis. The dielectric constant value of ~155 was estimated for surface passivated BCZT NPs through the slurry technique; however, due to higher innate surface conductivity, the dielectric permittivity of pristine BCZT nanopowder could not be measured. The PNC films were dispersed by ceramic fillers via the solution casting method, which was examined by SEM. Nano BCZT/PVDF-HFP films modified by NPh ligands exhibited maximum energy storage capacity at 5% filler concentration than untreated nano BCZT-P (VDF-HFP) and even pure polymer films. 8.5 J cm^{-3} was found as the maximum energy storage density value at 10% filler concentration for the PNC films of thickness 10 μm [42].

15.3 Conclusion and Future Perspective

In conclusion, this chapter puts the limelight on the significance of polymer nanodielectrics in the development of cost-effective and innovative high-voltage storage capacitors. Fabrication of metal or metal oxide NPs as nanofillers greatly increased the dielectric constant and dielectric permittivity of the polymeric film, which was rather difficult to attain with ceramic material. In some cases, the nanofiller functionalization was done to achieve high dispersion quality and to get better interaction between the polymer and the nanofiller. It was also found to enhance the breakdown strength of polymeric material with low dielectric loss, thus providing stability by increasing the energy density of the film. Though the results are quite satisfactory, the research work still needs more polish. Rather than using the chemical preparation method of PNC films, the scientific community should adopt a greener and atom economical approach to designing and developing PNC films. Also, rather than using the metal or metal oxide NPs to fabricate over PNC films, people should use the one-step functionalization of the polymer nanocomposite by a suitable functionalizing agent to attain flexibility and high dielectric constant value.

Acknowledgements PP is thankful to the Department of Basic Sciences, RIE, Bhubaneswar, Odisha, India. SC is grateful to the School of Sciences, IES University, Bhopal, Madhya Pradesh, India.

Conflict of Interest The authors declare no conflict of interest.

References

1. Dang Z-M, Yuan J-K, Yao S-H, Liao R-J (2013) Flexible nanodielectric materials with high permittivity for power energy storage. Adv Mater 25:6334–6365
2. Siwal SS, Zhang Q, Devi N, Thakur VK (2020) Carbon-based polymer nanocomposite for high-performance energy storage applications. Polymers 12:505–535
3. Zhang G, Li Q, Allahyarov E, Li Y, Zhu L (2021) Challenges and opportunities of polymer nanodielectrics for capacitive energy storage. ACS Appl Mater Interfaces 13:37939–37960
4. Tan DQ (2020) The search for enhanced dielectric strength of polymer-based dielectrics: a focused review on polymer nanocomposites. J Appl Polym Sci 137:49379
5. Cheng L, Chi X, Yan C, Xie D, Liu X, Wen Y, Liu W, Li S (2018) Polypropylene nanocomposite for power equipment: a review. IET Nanodielectrics 1:92–103

6. Barber P, Balasubramanian S, Anguchamy Y, Gong S, Wibowo A, Gao H, Ploehn HJ, zur Loye H-C (2009) Polymer composite and nanocomposite dielectric materials for pulse power energy storage. Materials 2:1697–1733
7. Hu J, Zhang S, Tang B (2021) 2D filler-reinforced polymer nanocomposite dielectrics for high-k dielectric and energy storage applications. Energy Storage Mater 34:260–281
8. Huang X, Sun B, Zhu Y, Li S, Jiang P (2019) High-k polymer nanocomposites with 1D filler for dielectric and energy storage applications. Prog Mater Sci 100:187–225
9. Zhang G, Allahyarov E, Zhu L (2018) Polymer nanodielectrics: current accomplishments and future challenges for electric energy storage. In: Nano/micro-structured materials for energy and biomedical applications. Springer, pp 1–48
10. Wei J, Zhu L (2020) Intrinsic polymer dielectrics for high energy density and low loss electric energy storage. Prog Polym Sci 106:101254
11. Pirzada BM, Sabir S (2018) Polymer-based nanocomposites for significantly enhanced dielectric properties and energy storage capability. In: Polymer-based nanocomposites for energy and environmental applications. Woodhead Publishing series in composites science and engineering, pp 131–183
12. Tan DQ (2020) Review of polymer-based nanodielectric exploration and film scale-up for advanced capacitors. Adv Func Mater 30:1808567
13. Luo H, Zhou X, Ellingfor C, Zhang Y, Chen S, Zhou K, Zhang D, Bowen CR, Wan C (2019) Interface design for high energy density polymer nanocomposites. Chem Soc Rev 48:4424–4465
14. Pal K, Chakroborty S, Panda P, Nath N, Soren S (2022) Environmental assessment of wastewater management via hybrid nanocomposite matrix implications—an organized review. Environ Sci Pollut Res 29:76626–76643
15. Panda P, Chakroborty S (2022) Optical sensor technology and its application in detecting environmental effluents: a review. J Environ Anal Chem. https://doi.org/10.1080/03067319.2022.2098480
16. Chakroborty S, Panda P (2022) Nanovaccinology against infectious disease. In: Nanovaccinology as targeted therapeutics. Wiley, pp 95–113
17. Panda P, Barik A, Unnamatla MVB, Chakroborty S (2021) Synthesis and antimicrobial abilities of metal oxide nanoparticles. In: Bio-manufactured nanomaterials. Springer, Cham, pp 41–58
18. Soren S, Panda P, Chakroborty S (2023) Nanotechnology in water and wastewater treatment. In: Agricultural and environmental nanotechnology. Springer, 127–143
19. Chakroborty S, Panda P, Unnamatla MVB, Chandra P, Varela-Guerreroc V (2023) Green nanotechnology research avenue in medicinal biology. In: Green nanoarchitectonics. Jenny Stanford Publishing, pp 167–195
20. Wu X, Chen X, Zhang QM, Tan DQ (2022) Advanced dielectric polymers for energy storage. Energy Storage Mater 44:29–47
21. Helal E, Pottier C, David E, Fréchette M, Demarquette NR (2018) Polyethylene/thermoplastic elastomer/zinc oxide nanocomposites for high voltage insulation applications: dielectric, mechanical and rheological behavior. Eur Polymer J 100:258–269
22. Fakhria P, Mahmood H, Jaleh B, Pegoretti A (2016) Improved electroactive phase content and dielectric properties of flexible PVDF nanocomposite films filled with Au- and Cu-doped graphene oxide hybrid nanofiller. Synth Met 220:653–660
23. Feng M, Li C, He M, Huang Y, Luo J (2020) Poly(arylene ether nitrile) ternary dielectric composites modulated via polydopamine-assisted $BaTiO_3$ decorating MoS_2 sheets. Ceram Int 46:19181–19190
24. Morsi MA, Asnag GM, Rajeh A, Awwad NS (2021) Nd:YAG nanosecond laser induced growth of Au nanoparticles within CMC/PVA matrix: multifunctional nanocomposites with tunable optical and electrical properties. Compos Commun 24:100662
25. Choudhary S (2018) Structural, optical, dielectric and electrical properties of (PEO–PVP)–ZnO nanocomposites. J Phys Chem Solids 121:196–209
26. Menazea AA, Ibrahium HA, Awwad NS, Moustapha ME, Farea MO, Bajaber MA (2022) Facile synthesis and high-performance dielectric properties of polyethylene oxide-chitosan-iron oxide nanocomposite for electrical applications. J Market Res 18:2273–2281

27. Alfonso MS, Lapeyronie C, Goubet M, Viala B, Tortai J-H (2021) Enhanced dielectric properties of epoxy-based photoresist nanocomposites using carbon-coated nickel nanoparticles for high voltage integrated capacitors. Compos Sci Technol 216:109063
28. Ul-Haq Y, Murtaza I, Mazhar S, Ullah R, Iqbal M, Zeeshan-ul-Huq, Qarni AA, Amin S (2020) Dielectric, thermal and mechanical properties of hybrid PMMA/RGO/Fe_2O_3 nanocomposites fabricated by in-situ polymerization. Ceram Int 46:5828–5840
29. Al-Harbi LM, Alsulami QA, Farea MO, Rajeh A (2023) Tuning optical, dielectric, and electrical properties of polyethylene oxide/carboxymethyl cellulose doped with mixed metal oxide nanoparticles for flexible electronic devices. J Mol Struct 1272:134244
30. Morsi MA, Abdelrazek EM, Ramadan RM, Elashmawi IS, Rajeh A (2022) Structural, optical, mechanical, and dielectric properties studies of carboxymethyl cellulose/polyacrylamide/lithium titanate nanocomposites films as an application in energy storage devices. Polym Testing 114:107705
31. Al-Muntaser AA, Pashameah RA, Sharma K, Alzahrani E, Hameed ST, Mors MA (2022) Boosting of structural, optical, and dielectric properties of PVA/CMC polymer blend using $SrTiO_3$ perovskite nanoparticles for advanced optoelectronic applications. Opt Mater 132:112799
32. Sengwa RJ, Dhatarwal P (2021) Polymer nanocomposites comprising PMMA matrix and ZnO, SnO_2, and TiO_2 nanofillers: a comparative study of structural, optical, and dielectric properties for multifunctional technological applications. Opt Mater 113:110837
33. Dhatarwal P, Sengwa RJ (2020) Tunable β-phase crystals, degree of crystallinity, and dielectric properties of three-phase PVDF/PEO/SiO_2 hybrid polymer nanocomposites. Mater Res Bull 129:110901
34. Chen C, Xie Y, Liu J, Li J, Wei X, Zhang Z (2020) Enhanced energy storage capability of P(VDF-HFP) nanodielectrics by HfO_2 passivation layer: preparation, performance and simulation. Compos Sci Technol 188:107968
35. Morsi MA, Rajeh A, Al-Muntaser AA (2019) Reinforcement of the optical, thermal and electrical properties of PEO based on MWCNTs/Au hybrid fillers: nanodielectric materials for organoelectronic devices. Compos B 173:106957
36. Sengwa RJ, Dhatarwal P (2022) Toward multifunctionality of PEO/PMMA/MMT hybrid polymer nanocomposites: promising morphological, nanostructural, thermal, broadband dielectric, and optical properties. J Phys Chem Solids 166:110708
37. Sengwa RJ, Dhatarwal P, Choudhary S (2020) A comparative study of different metal oxide nanoparticles dispersed PVDF/PEO blend matrix-based advanced multifunctional nanodielectrics for flexible electronic devices. Mater Today Commun 25:101380
38. Dhatarwal P, Sengwa RJ, Choudhary S (2022) Broadband radio frequency dielectric permittivity and electrical conductivity of dispersed tin oxide and silica nanoparticles in poly(ethylene oxide)/poly(methyl methacrylate) blend matrix-based nanocomposites for nanodielectric applications. J Macromol Sci 61:111–120
39. Sengwa RJ, Dhatarwal P (2022) Crystalline phases thermal behaviour and radio frequencies dielectric properties of PVDF/PEO/metal oxides hybrid polymer nanocomposite films. J Polym Res 5
40. Badi N, Mekala R, Herdandez FR (2013) Synthesis of Al–Al_2O_3/PVDF core shell nanodielectrics for energy storage applications. In: Technical proceedings of the 2013 clean technology conference and trade show. TechConnect Briefs, pp 330–333
41. Al-Muntaser AA, AlSaidi RAM, Sharma K, Alamri HR, Makhlouf MM (2022) Structural, optical, electrical, and DFT studies on polyvinyl pyrrolidone/polyethylene oxide polymer blend filled with MoO_3 nanoplates for flexible energy-storage devices. Int J Energy Res 46:13832–13843
42. Sadhu SPP, Siddabattuni S, Muthukumar SV, Varma KBR (2018) Enhanced dielectric properties and energy storage density of surface engineered BCZT/PVDF-HFP nanodielectrics. J Mater Sci: Mater Electron 29:6174–6182

Index

S. Moharana et al. (eds.), *Emerging Nanodielectric Materials for Energy Storage*, Nanostructure Science and Technology, https://doi.org/10.1007/978-3-031-40938-7

MIX
Papier aus verantwortungsvollen Quellen
Paper from responsible sources
FSC® C105338
www.fsc.org

If you have any concerns about our products,
you can contact us on
ProductSafety@springernature.com

In case Publisher is established outside the EU,
the EU authorized representative is:
Springer Nature Customer Service Center GmbH
Europaplatz 3, 69115 Heidelberg, Germany

Printed by Libri Plureos GmbH
in Hamburg, Germany